Nonlinear Physical Science

Nonlinear Physical Science

Nonlinear Physical Science focuses on recent advances of fundamental theories and principles, analytical and symbolic approaches, as well as computational techniques in nonlinear physical science and nonlinear mathematics with engineering applications.

Topics of interest in *Nonlinear Physical Science* include but are not limited to:

- New findings and discoveries in nonlinear physics and mathematics
- Nonlinearity, complexity and mathematical structures in nonlinear physics
- Nonlinear phenomena and observations in nature and engineering
- Computational methods and theories in complex systems
- Lie group analysis, new theories and principles in mathematical modeling
- Stability, bifurcation, chaos and fractals in physical science and engineering
- Nonlinear chemical and biological physics
- Discontinuity, synchronization and natural complexity in the physical sciences

Series editors

Albert C.J. Luo
Department of Mechanical and Industrial Engineering
Southern Illinois University Edwardsville
Edwardsville, IL 62026-1805, USA
e-mail: aluo@siue.edu

Nail H. Ibragimov
Department of Mathematics and Science
Blekinge Institute of Technology
S-371 79 Karlskrona, Sweden
e-mail: nib@bth.se

International Advisory Board

More information about this series at http://www.springer.com/series/8389

Marat Akhmet · Mehmet Onur Fen

Replication of Chaos in Neural Networks, Economics and Physics

Higher Education Press

Springer

Marat Akhmet
Department of Mathematics
Middle East Technical University
Ankara
Turkey

Mehmet Onur Fen
Neuroscience Institute
Georgia State University
Atlanta, GA
USA

ISSN 1867-8440 ISSN 1867-8459 (electronic)
Nonlinear Physical Science
ISBN 978-3-662-47499-0 ISBN 978-3-662-47500-3 (eBook)
DOI 10.1007/978-3-662-47500-3

Jointly published with Higher Education Press, Beijing
ISBN: 978-7-04-043102-5 Higher Education Press, Beijing

Library of Congress Control Number: 2015944467

Springer Heidelberg New York Dordrecht London

Printed on acid-free paper

Springer-Verlag GmbH Berlin Heidelberg is part of Springer Science+Business Media
(www.springer.com)

To our beloved families

Preface

The main novelty of this book is the consideration of chaos as an *input* for differential and hybrid equations. More precisely, we insert chaos on the right-hand side of the equations and investigate the results of perturbation. Moreover, we investigate many possible consequences of the input–output analysis in systems with many compartments. This is what makes our book on chaos unique among all others.

Let us give some arguments toward the importance of the input–output analysis of chaos for both theory and applications:

1. In the theory of dynamical systems, a large number of results use the input–output analysis. For example, there are many theorems that can be loosely formulated as follows: if a perturbation is periodic (bounded, almost periodic), then there is a unique periodic (bounded, almost periodic) solution. Generally speaking, our results can be formulated in the following way: if a perturbation is chaotic, then there exists a chaos in the set of solutions. Thus, one can say that our main proposal is to return investigation of chaos into the mainstream of classical differential/difference equations theory and, consequently, a huge number of rigorous mathematical methods, numerical instruments, and applications that rely on the input–output analysis will be involved for the investigation of chaotic processes.
2. Despite the fact that many distinguished specialists in the chaos theory and mathematics have been involved in the investigation, there are still many challenging problems related to the origin of the chaos theory. For instance, we do not have a rigorously approved chaos in Lorenz systems, Duffing equations, and other systems. Moreover, there is no universal method to detect chaos in multidimensional systems. Hopefully, the input–output analysis will give new opportunities for the analyses of the basic models and help to unify the knowledge of chaos. We believe that the exploitation of the mechanism in the considered models can give mathematical clarity there.

3. The input–output analysis can become a strong instrument in applications to real-world problems through the modeling of chaos expansion. We hope that unpredictability of weather, economical unpredictability, and irregularity as a global phenomenon will be reflected in mathematical investigations more comprehensively through this machinery. This is true not only for atmospheric or economic processes, but also for any large systems in biology, neural networks, and computer sciences. Utilization of the input–output analysis in cryptography and deciphering may also give effective results. The input–output analysis is very popular, for instance, in mechanics, chemistry, biology, cryptography, etc. Consequently, one can suppose that what we have suggested has to be realized for real-world problems of various natures.
4. We describe the expansion of chaos on the basis of the input–output mechanism using the concept of *morphogenesis* to emphasize that the expansion keeps the geometrical properties of chaos. Furthermore, it is not surprising that the replication of chaos, introduced in the book, relates to concepts of science with broad applications: *self-organization, synergetics, chaos-order relations, thermodynamics, biological patterns*.

The book is attractive in the mathematical sense, since we have introduced rigorous description of chaos for systems with continuous time for the first time. This may give a push for the functional analysis of chaos to involve the operator theory results, etc. Hopefully, our approach will give a basis for deeper comprehension and the possibility to unite different appearances of chaos. In this framework, we also hope that the results can be developed for partial differential equations, integro-differential equations, functional differential equations, evolution systems, etc.

A part of the book is devoted to problems of economics. We have analyzed chaos extension in economic models. Unpredictability in economics as sensitivity in dynamical models is considered, and on that basis, global extension of unpredictability is discussed.

The presence of chaos in neural networks is indispensable, and as applications of our results, replication of chaos by neural networks is presented in a separate chapter in this book.

We pay great attention to expansion of chaos through Lorenz models in meteorology. A special mathematical analysis has been made, since only dissipativeness property of a system is used to prove the chaos presence in perturbed systems.

Entrainment of limit cycles by chaos is discovered numerically through specially designed unidirectional coupling of two glow discharge-semiconductor systems. The result demonstrates that the input–output machinery is working effectively for partial differential equations. Chaotic control is through the external circuit equation and governs the electrical potential on the boundary. The expandability of the theory to collectives of glow discharge systems is discussed, and this increases the potential of applications of the results.

The content of the book is a good background for applications in mechanics, biology, molecular biology, physiology, pharmacology, secure communications, neural networks, and other real-world problems involving complex behavior of models. Since chaos is present everywhere, we can say that our results are applicable in any field, where differential and difference equations are utilized as models.

The authors would like to express their gratitude to those who contributed to the preparation of this book, Zhanar Akhmetova and Ismail Rafatov for the joined results, the Series Editor Prof. Albert Luo and Editor of HEP Liping Wang for their interest in the monograph and patience during the publication of the book.

Ankara, Turkey
Atlanta, GA, USA

Marat Akhmet
Mehmet Onur Fen

Contents

Chapter 1
Introduction

The theory of dynamical systems starts with H. Poincaré, who studied nonlinear differential equations by introducing qualitative techniques to discuss the global properties of solutions [1]. His discovery of the homoclinic orbits figures prominently in the studies of chaotic dynamical systems. Poincaré first encountered the presence of homoclinic orbits in the three-body problem of celestial mechanics [2]. A Poincaré homoclinic orbit is an orbit of intersection of the stable and unstable manifolds of a saddle periodic orbit. It is called structurally stable if the intersection is transverse, and structurally unstable or a homoclinic tangency if the invariant manifolds are tangent along the orbit [3]. In any neighborhood of a structurally stable Poincaré homoclinic orbit, there exist nontrivial hyperbolic sets containing a countable number of saddle periodic orbits and continuum of non-periodic Poisson stable orbits [3–5]. For this reason, the presence of a structurally stable Poincaré homoclinic orbit can be considered as a criterion for the presence of complex dynamics [3].

The first mathematically rigorous definition of chaos is introduced by Li and Yorke [6] for one-dimensional difference equations. According to [6], a continuous map $F : J \to J$, where $J \subset \mathbb{R}$ is an interval, exhibits chaos if: (i) For every natural number p, there exists a $p-$periodic point of F in J; (ii) There is an uncountable set $S \subset J$ containing no periodic points such that for every $s_1, s_2 \in S$ with $s_1 \neq s_2$ we have $\limsup_{k\to\infty} \left|F^k(s_1) - F^k(s_2)\right| > 0$ and $\liminf_{k\to\infty} \left|F^k(s_1) - F^k(s_2)\right| = 0$; (iii) For every $s \in S$ and periodic point $\sigma \in J$ we have $\limsup_{k\to\infty} \left|F^k(s) - F^k(\sigma)\right| > 0$. In the paper [6], it was proved that if a map on an interval has a point of period three, then it is chaotic.

Generalizations of Li-Yorke chaos to high-dimensional difference equations were provided in [7–10]. According to Marotto [10], if a repelling fixed point of a differentiable map has an associated homoclinic orbit that is transversal in some sense, then the map must exhibit chaotic behavior. More precisely, if a multidimensional differentiable map has a snap-back repeller, then it is chaotic. In the paper [9], Marotto's Theorem was used to prove rigorously the existence of Li-Yorke chaos in a spatiotemporal chaotic system. Furthermore, the notion of Li-Yorke sensitivity, which

M. Akhmet and M.O. Fen, *Replication of Chaos in Neural Networks, Economics and Physics*, Nonlinear Physical Science,
DOI 10.1007/978-3-662-47500-3_1

links the Li-Yorke chaos with the notion of sensitivity, was studied in [7], and generalizations of Li-Yorke chaos to mappings in Banach spaces and complete metric spaces were considered in [8].

Another mathematical definition of chaos for discrete-time dynamics was introduced by Devaney [1]. It is mentioned in [1] that a map $F : J \to J$, where $J \subset \mathbb{R}$ is an interval, has sensitive dependence on initial conditions if there exists $\delta > 0$ such that for any $x \in J$ and any neighborhood N of x there exists $y \in J$ and a positive integer k such that $\left|F^k(x) - F^k(y)\right| > \delta$. On the other hand, F is said to be topologically transitive if for any pair of open sets $U, V \subset J$ there exists a positive integer k such that $F^k(U) \cap V \neq \emptyset$. According to Devaney, a map $F : J \to J$ is chaotic on J if: (i) F has sensitive dependence on initial conditions; (ii) F is topologically transitive; (iii) Periodic points of F are dense in J. In other words, a chaotic map possesses three ingredients: unpredictability, indecomposability, and an element of regularity.

Symbolic dynamics, whose earliest examples were constructed by Hadamard [11] and Morse [12], is one of the oldest techniques for the study of chaos. Symbolic dynamical systems are systems whose phase space consists of one-sided or two-sided infinite sequences of symbols chosen from a finite alphabet. Such dynamics arises in a variety of situations such as in horseshoe maps and the logistic map. The set of allowed sequences is invariant under the shift map, which is the most important ingredient in symbolic dynamics [1, 13–17]. Moreover, it is known that the symbolic dynamics admits the chaos in the sense of both Devaney and Li-Yorke [1, 18–21].

The Smale Horseshoe map is first studied by Smale [22] and it is an example of a diffeomorphism which is structurally stable and possesses a chaotic invariant set [1, 15, 17]. The horseshoe arises whenever one has transverse homoclinic orbits, as in the case of the Duffing equation [23]. People used the symbolic dynamics to discover chaos, but we suppose that it can serve as an "embryo" for the morphogenesis of chaos.

From the mathematical point of view, chaotic systems are characterized by local instability and uniform boundedness of the trajectories. Since local instability of a linear system implies unboundedness of its solutions, chaotic system should be necessarily nonlinear [24]. Chaos in dynamical systems is commonly associated with the notion of a strange attractor, which is an attractive limit set with a complicated structure of orbit behavior. This term was introduced by Ruelle and Takens [25] in the sense where the word strange means the limit set has a fractal structure [3]. The dynamics of chaotic systems are sensitive to small perturbations of initial conditions. This means that if we take two close but different points in the phase space and follow their evolution, then we see that the two phase trajectories starting from these points eventually diverge [1, 26]. The sensitive dependence on the initial condition is used both to stabilize the chaotic behavior in periodic orbits and to direct trajectories to a desired state [27].

It was Lorenz [28] who discovered that the dynamics of an infinite-dimensional system being reduced to three-dimensional equation can be next analyzed in its chaotic appearances by application of the simple unimodal one-dimensional map. Smale [22] explained that the geometry of the horseshoe map is underneath of the

Van der Pol equation's complex dynamics which was investigated by Cartwright and Littlewood [29] and later by Levinson [30]. Nowadays, the Smale horseshoes with its chaotic dynamics is one of the basic instruments when one tries to recognize a chaos in a process. Guckenheimer and Williams [31] gave a geometric description of the flow of Lorenz attractor to show the structural stability of codimension 2. In addition to this, it was found out that the topology of the Lorenz attractor is considerably more complicated than the topology of the horseshoe [23]. Moreover, Levi [32] used a geometric approach for a simplified version of the Van der pol equation to show existence of horseshoes embedded within the Van der Pol map and how the horseshoes fit in the phase plane.

It is natural *to discover a chaos* [6, 10, 25, 28, 33–42] and proceed by producing basic definitions and creating the theory. On the other hand, one can *shape* an irregular process by inserting chaotic elements in a system which has regular dynamics (let us say comprising an asymptotically stable equilibrium, a global attractor, etc.). This approach to the problem also deserves consideration as it may allow for a more rigorous treatment of the phenomenon, and helps develop new methods of investigation. Our results are of this type.

In this book, we use the idea that chaos can be utilized as input in systems of equations. To explain the input–output procedure realized in our book, let us introduce examples of systems called as *the base-system, the replicator*, and *the generator*, which will be intensively used in the manuscript. Consider the following system of differential equations,

$$\frac{dz}{dt} = B(z). \tag{1.1}$$

The system (1.1) is called *the base-system*. We assume that the system admits a regular property. For example, there is a globally asymptotically stable equilibrium of (1.1). Next we apply to the system a perturbation, $I(t)$, which will be called an *input* and obtain the following system,

$$\frac{dy}{dt} = B(y) + I(t), \tag{1.2}$$

which will be called as *the replicator*.

Suppose that the input I admits a certain property, let us say, it is a bounded function. We assume then that there exists a unique solution, $y(t)$, of the last equation, the replicator, with the same property of boundedness. This solution is considered as an *output*. The process for obtaining the solution $y(t)$ of the replicator system by applying perturbation $I(t)$ to the base-system (1.1) is called the *input–output mechanism*, and sometimes we shall call it the *machinery*. It is known that for certain base-systems, if the input is periodic, almost periodic, bounded, then there exists an output, which is also periodic, almost periodic, bounded, respectively. In our book, we consider inputs of the new nature: chaotic sets and chaotic functions. The motions which are in the chaotic attractor of the Lorenz system considered altogether provide

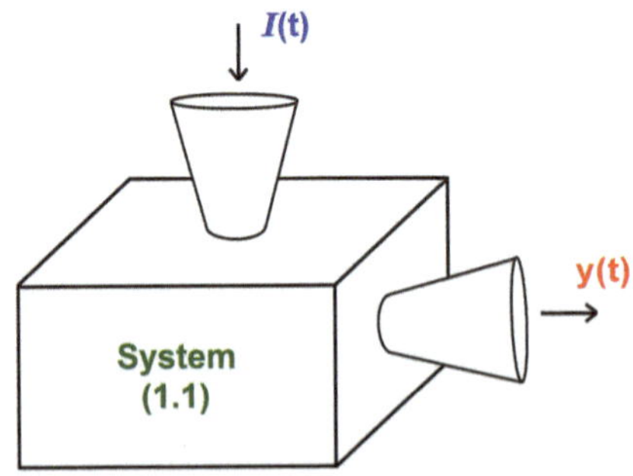

Fig. 1.1 The input–output mechanism

us an example of a chaotic set of functions. Any element of this set is considered as a chaotic function. Both of these types of inputs will be used effectively. For example, to prove rigorously, by verification of all ingredients, that there exists a certain type of chaos generated by the input–output procedure, we use the concept of the chaotic set. For simulations we shall use inputs in the form of chaotic functions. The diagram in Fig. 1.1 illustrates the input–output procedure schematically. We have to say that in the figure the input I can be a set of functions as well as a single function. The same is true for the output, $y(t)$.

The main sources of chaos in theory are difference and differential equations. For this reason, we consider inputs which are solutions of some systems of differential equations or discrete equations. These systems will be called *generators* in this book.

Thus, we can consider the following system of differential equations,

$$\frac{dx}{dt} = G(x), \tag{1.3}$$

and it is assumed that this system possesses chaos. We shall call this system a *generator*. If $x(t)$ is a solution of the system from the chaotic attractor, that is, it is a chaotic solution, then we notate $I(t) = x(t)$ and use the function $I(t)$ in Eq. (1.2). Types of the Eqs. (1.1)–(1.3) considered as a base-system, replicator system, and a generator system, respectively, can be varied in the future. For example, the systems may be non-autonomous and an input may be involved non-linearly.

In this book, we have proved rigorously that the output possesses the same type of chaos as the input if base-systems are with globally asymptotically stable equilibriums, with limit cycles and dissipative systems.

The term *morphogenesis* is used issuing from the sense of the words *morph* meaning "form" and *genesis* meaning "creation." This is similar to the idea such that morphogenesis is used in fields such as urban studies [43], architecture [44], mechanics [45], computer science [46], linguistics [47], and sociology [48, 49]. In the present book, morphogenesis is understood in this weak sense, and the mechanism of the replication is simple. In discussion form, we consider inheritance of intermittency, the double-scroll Chua's attractor, and quasiperiodical motions as a possible skeleton of a chaotic attractor. We use the concept of morphogenesis for two reasons. First of all, morphogenesis is convenient to describe how the input–output analysis works if chaos is an input. Second, it provides information about structure of the chaos-output, if one knows structure of the chaos-input. We give a full description of the chaos expansion as morphogenesis, if base-systems are linear and with constant matrices of coefficients. We make comparison of the main concept of our book with Turing's morphogenesis [50] and John von Neumann automata [51], considering that this may not be only a formal comparison, but will also give ideas for the chaos development in the morphogenesis of Turing and for self-replicating machines.

We propose a rigorous identification method for replication of chaos from a prior one to systems with large dimensions. Extension of the formal properties and features of a complex motion can be observed such that ingredients of chaos united as known types of chaos, Devaney's, Li-Yorke, and others. This is true for other appearances of chaos: intermittency, structure of the chaotic attractor, its fractal dimension, form of the bifurcation diagram, the spectra of Lyapunov exponents, etc.

In our theoretical results of chaos extension, we use coupled systems in which the generator influences the replicator in a unidirectional way, that is, the generator affects the behavior of the replicator, but not the converse. The possibility of making use of more than one replicators and non-identical systems in the machinery is an advantage of the procedure. On the other hand, contrary to the method that we present, in the synchronization of chaotic systems, one does not consider the type of the chaos that the master and slave systems admit. The problem that whether the synchronization of systems implies the same type of chaos for both master and slave has not been taken into account yet.

The concept of morphogenesis is considered carefully only in the second chapter of the book, for systems with stable equilibrium, since for other base-systems the discussion can be done very similarly. Another reason is that namely for systems with stable equilibriums all the known ingredients of chaos are proper, when for other types of base-systems the number of ingredients, which can be saved in the machinery, gradually decreases. We shall discuss this phenomenon in the main text of the book.

The significant theoretical meaning, in our opinion, can be given to the results on entrainment of chaos. Entrainment is a very general concept and it has been used in various fields of science. Mainly, the concept of entrainment is related to cyclic motions. We understand the entrainment of chaos as the seizure of irregularity of inputs by the vector field around the limit cycle such that the resultant vector field of the replicator behaves chaotically in a certain region. In theoretical sense, the analysis of chaos is more complex, since it requests the elaboration of the proof of Andronov-Witt Theorem [52] on orbital stability of a limit cycle, as well as the utilization of the theorem of Massera [53] for the existence of periodic solutions of nonlinear systems.

Finally, we have found that the base-systems can be dissipative systems, which do not necessarily possess a stable equilibrium or a limit cycle. This theoretical result immediately increases the number of models. For example, Lorenz systems are of these type of equations. In the same time, one has to say that by making the nature of attractors for base-systems more general and less obscure in their description (from equilibriums to dissipative systems), we have lost some ingredients of chaos for the resulting chaos-outputs. For example, in the last type of base-systems with dissipativeness, we can not prove the transitivity feature. The same is true for chaos on the basis of a limit cycle. Nevertheless, in these systems we have proved the presence of sensitivity as well as the existence of infinitely many unstable periodic solutions.

Taking into account that our results provide strong evidence of chaos expansion, we become able to interpret them for real-world problems. In this book, we consider models from mechanics and electronics by means of Duffing and Van der Pol equations, and double-scroll Chua's oscillator. Two subjects, economical models and atmospheric dynamics, are discussed in Chaps. 7 and 9, respectively. For both of them we provide a full range analysis based on our theoretical results. Doubtless, the main keyword for these applications is the word chaos itself, since it has to be interpreted as irregularity, which is a very natural feature for economic processes as well as for the dynamics of the weather. In Chap. 10 spatiotemporal chaos in glow discharge-semiconductor systems is verified. Nevertheless, by our opinion, the main ingredient of chaos that is important for the models is the sensitivity, since this property reflects unpredictability for atmospheric dynamics and it is important for economics, too. Thus, one can say that as a consequence of our theoretical results in the theory of dynamical systems, we have made the first step in the globalization of the unpredictability phenomenon for both economics and weather dynamics.

1.1 Synchronization of Chaotic Systems

One of the usage areas of master-slave systems is the study of synchronization of chaotic systems [54–60]. In 1990, Pecora and Carroll [59] realized that two identical chaotic systems can be synchronized under appropriate unidirectional coupling schemes. Consider the system

$$x' = G(x), \tag{1.1.4}$$

as the master, where $x \in \mathbb{R}^d$, such that the steady evolution of the system occurs in a chaotic attractor. The dynamics of the slave system is governed by the equation

$$y' = H(x, y). \tag{1.1.5}$$

When the unidirectional drive is established, suppose that the right hand side of Eq. (1.1.5) satisfies that

$$H(x, y) = G(x), \tag{1.1.6}$$

for $y = x$, and the slave system takes the form

$$y' = G(y), \tag{1.1.7}$$

which is a copy of system (1.1.4), in the absence of driving. In unidirectional couplings, the signals of the master system acts on the slave system, but the converse is not true. Moreover, this action becomes null when the two systems follow identical trajectories [56]. The continuous control scheme [61, 62] and the method of replacement of variables [63, 64] can be used to obtain couplings in the form of the system (1.1.4)+(1.1.5). Synchronization of a slave system to a master system, under the condition (1.1.6), is known as identical synchronization, and it occurs when there are sets of initial data $\mathscr{B}_x \subset \mathbb{R}^d$ and $\mathscr{B}_y \subset \mathbb{R}^d$ for the master and slave systems, respectively, such that the equation $\lim_{t\to\infty} \|x(t) - y(t)\| = 0$ holds, where $(x(t), y(t))$ is a solution of system (1.1.4)+(1.1.5) with initial data $(x(0), y(0)) \in \mathscr{B}_x \times \mathscr{B}_y$.

In paper [65], Afraimovich et al. proposed the synchronization of chaotic systems that are different and not restricted in coupling. To realize this proposal, Rulkov et al. [60] considered the concept of generalized synchronization for unidirectionally coupled systems.

Consider the unidirectionally coupled system (1.1.4)+(1.1.5) such that the dimensions of the master and slave systems are d and r, respectively. Generalized synchronization [54–58, 60] is said to occur if there exist sets $\mathscr{B}_x \subset \mathbb{R}^d$, $\mathscr{B}_y \subset \mathbb{R}^r$ of initial conditions and a transformation ψ, defined on the chaotic attractor of (1.1.4), such that for all $x(0) \in \mathscr{B}_x$, $y(0) \in \mathscr{B}_y$ the relation

$$\lim_{t\to\infty} \|y(t) - \psi(x(t))\| = 0 \tag{1.1.8}$$

holds. In this case, a motion that starts on $\mathscr{B}_x \times \mathscr{B}_y$ collapses onto a manifold $M \subset \mathscr{B}_x \times \mathscr{B}_y$ of synchronized motions. The transformation ψ is not required to exist for the transient trajectories. Generalized synchronization includes the identical synchronization as a particular case. That is, if ψ is the identity transformation, then identical synchronization takes place. The paper [57] deals with the case when the transformation ψ is differentiable.

According to Kocarev and Parlitz [58], generalized synchronization occurs in the dynamics of the coupled system (1.1.4)+(1.1.5) if and only if for all $x_0 \in \mathscr{B}_x$, $y_{10}, y_{20} \in \mathscr{B}_y$, the criterion

$$\lim_{t\to\infty} \|y(t, x_0, y_{10}) - y(t, x_0, y_{20})\| = 0 \tag{1.1.9}$$

holds, where $y(t, x_0, y_{10}), y(t, x_0, y_{20})$ denote the solutions of the slave system (1.1.5) with the initial data $y(0, x_0, y_{10}) = y_{10}$, $y(0, x_0, y_{20}) = y_{20}$ and the same $x(t)$, $x(0) = x_0$.

As a consequence of generalized synchronization, the behavior of the slave system (1.1.5) can be predicted by the knowledge of the trajectories of the master system (1.1.4) and the transformation ψ. The master system is also predictable from the slave system, if ψ is invertible [58].

Let us consider the synchronization concept simply as a chaos generation. In this case, one can guess that it is not possible to describe the output precisely as we have done, and for that reason in the first papers, only identical master and slave systems were discussed. It was assumed that if the two systems admit asymptotically close solutions, then it means that the solutions of the slave system behave chaotically. Nevertheless, the fact that the asymptotic closeness implies the chaoticity of the slave system has not been verified. That is, the authors decided that this is seen intuitively, and does not deserve to be checked. Our objection is that synchronization of chaos has not been approved in the sense that the chaos of the slave system is not verified. Moreover, we have to recognize that only identical systems as the master and the slave system have been considered, and this is a very strong restriction. Later, the synchronization was extended to arbitrary systems (not only identical). The presence of a transformation between the solutions of a master and slave system is required, and asymptotic property has been generalized. However, both of them are practically impossible to check in an analytical way and only numerical methods have been used to indicate the conditions. Our approach is free of the asymptotic condition and it is verifiable. It can be developed not only by means of linear stability, but also the Lyapunov functions can be applied to arrange the generation of chaos. The main advantage of our results is that we assure the structure of the generated chaos in many senses: the type of chaos, bifurcation diagrams, Lyapunov exponents, presence of periodic motions (periods), or almost periodic solutions (their spectra). Definitely, this is very important for the development of security of information, neural networks activities, synchronization problems of mechanics, chemistry, biology, etc. We do hope that the developed theory of synchronization can adopt our results to strengthen the theoretical and practical power of the theory.

1.2 Control of Chaos

The idea of chaos control is based on the fact that chaotic attractors have a skeleton made of an infinite number of unstable periodic orbits [1, 26, 56, 66, 67]. Stability can be described as the ability of a system to keep itself working properly even when perturbations act on it, and this is the main goal to be achieved by the control strategy that is embedded in the system [67]. In other words, the aim of chaos control is to stabilize a previously chosen unstable periodic orbit by means of small perturbations applied to the system, so the chaotic dynamics is substituted by a periodic one chosen at will among the several available [56]. That is, when control is present, a chaotic trajectory transforms into a periodic one [24]. Experimental demonstrations of chaos control methods were presented in the papers [68–75].

Small perturbations applied to control parameters can be used to stabilize chaos, keeping the parameters in the neighborhood of their nominal values, and this idea is first introduced by Ott et al. [76]. Experimental applications of the Ott-Grebogi-Yorke (OGY) control method require a permanent computer analysis of the state of the system. The method deals with a Poincaré map and therefore, the parameter changes are discrete in time. Using this method, one can stabilize only those periodic orbits whose maximal Lyapunov exponent is small compared to the reciprocal of the time interval between parameter changes [77]. Another control method has been developed by Pyragas [77] to stabilize unstable periodic orbits applying small time continuous control to a parameter of a system while it evolves in continuous time, instead of a discrete control at the crossing of a surface [56]. Pyragas control method uses a delayed feedback employing a suitably amplified difference of an output measurement of the chaotic system and the respectively delayed measurement for control. The control signal vanishes in the post-transient behavior for the stabilized orbit. For this reason, the delay time has to be the exact value of the period of the unstable periodic orbit that will be stabilized [78]. Both of the OGY and Pyragas control methods will be utilized throughout the book.

Scientists are interested in the chaos theory because of the fact that it can offer new controlling strategies which have some particularly interesting insights for economic policies. There was opinion among economists that dynamics of chaos is neither predictable nor controllable due to sensitivity. Results of Ott et al. [76] showed that control of a chaos can be made by very small corrections of parameters [56, 79]. This achievement has been widely used in economics by Kaas, Kopel, Holyst, Urbanovwics [80–83], and many others. Our results demonstrate that the control may not be local (applied to an isolated model) but a global phenomenon with strong effectiveness. A control applied to a model, which is realizable easily (for example, the logistic map or Feichtinger's generic model [84]), can be sufficient to rule the process in all models joined with the controlled one. Another benefit of our studies is that in the literature controls are applied to those systems which are simple and low-dimensional. It is worth noting that control of chaos (unstable periodic motions) becomes difficult if dimensions of systems increase and the construction of Poincar sections is complexified. For this reason, the idea to control the generated chaos by controlling the exogenous shocks is useful for applications. In the present book, the control of an economic system through the application of the OGY control to the logistic map is demonstrated. A chaos control can not be realized if we do not know the period of unstable the motion to be controlled. In our case, the control is applicable to models with arbitrary dimension if just the basic period of the generator is known. It is obvious that our methods provide us a scheme of investigations, which can be accompanied with detailed studies in the future. Control of chaos is a synonym to the suppression of chaos nowadays. Thus, our results give another way of suppression of chaos. If we find the controllable link (member) in a chain (collection) of connected chaotic systems, then we can suppress chaos in the whole chain. This is an effective consequence of our studies.

1.3 Neural Networks and Chaos

The chaos phenomenon has been observed in the dynamics of neural networks [85–96], and chaotic dynamics applying as external inputs are useful for separating image segments [93], information processing [90, 91] and synchronization of neural networks [97–99]. Aihara et al. [86] proposed a model of a single neuron with chaotic dynamics by considering graded responses, relative refractoriness, and spatio-temporal summation of inputs. Chaotic solutions of both the single chaotic neuron and the chaotic neural network composed of such neurons were demonstrated numerically in [86]. Focusing on the model proposed in [86], dynamical properties of a chaotic neural network in chaotic wandering state were studied concerning sensitivity to external inputs in [89]. On the other hand, in the paper [93], Aihara's chaotic neuron model is used as the fundamental model of elements in a network, and the synchronization characteristics in response to external inputs in a coupled lattice based on a Newman-Watts model are investigated. Besides, in the studies [90, 91], a network consisting of binary neurons which do not display chaotic behavior is considered, and by means of the reduction of synaptic connectivities it is shown that the state of the network in which cycle memories are embedded reveals chaotic wandering among memory attractor basins. Moreover, it is mentioned that chaotic wandering among memories is considerably intermittent. Chaotic solutions to the Hodgkin-Huxley equations with periodic forcing have been discovered in [85]. The paper [88] indicates the existence of chaotic solutions in the Hodgkin-Huxley model with its original parameters. An analytical proof for the existence of chaos through period-doubling cascade in a discrete-time neural network is given in [96], and the problem of creating a robust chaotic neural network is handled in [92]. Generally speaking, it is recognized that chaos is a friend of mind.

1.4 Extension of Chaos

To have a comprehensible discussion in this introduction, let us give an outline of a consequence of our results for collectives of systems. Suppose that there is a system, S_1, which is autonomous and possesses chaos. That is, a chaotic attractor of the system exists and the presence of chaos is proved by applying one of the definitions of chaos: Li-Yorke chaos, Devaney chaos, chaos through period-doubling cascade and sensitivity, etc. We call S_1 as the generator system (generator of chaos). Assume that there are other systems, $S_2, S_3, \ldots, S_n$, which are all interconnected in the unidirectional fashion. In Fig. 1.2, an example of the connection for the case $n = 16$ is depicted. The connection's nature is very simple. Solutions of system S_1 are utilized as perturbations for systems S_2, S_5, S_7 and S_{11}. Next, solutions of the perturbed systems are utilized to perturb other adjoint systems and so on.

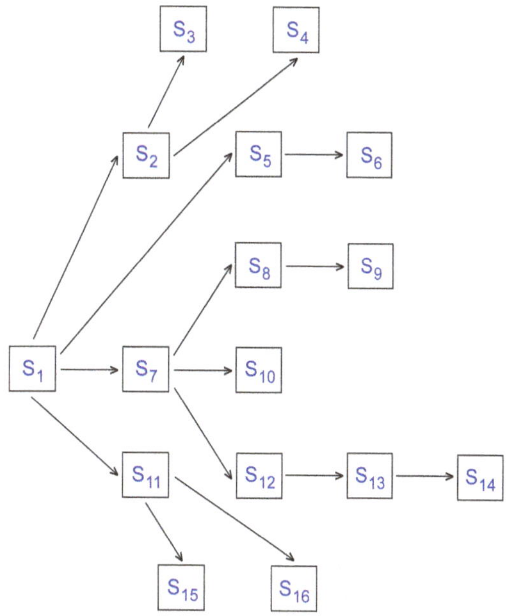

Fig. 1.2 Chaos extension mechanism in the net

That is, in their own turn systems S_2, S_5, S_7, and S_{11} become generator systems for all others, except S_1, etc. So, these connections will continue, while all the systems are connected in the net, which is seen in Fig. 1.2. Systems S_i, $i > 1$, admit globally stable equilibriums, if they are isolated (unperturbed). In this unconnected state, they are called base-systems. It implies from our results that when the connections are valid, they are all chaotic under certain conditions. Thus, all of the cells shown in Fig. 1.3 are chaotic such that the whole system S, the union of subsystems, S_i, $i = 1, 2, \ldots, 16$, is chaotic with the same type of chaos as S_1. Thus, in what follows we shall refer to the collection S of the chaotified systems to illustrate concepts of the discussion.

Fig. 1.3 Collection of chaotified systems

S_{10}	S_4	S_2	S_3
S_8	S_7	S_1	S_{15}
S_9	S_{12}	S_5	S_{11}
S_{13}	S_{14}	S_6	S_{16}

1.5 Ordering Chaos

One of the most interesting questions in science is as follows: How do chaos and order act together in the universe and relate to each other? It was ancient Greeks who accept that chaos is primary with respect to order. For example, Hesiod (eighth century B.C.) considers chaos as "a uniform disorderly mass, heavy and obscure, a mixture of water, earth, fire and air" [100, 101]. Similarly, Ovid in his Metamorphoses understands chaos as the raw and disordered formless mass from which the ordered universe is created. The antithesis of chaos was cosmos, understood as world or universe. This was described by Heraclitus, Plato, and Aristotle. In classical science, represented by the mechanics of Galileo and Newton, the order is primary, while chaos is a deviation from regularities. Moreover, chaos is regarded as the violation of order, not as the absence of order [101]. In their book [102], Prigogine and Stengers wrote that "Our scientific heritage includes two basic questions to which till now no answer was provided. One is the relation between disorder and order. The famous law of increase of entropy describes the world as evolving from order to disorder ..." In the same time, "... biological or social evolution shows us the complex emerging from the simple. How is this possible? How can structure arise from disorder? ... We know now that nonequilibrium, the flow of matter and energy, may be source of order." [102]. It was desired to say that chaos and order are not separated in time or space, the chaos is not something a formless, but one in which "there are underlying patterns, fractal structures, governed by a new view of our 'orderly' world" [103].

In his study [104], the German philosopher and scientist Immanuel Kant raised the idea which supports that chaos and order are antitheses considering the universe, and consequently, one cannot say which one is primary or secondary in the couple. He expressed the hypothesis of the wave-like character of the evolution of the universe as the sequential replacement of chaos and order, which goes on from the infinite past to the infinite future [101]. What we have done in the present book is a contribution to the point of view of Kant in the following sense. It is clear that chaos (better to say disorder) and order are antitheses, and in general philosophical sense, they are proper not only for the universe evolution, but also in every place where these concepts can be met. This is true for some abstract constructions too. So, it is worth noting that, in the present book, flows of unperturbed systems (the base-systems) present order, since they admit asymptotically stable equilibria (globally stable). When they are perturbed with chaotic functions (chaotic solutions of the generator system), they in turn produce a new chaos. That is, perturbations present chaos in our investigations. Thus, considering several systems in chain, we have the wave-like expansion of chaos represented in Fig. 1.4.

Fig. 1.4 The wave-like expansion of chaos

In other words, the *chaos-order* waves happen in the set of perturbation-flow models. Thus, one can observe an interaction of chaos and order in a new way. These waves are expanding not in time, but rather in the space of phase variables such that different coordinates are subdued to different systems (models). They are presumably finite, since a number of connected models (systems) are finite. Nevertheless, the wave exists. So, we think that our results make the Kant's hypothesis more realistic, and they support that chaos and order are antitheses and complimentary in the universe.

Our analysis implies that chaos can be ordered in the following way. It is possible to construct a system with subsystems—*cells*, having the same type of chaos (the subsystems can be identical). This is another reason to consider the morphogenesis concept by applying replication of chaos. The morphogenesis is expressed not only in the type of chaos, but also through the arrangement of similar bifurcation diagrams, fractals, sensitivity, shadowing property, structures of chaotic attractor's images by the stroboscopic method. We are very confident that it will take its place in all appearances of chaos in next investigations.

The main idea of the hypothesis is that chaos and order are equal in the relation to each other, they are antitheses. Does chaos deserve analysis against regular dynamics? It seems that the answer is rather positive than negative. Moreover, can we say that deterministic chaos is disorder? It is obvious that the answer has to be negative. Order is discovered in chaos: there are skeletons of chaos consisting of periodic, almost periodic, recurrent motions. Bifurcation diagrams, Feigenbaum number, and Lyapunov exponents are widely applied to give a proper description of irregular behavior. We can compare chaos of two systems through synchronization. From this point of view, possibly, it is better to say that we are on the way to give a more precise picture of order/disorder. The chaos theory considered in papers of the last half century is just one step in this direction. Possibly, it is better to call the theory of chaos investigated last century as *theory of complexity* as it is done in [105].

Another remark is as follows. The expansion of order, which was started in the studies of Newton, Galileo, continues by results in the theory of chaos. This is what was written by Durrenmatt in his play The Physicists "... science progresses by reducing the complexity of reality to a hidden simplicity" [106]. In other words, this idea for dynamical systems was expressed by M.W. Hirsh in [107], "The development of mathematical dynamical systems theory can be viewed as the simultaneous pursuit of two lines of research: on the one hand, the quest for simplicity, comprehensibility, stability; on the other hand, the discovery of complexity, instability, chaos. When new complexities are discovered we try to tame them through analysis and classification." In 1969, it was said by R. Abraham in [108] that "Some large (yin) sets of differential equations with generic properties are known, some small (yang) sets which can be classified are known, but in general the two domains have not yet met." In that sense, we suppose that the replication of chaos is a convenient instrument to give points of meeting for those differential equations and the methods of analyses, which are known in theory of differential equations (not only ordinary differential equations) and chaos appearances. This is true not only for enlarging the sets of equations, but also for enlarging the dimensions of equations, which admit irregular motions. R. Thom in his book [109] emphasizes the role of attractors as models in science by declaring that "Every object, or every physical form, can be represented by an attractor C of a dynamical system in a space M of internal variables." In his paper [107], M. Hirsh recognized that "It is in fact exceedingly difficult to decide, either theoretically or practically, whether a particular system has a strange attractor." Apparently, our results will enlighten the problem of indication of strange (chaotic) attractors. At least, we are now able to construct strange attractors of arbitrary dimension being confident that they are chaotic, which type of chaos is developing in the sets, whether a dense trajectory is present, whether infinitely many unstable periodic (almost periodic) trajectories exist, etc. From this point of view, we hope that analyses of particular models and their collectives will be improved.

What we have written is not something original. It is just an interpretation of dialectical laws [110].

1.6 Self-organization of Chaos

The idea of the transition of chaos from one system to another as well as the arrangement of chaos in an ordered way can be considered as another level of self-organization [105, 111]. In [106] it is described that "... a system is self-organizing if it acquires a spatial, temporal or functional structure without specific interference from the outside. By "specific" we mean that the structure of functioning is not impressed on the system, but the system is acted upon from the outside in a nonspecific fashion." here are three approaches to self-organization, namely thermodynamic (dissipative structures), synergetic and the autowave [101]. For the theory of dynamical systems (differential equations, e.g.), the phenomenon means that an autonomous system of equations admits a regular and stable motion such as periodic, quasi-periodic, almost periodic. This is what in literature is called autowave processes [112] or self-excited oscillations [103]. We inclined to add to the list another phenomenon, which is a consequence of replication of chaos. Consider the collection of systems $S_1, S_2, S_3, \ldots, S_n$ introduced in Sect. 1.4 once again, and assume that the first system, S_1, is autonomous and admits a chaos, let us say, of Devaney type. Because of the connections and conditions discovered in our analysis, all other subsystems, S_i, $i = 2, \ldots, n$, will have the same type of chaos. We suppose that one can call this as a self-organization phenomenon. That is, sustained chaotic motion is present in a multidimensional system (with several, if not infinite subsystems). The chaos is homogeneous such that at each of the subsystems the chaos type is the same with similar characteristics (bifurcation diagrams, Lyapunov exponents, sensitivity, quantitative characteristics, etc.). This phenomenon can be restricted only for autonomous systems or it can be even interpreted for non-autonomous systems too. If needed, we may also give arguments for periodicity by the consideration of chaos-order waves (see Sect. 1.5) as a strong argument of self-organization. Not only the community of systems with the same chaos type is large, participating in the replication, but also a chaos considered in an isolated model is a "thicker" object than let us say a cycle, the trajectory of a periodic motion. So, we can say that replication of chaos is an example of self-organization phenomenon, that is a coherent behavior of a large number of systems and even a single system [111].

To observe self-organization in models, it is useful to find a cooperation of dissipativeness with forces, which repel from an equilibrium. This is true for morphogenesis. In his famous paper [50], the English mathematician A. Turing constructed a model of morphogenesis, a reaction-diffusion model, where formation of organs in organisms appears, by considering transformation in the couple stability–instability. In our investigations, we have similar effect of cooperation of stability and instability. Our base-systems are assumed to be asymptotically stable and chaotic perturbations are representatives of strong instability (analogues of the diffusion-driving in the Turing's model). Thus, if in morphogenesis one says about Turing's instability, our results may give the new concept of chaotic instability. Self-organization is a process of formation of dissipative systems [113]. According to Glansdorff and Prigogine [113], "Classical thermodynamics had solved the problem of the competition between randomness and organization for equilibrium systems. But what happens far from equilibrium? Can we find there new organizations, new structures stabilized through the interaction with the outside world?" and "Dissipative structures are formed and maintained through the effect of exchange of energy and matter in non-equilibrium conditions. The formation of cell patterns at the onset of free connection is a typical example of a dissipative structure. We may consider a convection cell as a giant fluctuation stabilized by the flow of energy and matter prescribed by the boundary conditions." The role of fluctuations is emphasized in the sentence [113] "... a new 'structure' is always the result of an instability. It originates from fluctuation." One can compare what we have done with the theory of dissipative structures to see that the role of fluctuations is prescribed in our case to chaotic perturbations, while the flow of base-systems pointed to equilibriums are similar to the flow of energy and matter present in the thermodynamical processes. One more thing to be compared is as follows. It is mentioned in [102] that "In far-from-equilibrium conditions we may have transformations from disorder, from thermal chaos, into order." It is reasonable to ask that "What is the analogue of the order in our case?." We have the following answer for the question. The chaotic "cells," S_i, $i = 1, \ldots, n$, constitute the new order of chaotified systems. It is similar to that we have in Bénard convection phenomenon or biological structures, which appear in morphogenesis. To demonstrate this, let us specify the systems S_i by considering $n = 9$ with the following differential equations,

$$
\begin{aligned}
x_1' &= x_2\\
x_2' &= -0.05x_2 - x_1^3 + 7.5\cos t\\
x_3' &= x_3 + 4x_4 + x_1\\
x_4' &= -2x_3 - 3x_4 + 0.001x_4^2 + x_2\\
x_5' &= x_5 + 4.01x_6 + 0.001x_6^2 + x_4\\
x_6' &= -2x_5 - 3x_6 + 0.6x_3 + 0.5\sin x_3\\
x_7' &= x_7 + 4.01x_8 + 0.005x_7^3 + x_5 + 0.0001x_5^3\\
x_8' &= -2.01x_7 - 3x_8 + \tan(x_6/10)\\
x_9' &= 1.01x_9 + 4x_{10} + 0.03\sin x_9 + 2x_8\\
x_{10}' &= -2x_9 - 3.01x_{10} + 4x_7 + 2\cos x_7\\
x_{11}' &= x_{11} + 4.01x_{12} + e^{x_9/2}\\
x_{12}' &= -2.01x_{11} - 3.01x_{12} + 0.001x_{11}^2 + x_{10}\\
x_{13}' &= 1.01x_{13} + 4x_{14} + 0.001x_{13}^2 + x_{12}\\
x_{14}' &= -2.01x_{13} - 3.01x_{14} + x_{11} + \tanh(x_{11})\\
x_{15}' &= 1.01x_{15} + 4.01x_{16} + x_{13}\\
x_{16}' &= -2x_{15} - 3.01x_{16} + 0.004x_{16}^3 + \arctan(x_{14})\\
x_{17}' &= 1.01x_{17} + 4.01x_{18} + 0.002x_{17}^3 + (x_{15}^2 + 3x_{15} - 1)/(x_{15} + 3)\\
x_{18}' &= -2.01x_{17} - 3x_{18} + 0.3x_{16}.
\end{aligned}
\tag{1.6.10}
$$

Since the first subsystem

$$
\begin{aligned}
x_1' &= x_2\\
x_2' &= -0.05x_2 - x_1^3 + 7.5\cos t
\end{aligned}
\tag{1.6.11}
$$

possesses a chaotic attractor [114] as it is the Duffing equation with "chaotic" coefficients, all remaining eight 2-dimensional subsystems are chaotic by replication of chaos, due to the connections. The chaotic attractor of the whole 18-dimensional system (1.6.10) is shown in Fig. 1.5. Thus, one can see that the scenario of self-organization is repeated in our studies, too. That is the reason why we observe "morphogenesis" in our simulations and this is true not only for geometrical objects which can be visualized in the collective flow, but also for bifurcation diagrams, quantitative characteristics of period-doubling cascade, etc. This is a new form of coherence.

Saying about chemical clocks in [102], authors indicate that "... a situation is no longer be described in terms of chaotic behavior. A new type of order has appeared. We can speak of a new coherence, of a mechanism of "communication" among molecules. But this type of communication can arise only in far-from-equilibrium conditions. It is quite interesting that such communication seems to be the rule in the world of biology." We suppose that chaos extension discovered through replication of chaos also has to be accepted as a new form of order. One can name it as an ordered net of chaotic "cells." The reader can find more about the architecture of the extension of chaos in the next section.

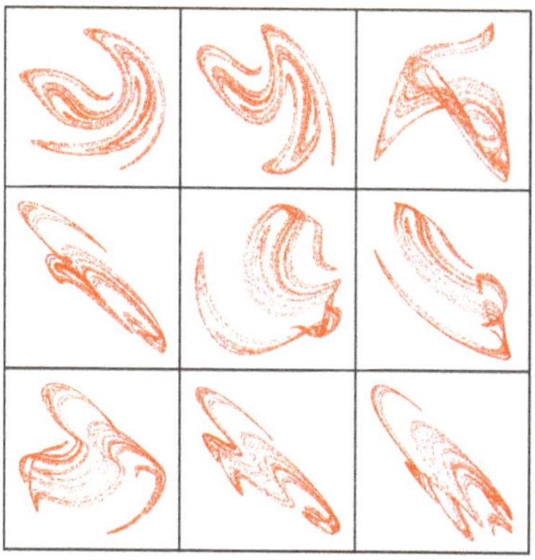

Fig. 1.5 Chaotic attractor of system (1.6.10)

There is no doubt that this type of order exists not only for models in the form of ordinary differential equations or difference equations, but also for partial differential equations, equations in Banach spaces and, in fact, for all other types of equations. We have to recognize that our theoretically approved results have not been seen in experiments yet, but only in simulations through numerical analysis. So, we hope that in very close future special experiments will be done, and one will see the results of replication of chaos in electrical devices, mechanisms, artificial intelligence, etc.

In fact, biological pattern formation, dissipative structures, and replication of chaos are results of interaction of dissipation and diffusion-driven instability, fluctuations, and chaotic perturbations. That is, there are forces acting from outside toward, let us say an attractor, and forces which "revolt" acting outward, called "instabilities." The sustained structures in morphogenesis, dissipative structures or ordered nets of chaos are results of balance of the forces. In order to give more visible impression of this type of self-organization, let us take in attention the life on the Earth. It is from one side result of the gravitational force, which is universally needed for life, and from another side of many forces which are acting "against" the force of gravitation and all together bring the life phenomenon to the globe. In [102] it is written that "... the type of dissipative structure depends critically on the conditions in which the structure is formed. External fields such as the gravitational field of Earth, as well as the magnetic field, may play an essential role in the selection mechanism of self-organization." An important idea that was first proposed by Turing [50] is that a system of chemicals, which is stable in the absence of diffusion, becomes unstable in the presence of diffusion. The same scenario of chaos replication is played by forces in two opposite directions. One force is toward to the equilibrium and another "revolts" chaotically such that as a result we have the chaotic attractor in the replicator system. Finally, we do not pretend that replication of chaos will be on the same level of universality as biological pattern formation or dissipative structures (first of all, we have to see the phenomenon in experiments), but we are sure that, at least, replication of chaos will be found later as a link in some universal chains of self-organization.

1.7 Morphogenesis of Chaos

In this book, we try to use the term morphogenesis issuing from the sense of the words *morph* meaning "form" and *genesis* meaning "creation" [115]. In other words, similar to the ideas of René Thom [116], we employ the word *morphogenesis* as its etymology indicates, to denote *processes creating forms*. One should understand *morphogenesis of chaos* as a form-generating mechanism emerging from a dynamical process which is based on replication of chaos. Here, we accept the form (morph) not only as a type of chaos, but also accompanying concepts as the structure of the chaotic attractor, its fractal dimension, form of the bifurcation diagram, the spectra of Lyapunov exponents, inheritance of intermittency, etc. To illustrate the point of view one can see the "chaotic attractor" of the 18-dimensional system in Fig. 1.5.

To understand the concept of this book better, let us consider morphogenesis of fractal structures [117, 118]. It is important to say that Mandelbrot fractal structures exhibit the appearance of fractal hierarchy looking *in*, that is, *a part is similar to the whole*. Examples for this are the Julia sets [119, 120] and the Sierpinski carpet [121]. In our morphogenesis both directions, *in* and *out*, are present. Indeed, the fractal structure of the prior chaos has hierarchy looking *in*, and the structure for the result-system is obtained considering hierarchy looking *out*, that is, when *the whole is similar to the part*. It is important to say that morphogenesis *out* is very important for biological evolution [122] and also for industry, urban, and computer developments.

In our results, we do not consider the chaos synchronization problem, but we say that the type of the chaos is kept *invariant* in the procedure. That is why the classes which can be considered with respect to this *invariance* is expectedly wider then those investigated for synchronization of chaos. Since we do not request strong relation and accordance between the solutions of the generator and the replicator in the asymptotic point of view, the terms *master* and *slave* as well as *drive* and *response* are not preferred to be used for the analyzed systems. On the other hand, contrary to the method that we present, in the synchronization of chaotic systems, one does not consider the type of the chaos that the master and slave systems admit. The problem that whether the synchronization of systems implies the same type of chaos for both master and slave has not been taken into account yet.

The phenomenon of the form recognition for chaotic processes has already begun in pioneering papers [22, 23, 28–32]. All these results say about chaos recognition, by reducing complex behavior to the structure with recognizable chaos. In [19, 20, 123–128], we provide a different and constructive way when a recognized chaos can be extended saving the form of chaos to a multidimensional system. The idea is extended to the morphogenesis of chaos in the present book.

Nowadays, one can consider the development of a multidimensional chaos from a low-dimensional one in different ways. One of them is the chaotic itinerancy [95, 129–135]. The itinerant motion among varieties of ordered states through high-dimensional chaotic motion can be observed and this behavior is named as chaotic itinerancy. In other words, chaotic itinerancy is a universal dynamics in high-dimensional dynamical systems, showing itinerant motion among varieties of low-dimensional ordered states through high dimensional chaos. This phenomenon occurs in different real-world processes: optical turbulence [129], globally coupled chaotic systems [130, 131], non-equilibrium neural networks [95, 135], analysis of brain activities [136] and ecological systems [137]. One can see that in its degenerated form chaotic itinerancy relates to intermittency [36, 103], since they both represent dynamical interchange of irregularity and regularity.

Likewise the itinerant chaos observed in brain activities, we have low-dimensional chaos in the subsystems considered and high dimensional chaos is obtained when one considers all subsystems as a whole. The main difference compared to our technique is in the elapsed time for the occurrence of the process. In our discussions, no itinerant motion is observable and all resultant chaotic subsystems process simultaneously, whereas the low-dimensional chaotic motions take place as time elapses in the case of chaotic itinerancy. Knowledge of the type of chaos is another difference between chaotic itinerancy and our procedure. Possibly the present way of replication of chaos will give a light to the solutions of problems about extension of irregular behavior (crises, collapses, etc.) in interrelated or multiple connected systems which can arise in problems of classical mechanics [103], electrical systems [138, 139], economic theory [140], and brain activity investigations [136].

In systems whose dimension is at least four, it is possible to observe chaotic attractors with at least two positive Lyapunov exponents and such systems are called hyperchaotic [141]. An example of a four-dimensional hyperchaotic system is discovered by Rössler [38]. Combining two or more chaotic, not necessarily identical, systems is a way of achieving hyperchaos [139, 142, 143]. However, we take into account exactly one chaotic system with a known type of chaos, and use this system as the generator to reproduce the same type of chaos in other systems. On the other hand, the crucial phenomenon in the hyperchaotic systems is the existence of two or more positive Lyapunov exponents and the type of chaos is not taken into account. In our way of morphogenesis, the critical situation is rather the replication of a known type of chaos.

The paper [50] was one of the first studies that consider mathematically the self-replicating forms using a set of reaction-diffusion equations [144]. Taking inspiration from the ideas of Turing, Smale [145] considers the problem of whether oscillations can be generated through coupling of identical systems which tend to an equilibrium. A similar question is also reasonable for the achievement of chaos in such systems and it is found out that, without using a chaotic input, it is possible to obtain coupled systems which exhibit chaotic behavior. The existence of strange attractors in a family of vector fields consisting of two Brusselators linearly coupled by diffusion is proved analytically in the paper [146] and numerical examples of such a chaotic behavior are provided in [147]. Such couplings display several cases of Hopf-pitchfork singularities of codimensions 2, 3, and 4. In all these cases, the corresponding bifurcation diagrams provide regions of parameters such that the system exhibits synchronization, regions where synchronization depends on the initial state and regions where orbits show infinitely many transients of synchronization [148]. Another example of a linearly coupled system which exhibit chaotic behavior can be found in [149]. According to the results of [149], a sufficiently large coupling coefficient used in a network of linearly coupled identical systems, where each node is located in a non-chaotic region, leads to the existence of a positive transversal Lyapunov exponent and makes the system behave chaotically. The Lorenz systems with stable equilibriums can be used in the construction of such a network of linearly coupled systems. Distinctively, we make use of coupled systems such that exactly one of them is chaotic with a known type of chaos and prove theoretically that the same type of chaos is extended. Moreover, in the presented mechanism, we are not restricted to use linear couplings as well as identical systems.

1.8 Chaos and Cellular Automata

The concept of self-replicating machines, in the abstract sense, starts with the ideas of von Neumann [51] and these ideas are supposed to be the origins of cellular automata theory [144]. Morphogenesis was deeply involved in mathematical discussions through Turing's investigations [50] as well as in the concept of structural stability [109]. In the book, the term "morphogenesis" is used in the meaning of "processes creating forms" where we accept the *form* not only as a type of chaos, but also accompanying concepts as the structure of the chaotic attractor, its fractal dimension, form of the bifurcation diagram, the spectra of Lyapunov exponents, inheritance of intermittency, etc. This is similar to the idea such that morphogenesis is used in fields such as urban studies [43], architecture [44], mechanics [45], computer science [46], linguistics [47], and sociology [48, 49].

According to von Neumann, it is feasible in principle to create a self-replicating machine, which he refers as an "automaton," by starting with a machine A, which has the ability to construct any other machine once it is furnished with a set of instructions, and then attaching to A another component B that can make a copy of any instruction supplied to it. Together with a third component labeled C, it is possible to create a machine, denoted by R, with components A, B, and C such that C is responsible to initiate A to construct a machine as described by the instructions, then make B to create a copy of the instructions, and supply the copy of the instructions to the entire apparatus. The component C is referred as "control mechanism." It is the resulting machine R', obtained by furnishing the machine R by instructions I_R, that is capable of replicating itself. Multiple usage of the set of instructions I_R is crucial in the mechanism of self-replication. First, the instructions must be fulfilled by the machine A, then they must be copied by B, and finally the copy must be attached to machine R to form the system R' once again [51, 144].

Our theory of replication of chaos relates the ideas of von Neumann about self-replicating machines in the following sense. Initially, we take into account a system of differential equations (the generator) which plays the role of machine A as in the ideas of von Neumann, and we use this system to influence in a unidirectional way, another system (the replicator) in the role of machine B, in such a manner that the replicator mimics the same ingredients of chaos furnished to the generator. In this book, we use such ingredients in the form of period-doubling cascade, Devaney's and Li-Yorke chaos. In conclusion, the generator system with the replicator counterpart together, that is, the result-system, admits ingredients of the generator. In other words, a known type of chaos is replicated.

1.9 Synergetics and Chaos

In his fascinating book [111], German theoretical physicist Hermann Haken introduced a new interdisciplinary field of science, synergetics, which studies the origins and the evolution of spatio-temporal structures. Profound part of synergetics is based on the dynamical systems theory. Depending on the mathematical discussion of our book, it is natural that we concentrate on the differential and difference equations. Everything that will be said below about synergetics concern first of all dynamical systems with mathematical approach. One of the main features of systems in synergetics is self-organization, which is discussed in Sect. 1.6, and we realize that the phenomenon is present in replication of chaos in strong or weak sense. According to Haken [111], the central question in synergetics is whether there are general principles that govern the self-organized formation of structures and/or functions. The main principles by the founder of the theory are instability, order parameters and slaving [111]. Instability is understood as the formation or collapse of structures (patterns). This is very common in fluid dynamics, lasers, chemistry, and biology [111, 122, 150–152]. A number of examples of instability can be found in the literature about morphogenesis. For instance, the pattern formation examples can be found in fluid dynamics. The phenomenon is called instability because the former state of the fluid transforms to a new one, loses its ability to persist and becomes unstable.

We see instability in the replication of chaos, first of all, through the chaotification of the base-system, which is originally non-chaotic. Secondly, if one considers real-world systems, then, for example, this chaotification can be evolved in time by a new replicator joined to the former generator. Finally, since we consider macroscopic processes, increasing the field of chaos by adding new subsystems can be accepted as instability.

The concepts of the order parameter and slaving are strongly related in synergetics. In the theory of differential equations, order parameters mean those dependent variables whose behavior formates the main properties of a macroscopic structure that dominates over all other variables in the formation such that they can even depend on the order parameters functionally. The mathematically proved (discovered) dependence is called as slaving [111]. It is not difficult to see that in the replication of chaos, considering a generator and several replicators, the variables of the generator are order parameters and they determine the chaotic behavior of replicators' variables. That is, the slaving principle is present there. We have to say that, generally, the slaving principle works in bidirectional connections, but we discuss unidirectionally interacting systems. Nevertheless, the slaving principle is in the basis of replication of chaos as well as in synergetics.

1.10 Mathematics in Chaos Theory

A very serious question may occur concerning the results of our investigations. Should we call the phenomenon discussed as *replication of chaos*? It seems that one can name the results as, let us say, *theorems on the existence of chaos* in a similar way to the theorems on the existence of periodic solutions for periodically perturbed linear systems. We have to say that this similarity is superficial. Indeed, the input–output analysis realized in differential equations theory to obtain the existence of periodic solutions, almost periodic solutions, etc. deals with separate, individual functions and solutions. One can say that it replicates periodic, almost periodic functions. However, in our case, the mechanism replicates a set of functions (a chaotic set), which comprises infinitely many elements. Moreover, we can say that *a phenomenon* is replicated. That is, we are considering the apparition. This makes the replication of chaos closer to those results obtained for reaction-diffusion equations and thermodynamics [102, 113] rather than finite-dimensional flows. Additionally, we fix the presence of chaos and say about all accompanying characteristics of irregular motions. This is a new theory, which can be developed in a similar way to the theory of periodic solutions, stability and other known ones for differential equations. In that sense, our results bring the investigation of chaos to the bosom of differential equations theory back, where it was originated. Thus, we somehow support Philip Holmes' opinion that the theory of chaos has to be a part of the theory of differential equations [153]. According to Holmes [153], much remains to be learned about the applicability and validity of the methods of data analysis for chaos such as dimension computations, Lyapunov exponents, phase-space reconstruction, spectra of dimensions, etc. Moreover, the author recognizes that "... I do not really see a "new science" here, in particular I do not see that "chaos theory" even exists as a coherent object, for example like the quantum and relativity theories." Since the main body of results on chaotic motions have been formulated in terms of differential and difference equations, we may suggest that all these achievements have to be embedded in the theory of dynamical systems or more specifically, in the theory of differential equations or hybrid systems. In this sense, our results, which use the methods developed in the theory of ordinary differential equations for quasilinear systems and realized on the basis of definitions obtained to introduce chaos in the set of continuous functions, can be considered as an attempt to see the results about chaos in the theory of differential equations.

Possibly, we will achieve more successes in the direction of investigation. These results will pull chaos analysis back to the classical dynamical systems theory. If the apparatus will be utilized to the problems in the whole of its might, one will definitely be able to response to what was written in [153]: "Certain persons seem to prefer to abandon hard won, detailed knowledge of problems like turbulence in boundary or shear layers in favor of metaphors, such as coupled map lattices, which have little obvious connection with underlying physics." According to L. Shil'nikov [154], quasi-minimal sets possibly constitute the basis of many chaos appearances. The sets that are closures of Poisson stable trajectories are in the content of classical dynamics theory [107]. Nowadays, not all mathematicians agree that chaos is present in the world as much as the regular motions. Chaos is considered by the majority of mathematicians as an exotic phenomenon. Nevertheless, specialists who are working with chaos have become more and more confident that the chaos presence extends in the real-world as much as regular motions. Possibly, it will be verified that the places where chaotic motions evolve in reality are much more than the ones with non-chaotic motions, as it is for rational and irrational numbers, for example. In that sense, we hope that our proposals based on the input–output analysis for the chaos extension may give more arguments and tools for further investigation of chaos in the mathematical theory as well as in applications. However, nowadays, we continue to consider chaos being present in the real-world problems less than it deserves even in the measure sense. Remember that the dimension of a chaotic attractor is not only a fractal, but also it is less than the dimension of a system, where the chaos is observed. The phase space of the system may admit partitions such as periodic, almost periodic trajectories and other regular alternatives. These regular orbits are also the sets with small measure, even zero. So, in that case, one can consider chaotic trajectory or chaotic set of orbits in the same way as periodic, quasi-periodic or almost periodic ones. This consideration is not something exceptional. Our investigation gives a light on the concept of chaos in that sense. That is, we try to treat chaotic motions (functions) as the ones in the row of functions: periodic, quasi-periodic, almost periodic, chaotic, bounded on the real axis. Moreover, we guess that chaotic functions can be investigated more deeply within the scope of theory of functions in the future.

1.11 Chaos Theory and Real World

A very interesting question is the following one: How do chaotic systems interact with their environment or other systems? More specifically, how can several chaotic and non-chaotic systems interact to give some desirable effect or to suppress harmful influences? Synchronization of chaos and suppression of chaos by control are two fields of the modern science, where this question is considered. Synchronization is very important for comprehension processes in the brain and to develop reliable security of communication systems. It is investigated for circadian rhythms, chemical oscillators, electrical circuits, plasma tubes, technology of communications, etc.

There are some observations, which confirm that if a chaotic oscillator is acted on by a periodic perturbation, then there is the locking of the phase of the oscillator to that of the force. Since periodic perturbations are very common in magneto-mechanical systems, chemical reactions, which have been observed in laboratories as well as for the earth and life sciences, the role of the suppression of chaos is obviously increasing with the development of interdisciplinary connections of dynamical systems theory. This is true not only for mechanical and electrical engineering with undesirable and harmful irregular oscillations, but also for economics, biology, and neural networks theory. There is another possibility of interaction of a chaotic system with environment when the latter is chaotic. This phenomenon is described in the literature as the interaction of a chaotic driving system, which is connected unidirectionally with another chaotic response system. Obviously, the problem is difficult for analysis, and it has been analyzed only within the scope of the synchronization theory. Exceptionally, it is difficult if the interconnected systems are not identical. If the drive and response systems are identical, then there are results such that the systems have solutions which are asymptotically close as the time evolves to infinity [56, 59]. Despite the fact that just sensitivity can be confirmed intuitively only, it is recognized that the asymptotic closeness indicates the presence of chaos for the response system. The asymptotic closeness of solutions can be easily proved by applying, let us say, the second Lyapunov method [155]. Nevertheless, as we have just said, chaoticity of the response system has not been discussed from the point of view of the rigorous mathematics of dynamical systems theory. In the case of non-identical drive and response systems (generalized synchronization), there is an additional request that there has to be a correspondence between solutions of the systems: a homeomorphism in the most restrictive case. That is the reason why applications of the generalized synchronization is not a so much obvious task. In order to detect the synchronization, conditional Lyapunov exponents, auxiliary system approach, and mutual false nearest neighbors techniques can be used [54, 56, 59, 60]. All these methods of indication are rather numerical. So, the problem has to be further considered by theoretical investigations.

What we do in our analysis in its particular case can be considered as an interaction of a chaotic system, which we call as *the generator*, with a regular system (for example, a system having a globally asymptotically stable equilibrium), which is called *the base-system*. When the latter is perturbed by solutions of the generator, (generally non-linearly), the obtained system is called *the replicator*. Thus, we can say that the subject is two systems: a chaotic system and a non-chaotic one. There is a unidirectional connection of the first one with the second. It is rigorously proved that the replicator admits chaos provided that the base-system is with stable equilibrium, limit cycle or torus. This problem has not been considered before in the literature. Our method is based on rigorous definitions of chaotic sets of functions and chaotic functions as elements of a chaotic set. This is also done for the first time in the literature. Thus, we consider a chaotic environment and prove that a non-chaotic system (an oscillator) with certain properties also becomes chaotic in the environment.

In the book [103], F.C. Moon states that "... engineers have always known about chaos—it was called noise or turbulence and huge factors or factors of safety were used to design around these apparent random unknowns that seem to crop up in every technical device ..." According to Moon [103], the recognition of chaos in nonlinear deterministic systems may be useful for understanding the source of random like noise. Moreover, the author declares that "the new discoveries in nonlinear dynamics bring with them new concepts and tools for detecting chaotic vibrations in physical systems and for quantifying this 'deterministic noise' with new measures such as fractal dimensions and Lyapunov exponents." Our investigations realize expectations from the chaos theory in both cases. We suggest that if one investigates the presence of chaos in a system with connections to its neighbor systems among which there is a system that is respectively simple for the chaos analysis, then the same characteristics of chaos can also be determined for the examined system. The characteristics of chaos can be considered as the chaos type, unstable periodic solutions in the skeleton of the chaos, bifurcation diagrams, and quantities such as Lyapunov exponents and fractal dimensions. We remind that it is standard to look for chaos through the construction of special low-dimensional maps. On the basis of the map, one can make conclusion on the multidimensional chaos. However, in our case, we reduce the analysis to more simple systems by looking at interconnections in a collection of systems. This is something new in the theoretical indication of chaos.

Is unpredictability a generic phenomenon in economics, meteorology, mechanical experiments and neural networks? This question is actual today when the world economy is shaken by the crisis, and it should push economists to wonder about the approach used to analyze economic phenomena. Another question is as follows: Is it possible to recognize an economic crisis before it occurs? Our results may give a slight light on the problem. On the basis of our results, we show that unpredictability is not avoidable in economics globally. The fact that chaos is a generic property of dynamics is not new. We will give more literature observation later. But now, let us describe why unpredictability is generic through our investigations. For that purpose let us consider a global economy as interconnected models, where each model is a system of differential, difference or hybrid equations. It implies from our results that if one of the models generates endogenous chaos (unpredictability), then it infects the "neighbors" through the connections, provided that some certain criteria are satisfied. That is why unpredictability is a global phenomenon. The topology of the expansion can be of various types. If one wants to study the globalization of unpredictability in economics, it has to be taken into account that our results consider the expansion among models with strict conditions, for instance, the Lipschitz condition with small coefficients. We consider also quasilinear systems which are with hyperbolic linear parts. Thus, there are opportunities to consider the generation of chaos in the global economy.

Detection of chaos is one of the main problems considered within the scope of applications of chaos. The methods of bifurcation diagrams, Lyapunov exponents, etc. do not always work. The next question is the determination of the chaos type that the irregular dynamics belong to. We suggest to consider a cluster of models with irregular behavior, confirm that they are interconnected, learn which of the models is the most simple to analyze for chaos, specify the chaos type and the values of all parameters such as Lyapunov exponents, etc., and then to verify that counterparts of the model satisfy conditions of our results. Finally, one can decide that the same chaos is proper for the whole system of models or those of models which satisfy the needed criteria. Hence, the problem of determination of chaos can be solved by the algorithm and it may give a new approach to complex system development and indication. Moreover, we can estimate the divergence of the output motions by using the knowledge about the input motions. That is what we can give for practice.

One can take into account the following source of chaotic shocks: the bio-economic synthesis (see Chap. 13 in [84]). More precisely, assuming that economic endogenous chaos is questionable, we accept that biological (population dynamics) chaos is approved. Then we can connect economy with fishery, etc., to obtain that chaos enters into economics through exogenous biological shocks. We may also support relay functions by seasonal harvesting, etc.

It is known that there are comparisons of economics with thermodynamics and in general physics [156]. There is another science to compare economics with meteorology and atmospheric physics. This is useful to imagine exceptionally if one talks about global processes.

1.12 Organization of the Book

The book is organized as follows.

We start with the descriptions of the chaotic sets of functions in Devaney and Li-Yorke sense as well as hyperbolic set of functions in Chap. 2. Next, replication of chaos on the basis of the input–output mechanism is provided with rigorous verification. Machinery for the extension of chaos in the sense of Devaney and Li-Yorke as well as the one obtained through period-doubling cascade is established. Control of chaos inserted in the input–output procedure is realized such that periodic motions, which are in the skeleton of chaos, are stabilized. In discussions of the main results of the chapter, morphogenesis of intermittency and the double-scroll Chua's attractor are considered as well as quasi-periodicity, Shil'nikov orbits and the case when the linear system admits non-negative eigenvalues are handled.

Chapter 3 is devoted to the subject of the second chapter in the case that the condition on the eigenvalues of the coefficient matrix is weakened. We consider the hyperbolic case such that some of the eigenvalues are with positive real parts. Extension of chaos is observed for this case too. Possibly, this chapter is first of all of theoretical interest.

In Chap. 4, we consider a limit cycle as the main qualitative feature of the base-system. The entrainment of limit cycles by chaos, which is understood as the deformation of limit cycles to chaotic cycles, is considered. The extension of sensitivity as well as the presence of infinitely many unstable periodic solutions and their control are realized through the input–output analysis. The entrainment of toroidal attractors by chaos is also considered. It is shown that the phenomenon cannot be considered as generalized synchronization of chaotic systems.

We take into account the base-system with impulsive actions in Chap. 5. The presence of chaotic dynamics is rigorously approved by implementing chaotic perturbations. The results are applied to mechanical systems by means of Duffing equations. Controllability of the obtained chaos is also proposed.

In Chap. 6, we handle relay functions and impulsive perturbations, which are the most specific inputs in the book. Again chaos in the sense of Devaney, Li-Yorke, and chaos on the basis of period-doubling cascade are generated, but this time inputs and outputs are of different natures. If the former are discontinuous, then the latter are continuous or discontinuous (for impulsive systems). The classical concepts of quasi-minimal sets, hyperbolic sets, and shadowing property are discussed through the input–output analysis. Applications of the theoretical results such as Duffing equation and others are investigated subdued to chaotic perturbations.

The last four chapters revolve around more applications of our theoretical results. Chapter 7 is concerned with economic models, where chaotic inputs are interpreted as exogenous shocks. The base-systems are considered with stable equilibrium points and limit cycles. The attractors have very realistic economical sense as equilibria, limit cycles or tori. Continuous and discontinuous shocks are applied as inputs for the base-systems. In the case of discontinuous shocks, discrete variables are the values of the economical variables as well as the moments of time, when the shocks happen. A comparison of exogenous chaos and endogenous one in the business cycles is performed through our results.

Chapter 8 is devoted to neural networks. In that chapter, we consider the problem of chaos generation in shunting inhibitory cellular neural networks and Hopfield neural networks by taking advantage of external inputs. Retarded and impulsive neural networks are studied with chaotic external inputs. The generation of cyclic and toroidal chaotic motions by neural networks are also discussed.

The question "Why is the weather unpredictability a global phenomenon?" is under discussion in Chap. 9. We specify it as the Lorenz unpredictability as it is studied on the basis of the famous model. While the base-systems are considered with equilibrium points and limit cycles in the previous chapters, in Chap. 9 the base-system is just a dissipative one. On the basis of the input–output analysis, we theoretically approve the concept of the global weather unpredictability in the toy case, considering it as sensitivity in the dynamical system.

Finally, Chap. 10 is entirely about spatiotemporal chaos in glow discharge-semiconductor systems. The results of that part of the book demonstrate that the extension of chaos can be observed in any type of models, this time partial differential equations.

References

1. R.L. Devaney, *An Introduction to Chaotic Dynamical Systems* (Addison-Wesley, Menlo Park, 1989)
2. K.G. Andersson, Poincaré's discovery of homoclinic points. Arch. Hist. Exact Sci. **48**, 133–147 (1994)
3. S.V. Gonchenko, L.P. Shil'nikov, D.V. Turaev, Dynamical phenomena in systems with structurally unstable Poincaré homoclinic orbits. Chaos **6**(1), 15–31 (1996)
4. L.P. Shil'nikov, On a Poincaré-Birkhoff problem. Math. USSR-Sbornik **3**(3), 353–371 (1967)
5. S. Smale, Diffeomorphisms with many periodic points, in *Differential and Combinatorial Topology: A Symposium in Honor of Marston Morse* (Princeton University Press, Princeton, 1965), pp. 63–70
6. T.Y. Li, J.A. Yorke, Period three implies chaos. Am. Math. Mon. **82**, 985–992 (1975)
7. E. Akin, S. Kolyada, Li-Yorke sensitivity. Nonlinearity **16**, 1421–1433 (2003)
8. P. Kloeden, Z. Li, Li-Yorke chaos in higher dimensions: a review. J. Differ. Equ. Appl. **12**, 247–269 (2006)
9. P. Li, Z. Li, W.A. Halang, G. Chen, Li-Yorke chaos in a spatiotemporal chaotic system. Chaos Solitons Fractals **33**(2), 335–341 (2007)

10. F.R. Marotto, Snap-back repellers imply chaos in $\mathbb{R}^n$. J. Math. Anal. Appl. **63**, 199–223 (1978)
11. J. Hadamard, Les surfaces à courbures opposées et leurs lignes géodésiques. J. Math. Pures et Appl. **4**, 27–74 (1898)
12. M. Morse, G.A. Hedlund, Symbolic dynamics. Am. J. Math. **60**, 815–866 (1938)
13. C. Grebogi, J.A. Yorke, *The Impact of Chaos on Science and Society* (United Nations University Press, Tokyo, 1997)
14. J. Hale, H. Koçak, *Dynamics and Bifurcations* (Springer, New York, 1991)
15. J. Kennedy, J.A. Yorke, Topological horseshoes. Trans. Am. Math. Soc. **353**, 2513–2530 (2001)
16. S. Wiggins, *Global Bifurcations and Chaos: Analytical Methods* (Springer, New York, 1988)
17. S. Wiggins, *Introduction to Applied Nonlinear Dynamical Systems and Chaos* (Springer, New York, 2003)
18. M. Akhmet, *Nonlinear Hybrid Continuous/Discrete-Time Models* (Atlantis Press, Paris, 2011)
19. M.U. Akhmet, Devaney's chaos of a relay system. Commun. Nonlinear Sci. Numer. Simul. **14**, 1486–1493 (2009)
20. M.U. Akhmet, Li-Yorke chaos in the system with impacts. J. Math. Anal. Appl. **351**, 804–810 (2009)
21. C. Robinson, *Dynamical Systems: Stability, Symbolic Dynamics, and Chaos* (CRC Press, Boca Raton, 1995)
22. S. Smale, Differentiable dynamical systems. Bull. Am. Math. Soc. **73**, 747–817 (1967)
23. J. Guckenheimer, P.J. Holmes, *Nonlinear Oscillations, Dynamical Systems and Bifurcations of Vector Fields* (Springer, New York, 1997)
24. A.L. Fradkov, *Cybernetical Physics: From Control of Chaos to Quantum Control* (Springer, Berlin, 2007)
25. D. Ruelle, F. Takens, On the nature of turbulence. Commun. Math. Phys. **20**, 167–192 (1971)
26. D. Gulick, *Encounters with Chaos* (University of Maryland, College Park, 1992)
27. H.G. Schuster, *Handbook of Chaos Control* (Wiley-VCH, Weinheim, 1999)
28. E.N. Lorenz, Deterministic nonperiodic flow. J. Atmos. Sci. **20**, 130–141 (1963)
29. M. Cartwright, J. Littlewood, On nonlinear differential equations of the second order I: the equation $\ddot{y} - k(1-y^2)'y + y = bkcos(\lambda t + a)$, k large. J. Lond. Math. Soc. **20**, 180–189 (1945)
30. N. Levinson, A second order differential equation with singular solutions. Ann. Math. **50**, 127–153 (1949)
31. J. Guckenheimer, R.F. Williams, Structural stability of Lorenz attractors. Publ. Math. **50**, 307–320 (1979)
32. M. Levi, *Qualitative Analysis of the Periodically Forced Relaxation Oscillations* (Memoirs of the American Mathematical Society, Providence, 1981)
33. L.O. Chua, M. Komuro, T. Matsumoto, The double scroll family, parts I and II. IEEE Trans. Circuit Syst. **CAS-33**, 1072–1118 (1986)
34. M. Hénon, A two-dimensional mapping with a strange attractor. Commun. Math. Phys. **50**(1), 69–77 (1976)
35. T. Matsumoto, L.O. Chua, M. Komuro, The double scroll. IEEE Trans. Circuit Syst. **CAS-32**, 797–818 (1985)
36. Y. Pomeau, P. Manneville, Intermittent transition to turbulence in dissipative dynamical systems. Commun. Math. Phys. **74**, 189–197 (1980)
37. O.E. Rössler, An equation for continuous chaos. Phys. Lett. **57A**, 397–398 (1976)
38. O.E. Rössler, An equation for hyperchaos. Phys. Lett. A **71**, 155–157 (1979)
39. E. Sander, J.A. Yorke, Period-doubling cascades for large perturbations of Hénon families. J. Fixed Point Theory Appl. **6**, 153–163 (2009)
40. S. Sato, M. Sano, Y. Sawada, Universal scaling property in bifurcation structure of Duffing's and of generalized Duffing's equations. Phys. Rev. A **28**, 1654–1658 (1983)
41. S.W. Shaw, A.G. Haddow, S.R. Hsieh, Properties of cross-well chaos in an impacting systems. Philos. Trans. R. Soc. Lond. A **347**, 391–410 (1994)
42. K. Thamilmaran, M. Lakshmanan, Rich variety of bifurcations and chaos in a variant of Murali-Lakshmanan-Chua circuit. Int. J. Bifurc. Chaos **10**, 1781–1785 (2000)

43. T. Courtat, C. Gloaguen, S. Douady, Mathematics and morphogenesis of cities: a geometrical approach. Phys. Rev. E **83**, 1–12 (2011)
44. S. Roudavski, Towards morphogenesis in architecture. Int. J. Archit. Comput. **7**, 345–374 (2009)
45. L.A. Taber, Towards a unified theory for morphomechanics. Philos. Trans. R. Soc. A **367**, 3555–3583 (2009)
46. P. Bourgine, A. Lesne, *Morphogenesis: Origins of Patterns and Shapes* (Springer, Berlin, 2011)
47. C. Hagége, *The Language Builder: An Essay on the Human Signature in Linguistic Morphogenesis* (John Benjamins Publishing Co., Amsterdam, 1993)
48. M.S. Archer, *Realistic Social Theory: The Morphogenetic Approach* (Cambridge University Press, Cambridge, 1995)
49. W. Buckley, *Sociology and Modern Systems Theory* (Prentice Hall, New Jersey, 1967)
50. A.M. Turing, The chemical basis of morphogenesis. Philos. Trans. R. Soc. Lond., Ser. B, Biol. Sci. **237**, 37–72 (1952)
51. J. Von Neumann, A.W. Burks (eds.), *The Theory of Self-Reproducing Automata* (University of Illinois Press, Urbana, 1966)
52. M. Farkas, *Periodic Motions* (Springer, New York, 2010)
53. J.L. Massera, The existence of periodic solutions of systems of differential equations. Duke Math. J. **17**, 457–475 (1950)
54. H.D.I. Abarbanel, N.F. Rulkov, M.M. Sushchik, Generalized synchronization of chaos: the auxiliary system approach. Phys. Rev. E **53**, 4528–4535 (1996)
55. V. Afraimovich, J.R. Chazottes, A. Cordonet, Nonsmooth functions in generalized synchronization of chaos. Phys. Lett. A **283**, 109–112 (2001)
56. J.M. Gonzáles-Miranda, *Synchronization and Control of Chaos* (Imperial College Press, London, 2004)
57. B.R. Hunt, E. Ott, J.A. Yorke, Differentiable generalized synchronization of chaos. Phys. Rev. E **55**(4), 4029–4034 (1997)
58. L. Kocarev, U. Parlitz, Generalized synchronization, predictability, and equivalence of unidirectionally coupled dynamical systems. Phys. Rev. Lett. **76**(11), 1816–1819 (1996)
59. L.M. Pecora, T.L. Carroll, Synchronization in chaotic systems. Phys. Rev. Lett. **64**, 821–825 (1990)
60. N.F. Rulkov, M.M. Sushchik, L.S. Tsimring, H.D.I. Abarbanel, Generalized synchronization of chaos in directionally coupled chaotic systems. Phys. Rev. E **51**(2), 980–994 (1995)
61. M. Ding, E. Ott, Enhancing synchronism of chaotic systems. Phys. Rev. E **49**, R945–R948 (1994)
62. T. Kapitaniak, Synchronization of chaos using continuous control. Phys. Rev. E **50**, 1642–1644 (1994)
63. K.M. Cuomo, A.V. Oppenheim, Circuit implementation of synchronized chaos with applications to communications. Phys. Rev. Lett. **71**, 65–68 (1993)
64. L.M. Pecora, T.L. Carroll, Driving systems with chaotic signals. Phys. Rev. A **44**, 2374–2383 (1991)
65. V.S. Afraimovich, N.N. Verichev, M.I. Rabinovich, Stochastic synchronization of oscillation in dissipative systems. Radiophys. Quantum Electron. **29**, 795–803 (1986)
66. T. Kapitaniak, *Controlling Chaos: Theoretical and Practical Methods in Non-linear Dynamics* (Butler and Tanner Ltd., Frome, 1996)
67. E. Schöll, H.G. Schuster, *Handbook of Chaos Control* (Wiley-VCH, Weinheim, 2008)
68. A. Azevedo, S.M. Rezende, Controlling chaos in spin-wave instabilities. Phys. Rev. Lett. **66**(10), 1342–1345 (1991)
69. G.L. Baker, Control of the chaotic driven pendulum. Am. J. Phys. **63**, 832–838 (1995)
70. S. Bielawski, D. Derozier, P. Glorieux, Controlling unstable periodic orbits by a delayed continuous feedback. Phys. Rev. E **49**, R971–R974 (1994)
71. W.L. Ditto, S.N. Tauseo, M.L. Spano, Experimental control of chaos. Phys. Rev. Lett. **65**(26), 3211–3214 (1990)

72. A. Garfinkel, M.L. Spano, W.L. Ditto, J.N. Weiss, Controlling cardiac chaos. Science **257**, 1230–1233 (1992)
73. S. Hayes, C. Grebogi, E. Ott, Communicating with chaos. Phys. Rev. Lett. **70**(20), 3031–3034 (1993)
74. R. Meucci, W. Gadomski, M. Ciofini, F.T. Arecchi, Experimental control of chaos by means of weak parametric perturbations. Phys. Rev. E **49**(4), R2528–R2531 (1994)
75. S.J. Schiff, K. Jerger, D.H. Duong, T. Chang, M.L. Spano, W.L. Ditto, Controlling chaos in the brain. Nature **370**, 615–620 (1994)
76. E. Ott, C. Grebogi, J.A. Yorke, Controlling chaos. Phys. Rev. Lett. **64**, 1196–1199 (1990)
77. K. Pyragas, Continuous control of chaos by self-controlling feedback. Phys. Rev. A **170**, 421–428 (1992)
78. G. Herrmann, A robust delay adaptation scheme for Pyragas' chaos control method. Phys. Lett. A **287**(3–4), 245–256 (2001)
79. J.A. Holyst, K. Urbanowicz, Chaos control in economical model by time delayed feedback method. Phys. A: Stat. Mech. Appl. **287**(3–4), 587–598 (2000)
80. M. Allais, The economic science of today and global disequilibrium, in *Global Disequilibrium in the World Economy*, ed. by M. Baldassarry, J. McCallum, R.A. Mundell (Macmillan, Basingstoke, 1992)
81. J.A. Holyst, T. Hagel, G. Haag, W. Weidlich, How to control a chaotic economy? J. Evol. Econ. **6**(1), 31–42 (1996)
82. L. Kaas, Stabilizing chaos in a dynamic macroeconomic model. J. Econ. Behav. Organ. **33**, 313–332 (1998)
83. M. Kopel, Improving the performance of an economic system: controlling chaos. J. Evol. Econ. **7**, 269–289 (1997)
84. G. Feichtinger, Nonlinear threshold dynamics: further examples for chaos in social sciences, in *Economic Evolution and Demographic Change*, ed. by G. Haag, U. Mueller, K.G. Troitzsh (Springer, Berlin, 1992)
85. K. Aihara, G. Matsumoto, Chaotic oscillations and bifurcations in squid giant axons, in *Chaos*, ed. by A. Holden (Manchester University Press, Manchester, 1986), pp. 257–269
86. K. Aihara, T. Takabe, M. Toyoda, Chaotic neural networks. Phys. Lett. A **144**, 333–340 (1990)
87. W.J. Freeman, Tutorial on neurobiology: from single neurons to brain chaos. Int. J. Bifurc. Chaos **2**(3), 451–482 (1992)
88. J. Guckenheimer, R.A. Oliva, Chaos in the Hodgkin-Huxley model. SIAM J. Appl. Dyn. Syst. **1**(1), 105–114 (2002)
89. J. Kuroiwa, N. Masutani, S. Nara, K. Aihara, Chaotic wandering and its sensitivity to external input in a chaotic neural network, in *Proceedings of the 9th International Conference on Neural Information Processing (ICONIP'O2)*, ed. by L. Wang, J.C. Rajapakse, K. Fukushima, S.Y. Lee, X. Yao (Orchid Country Club, Singapore, 2002), pp. 353–357
90. S. Nara, P. Davis, Chaotic wandering and search in a cycle-memory neural network. Prog. Theor. Phys. **88**(5), 845–855 (1992)
91. S. Nara, P. Davis, M. Kawachi, H. Totsuji, Chaotic memory dynamics in a recurrent neural network with cycle memories embedded by pseudo-inverse method. Int. J. Bifurc. Chaos **5**(4), 1205–1212 (1995)
92. A. Potapov, M.K. Ali, Robust chaos in neural networks. Phys. Lett. A **277**(6), 310–322 (2000)
93. M. Shibasaki, M. Adachi, Response to external input of chaotic neural networks based on Newman-Watts model, in *The 2012 International Joint Conference on Neural Networks*, ed. by J. Liu, C. Alippi, B. Bouchon-Meunier, G.W. Greenwood, H.A. Abbass (Brisbane, 2012), pp. 1–7
94. C.A. Skarda, W.J. Freeman, How brains make chaos in order to make sense of the world. Behav. Brain Sci. **10**(2), 161–173 (1987)
95. I. Tsuda, Chaotic itinerancy as a dynamical basis of hermeneutics in brain and mind. World Futures **32**, 167–184 (1991)
96. X. Wang, Period-doublings to chaos in a simple neural network: an analytical proof. Complex Syst. **5**, 425–441 (1991)

97. Q. Liu, S. Zhang, Adaptive lag synchronization of chaotic Cohen-Grossberg neural networks with discrete delays. Chaos **22**(3), 033123 (2012)
98. W. Lu, T. Chen, Synchronization of coupled connected neural networks with delays. IEEE Trans. Circuits Syst.-I: Regul. Pap. **51**(12), 2491–2503 (2004)
99. W. Yu, J. Cao, W. Lu, Synchronization control of switched linearly coupled neural networks with delay. Neurocomputing **73**(4–6), 858–866 (2010)
100. R. Graves, *New Larousse Encyclopedia of Mythology* (Prometheus Press/Hamlyn, New York, 1968)
101. M. Bushev, *Synergetics: Chaos, Order Self-Organization* (World Scientific, Singapore, 1994)
102. I. Prigogine, I. Stengers, *Order Out of Chaos: Man's Dialogue with Nature* (Bantam Books, Toronto, 1984)
103. F.C. Moon, *Chaotic Vibrations: An Introduction for Applied Scientists and Engineers* (Wiley, Hoboken, 2004)
104. I. Kant, *Universal Natural History and Theory of the Heavens* (Richer Resources Publications, Virginia, 2008)
105. G. Nicolis, I. Prigogine, *Exploring Complexity: An Introduction* (W.H. Freeman, New York, 1989)
106. F. Durrenmatt, *The Physicists* (Grove, New York, 1964)
107. M.W. Hirsh, The dynamical systems approach to differential equations. Bull. Am. Math. Soc. **11**, 1–64 (1984)
108. R. Abraham, *Predictions for the Future of Differential Equations*. Lecture Notes in Mathematics, vol. 206 (Springer, Berlin, 1971)
109. R. Thom, *Stabilité Structurelle et Morphogénèse* (W.A. Benjamin, New York, 1972)
110. C.W. Hegel, *Lectures on the History of Philosophy* (Humanities Press, New York, 1974)
111. H. Haken, *Advanced Synergetics: Instability Hierarchies of Self-Organizing Systems and Devices* (Springer, Berlin, 1983)
112. A.A. Andronov, A.A. Vitt, C.E. Khaikin, *Theory of Oscillations* (Pergamon Press, Oxford, 1966)
113. P. Glansdorff, I. Prigogine, *Thermodynamic Theory of Structure, Stability and Fluctuations* (Wiley, London, 1971)
114. J.M.T. Thompson, H.B. Stewart, *Nonlinear Dynamics and Chaos* (Wiley, Chichester, 2002)
115. J.A. Davies, *Mechanisms of Morphogenesis: The Creation of Biological Form* (Elsevier Academic Press, Amsterdam, 2005)
116. R. Thom, *Mathematical Models of Morphogenesis* (Ellis Horwood Limited, Chichester, 1983)
117. B.M. Mandelbrot, *The Fractal Geometry of Nature* (Freeman, San Francisco, 1982)
118. B.M. Mandelbrot, *Fractals: Form, Chance and Dimension* (Freeman, San Francisco, 1977)
119. B. Branner, L. Keen, A. Douady, P. Blanchard, J.H. Hubbard, D. Schleicher, R.L. Devaney, in *Complex Dynamical Systems: The Mathematics Behind Mandelbrot and Julia Sets*, ed. by R.L. Devaney (American Mathematical Society, Providence, 1994)
120. J. Milnor, *Dynamics in One Complex Variable* (Princeton University Press, Princeton, 2006)
121. H.O. Peitgen, H. Jürgens, D. Saupe, *Chaos and Fractals: New Frontiers of Science* (Springer, New York, 2004)
122. J.D. Murray, *Mathematical Biology II: Spatial Models and Biomedical Applications* (Springer, New York, 2003)
123. M. Akhmet, *Principles of Discontinuous Dynamical Systems* (Springer, New York, 2010)
124. M.U. Akhmet, Creating a chaos in a system with relay. Int. J. Qual. Theory Differ. Equ. Appl. **3**, 3–7 (2009)
125. M.U. Akhmet, Dynamical synthesis of quasi-minimal sets. Int. J. Bifur. Chaos **19**, 2423–2427 (2009)
126. M.U. Akhmet, Shadowing and dynamical synthesis. Int. J. Bifur. Chaos **19**, 3339–3346 (2009)
127. M.U. Akhmet, M.O. Fen, Chaotic period-doubling and OGY control for the forced Duffing equation. Commun. Nonlinear Sci. Numer. Simul. **17**, 1929–1946 (2012)
128. M.U. Akhmet, M.O. Fen, The period-doubling route to chaos in the relay system, in *Proceedings of Dynamic Systems and Applications*, vol. 6, ed. by G.S. Ladde, N.G. Medhin, C. Peng, M. Sambandham (Dynamic Publisher Inc., Atlanta, 2012), pp. 22–26

129. K. Ikeda, K. Matsumoto, K. Otsuka, Maxwell-Bloch turbulence. Prog. Theor. Phys. Suppl. **99**, 295 (1989)
130. K. Kaneko, Clustering, coding, switching, hierarchical ordering, and control in network of chaotic elements. Phys. D **41**, 137–172 (1990)
131. K. Kaneko, Globally coupled circle maps. Phys. D **54**, 5–19 (1991)
132. K. Kaneko, I. Tsuda, *Complex Systems: Chaos and Beyond, A Constructive Approach with Applications in Life Sciences* (Springer, Berlin, 2000)
133. K. Kaneko, I. Tsuda, Chaotic itinerancy. Chaos **13**, 926–936 (2003)
134. T. Sauer, Chaotic itinerancy based on attractors of one-dimensional maps. Chaos **13**, 947–952 (2003)
135. I. Tsuda, Dynamic link of memory–chaotic memory map in nonequilibrium neural networks. Neural Netw. **5**, 313–326 (1992)
136. W.J. Freeman, J.M. Barrie, Chaotic oscillations and the genesis of meaning in cerebral cortex, in *Temporal Coding in the Brain*, ed. by G. Buzsáki, R. Llinás, W. Singer, A. Berthoz, Y. Christen (Springer, Berlin, 1994), pp. 13–37
137. P. Kim, T. Ko, H. Jeong, K.J. Lee, S.K. Han, Emergence of chaotic itinerancy in simple ecological systems. Phys. Rev. E **76**, 1–4 (2007)
138. L.O. Chua, Chua's circuit: ten years later. IEICE Trans. Fundam. Electron. Commun. Comput. Sci. **E77-A**, 1811–1822 (1994)
139. T. Kapitaniak, L.O. Chua, Hyperchaotic attractors of unidirectionally-coupled Chua's circuits. Int. J. Bifurc. Chaos Appl. Sci. Eng. **4**, 477–482 (1994)
140. H.W. Lorenz, *Nonlinear Dynamical Economics and Chaotic Motion* (Springer, New York, 1993)
141. J.C. Sprott, *Elegant Chaos: Algebraically Simple Chaotic Flows* (World Scientific Publishing, Singapore, 2010)
142. T. Kapitaniak, Transition to hyperchaos in chaotically forced coupled oscillators. Phys. Rev. E **47**, R2975–R2978 (1993)
143. T. Kapitaniak, L.O. Chua, G. Zhong, Experimental hyperchaos in coupled Chua's circuits. IEEE Trans. Circuits Syst.-I: Fundam. Theory Appl. **41**, 499–503 (1994)
144. J.L. Schiff, *Cellular Automata: A Discrete View of the World* (Wiley, Hoboken, 2008)
145. S. Smale, A mathematical model of two cells via Turing's equation, in *Some Mathematical Questions in Biology, V, Proceedings of the Seventh Symposium on Mathematical Biology, Mathematics and Life Sciences*, vol. 6 (American Mathematical Society, Mexico, 1973), pp. 15–26
146. F. Drubi, S. Ibáñez, J.A. Rodriguez, Coupling leads to chaos. J. Differ. Equ. **239**(2), 371–385 (2007)
147. F. Drubi, S. Ibáñez, J.A. Rodriguez, Singularities and chaos in coupled systems. Bull. Belg. Math. Soc. Simon Stevin **15**(5), 797–808 (2008)
148. F. Drubi, S. Ibáñez, J.A. Rodriguez, Hopf-pitchfork singularities in coupled systems. Phys. D **240**(9–10), 825–840 (2011)
149. W.-J. Yuan, X.-S. Luo, P.-Q. Jiang, B.-H. Wang, J.-Q. Fang, Transition to chaos in small-world dynamical network. Chaos Solitons Fractals **37**(3), 799–806 (2008)
150. H. Haken, *Information and Self-Organization: A Macroscopic Approach to Complex Systems* (Springer, Berlin, 1988)
151. H. Haken, *Brain Dynamics, Synchronization and Activity Patterns in Pulse-Coupled Neural Nets with Delays and Noise* (Springer, Berlin, 2002)
152. M.A. Vorontsov, W.B. Miller, *Self-Organization in Optical Systems and Applications in Information Technology* (Springer, Berlin, 1998)
153. P. Holmes, Poincaré, celestial mechanics, dynamical-systems theory and "chaos". Phys. Rep., Rev. Sect. Phys. Lett. **193**, 137–163 (1990)
154. L. Shilnikov, Bifurcations and strange attractors, in *Proceedings of the International Congress of Mathematicians*, vol. III (Higher Education Press, Beijing, 2002), pp. 349–372
155. T. Yoshizawa, *Stability Theory and the Existence of Periodic Solutions and Almost Periodic Solutions* (Springer, New York, 1975)
156. J.B. Rosser Jr., *From Catastrophe to Chaos: A General Theory of Economic Discontinuities*, 2nd edn. (Kluwer Academic Publishers, Norwell, 2000)

Chapter 2
Replication of Continuous Chaos About Equilibria

To approve stable chaotic motions by utilizing the input–output analysis one needs base-systems with attractors. The simplest attractors are equilibriums. This is why we start with perturbation of linear systems with constant coefficients and globally asymptotically stable equilibriums. In this chapter, we introduce chaotic sets of functions, the generator and replicator of chaos, precise description of ingredients for Devaney and Li–Yorke chaos in continuous dynamics. Moreover, we shall discuss morphogenesis phenomenon, hyperbolic set of functions, intermittency, chaos control, the double-scroll Chua's attractor, and quasiperiodicity. Appropriate simulations which confirm the theoretical results are provided. We consider the morphogenesis concept, since it helps us to describe in the most general form the expansion of chaos, which is not only an enlargement of the dimension of chaotic systems, but also saving properties of chaos during the extension.

2.1 Introduction

It is known that if one considers the evolution equation $u' = L[u] + I(t)$, where $L[u]$ is a linear operator with spectra placed in the left half of the complex plane, then a function $I(t)$ being considered as an *input* with a certain property (boundedness, periodicity, almost periodicity) produces through the equation the *output*, a solution with a similar property, boundedness/periodicity/almost periodicity [1–4].

A reasonable question appears whether it is possible to use as input a chaotic motion and to obtain output also as a chaos of certain type. The present chapter is devoted to answer this question even if the input is inserted nonlinearly. One must say that we consider as an input first of all a single function, a member of a chaotic set to obtain a solution, which is a member of another chaotic set. Besides that we consider the chaotic sets as the input and the output. We have been forced to

M. Akhmet and M.O. Fen, *Replication of Chaos in Neural Networks, Economics and Physics*, Nonlinear Physical Science,
DOI 10.1007/978-3-662-47500-3_2

consider sets of functions as inputs and outputs, since Devaney or Li–Yorke chaos are indicated through relation of motions (sensitivity, transitiveness, proximality). Thus, we consider the input and the output not only as single functions, but also as collections of functions. The way of our investigation is arranged in the well-accepted traditional mathematical fashion, but with a new and a more complex way of arrangement of the connections between the input and the output.

Since the concept of chaos is much more complex than just single periodic or almost periodic solutions, we have to use a special terminology for the chaos generation through the input–output mechanism, *replication of chaos*.

The technique of the replication used in this chapter is as follows. We need a source of chaotic inputs, but mostly chaos can be obtained through solving differential or difference equations. For this reason, we use special generator systems as the source of chaos or chaotic functions. Nevertheless, we emphasize that the generator is not necessarily the element of the replication procedure since it can be replaced by another source of a chaotic input, and in applications present result may be considered with, for example, chaotic inputs obtained from experimental activity. So, initially, we take into account a system of differential equations (the generator) which produces chaos, and we use this system to influence in a unidirectional way, another system (the replicator) in such a manner that the replicator mimics the same ingredients of chaos provided to the generator. In the present chapter, we use such ingredients in the form of period-doubling cascade, Devaney and Li–Yorke chaos. For the study of the subject, we introduce new definitions such as chaotic sets of functions, the generator and replicator of chaos, and precise description of ingredients for Devaney and Li–Yorke chaos in continuous dynamics.

Throughout the chapter, *the generator* will be considered as a system of the form

$$x' = F(t, x), \tag{2.1.1}$$

where $F : \mathbb{R} \times \mathbb{R}^m \to \mathbb{R}^m$ is a continuous function in all its arguments, and *the replicator* is assumed to have the form

$$y' = Ay + g(x(t), y), \tag{2.1.2}$$

where $g : \mathbb{R}^m \times \mathbb{R}^n \to \mathbb{R}^n$ is a continuous function in all its arguments, the constant $n \times n$ real-valued matrix A has real parts of eigenvalues all negative and the function $x(t)$ is a solution of system (2.1.1).

We consider, in this chapter, the linear equation

$$z' = Az \tag{2.1.3}$$

as *the base-system*. The condition on eigenvalues of matrix A implies that the base-system admits asymptotically stable equilibrium. The generator–replicator couple, (2.1.1)+(2.1.2), will be called in the remaining parts of the chapter as the *result-system*.

Now, to illustrate the replication mechanism discussed in this chapter, let us consider the following example. For our purposes, as the generator we shall take into account the Duffing's oscillator represented by the differential equation

$$x'' + 0.05x' + x^3 = 7.5 \cos t. \tag{2.1.4}$$

It is known that Eq. (2.1.4) possesses a chaotic attractor [5]. Defining the variables $x_1 = x$ and $x_2 = x'$, Eq. (2.1.4) can be reduced to the system

$$\begin{aligned} x_1' &= x_2 \\ x_2' &= -0.05x_2 - x_1^3 + 7.5 \cos t. \end{aligned} \tag{2.1.5}$$

Next, let us consider the following system:

$$\begin{aligned} x_3' &= x_4 + x_1(t) \\ x_4' &= -3x_3 - 2x_4 - 0.008x_3^3 + x_2(t). \end{aligned} \tag{2.1.6}$$

In this form system (2.1.6) is a replicator. One has to emphasize that the linear part of the associated with (2.1.6) non-perturbed system

$$\begin{aligned} x_3' &= x_4 \\ x_4' &= -3x_3 - 2x_4 - 0.008x_3^3, \end{aligned} \tag{2.1.7}$$

has eigenvalues with negative real parts and does not admit chaos.

Figure 2.1 shows the trajectory of system (2.1.7) with $x_3(0) = -2$ and $x_4(0) = 1$. It is seen in the figure that the behavior of the solution is non-chaotic.

To visualize the process of replication by the result-system, (2.1.5) + (2.1.6), let us consider the Poincaré sections of the both. By marking the trajectory of this system with the initial data $x_1(0) = 2$, $x_2(0) = 3$, $x_3(0) = -1$, $x_4(0) = 1$ stroboscopically

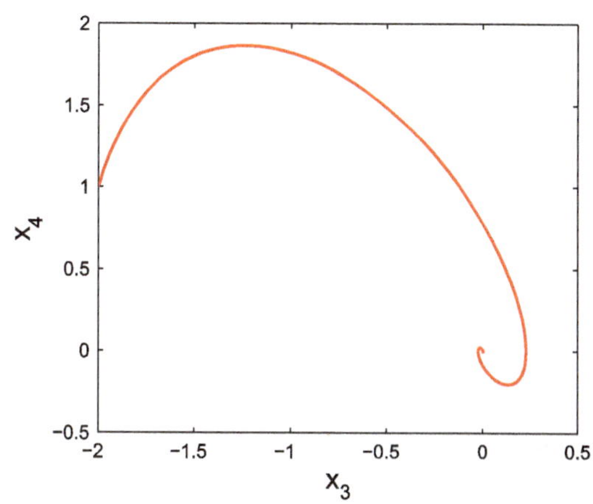

Fig. 2.1 The trajectory of system (2.1.7) with $x_3(0) = -2$ and $x_4(0) = 1$

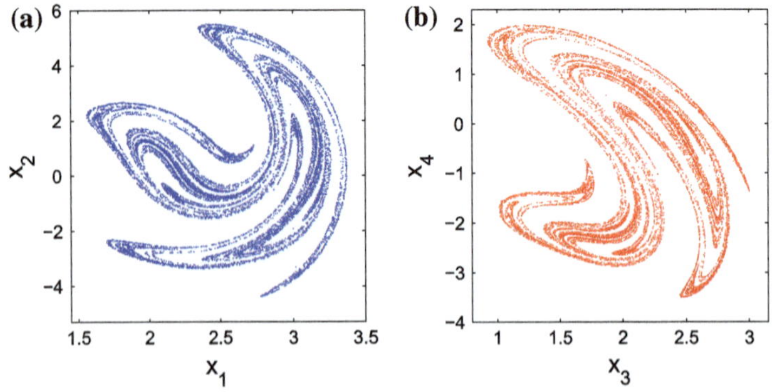

Fig. 2.2 The picture in **a** not only represents the projection of the whole attractor on the $x_1 - x_2$ plane, but also the strange attractor of the generator. In a similar way, the picture introduced in **b** represents the chaotic attractors of the replicator. The presented chaotic attractors of the generator and the replicator systems reveal that the chaos replication mechanism works consummately

at times that are integer multiples of 2π, we obtain the Poincaré section and in Fig. 2.2, where the chaos replication is apparent, we illustrate its 2-dimensional projections. Figure 2.2a represents the projection of the Poincaré section on the $x_1 - x_2$ plane, and we note that this projection is in fact the strange attractor of the generator system (2.1.5). On the other hand, the projection on the $x_3 - x_4$ plane presented in Fig. 2.2b is the attractor corresponding to the replicator system (2.1.6). One can see that the attractor indicated in Fig. 2.2b repeated the structure of the attractor shown in Fig. 2.2a and this result is a manifestation of the replication of chaos. One has to think about mathematical aspects of this phenomena and in this chapter we handle this problem.

In our theoretical results, we use coupled systems in which the generator influences the replicator in a unidirectional way. In other words, the generator affects the behavior of the replicator counterpart in such a way that the solutions of the generator are used as an input for the latter. The possibility of making use of more than one replicator systems with arbitrarily high dimensions in the extension mechanism is an advantage of our procedure. Moreover, we are describing a process involving the replication of chaos which does not occur in the course of time, but instead an **instantaneous** one. In other words, the prior chaos is mimicked in all existing replicators such that the generating mechanism works through arranging connections between systems not with the lapse of time.

Since we do not restrict ourselves in this chapter with a simple couple *the generator–the replicator*, but get them in different combinations and numbers, having the geometric features of chaos saved, we shall call the extension of chaos as *morphogenesis*.

In the next section we will present assumptions for systems (2.1.1) and (2.1.2) which are needed for the chaos replication, and introduce the chaotic attractors of these systems in the functional sense.

2.2 Preliminaries

Throughout the chapter, $\mathbb{R}$ and $\mathbb{N}$ will denote the sets of real numbers and natural numbers, respectively. We will make use of the usual Euclidean norm for vectors and the norm induced by the Euclidean norm for square matrices [6], that is,

$$\|\Gamma\| = \max\left\{\sqrt{\varsigma} : \varsigma \text{ is an eigenvalue of } \Gamma^T\Gamma\right\}$$

for any square matrix Γ with real entries, and Γ^T denotes the transpose of the matrix Γ.

Since the matrix A, which is aforementioned in system (2.1.2), is supposed to admit eigenvalues all with negative real parts, it is easy to verify the existence of positive numbers N and ω such that $\left\|e^{At}\right\| \le Ne^{-\omega t}$, $t \ge 0$. These numbers will be used in the last condition below.

The following assumptions on systems (2.1.1) and (2.1.2) are needed throughout the chapter:

(A1) There exists a positive number T such that the function $F(t,x)$ satisfies the periodicity condition

$$F(t+T,x) = F(t,x),$$

for all $t \in \mathbb{R}$, $x \in \mathbb{R}^m$;

(A2) There exists a positive number L_0 such that

$$\|F(t,x_1) - F(t,x_2)\| \le L_0 \|x_1 - x_2\|,$$

for all $t \in \mathbb{R}$, $x_1, x_2 \in \mathbb{R}^m$;

(A3) There exists a positive number $H_0 < \infty$ such that

$$\sup_{t\in\mathbb{R},x\in\mathbb{R}^m} \|F(t,x)\| \le H_0;$$

(A4) There exists a positive number L_1 such that

$$\|g(x_1,y) - g(x_2,y)\| \ge L_1 \|x_1 - x_2\|,$$

for all $x_1, x_2 \in \mathbb{R}^m$, $y \in \mathbb{R}^n$;

(A5) There exist positive numbers L_2 and L_3 such that

$$\|g(x_1,y) - g(x_2,y)\| \le L_2 \|x_1 - x_2\|,$$

for all $x_1, x_2 \in \mathbb{R}^m$, $y \in \mathbb{R}^n$, and

$$\|g(x,y_1) - g(x,y_2)\| \le L_3 \|y_1 - y_2\|,$$

for all $x \in \mathbb{R}^m$, $y_1, y_2 \in \mathbb{R}^n$;

(A6) There exists a positive number $M_0 < \infty$ such that

$$\sup_{x \in \mathbb{R}^m, y \in \mathbb{R}^n} \|g(x, y)\| \leq M_0;$$

(A7) $NL_3 - \omega < 0$.

Remark 2.1 The results presented in the remaining parts are also true even if we replace the nonautonomous system (2.1.1) by the autonomous equation

$$x' = \overline{F}(x), \tag{2.2.8}$$

where the function $\overline{F} : \mathbb{R}^m \to \mathbb{R}^m$ is continuous with conditions which are counterparts of $(A2)$ and $(A3)$.

At the present time, systems of differential equations which are known to exhibit chaotic behavior are either nonautonomous and periodic in time such as the Duffing and Van der Pol oscillators or autonomous such as the Lorenz, Chua and Rössler systems. In a similar way, in our investigations of chaos generation, we take advantage of periodic nonautonomous systems as well as autonomous ones as generators.

Using the theory of quasilinear equations [7], one can verify that for a given solution $x(t)$ of system (2.1.1), there exists a unique bounded on $\mathbb{R}$ solution $y(t)$ of the system $y' = Ay + g(x(t), y)$, denoted by $y(t) = \phi_{x(t)}(t)$, which satisfies the integral equation

$$y(t) = \int_{-\infty}^{t} e^{A(t-s)} g(x(s), y(s)) ds. \tag{2.2.9}$$

Our main assumption is the existence of a nonempty set $\mathscr{A}_x$ of all solutions of system (2.1.1), uniformly bounded on $\mathbb{R}$. That is, there exists a positive number H such that $\sup_{t \in \mathbb{R}} \|x(t)\| \leq H$, for all $x(t) \in \mathscr{A}_x$.

Let us introduce the sets of functions

$$\mathscr{A}_y = \left\{ \phi_{x(t)}(t) \mid x(t) \in \mathscr{A}_x \right\}, \tag{2.2.10}$$

and

$$\mathscr{A} = \left\{ (x(t), \phi_{x(t)}(t)) \mid x(t) \in \mathscr{A}_x \right\}. \tag{2.2.11}$$

We note that for all $y(t) \in \mathscr{A}_y$ one has $\sup_{t \in \mathbb{R}} \|y(t)\| \leq M$, where $M = \dfrac{NM_0}{\omega}$.

Next, we reveal that if the set $\mathscr{A}_x$ is an attractor with basin $\mathscr{U}_x$, that is, for each $x(t) \in \mathscr{U}_x$ there exists $\overline{x}(t) \in \mathscr{A}_x$ such that $\|x(t) - \overline{x}(t)\| \to 0$ as $t \to \infty$, then the set $\mathscr{A}_y$ is also an attractor in the same sense. Denote by $\mathscr{U}_y$ the set consisting of all solutions of system $y' = Ay + g(x(t), y)$, where $x(t) \in \mathscr{U}_x$. In the next lemma we specify the basin of attraction of $\mathscr{A}_y$.

Lemma 2.1 *$\mathscr{U}_y$ is a basin of $\mathscr{A}_y$.*

Proof Fix an arbitrary positive number ε and let $y(t) \in \mathscr{U}_y$ be a given solution of the system $y' = Ay + g(x(t), y)$ for some $x(t) \in \mathscr{U}_x$. In this case, there exists $\overline{x}(t) \in \mathscr{A}_x$ such that $\|x(t) - \overline{x}(t)\| \to 0$ as $t \to \infty$. Let $\alpha = \dfrac{\omega - NL_3}{\omega - NL_3 + NL_2}$ and $\overline{y}(t) = \phi_{\overline{x}(t)}(t)$. Condition $(A7)$ implies that the number α is positive. Under the circumstances, one can find $R_0 = R_0(\varepsilon) > 0$ such that if $t \geq R_0$, then $\|x(t) - \overline{x}(t)\| < \alpha\varepsilon$ and $N \|y(R_0) - \overline{y}(R_0)\| e^{(NL_3-\omega)t} < \alpha\varepsilon$. The functions $y(t)$ and $\overline{y}(t)$ satisfy the relations

$$y(t) = e^{A(t-R_0)}y(R_0) + \int_{R_0}^{t} e^{A(t-s)}g(x(s), y(s))ds,$$

and

$$\overline{y}(t) = e^{A(t-R_0)}\overline{y}(R_0) + \int_{R_0}^{t} e^{A(t-s)}g(\overline{x}(s), \overline{y}(s))ds,$$

respectively. Making use of these relations, one can verify that

$$\begin{aligned} y(t) - \overline{y}(t) &= e^{A(t-R_0)}(y(R_0) - \overline{y}(R_0)) \\ &+ \int_{R_0}^{t} e^{A(t-s)} [g(x(s), y(s)) - g(x(s), \overline{y}(s))] \, ds \\ &+ \int_{R_0}^{t} e^{A(t-s)} [g(x(s), \overline{y}(s)) - g(\overline{x}(s), \overline{y}(s))] \, ds. \end{aligned}$$

Therefore, we have

$$\|y(t) - \overline{y}(t)\| \leq Ne^{-\omega(t-R_0)} \|y(R_0) - \overline{y}(R_0)\| + \frac{NL_2\alpha\varepsilon}{\omega} e^{-\omega t} \left(e^{\omega t} - e^{\omega R_0}\right)$$
$$+NL_3 \int_{R_0}^{t} e^{-\omega(t-s)} \|y(s) - \overline{y}(s)\| \, ds.$$

Let $u : [R_0, \infty) \to [0, \infty)$ be a function defined as $u(t) = e^{\omega t} \|y(t) - \overline{y}(t)\|$. By means of this definition, we reach the inequality

$$u(t) \leq Ne^{\omega R_0} \|y(R_0) - \overline{y}(R_0)\| + \frac{NL_2\alpha\varepsilon}{\omega} \left(e^{\omega t} - e^{\omega R_0}\right) + NL_3 \int_{R_0}^{t} u(s)ds.$$

Now, let $\psi(t) = \dfrac{NL_2\alpha\varepsilon}{\omega} e^{\omega t}$ and $\phi(t) = \psi(t) + c$, where

$$c = Ne^{\omega R_0} \|y(R_0) - \overline{y}(R_0)\| - \frac{NL_2\alpha\varepsilon}{\omega} e^{\omega R_0}.$$

Using these functions we get

$$u(t) \leq \phi(t) + NL_3 \int_{R_0}^{t} u(s)ds.$$

Applying Gronwall's Lemma [8] to the last inequality for $t \geq R_0$, we attain the inequality

$$u(t) \leq c + \psi(t) + NL_3 \int_{R_0}^{t} e^{NL_3(t-s)} cds + NL_3 \int_{R_0}^{t} e^{NL_3(t-s)} \psi(s)ds$$

and hence,

$$\begin{aligned}
&u(t) \leq c + \psi(t) + c\left(e^{NL_3(t-R_0)} - 1\right) \\
&+\frac{N^2L_2L_3\alpha\varepsilon}{\omega(\omega - NL_3)} e^{\omega t}\left(1 - e^{(NL_3-\omega)(t-R_0)}\right) \\
&= \frac{NL_2\alpha\varepsilon}{\omega} e^{\omega t} + N\|y(R_0) - \overline{y}(R_0)\| e^{\omega R_0} e^{NL_3(t-R_0)} \\
&-\frac{NL_2\alpha\varepsilon}{\omega} e^{\omega R_0} e^{NL_3(t-R_0)} + \frac{N^2L_2L_3\alpha\varepsilon}{\omega(\omega - NL_3)} e^{\omega t}\left(1 - e^{(NL_3-\omega)(t-R_0)}\right).
\end{aligned}$$

Thus,

$$\begin{aligned}
&\|y(t) - \overline{y}(t)\| \leq \frac{NL_2\alpha\varepsilon}{\omega} + N\|y(R_0) - \overline{y}(R_0)\| e^{(NL_3-\omega)(t-R_0)} \\
&-\frac{NL_2\alpha\varepsilon}{\omega} e^{(NL_3-\omega)(t-R_0)} + \frac{N^2L_2L_3\alpha\varepsilon}{\omega(\omega - NL_3)}\left(1 - e^{(NL_3-\omega)(t-R_0)}\right) \\
&< N\|y(R_0) - \overline{y}(R_0)\| e^{(NL_3-\omega)(t-R_0)} + \frac{NL_2\alpha\varepsilon}{\omega - NL_3}.
\end{aligned}$$

Consequently, for $t \geq 2R_0$, we have that

$$\|y(t) - \overline{y}(t)\| < \left(1 + \frac{NL_2}{\omega - NL_3}\right)\alpha\varepsilon = \varepsilon,$$

and hence $\|y(t) - \overline{y}(t)\| \to 0$ as $t \to \infty$.

The proof of the lemma is completed. □

Now, let us define the set $\mathscr{U}$ consisting the solutions $(x(t), y(t))$ of system (2.1.1)+(2.1.2), where $x(t) \in \mathscr{U}_x$. Next, we state the following corollary of Lemma 2.1.

Corollary 2.1 *$\mathscr{U}$ is a basin of $\mathscr{A}$.*

Proof Let $(x(t), y(t)) \in \mathscr{U}$ be a given solution of system (2.1.1)+(2.1.2). According to Lemma 2.1, there exists a function $(\overline{x}(t), \overline{y}(t)) \in \mathscr{A}$ such that both $\|x(t) - \overline{x}(t)\|$ and $\|y(t) - \overline{y}(t)\|$ tend to 0 as t tends to ∞. Consequently,

$$\|(x(t), y(t)) - (\overline{x}(t), \overline{y}(t))\| \to 0 \text{ as } t \to \infty.$$

The proof is finalized. □

2.3 Chaotic Sets of Functions

In this section, the descriptions for the chaotic sets of continuous functions will be introduced and the definitions of the chaotic features will be presented, both in the Devaney's sense and in the sense of Li–Yorke.

Let us denote by

$$\mathscr{B} = \{\psi(t) \mid \psi : \mathbb{R} \to K \text{ is continuous}\} \tag{2.3.12}$$

a collection of functions, where $K \subset \mathbb{R}^q$, $q \in \mathbb{N}$, is a bounded region.

We start with the description of chaotic sets of functions in Devaney's sense and then continue with the Li–Yorke counterpart.

2.3.1 Devaney Set of Functions

In this part, we shall elucidate the ingredients of the chaos in Devaney's sense for the set $\mathscr{B}$, which is introduced by (2.3.12), and the first definition is about the sensitivity of chaotic sets of functions.

Definition 2.1 $\mathscr{B}$ is called sensitive if there exist positive numbers ε and Δ such that for every $\psi(t) \in \mathscr{B}$ and for arbitrary $\delta > 0$ there exist $\overline{\psi}(t) \in \mathscr{B}$, $t_0 \in \mathbb{R}$ and an interval $J \subset [t_0, \infty)$, with length not less than Δ, such that $\left\|\psi(t_0) - \overline{\psi}(t_0)\right\| < \delta$ and $\left\|\psi(t) - \overline{\psi}(t)\right\| > \varepsilon$, for all $t \in J$.

Definition 2.1 considers the inequality ($>\varepsilon$) over the interval J. In the Devaney's chaos definition for the map, the inequality is assumed for discrete moments. Let us reveal how one can extend the inequality from a discrete point to an interval by considering continuous flows.

In [9], it is indicated that a continuous map $\varphi : \Lambda \to \Lambda$, with an invariant domain $\Lambda \subset \mathbb{R}^k$, $k \in \mathbb{N}$, has sensitive dependence on initial conditions if there exists $\overline{\varepsilon} > 0$ such that for any $x \in \Lambda$ and any neighborhood $\mathscr{U}$ of x, there exist $y \in \mathscr{U}$ and a natural number n such that $\|\varphi^n(x) - \varphi^n(y)\| > \overline{\varepsilon}$.

Suppose that the set $\mathscr{A}_x$ satisfies the definition of sensitivity in the following sense: There exists $\overline{\varepsilon} > 0$ such that for every $x(t) \in \mathscr{A}_x$ and arbitrary $\delta > 0$, there exist

$\overline{x}(t) \in \mathscr{A}_x$, $t_0 \in \mathbb{R}$ and a real number $\zeta \geq t_0$ such that $\|x(t_0) - \overline{x}(t_0)\| < \delta$ and $\|x(\zeta) - \overline{x}(\zeta)\| > \overline{\varepsilon}$.

In this case, for given $x(t) \in \mathscr{A}_x$ and $\delta > 0$, one can find $\overline{x}(t) \in \mathscr{A}_x$ and $\zeta \geq t_0$ such that $\|x(t_0) - \overline{x}(t_0)\| < \delta$ and $\|x(\zeta) - \overline{x}(\zeta)\| > \overline{\varepsilon}$. Let $\Delta = \dfrac{\overline{\varepsilon}}{8HL_0}$ and take a number Δ_1 such that $\Delta \leq \Delta_1 \leq \dfrac{\overline{\varepsilon}}{4HL_0}$. Using appropriate integral equations for $t \in [\zeta, \zeta + \Delta_1]$, it can be verified that

$$\begin{aligned}\|x(t) - \overline{x}(t)\| &\geq \|x(\zeta) - \overline{x}(\zeta)\| - \left\| \int_\zeta^t [F(s, x(s)) - F(s, \overline{x}(s))]ds \right\| \\ &> \overline{\varepsilon} - 2HL_0\Delta_1 \\ &\geq \frac{\overline{\varepsilon}}{2}.\end{aligned}$$

The last inequality confirms that $\mathscr{A}_x$ satisfies Definition 2.1 with $\varepsilon = \overline{\varepsilon}/2$ and $J = [\zeta, \zeta + \Delta_1]$. So the definition is a natural one. It provides more information then discrete moments and for us it is important that the extension on the interval is useful to prove the property for chaos extension.

In the next two definitions, we continue with the existence of a dense function in the set of chaotic functions followed by the transitivity property.

Definition 2.2 $\mathscr{B}$ possesses a dense function $\psi^*(t) \in \mathscr{B}$ if for every function $\psi(t) \in \mathscr{B}$, arbitrary small $\varepsilon > 0$ and arbitrary large $E > 0$, there exist a number $\xi > 0$ and an interval $J \subset \mathbb{R}$, with length E, such that $\|\psi(t) - \psi^*(t + \xi)\| < \varepsilon$, for all $t \in J$.

Definition 2.3 $\mathscr{B}$ is called transitive if it possesses a dense function.

Now, let us recall the definition of transitivity for maps [9]. A continuous map φ with an invariant domain $\Lambda \subset \mathbb{R}^k, k \in \mathbb{N}$, possesses a dense orbit if there exists $c^* \in \Lambda$ such that for each $c \in \Lambda$ and arbitrary number $\varepsilon > 0$, there exist natural numbers k_0 and l_0 such that $\left\|\varphi^{l_0}(c) - \varphi^{l_0+k_0}(c^*)\right\| < \varepsilon$, and maps which have dense orbits are called transitive.

Suppose that $\mathscr{A}_x$ satisfies the transitivity property in the following sense. There exists a function $x^*(t) \in \mathscr{A}_x$ such that for each $x(t) \in \mathscr{A}_x$ and arbitrary positive number ε, there exist a real number ζ_0 and a natural number m_0 such that $\|x(\zeta_0) - x^*(\zeta_0 + m_0T)\| < \varepsilon$.

Fix an arbitrary function $x(t) \in \mathscr{A}_x$, arbitrary small $\varepsilon > 0$ and arbitrary large $E > 0$. Under the circumstances, one can find $\zeta_0 \in \mathbb{R}$ and $m_0 \in \mathbb{N}$ such that $\|x(\zeta_0) - x^*(\zeta_0 + m_0T)\| < \varepsilon e^{-L_0E}$.

Using the condition $(A2)$ together with the convenient integral equations that $x(t)$ and $x^*(t)$ satisfy, it is easy to obtain for $t \in [\zeta_0, \zeta_0 + E]$ that

$$\begin{aligned}&\left\|x(t) - x^*(t + m_0T)\right\| \leq \left\|x(\zeta_0) - x^*(\zeta_0 + m_0T)\right\| \\ &+ \int_{\zeta_0}^t L_0 \left\|x(s) - x^*(s + m_0T)\right\| ds,\end{aligned}$$

and by the help of the Gronwall–Bellman inequality [2], we get

$$\left\| x(t) - x^*(t + m_0 T) \right\| \le \left\| x(\zeta_0) - x^*(\zeta_0 + m_0 T) \right\| e^{L_0(t-\zeta_0)} < \varepsilon.$$

The last inequality shows that the set $\mathscr{A}_x$ satisfies Definition 2.2 with $\xi = k_0 T$ and is transitive in accordance with Definition 2.3.

The following definition describes the density of periodic functions inside $\mathscr{B}$.

Definition 2.4 $\mathscr{B}$ admits a dense collection $\mathscr{G} \subset \mathscr{B}$ of periodic functions if for every function $\psi(t) \in \mathscr{B}$, arbitrary small $\varepsilon > 0$ and arbitrary large $E > 0$, there exist $\widetilde{\psi}(t) \in \mathscr{G}$ and an interval $J \subset \mathbb{R}$, with length E, such that $\left\| \psi(t) - \widetilde{\psi}(t) \right\| < \varepsilon$, for all $t \in J$.

Let us remind the definition of density of periodic orbits for maps [9]. The set of periodic orbits of a continuous map φ with an invariant domain $\Lambda \subset \mathbb{R}^k, k \in \mathbb{N}$, is called dense in Λ if for each $c \in \Lambda$, arbitrary positive number ε, there exist a natural number l_0 and a point $\widetilde{c} \in \Lambda$ such that the sequence $\left\{\varphi^i\left(\widetilde{c}\right)\right\}$ is periodic and $\left\| \varphi^{l_0}(c) - \varphi^{l_0}\left(\widetilde{c}\right) \right\| < \varepsilon$.

Let us denote by $\mathscr{G}_x$ the set of all periodic functions inside $\mathscr{A}_x$. Suppose that $\mathscr{A}_x$ satisfies density of periodic solutions as follows. For an arbitrary function $x(t) \in \mathscr{A}_x$ and arbitrary small $\varepsilon > 0$ there exist a periodic function $\widetilde{x}(t) \in \mathscr{G}_x$ and a number $\zeta_0 \in \mathbb{R}$ such that $\|x(\zeta_0) - \widetilde{x}(\zeta_0)\| < \varepsilon$.

Let us fix an arbitrary function $x(t) \in \mathscr{A}_x$, arbitrary small $\varepsilon > 0$ and arbitrary large $E > 0$. In that case, there exist a periodic function $\widetilde{x}(t) \in \mathscr{G}_x$ and $\zeta_0 \in \mathbb{R}$ such that $\|x(\zeta_0) - \widetilde{x}(\zeta_0)\| < \varepsilon e^{-L_0 E}$.

It can be easily verified for $t \in [\zeta_0, \zeta_0 + E]$ that the inequality

$$\|x(t) - \widetilde{x}(t)\| \le \|x(\zeta_0) - \widetilde{x}(\zeta_0)\| + \int_{\zeta_0}^{t} L_0 \|x(s) - \widetilde{x}(s)\| \, ds,$$

holds, and therefore for each t from the same interval of time we have

$$\|x(t) - \widetilde{x}(t)\| \le \|x(\zeta_0) - \widetilde{x}(\zeta_0)\| \, e^{L_0(t-\zeta_0)} < \varepsilon.$$

Consequently, the set $\mathscr{A}_x$ satisfies Definition 2.4 with $J = [\zeta_0, \zeta_0 + E]$.

Finally, we introduce in the next definition the chaotic set of functions in Devaney's sense.

Definition 2.5 The collection $\mathscr{B}$ of functions is called a Devaney's chaotic set if

(D1) $\mathscr{B}$ is sensitive;
(D2) $\mathscr{B}$ is transitive;
(D3) $\mathscr{B}$ admits a dense collection of periodic functions.

In the next subsection, the chaotic properties of the set $\mathscr{B}$ will be imposed in the sense of Li–Yorke.

2.3.2 *Li–Yorke Set of Functions*

The ingredients of Li–Yorke chaos for the collection of functions $\mathscr{B}$, which is defined by (2.3.12), will be described in this part. Making use of discussions similar to the ones made in the previous subsection, we extend, below, the definitions for the ingredients of Li–Yorke chaos from maps [10–13] to continuous flows and we just omit these indications here.

Definition 2.6 A couple of functions $\left(\psi(t), \overline{\psi}(t)\right) \in \mathscr{B} \times \mathscr{B}$ is called proximal if for arbitrary small $\varepsilon > 0$ and arbitrary large $E > 0$, there exist infinitely many disjoint intervals of length not less than E such that $\left\|\psi(t) - \overline{\psi}(t)\right\| < \varepsilon$, for each t from these intervals.

Definition 2.7 A couple of functions $\left(\psi(t), \overline{\psi}(t)\right) \in \mathscr{B} \times \mathscr{B}$ is frequently (ε_0, Δ)-separated if there exist positive numbers ε_0, Δ and infinitely many disjoint intervals of length no less than Δ, such that $\left\|\psi(t) - \overline{\psi}(t)\right\| > \varepsilon_0$, for each t from these intervals.

Remark 2.2 The numbers ε_0 and Δ taken into account in Definition 2.7 depend on the functions $\psi(t)$ and $\overline{\psi}(t)$.

Definition 2.8 A couple of functions $\left(\psi(t), \overline{\psi}(t)\right) \in \mathscr{B} \times \mathscr{B}$ is a Li–Yorke pair if they are proximal and frequently (ε_0, Δ)-separated for some positive numbers ε_0 and Δ.

Definition 2.9 An uncountable set $\mathscr{C} \subset \mathscr{B}$ is called a scrambled set if $\mathscr{C}$ does not contain any periodic functions and each couple of different functions inside $\mathscr{C} \times \mathscr{C}$ is a Li–Yorke pair.

Definition 2.10 $\mathscr{B}$ is called a Li–Yorke chaotic set if

(LY1) There exists a positive number T_0 such that $\mathscr{B}$ admits a periodic function of period kT_0, for any $k \in \mathbb{N}$;
(LY2) $\mathscr{B}$ possesses a scrambled set $\mathscr{C}$;
(LY3) For any function $\psi(t) \in \mathscr{C}$ and any periodic function $\overline{\psi}(t) \in \mathscr{B}$, the couple $\left(\psi(t), \overline{\psi}(t)\right)$ is frequently (ε_0, Δ)—separated for some positive numbers ε_0 and Δ.

2.4 Hyperbolic Set of Functions

The definitions of stable and unstable sets of hyperbolic periodic orbits of autonomous systems are given in [14], and information about such sets of solutions of perturbed nonautonomous systems can be found in [15]. Moreover, homoclinic structures in almost periodic systems were studied in [16–18]. In this section, we give a definition for hyperbolic collection of uniformly bounded functions and before this, we start with the descriptions of stable and unstable sets of a function.

We define the stable set of a function $\psi(t) \in \mathscr{B}$, where the collection $\mathscr{B}$ is defined by (2.3.12), as the set of functions

$$W^s(\psi(t)) = \{u(t) \in \mathscr{B} \mid \|u(t) - \psi(t)\| \to 0 \text{ as } t \to \infty\}, \tag{2.4.13}$$

and, similarly, we define the unstable set of a function $\psi(t) \in \mathscr{B}$ as the set of functions

$$W^u(\psi(t)) = \{v(t) \in \mathscr{B} \mid \|v(t) - \psi(t)\| \to 0 \text{ as } t \to -\infty\}. \tag{2.4.14}$$

Definition 2.11 The set of functions $\mathscr{B}$ is called hyperbolic if the stable and unstable sets of each function $\psi(t) \in \mathscr{B}$ possess at least one element different from $\psi(t)$.

Theorem 2.1 *If $\mathscr{A}_x$ is hyperbolic, then the same is true for $\mathscr{A}_y$.*

Proof Fix an arbitrary positive number ε and a function $y(t) = \phi_{x(t)}(t) \in \mathscr{A}_y$. Let $\alpha = \dfrac{\omega - NL_3}{\omega - NL_3 + NL_2}$ and $\beta = \dfrac{\omega - NL_3}{1 + NL_2}$. By condition $(A7)$, one can verify that the numbers α and β are both positive.

Due to hyperbolicity of $\mathscr{A}_x$, the function $x(t)$ has a nonempty stable set $W^s(x(t))$ and a nonempty unstable set $W^u(x(t))$.

Let us take an arbitrary function $u(t) \in W^s(x(t))$. Since $\|x(t) - u(t)\| \to 0$ as $t \to \infty$ and $NL_3 - \omega < 0$, there exists a positive number R_1, which depends on ε, such that $\|x(t) - u(t)\| < \alpha\varepsilon$ and $e^{(NL_3-\omega)t} < \dfrac{\omega\alpha\varepsilon}{2M_0N}$ for $t \ge R_1$. Let $\overline{y}(t) = \phi_{u(t)}(t)$. We shall prove that the function $\overline{y}(t)$ belongs to the stable set of $y(t)$.

The bounded on $\mathbb{R}$ functions $y(t)$ and $\overline{y}(t)$ satisfy the relations

$$y(t) = \int_{-\infty}^{t} e^{A(t-s)} g(x(s), y(s))ds,$$

and

$$\overline{y}(t) = \int_{-\infty}^{t} e^{A(t-s)} g(u(s), \overline{y}(s))ds,$$

respectively, for $t \ge R_1$.

Therefore, one can easily reach up the equation

$$y(t) - \overline{y}(t) = \int_{-\infty}^{R_1} e^{A(t-s)}[g(x(s), y(s)) - g(u(s), \overline{y}(s))]ds$$
$$+ \int_{R_1}^{t} e^{A(t-s)}\Big\{[g(x(s), y(s)) - g(x(s), \overline{y}(s))]$$
$$+ [g(x(s), \overline{y}(s)) - g(u(s), \overline{y}(s))]\Big\}ds,$$

which implies that

$$\|y(t) - \overline{y}(t)\| \le \int_{-\infty}^{R_1} 2M_0Ne^{-\omega(t-s)}ds$$
$$+ \int_{R_1}^{t} e^{-\omega(t-s)}\left(NL_3\|y(s) - \overline{y}(s)\| + NL_2\|x(s) - u(s)\|\right)ds$$
$$\le \frac{2M_0N}{\omega}e^{-\omega(t-R_1)} + \int_{R_1}^{t} e^{-\omega(t-s)}\left(NL_3\|y(s) - \overline{y}(s)\| + NL_2\alpha\varepsilon\right)ds.$$

Using the Gronwall type inequality indicated in [19], we obtain for $t \ge R_1$ that

$$\|y(t) - \overline{y}(t)\| \le \frac{2M_0N}{\omega}e^{(NL_3-\omega)(t-R_1)} + \frac{NL_2\alpha\varepsilon}{\omega - NL_3}[1 - e^{(NL_3-\omega)(t-R_1)}].$$

For this reason, for all $t \ge 2R_1$, one has

$$\|y(t) - \overline{y}(t)\| \le \frac{2M_0N}{\omega}e^{(NL_3-\omega)R_1} + \frac{NL_2\alpha\varepsilon}{\omega - NL_3} < \left(1 + \frac{NL_2}{\omega - NL_3}\right)\alpha\varepsilon = \varepsilon.$$

The last inequality implies that $\|y(t) - \overline{y}(t)\| \to 0$ as $t \to \infty$. Hence, the function $\overline{y}(t)$ belongs to the stable set $W^s(y(t))$ of $y(t)$.

On the other hand, let $v(t)$ be a function inside the unstable set $W^u(x(t))$. Since $\|x(t) - v(t)\|$ tends to 0 as $t \to -\infty$, there exists a negative number $R_2(\varepsilon)$ such that $\|x(t) - v(t)\| < \beta\varepsilon$ for $t \le R_2$. Let $\widetilde{y}(t) = \phi_{v(t)}(t)$. Now, our purpose is to show that $\widetilde{y}(t)$ belongs to the unstable set of $y(t)$.

By the help of the integral equations

$$y(t) = \int_{-\infty}^{t} e^{A(t-s)}g(x(s), y(s))ds,$$

and

$$\widetilde{y}(t) = \int_{-\infty}^{t} e^{A(t-s)}g(v(s), \widetilde{y}(s))ds,$$

we obtain that

$$y(t) - \widetilde{y}(t) = \int_{-\infty}^{t} e^{A(t-s)}[g(x(s), y(s)) - g(v(s), y(s))]ds$$
$$+ \int_{-\infty}^{t} e^{A(t-s)}[g(v(s), y(s)) - g(v(s), \widetilde{y}(s))]ds.$$

Therefore, for $t \le R_2$, one has

$$\|y(t) - \widetilde{y}(t)\| \le \int_{-\infty}^{t} NL_2 e^{-\omega(t-s)} \|x(t) - v(t)\| \, ds$$
$$+ \int_{-\infty}^{t} e^{-\omega(t-s)} NL_3 \|y(s) - \widetilde{y}(s)\| \, ds$$
$$\le \frac{NL_2\beta\varepsilon}{\omega} + \frac{NL_3}{\omega} \sup_{t \le R_2} \|y(t) - \widetilde{y}(t)\| .$$

Hence,

$$\sup_{t \le R_2} \|y(t) - \widetilde{y}(t)\| \le \frac{NL_2\beta\varepsilon}{\omega} + \frac{NL_3}{\omega} \sup_{t \le R_2} \|y(t) - \widetilde{y}(t)\|$$

and

$$\sup_{t \le R_2} \|y(t) - \widetilde{y}(t)\| \le \frac{NL_2\beta\varepsilon}{\omega - NL_3} < \varepsilon.$$

The last inequality confirms that $\|y(t) - \widetilde{y}(t)\| \to 0$ as $t \to -\infty$. Therefore $\widetilde{y}(t) \in W^u(y(t))$.

Consequently, $\mathscr{A}_y$ is hyperbolic since $y(t)$ possesses both nonempty stable and unstable sets, denoted by $W^s(y(t))$ and $W^u(y(t))$, respectively. The theorem is proved. □

Theorem 2.1 implies the following corollary:

Corollary 2.2 *If $\mathscr{A}_x$ is hyperbolic, then the same is true for $\mathscr{A}$.*

Next, we continue with another corollary of Theorem 2.1, following the definitions of homoclinic and heteroclinic functions.

A function $\psi_1(t) \in \mathscr{B}$ is said to be homoclinic to the function $\psi_0(t) \in \mathscr{B}$, $\psi_0(t) \neq \psi_1(t)$, if $\psi_1(t) \in W^s(\psi_0(t)) \cap W^u(\psi_0(t))$.

On the other hand, a function $\psi_2(t) \in \mathscr{B}$ is called heteroclinic to the functions $\psi_0(t), \psi_1(t) \in \mathscr{B}$, $\psi_0(t) \neq \psi_2(t)$, $\psi_1(t) \neq \psi_2(t)$, if $\psi_2(t) \in W^s(\psi_0(t)) \cap W^u(\psi_1(t))$.

Corollary 2.3 *If $x_1(t) \in \mathscr{A}_x$ is homoclinic to the function $x_0(t) \in \mathscr{A}_x$, $x_0(t) \neq x(t)$, then $\phi_{x_1(t)}(t)$ is homoclinic to the function $\phi_{x_0(t)}(t)$.*

Similarly, if $x_2(t) \in \mathscr{A}_x$ *is heteroclinic to the functions* $x_0(t), x_1(t) \in \mathscr{A}_x$, $x_0(t) \neq x_2(t)$, $x_1(t) \neq x_2(t)$, *then* $\phi_{x_2(t)}(t)$ *is heteroclinic to the functions* $\phi_{x_0(t)}(t), \phi_{x_1(t)}(t)$.

In the next section, we theoretically prove that the set $\mathscr{A}_y$ replicates the ingredients of Devaney's chaos provided to the set $\mathscr{A}_x$, and as a consequence the same is valid also for the set $\mathscr{A}$. The same problem for the chaos in the sense of Li–Yorke will be handled in Sect. 2.6.

2.5 Replication of Devaney's Chaos

In this part, we will prove theoretically that the ingredients of Devaney's chaos furnished to the set $\mathscr{A}_x$ are all replicated by the set $\mathscr{A}_y$.

Suppose that the function $g(x, y)$ which is used in the right hand side of system (2.1.2) has component functions $g_j(x, y)$, $j = 1, 2, \ldots, n$. That is,

$$g(x, y) = \begin{pmatrix} g_1(x, y) \\ g_2(x, y) \\ \vdots \\ g_n(x, y) \end{pmatrix},$$

where each $g_j(x, y)$, $j = 1, 2, \ldots, n$, is a real valued function.

We start with the following assertion, which will be needed in the proof of Lemma 2.3.

Lemma 2.2 *The set of functions*

$$\mathscr{F} = \left\{ g_j(x(t), \phi_{x(t)}(t)) - g_j(\overline{x}(t), \phi_{x(t)}(t)) \mid 1 \leq j \leq n,\ x(t) \in \mathscr{A}_x,\ \overline{x}(t) \in \mathscr{A}_x \right\}$$

is an equicontinuous family on $\mathbb{R}$.

Proof Let us define a function $h : \mathbb{R}^m \times \mathbb{R}^m \times \mathbb{R}^n \to \mathbb{R}^n$ by the formula

$$h(x_1, x_2, x_3) = g(x_1, x_3) - g(x_2, x_3).$$

Being continuous on the compact region

$$\mathscr{D} = \left\{ (x_1, x_2, x_3) \in \mathbb{R}^m \times \mathbb{R}^m \times \mathbb{R}^n \mid \|x_1\| \leq H, \|x_2\| \leq H, \|x_3\| \leq M \right\},$$

the function $h(x_1, x_2, x_3)$ is uniformly continuous on $\mathscr{D}$.

Fix an arbitrary positive number ε. There exists a number $\delta_1 = \delta_1(\varepsilon) > 0$ such that for all $\left(x_1^0, x_2^0, x_3^0\right), \left(x_1^1, x_2^1, x_3^1\right) \in \mathbb{R}^m \times \mathbb{R}^m \times \mathbb{R}^n$ with

$$\left\| \left(x_1^0, x_2^0, x_3^0\right) - \left(x_1^1, x_2^1, x_3^1\right) \right\| < \delta_1,$$

the inequality

$$\left\| h\left(x_1^0, x_2^0, x_3^0\right) - h\left(x_1^1, x_2^1, x_3^1\right) \right\| < \varepsilon$$

holds.

Since $\|x'(t)\| \le H_0$ for each $x(t) \in \mathscr{A}_x$, the set $\mathscr{A}_x$ is an equicontinuous family on $\mathbb{R}$. Therefore, there exists a number $\delta_2 = \delta_2(\delta_1) > 0$ such that for all $t_1, t_2 \in \mathbb{R}$ satisfying $|t_1 - t_2| < \delta_2$ we have $\|x(t_1) - x(t_2)\| < \delta_1/3$ for all $x(t) \in \mathscr{A}_x$.

Similarly, the set $\mathscr{A}_y$ is also an equicontinuous family on $\mathbb{R}$, since $\|y'(t)\| \le \|A\|\, M + M_0$ for each $y(t) \in \mathscr{A}_y$. Thus, one can find a number $\delta_3 = \delta_3(\delta_1) > 0$ such that for all $t_1, t_2 \in \mathbb{R}$ with $|t_1 - t_2| < \delta_3$, the inequality $\|y(t_1) - y(t_2)\| < \delta_1/3$ is valid for all $y(t) \in \mathscr{A}_y$.

In this case, for all $t_1, t_2 \in \mathbb{R}$ with $|t_1 - t_2| < \min\{\delta_2, \delta_3\}$, one has

$$\begin{aligned}
&\left\| \left(x(t_1), \overline{x}(t_1), \phi_{x(t)}(t_1)\right) - \left(x(t_2), \overline{x}(t_2), \phi_{x(t)}(t_2)\right) \right\| \\
&\le \|x(t_1) - x(t_2)\| + \|\overline{x}(t_1) - \overline{x}(t_2)\| + \left\|\phi_{x(t)}(t_1) - \phi_{x(t)}(t_2)\right\| \\
&< \delta_1,
\end{aligned}$$

for all $x(t), \overline{x}(t) \in \mathscr{A}_x$.

Hence, taking $\delta = \min\{\delta_2, \delta_3\}$, one can confirm for all $t_1, t_2 \in \mathbb{R}$ with $|t_1 - t_2| < \delta$ that the inequality

$$\begin{aligned}
&\Big\| \left(g_j(x(t_1), \phi_{x(t)}(t_1)) - g_j(\overline{x}(t_1), \phi_{x(t)}(t_1))\right) \\
&- \left(g_j(x(t_2), \phi_{x(t)}(t_2)) - g_j(\overline{x}(t_2), \phi_{x(t)}(t_2))\right) \Big\| \\
&\le \left\| h\left(x(t_1), \overline{x}(t_1), \phi_{x(t)}(t_1)\right) - h\left(x(t_2), \overline{x}(t_2), \phi_{x(t)}(t_2)\right) \right\| \\
&< \varepsilon
\end{aligned}$$

holds for each $j = 1, 2, \ldots, n$ and $x(t), \overline{x}(t) \in \mathscr{A}_x$. Consequently, the family $\mathscr{F}$ is equicontinuous on $\mathbb{R}$. □

We continue with replication of sensitivity in the next lemma.

Lemma 2.3 *Sensitivity of the set $\mathscr{A}_x$ implies the same feature for the set $\mathscr{A}_y$.*

Proof Fix an arbitrary $\delta > 0$ and let $y(t) \in \mathscr{A}_y$ be a given solution of system (2.1.2). In this case, there exists $x(t) \in \mathscr{A}_x$ such that $y(t) = \phi_{x(t)}(t)$.

Let us choose a number $\overline{\varepsilon} = \overline{\varepsilon}(\delta) > 0$ small enough which satisfies the inequality

$$\left(1 + \frac{NL_2}{\omega - NL_3}\right)\overline{\varepsilon} < \delta.$$

Then take $R = R(\overline{\varepsilon}) < 0$ sufficiently large in absolute value such that

$$\frac{2M_0N}{\omega} e^{(\omega - NL_3)R} < \overline{\varepsilon},$$

and let $\delta_1 = \delta_1(\overline{\varepsilon}, R) = \overline{\varepsilon} e^{L_0 R}$. Since the set of functions $\mathscr{A}_x$ is sensitive, there exist positive numbers ε_0 and Δ such that the inequalities $\|x(t_0) - \overline{x}(t_0)\| < \delta_1$ and $\|x(t) - \overline{x}(t)\| > \varepsilon_0,\ t \in J$, hold for some solution $\overline{x}(t) \in \mathscr{A}_x$, a number $t_0 \in \mathbb{R}$ and an interval $J \subset [t_0, \infty)$ whose length is not less than Δ.

Using the couple of integral equations

$$x(t) = x(t_0) + \int_{t_0}^{t} F(s, x(s))ds,$$

$$\overline{x}(t) = \overline{x}(t_0) + \int_{t_0}^{t} F(s, \overline{x}(s))ds$$

together with condition $(A2)$, one can see that the inequality

$$\|x(t) - \overline{x}(t)\| \leq \|x(t_0) - \overline{x}(t_0)\| + \left| \int_{t_0}^{t} L_0 \|x(s) - \overline{x}(s)\| \, ds \right|$$

holds for $t \in [t_0 + R, t_0]$. Applying the Gronwall–Bellman inequality [2], we obtain that

$$\|x(t) - \overline{x}(t)\| \leq \|x(t_0) - \overline{x}(t_0)\| \, e^{L_0 |t - t_0|}$$

and therefore $\|x(t) - \overline{x}(t)\| < \overline{\varepsilon}$ for $t \in [t_0 + R, t_0]$.

Let us denote $\overline{y}(t) = \phi_{\overline{x}(t)}(t)$. First, we will show that $\|y(t_0) - \overline{y}(t_0)\| < \delta$.

The functions $y(t)$ and $\overline{y}(t)$ satisfy the relations

$$y(t) = \int_{-\infty}^{t} e^{A(t-s)} g(x(s), y(s))ds$$

and

$$\overline{y}(t) = \int_{-\infty}^{t} e^{A(t-s)} g(\overline{x}(s), \overline{y}(s))ds,$$

respectively. Therefore,

$$y(t) - \overline{y}(t) = \int_{-\infty}^{t} e^{A(t-s)} [g(x(s), y(s)) - g(\overline{x}(s), \overline{y}(s))]ds$$

and hence we obtain that

$$\|y(t)-\overline{y}(t)\| \leq \int_{t_0+R}^{t} Ne^{-\omega(t-s)} \|g(x(s),y(s))-g(x(s),\overline{y}(s))\|\, ds$$
$$+\int_{t_0+R}^{t} Ne^{-\omega(t-s)} \|g(x(s),\overline{y}(s))-g(\overline{x}(s),\overline{y}(s))\|\, ds$$
$$+\int_{-\infty}^{t_0+R} Ne^{-\omega(t-s)} \|g(x(s),y(s))-g(\overline{x}(s),\overline{y}(s))\|\, ds.$$

Since $\|x(t)-\overline{x}(t)\| < \overline{\varepsilon}$ for $t \in [t_0+R, t_0]$, one has

$$\|y(t)-\overline{y}(t)\| \leq NL_3 \int_{t_0+R}^{t} e^{-\omega(t-s)} \|y(s)-\overline{y}(s)\|\, ds$$
$$+\frac{NL_2\overline{\varepsilon}}{\omega} e^{-\omega t}(e^{\omega t}-e^{\omega(t_0+R)}) + \frac{2M_0N}{\omega} e^{-\omega(t-t_0-R)}.$$

Now, let us introduce the functions $u(t)=e^{\omega t}\|y(t)-\overline{y}(t)\|$, $k(t)=\frac{NL_2\overline{\varepsilon}}{\omega}e^{\omega t}$ and $h(t)=c+k(t)$, where $c=\left(\frac{2M_0N}{\omega}-\frac{NL_2\overline{\varepsilon}}{\omega}\right)e^{\omega(t_0+R)}$.

These definitions give us the inequality

$$u(t) \leq h(t) + \int_{t_0+R}^{t} NL_3 u(s) ds.$$

Applying Lemma 2.2 [20] to the last inequality, we achieve that

$$u(t) \leq h(t) + NL_3 \int_{t_0+R}^{t} e^{NL_3(t-s)} h(s) ds.$$

Therefore, on the time interval $[t_0+R, t_0]$, the inequality

$$u(t) \leq c + k(t) + c\left(e^{NL_3(t-t_0-R)}-1\right)$$
$$+\frac{N^2L_2L_3\overline{\varepsilon}}{\omega} e^{NL_3 t} \int_{t_0+R}^{t} e^{(\omega-NL_3)s} ds$$
$$=\frac{NL_2\overline{\varepsilon}}{\omega} e^{\omega t} + \left(\frac{2M_0N}{\omega}-\frac{NL_2\overline{\varepsilon}}{\omega}\right) e^{\omega R} e^{NL_3(t-t_0-R)}$$
$$+\frac{N^2L_2L_3\overline{\varepsilon}}{\omega(\omega-NL_3)} e^{\omega t}\left[1-e^{(NL_3-\omega)(t-t_0-R)}\right]$$

holds.

The last inequality leads to

$$\|y(t) - \overline{y}(t)\| \le \frac{NL_2\overline{\varepsilon}}{\omega - NL_3} + \frac{2M_0N}{\omega} e^{(NL_3-\omega)(t-t_0-R)},$$

and consequently we obtain that

$$\begin{aligned}\|y(t_0) - \overline{y}(t_0)\| &\le \frac{NL_2\overline{\varepsilon}}{\omega - NL_3} + \frac{2M_0N}{\omega} e^{(\omega-NL_3)R} \\ &< \left(1 + \frac{NL_2}{\omega - NL_3}\right)\overline{\varepsilon} \\ &< \delta.\end{aligned}$$

In the remaining part of the proof, we will show the existence of a positive number ε_1 and an interval $J^1 \subset J$, with a fixed length which is independent of $y(t), \overline{y}(t) \in \mathscr{A}_y$, such that the inequality $\|y(t) - \overline{y}(t)\| > \varepsilon_1$ holds for all $t \in J^1$.

According to Lemma 2.2, there exists a positive number $\tau < \Delta$, independent of the functions $x(t), \overline{x}(t) \in \mathscr{A}_x$, $y(t), \overline{y}(t) \in \mathscr{A}_y$, such that for any $t_1, t_2 \in \mathbb{R}$ with $|t_1 - t_2| < \tau$ the inequality

$$\begin{aligned}&\Big| \left(g_j\left(x(t_1), y(t_1)\right) - g_j\left(\overline{x}(t_1), y(t_1)\right)\right) \\ &- \left(g_j\left(x(t_2), y(t_2)\right) - g_j\left(\overline{x}(t_2), y(t_2)\right)\right) \Big| \\ &< \frac{L_1\varepsilon_0}{2n}\end{aligned} \tag{2.5.15}$$

holds, for all $1 \le j \le n$.

Condition $(A4)$ implies that, for all $t \in J$, the inequality

$$\|g(x(t), y(t)) - g(\overline{x}(t), y(t))\| \ge L_1 \|x(t) - \overline{x}(t)\|$$

is satisfied. Therefore, for each $t \in J$, there exists an integer $j_0 = j_0(t)$, $1 \le j_0 \le n$, such that

$$\left|g_{j_0}(x(t), y(t)) - g_{j_0}(\overline{x}(t), y(t))\right| \ge \frac{L_1}{n} \|x(t) - \overline{x}(t)\|.$$

Otherwise, if there exists $s \in J$ such that for all $1 \le j \le n$, the inequality

$$\left|g_j\left(x\left(s\right), y\left(s\right)\right) - g_j(\overline{x}(s), y(s))\right| < \frac{L_1}{n} \|x(s) - \overline{x}(s)\|$$

holds, then one encounters with a contradiction since

$$\begin{aligned}\|g(x(s), y(s)) - g(\overline{x}(s), y(s))\| &\le \sum_{j=1}^{n} \left|g_j(x(s), y(s)) - g_j(\overline{x}(s), y(s))\right| \\ &< L_1 \|x(s) - \overline{x}(s)\|.\end{aligned}$$

Now, let s_0 be the midpoint of the interval J and $\theta = s_0 - \tau/2$. One can find an integer $j_0 = j_0(s_0)$, $1 \le j_0 \le n$, such that

$$\left|g_{j_0}(x(s_0), y(s_0)) - g_{j_0}(\overline{x}(s_0), y(s_0))\right| \ge \frac{L_1}{n} \left\|x(s_0) - \overline{x}(s_0)\right\| > \frac{L_1 \varepsilon_0}{n}. \tag{2.5.16}$$

On the other hand, making use of inequality (2.5.15), for all $t \in [\theta, \theta + \tau]$ we have

$$\begin{aligned}
&\left|g_{j_0}(x(s_0), y(s_0)) - g_{j_0}(\overline{x}(s_0), y(s_0))\right| - \left|g_{j_0}(x(t), y(t)) - g_{j_0}(\overline{x}(t), y(t))\right| \\
&\le \left|\left(g_{j_0}(x(t), y(t)) - g_{j_0}(\overline{x}(t), y(t))\right) - \left(g_{j_0}(x(s_0), y(s_0)) - g_{j_0}(\overline{x}(s_0), y(s_0))\right)\right| \\
&< \frac{L_1 \varepsilon_0}{2n}.
\end{aligned}$$

Therefore, by means of (2.5.16), we obtain that the inequality

$$\begin{aligned}
&\left|g_{j_0}(x(t), y(t)) - g_{j_0}(\overline{x}(t), y(t))\right| \\
&> \left|g_{j_0}(x(s_0), y(s_0)) - g_{j_0}(\overline{x}(s_0), y(s_0))\right| - \frac{L_1 \varepsilon_0}{2n} \\
&> \frac{L_1 \varepsilon_0}{2n}
\end{aligned} \tag{2.5.17}$$

holds for all $t \in [\theta, \theta + \tau]$.

By applying the mean value theorem for integrals, one can find $s_1, s_2, \ldots, s_n \in [\theta, \theta + \tau]$ such that

$$\begin{aligned}
&\int_\theta^{\theta+\tau} \left[g(x(s), y(s)) - g(\overline{x}(s), y(s))\right] ds \\
&= \begin{pmatrix}
\tau \left[g_1(x(s_1), y(s_1)) - g_1(\overline{x}(s_1), y(s_1))\right] \\
\tau \left[g_2(x(s_2), y(s_2)) - g_2(\overline{x}(s_2), y(s_2))\right] \\
\vdots \\
\tau \left[g_n(x(s_n), y(s_n)) - g_n(\overline{x}(s_n), y(s_n))\right]
\end{pmatrix}.
\end{aligned}$$

Thus, using (2.5.17), one can verify that

$$\begin{aligned}
&\left\|\int_\theta^{\theta+\tau} \left[g(x(s), y(s)) - g(\overline{x}(s), y(s))\right] ds\right\| \\
&\ge \tau \left|g_{j_0}(x(s_{j_0}), y(s_{j_0})) - g_{j_0}(\overline{x}(s_{j_0}), y(s_{j_0}))\right| \\
&> \frac{\tau L_1 \varepsilon_0}{2n}.
\end{aligned} \tag{2.5.18}$$

It is clear that, for $t \in [\theta, \theta + \tau]$, the solutions $y(t)$ and $\overline{y}(t)$ satisfy the integral equations

$$y(t) = y(\theta) + \int_{\theta}^{t} Ay(s)ds + \int_{\theta}^{t} g(x(s), y(s))ds,$$

and

$$\overline{y}(t) = \overline{y}(\theta) + \int_{\theta}^{t} A\overline{y}(s)ds + \int_{\theta}^{t} g(\overline{x}(s), \overline{y}(s))ds,$$

respectively, and herewith the equation

$$\begin{aligned} y(t) - \overline{y}(t) &= (y(\theta) - \overline{y}(\theta)) + \int_{\theta}^{t} A(y(s) - \overline{y}(s))ds \\ &+ \int_{\theta}^{t} [g(x(s), y(s)) - g(\overline{x}(s), y(s))]ds \\ &+ \int_{\theta}^{t} [g(\overline{x}(s), y(s)) - g(\overline{x}(s), \overline{y}(s))]ds \end{aligned}$$

holds. Hence, we have the inequality

$$\begin{aligned} \|y(\theta+\tau) - \overline{y}(\theta+\tau)\| &\geq \left\| \int_{\theta}^{\theta+\tau} [g(x(s), y(s)) - g(\overline{x}(s), y(s))]ds \right\| \\ &- \|y(\theta) - \overline{y}(\theta)\| - \int_{\theta}^{\theta+\tau} \|A\| \, \|y(s) - \overline{y}(s)\| \, ds \\ &- \int_{\theta}^{\theta+\tau} L_3 \, \|y(s) - \overline{y}(s)\| \, ds. \end{aligned} \tag{2.5.19}$$

Now, assume that $\max_{t \in [\theta, \theta+\tau]} \|y(t) - \overline{y}(t)\| \leq \dfrac{\tau L_1 \varepsilon_0}{2n[2 + \tau(L_3 + \|A\|)]}$. In the present case, one encounters with a contradiction since, by means of the inequalities (2.5.18) and (2.5.19), we have

$$\begin{aligned} \max_{t \in [\theta, \theta+\tau]} \|y(t) - \overline{y}(t)\| &\geq \|y(\theta+\tau) - \overline{y}(\theta+\tau)\| \\ &> \frac{\tau L_1 \varepsilon_0}{2n} - [1 + \tau(L_3 + \|A\|)] \max_{t \in [\theta, \theta+\tau]} \|y(t) - \overline{y}(t)\| \\ &\geq \frac{\tau L_1 \varepsilon_0}{2n[2 + \tau(L_3 + \|A\|)]}. \end{aligned}$$

Therefore, one can see that the inequality

$$\max_{t \in [\theta, \theta+\tau]} \|y(t) - \overline{y}(t)\| > \frac{\tau L_1 \varepsilon_0}{2n[2 + \tau(L_3 + \|A\|)]}$$

is valid.

Suppose that at a point $\eta \in [\theta, \theta + \tau]$, the real valued function $\|y(t) - \overline{y}(t)\|$ takes its maximum on the interval $[\theta, \theta + \tau]$. That is,

$$\max_{t \in [\theta, \theta+\tau]} \|y(t) - \overline{y}(t)\| = \|y(\eta) - \overline{y}(\eta)\|.$$

For $t \in [\theta, \theta + \tau]$, by virtue of the integral equations

$$y(t) = y(\eta) + \int_\eta^t Ay(s)ds + \int_\eta^t g(x(s), y(s))ds,$$

and

$$\overline{y}(t) = \overline{y}(\eta) + \int_\eta^t A\overline{y}(s)ds + \int_\eta^t g(\overline{x}(s), \overline{y}(s))ds,$$

we obtain

$$y(t) - \overline{y}(t) = (y(\eta) - \overline{y}(\eta)) + \int_\eta^t A(y(s) - \overline{y}(s))ds$$
$$+ \int_\eta^t [g(x(s), y(s)) - g(\overline{x}(s), \overline{y}(s))]ds.$$

Define

$$\tau^1 = \min\left\{\frac{\tau}{2}, \frac{\tau L_1 \varepsilon_0}{8n(M\|A\| + M_0)[2 + \tau(L_3 + \|A\|)]}\right\}$$

and let

$$\theta^1 = \begin{cases} \eta, & \text{if } \eta \le \theta + \tau/2 \\ \eta - \tau^1, & \text{if } \eta > \theta + \tau/2 \end{cases}.$$

We note that the interval $J^1 = [\theta^1, \theta^1 + \tau^1]$ is a subset of $[\theta, \theta + \tau]$ and hence of J.

For $t \in J^1$, we have that

$$\|y(t) - \overline{y}(t)\| \ge \|y(\eta) - \overline{y}(\eta)\| - \left|\int_\eta^t \|A\| \, \|y(s) - \overline{y}(s)\| \, ds\right|$$
$$- \left|\int_\eta^t \|g(x(s), y(s)) - g(\overline{x}(s), \overline{y}(s))\| \, ds\right|$$
$$> \frac{\tau L_1 \varepsilon_0}{2n[2 + \tau(L_3 + \|A\|)]} - 2\tau^1(M\|A\| + M_0)$$
$$\ge \frac{\tau L_1 \varepsilon_0}{4n[2 + \tau(L_3 + \|A\|)]}.$$

Consequently, the inequality $\|y(t) - \overline{y}(t)\| > \varepsilon_1$ holds for $t \in J^1$, where

$$\varepsilon_1 = \frac{\tau L_1 \varepsilon_0}{4n[2 + \tau(L_3 + \|A\|)]},$$

and the length of the interval J^1 does not depend on the functions $x(t), \overline{x}(t) \in \mathscr{A}_x$.

The proof of the lemma is finalized. □

Through Lemma 2.3, we mention the replication of sensitivity feature from the set of functions $\mathscr{A}_x$ to $\mathscr{A}_y$, that is, from the generator system to the replicator counterpart. In a similar way, it is reasonable to analyze the sensitivity of the set of functions $\mathscr{A}$, which is defined through Eq. (2.2.11). In the present case, we shall say that the set $\mathscr{A}$ is sensitive provided that $\mathscr{A}_y$ is sensitive. This description is a natural one since, otherwise, the inequality $\|x(t) - \overline{x}(t)\| > \varepsilon_0$ implies that $\left\| \big(x(t), \phi_{x(t)}(t)\big) - \big(\overline{x}(t), \phi_{\overline{x}(t)}(t)\big) \right\| > \varepsilon_0$ in the same interval of time, which already signifies sensitivity of $\mathscr{A}$. But in replication of chaos, the crucial idea is the extension of sensitivity not only by the result-system, but also by the replicator, and one should understand sensitivity of the result-system as a property which is equivalent to the sensitivity of the replicator. According to this explanation, we note that if $\mathscr{A}_x$ is sensitive, then Lemma 2.3 implies the same feature for the set $\mathscr{A}_y$, and hence for the set $\mathscr{A}$.

Now, let us illustrate the replication of sensitivity through an example. It is known that the Lorenz system

$$\begin{aligned} x_1' &= \sigma(-x_1 + x_2) \\ x_2' &= -x_2 + r x_1 - x_1 x_3 \\ x_3' &= -b x_3 + x_1 x_2, \end{aligned} \tag{2.5.20}$$

with the coefficients $\sigma = 10$, $b = 8/3$, $r = 28$ admits sensitivity [21]. We use system (2.5.20) with the specified coefficients as the generator and constitute the 6-dimensional result-system

$$\begin{aligned} x_1' &= 10(-x_1 + x_2) \\ x_2' &= -x_2 + 28x_1 - x_1 x_3 \\ x_3' &= -\tfrac{8}{3}x_3 + x_1 x_2 \\ x_4' &= -5x_4 + x_3 \\ x_5' &= -2x_5 + 0.0002(x_2 - x_5)^3 + 4x_2 \\ x_6' &= -3x_6 - 3x_1. \end{aligned} \tag{2.5.21}$$

When system (2.5.21) is considered in the form of system (2.1.1)+(2.1.2), one can see that the diagonal matrix A whose entries on the diagonal are $-5, -2, -3$ satisfies the inequality $\left\|e^{At}\right\| \le N e^{-\omega t}$ with the coefficients $N = 1$ and $\omega = 2$. We note that the function $g : \mathbb{R}^3 \times \mathbb{R}^3 \to \mathbb{R}^3$ defined as

$$g(x_1, x_2, x_3, x_4, x_5, x_6) = \Big(x_3, 0.0002(x_2 - x_5)^3 + 4x_2, -3x_1\Big)$$

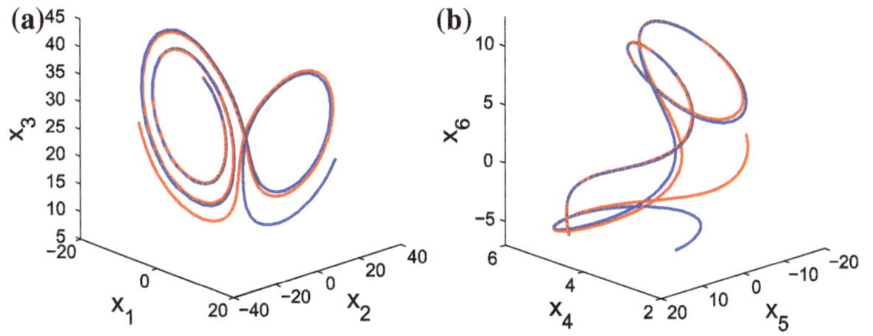

Fig. 2.3 Replication of sensitivity in the result-system (2.5.21). **a** 3-dimensional projection in the $x_1 - x_2 - x_3$ space. **b** 3-dimensional projection in the $x_4 - x_5 - x_6$ space. The sensitivity property is observable both in (**a**) and (**b**) such that the trajectories presented by *blue* and *red* colors move together in the first stage and then diverge. In other words, the sensitivity property of the generator system is mimicked by the replicator counterpart

provides the conditions $(A4)$ and $(A5)$ with constants $L_1 = 1/\sqrt{3}$, $L_2 = 11\sqrt{3}/2$ and $L_3 = 3/2$ since the chaotic attractor of system (2.5.21) is inside a compact region such that $|x_2| \leq 30$ and $|x_5| \leq 50$. Consequently, system (2.5.21) satisfies the condition $(A7)$.

In Fig. 2.3, one can see the 3-dimensional projections in the $x_1 - x_2 - x_3$ and $x_4 - x_5 - x_6$ spaces of two different trajectories of the result-system (2.5.21) with adjacent initial conditions, such that one of them is in blue color and the other in red color. For the trajectory with blue color, we make use of the initial data $x_1(0) = -8.57$, $x_2(0) = -2.39$, $x_3(0) = 33.08$, $x_4(0) = 5.32$, $x_5(0) = 10.87$, $x_6(0) = -6.37$ and for the one with red color, we use the initial data $x_1(0) = -8.53$, $x_2(0) = -2.47$, $x_3(0) = 33.05$, $x_4(0) = 5.33$, $x_5(0) = 10.86$, $x_6(0) = -6.36$. In the simulation, the trajectories move on the time interval $[0, 3]$. The results seen in Fig. 2.3 supports our theoretical results indicated in Lemma 2.3 such that the replicator system, likewise the generator counterpart, admits the sensitivity feature. That is, the solutions of both the generator and the replicator given by blue and red colors diverge, even though they start and move close to each other in the first stage.

In the next assertion we continue with the replication of transitivity.

Lemma 2.4 *Transitivity of $\mathscr{A}_x$ implies the same feature for $\mathscr{A}_y$.*

Proof Fix an arbitrary small $\varepsilon > 0$, an arbitrary large $E > 0$ and let $y(t) \in \mathscr{A}_y$ be a given function. Arising from the description (2.2.10) of the set $\mathscr{A}_y$, there exists a function $x(t) \in \mathscr{A}_x$ such that $y(t) = \phi_{x(t)}(t)$. Let $\gamma = \dfrac{\omega(\omega - NL_3)}{2M_0N(\omega - NL_3) + NL_2\omega}$. Condition $(A7)$ guarantees that γ is positive. Since there exists a dense solution $x^*(t) \in \mathscr{A}_x$, one can find $\xi > 0$ and an interval $J \subset \mathbb{R}$ with length E such that $\|x(t) - x^*(t + \xi)\| < \gamma\varepsilon$ for all $t \in J$. Without loss of generality, assume that J is a closed interval, that is, $J = [a, a + E]$ for some real number a.

Let $y^*(t) = \phi_{x^*(t)}(t)$. For $t \in J$, the bounded on $\mathbb{R}$ solutions $y(t)$ and $y^*(t)$ satisfy the relations

$$y(t) = \int_{-\infty}^{t} e^{A(t-s)} g(x(s), y(s))ds,$$

and

$$y^*(t) = \int_{-\infty}^{t} e^{A(t-s)} g(x^*(s), y^*(s))ds,$$

respectively. The second equation above implies that

$$y^*(t+\xi) = \int_{-\infty}^{t+\xi} e^{A(t+\xi-s)} g(x^*(s), y^*(s))ds.$$

Using the transformation $\overline{s} = s - \xi$, and replacing $\overline{s}$ by s again, it is easy to verify that

$$y^*(t+\xi) = \int_{-\infty}^{t} e^{A(t-s)} g(x^*(s+\xi), y^*(s+\xi))ds.$$

Therefore, for $t \in J$, we have that

$$\begin{aligned} y(t) - y^*(t+\xi) &= \int_{-\infty}^{a} e^{A(t-s)} [g(x(s), y(s)) - g(x^*(s+\xi), y^*(s+\xi))]ds \\ &+ \int_{a}^{t} e^{A(t-s)} [g(x(s), y(s)) - g(x(s), y^*(s+\xi))]ds \\ &+ \int_{a}^{t} e^{A(t-s)} [g(x(s), y^*(s)) - g(x^*(s+\xi), y^*(s+\xi))]ds, \end{aligned}$$

which implies the inequality

$$\begin{aligned} \|y(t) - y^*(t+\xi)\| &\le \int_{-\infty}^{a} 2M_0 N e^{-\omega(t-s)} ds \\ &+ \int_{a}^{t} N L_3 e^{-\omega(t-s)} \|y(s) - y^*(s+\xi)\| ds \\ &+ \int_{a}^{t} N L_2 e^{-\omega(t-s)} \|x(s) - x^*(s+\xi)\| ds \\ &\le \frac{2M_0 N}{\omega} e^{-\omega(t-a)} + \frac{N L_2 \gamma \varepsilon}{\omega} e^{-\omega t} \left(e^{\omega t} - e^{\omega a}\right) \\ &+ \int_{a}^{t} N L_3 e^{-\omega(t-s)} \|y(s) - y^*(s+\xi)\| ds. \end{aligned}$$

Hence, we get

$$e^{\omega t}\left\|y(t)-y^*(t+\xi)\right\| \le \frac{2M_0N}{\omega}e^{\omega a}+\frac{NL_2\gamma\varepsilon}{\omega}\left(e^{\omega t}-e^{\omega a}\right)$$
$$+\int_a^t NL_3e^{\omega s}\left\|y(s)-y^*(s+\xi)\right\| ds.$$

Through the implementation of Lemma 2.2 [20] to the last inequality, we obtain

$$e^{\omega t}\left\|y(t)-y^*(t+\xi)\right\| \le \frac{2M_0N}{\omega}e^{\omega a}+\frac{NL_2\gamma\varepsilon}{\omega}\left(e^{\omega t}-e^{\omega a}\right)$$
$$+\int_a^t NL_3\left[\frac{2M_0N}{\omega}e^{\omega a}+\frac{NL_2\gamma\varepsilon}{\omega}\left(e^{\omega s}-e^{\omega a}\right)\right]e^{NL_3(t-s)}ds$$
$$=\frac{NL_2\gamma\varepsilon}{\omega}e^{\omega t}+\left(\frac{2M_0N}{\omega}-\frac{NL_2\gamma\varepsilon}{\omega}\right)e^{\omega a}e^{NL_3(t-a)}$$
$$+\frac{N^2L_2L_3\gamma\varepsilon}{\omega(\omega-NL_3)}e^{NL_3t}\left(e^{(\omega-NL_3)t}-e^{(\omega-NL_3)a}\right).$$

Multiplying both sides by $e^{-\omega t}$, one can attain that

$$\left\|y(t)-y^*(t)\right\| \le \frac{2M_0N}{\omega}e^{(NL_3-\omega)(t-a)}$$
$$+\left(\frac{NL_2\gamma\varepsilon}{\omega}+\frac{N^2L_2L_3\gamma\varepsilon}{\omega(\omega-NL_3)}\right)\left(1-e^{(NL_3-\omega)(t-a)}\right)$$
$$=\frac{2M_0N}{\omega}e^{(NL_3-\omega)(t-a)}+\frac{NL_2\gamma\varepsilon}{\omega-NL_3}\left(1-e^{(NL_3-\omega)(t-a)}\right).$$

Now, suppose that the number E is sufficiently large such that

$$E>\frac{2}{\omega-NL_3}\ln\left(\frac{1}{\gamma\varepsilon}\right).$$

If $t\in[a+E/2,a+E]$, then it is true that

$$e^{(NL_3-\omega)(t-a)}\le e^{(NL_3-\omega)\frac{E}{2}}<\gamma\varepsilon.$$

As a result, we have

$$\left\|y(t)-y^*(t+\xi)\right\|<\left[\frac{2M_0N}{\omega}+\frac{NL_2}{\omega-NL_3}\right]\gamma\varepsilon=\varepsilon,$$

for $t\in J_1=[a_1,a_1+E_1]$, where $a_1=a+E/2$ and $E_1=E/2$. Consequently, the set $\mathscr{A}_y$ is transitive in compliance with Definition 2.3.

The lemma is proved. □

The extension of the last ingredient of chaos in the sense of Devaney is presented in the following lemma:

Lemma 2.5 *If $\mathscr{A}_x$ admits a dense collection of periodic functions, then the same is true for $\mathscr{A}_y$.*

Proof Fix a function $y(t) = \phi_{x(t)}(t) \in \mathscr{A}_y$, an arbitrary small number $\varepsilon > 0$ and an arbitrary large number $E > 0$. Let $\gamma = \dfrac{\omega(\omega - NL_3)}{2M_0N(\omega - NL_3) + NL_2\omega}$, which is a positive number by condition $(A7)$. Suppose that $\mathscr{G}_x$ is a dense collection of periodic functions inside $\mathscr{A}_x$. In this case, there exist $\widetilde{x}(t) \in \mathscr{G}_x$ and an interval $J \subset \mathbb{R}$ with length E such that $\|x(t) - \widetilde{x}(t)\| < \gamma\varepsilon$, for all $t \in J$. Without loss of generality, assume that J is a closed interval, that is, $J = [a, a + E]$ for some $a \in \mathbb{R}$.

We note that by condition $(A4)$ there is a one-to-one correspondence between the sets $\mathscr{G}_x$ and

$$\mathscr{G}_y = \left\{\phi_{x(t)}(t) \mid x(t) \in \mathscr{G}_x\right\}, \tag{2.5.22}$$

such that if $x(t) \in \mathscr{G}_x$ is periodic then $\phi_{x(t)}(t) \in \mathscr{G}_y$ is also periodic with the same period, and vice versa. Therefore, $\mathscr{G}_y \subset \mathscr{A}_y$ is a collection of periodic functions and in the proof our aim is to verify that the set $\mathscr{G}_y$ is dense in $\mathscr{A}_y$.

Let $\widetilde{y}(t) = \phi_{\widetilde{x}(t)}(t)$, which clearly belongs to the set $\mathscr{G}_y$. Making use of the relations

$$y(t) = \int_{-\infty}^{t} e^{A(t-s)} g(x(s), y(s))ds,$$

and

$$\widetilde{y}(t) = \int_{-\infty}^{t} e^{A(t-s)} g(\widetilde{x}(s), \widetilde{y}(s))ds,$$

for $t \in J$, we attain that

$$\begin{aligned} y(t) - \widetilde{y}(t) = & \int_{-\infty}^{a} e^{A(t-s)}[g(x(s), y(s)) - g(\widetilde{x}(s), \widetilde{y}(s))]ds \\ & + \int_{a}^{t} e^{A(t-s)}[g(x(s), y(s)) - g(x(s), \widetilde{y}(s))]ds \\ & + \int_{a}^{t} e^{A(t-s)}[g(x(s), \widetilde{y}(s)) - g(\widetilde{x}(s), \widetilde{y}(s))]ds. \end{aligned}$$

The last equation implies that

$$\|y(t) - \widetilde{y}(t)\| \leq \int_{-\infty}^{a} 2M_0 N e^{-\omega(t-s)} ds$$

$$+ \int_a^t N L_3 e^{-\omega(t-s)} \|y(s) - \widetilde{y}(s)\| \, ds$$

$$+ \int_a^t N L_2 e^{-\omega(t-s)} \|x(s) - \widetilde{x}(s)\| \, ds$$

$$\leq \frac{2M_0 N}{\omega} e^{-\omega(t-a)} + \frac{N L_2 \gamma \varepsilon}{\omega} e^{-\omega t} \left(e^{\omega t} - e^{\omega a}\right)$$

$$+ \int_a^t N L_3 e^{-\omega(t-s)} \|y(s) - \widetilde{y}(s)\| \, ds.$$

Hence, we have

$$e^{\omega t} \|y(t) - \widetilde{y}(t)\| \leq \frac{2M_0 N}{\omega} e^{\omega a} + \frac{N L_2 \gamma \varepsilon}{\omega} \left(e^{\omega t} - e^{\omega a}\right)$$

$$+ \int_a^t N L_3 e^{\omega s} \|y(s) - \widetilde{y}(s)\| \, ds.$$

Application of Lemma 2.2 [20] to the last inequality yields

$$e^{\omega t} \|y(t) - \widetilde{y}(t)\| \leq \frac{2M_0 N}{\omega} e^{\omega a} + \frac{N L_2 \gamma \varepsilon}{\omega} \left(e^{\omega t} - e^{\omega a}\right)$$

$$+ \int_a^t N L_3 \left[\frac{2M_0 N}{\omega} e^{\omega a} + \frac{N L_2 \gamma \varepsilon}{\omega} \left(e^{\omega s} - e^{\omega a}\right)\right] e^{N L_3 (t-s)} ds$$

$$= \frac{N L_2 \gamma \varepsilon}{\omega} e^{\omega t} + \left(\frac{2M_0 N}{\omega} - \frac{N L_2 \gamma \varepsilon}{\omega}\right) e^{\omega a} e^{N L_3 (t-a)}$$

$$+ \frac{N^2 L_2 L_3 \gamma \varepsilon}{\omega(\omega - N L_3)} e^{N L_3 t} \left(e^{(\omega - N L_3) t} - e^{(\omega - N L_3) a}\right).$$

Multiplying both sides by $e^{-\omega t}$, we obtain that

$$\|y(t) - \widetilde{y}(t)\| \leq \frac{2M_0 N}{\omega} e^{(N L_3 - \omega)(t-a)}$$

$$+ \left(\frac{N L_2 \gamma \varepsilon}{\omega} + \frac{N^2 L_2 L_3 \gamma \varepsilon}{\omega(\omega - N L_3)}\right) \left(1 - e^{(N L_3 - \omega)(t-a)}\right)$$

$$= \frac{2M_0 N}{\omega} e^{(N L_3 - \omega)(t-a)} + \frac{N L_2 \gamma \varepsilon}{\omega - N L_3} \left(1 - e^{(N L_3 - \omega)(t-a)}\right).$$

Suppose that the number E is sufficiently large such that

$$E > \frac{2}{\omega - N L_3} \ln\left(\frac{1}{\gamma \varepsilon}\right).$$

If $a + \frac{E}{2} \leq t \leq a + E$, then one has $e^{(NL_3-\omega)(t-a)} \leq e^{(NL_3-\omega)E/2} < \gamma\varepsilon$. Consequently, the inequality

$$\|y(t) - \widetilde{y}(t)\| < \left(\frac{2M_0N}{\omega} + \frac{NL_2}{\omega - NL_3}\right)\gamma\varepsilon = \varepsilon,$$

holds for $t \in J_1 = [a_1, a_1 + E_1]$, where $a_1 = a + E/2$ and $E_1 = E/2$.

The proof of the lemma is accomplished. □

We end up the present part by stating the following theorem and its immediate corollary, which can be verified as consequences of Lemmas 2.3, 2.4, and 2.5.

Theorem 2.2 *If the set $\mathscr{A}_x$ is Devaney's chaotic, then the same is true for the set $\mathscr{A}_y$.*

Corollary 2.4 *If the set $\mathscr{A}_x$ is Devaney's chaotic, then $\mathscr{A}$ is chaotic in the same way.*

In the next part, the replication of chaos in the Li–Yorke sense is taken into account.

2.6 Extension of Li–Yorke Chaos

Our aim in this section is to prove that if $\mathscr{A}_x$ is chaotic in the sense of Li–Yorke, then the same is valid for the set $\mathscr{A}_y$, and consequently for the set $\mathscr{A}$.

We start by indicating the following assertion, which presents the replication of proximality feature in accordance with Definition 2.6.

Lemma 2.6 *If a couple of functions $(x(t), \overline{x}(t)) \in \mathscr{A}_x \times \mathscr{A}_x$ is proximal, then the same is true for the couple $\left(\phi_{x(t)}(t), \phi_{\overline{x}(t)}(t)\right) \in \mathscr{A}_y \times \mathscr{A}_y$.*

Proof Fix an arbitrary small positive number ε and an arbitrary large positive number E. Define $\gamma = \dfrac{\omega(\omega - NL_3)}{2M_0N(\omega - NL_3) + NL_2\omega}$. Condition $(A7)$ implies that γ is positive. Because a given couple of functions $(x(t), \overline{x}(t)) \in \mathscr{A}_x \times \mathscr{A}_x$ is proximal, one can find a sequence of real numbers $\{E_i\}$ satisfying $E_i \geq E$ for each $i \in \mathbb{N}$, and a sequence $\{a_i\}$, $a_i \to \infty$ as $i \to \infty$, such that we have $\|x(t) - \overline{x}(t)\| < \gamma\varepsilon$, for each t from the intervals $J_i = [a_i, a_i + E_i]$, $i \in \mathbb{N}$, and $J_i \cap J_j = \emptyset$ whenever $i \neq j$.

Let us fix an arbitrary natural number i. Since the functions $y(t) = \phi_{x(t)}(t) \in \mathscr{A}_y$ and $\overline{y}(t) = \phi_{\overline{x}(t)}(t) \in \mathscr{A}_y$ satisfy the relations

$$y(t) = \int_{-\infty}^{t} e^{A(t-s)} g(x(s), y(s))ds,$$

and

$$\overline{y}(t) = \int_{-\infty}^{t} e^{A(t-s)} g(\overline{x}(s), \overline{y}(s))ds,$$

respectively, for $t \in J_i$, we have that

$$y(t) - \overline{y}(t) = \int_{-\infty}^{a_i} e^{A(t-s)}[g(x(s), y(s)) - g(\overline{x}(s), \overline{y}(s))]ds$$
$$+ \int_{a_i}^{t} e^{A(t-s)}[g(x(s), y(s)) - g(x(s), \overline{y}(s))]ds$$
$$+ \int_{a_i}^{t} e^{A(t-s)}[g(x(s), \overline{y}(s)) - g(\overline{x}(s), \overline{y}(s))]ds.$$

This implies that the inequality

$$\|y(t) - \overline{y}(t)\| \leq \int_{-\infty}^{a_i} 2M_0 N e^{-\omega(t-s)} ds$$
$$+ \int_{a_i}^{t} N L_3 e^{-\omega(t-s)} \|y(s) - \overline{y}(s)\| ds$$
$$+ \int_{a_i}^{t} N L_2 e^{-\omega(t-s)} \|x(s) - \overline{x}(s)\| ds$$
$$\leq \frac{2M_0 N}{\omega} e^{-\omega(t-a)} + \frac{N L_2 \gamma \varepsilon}{\omega} e^{-\omega t} \left(e^{\omega t} - e^{\omega a}\right)$$
$$+ \int_{a_i}^{t} N L_3 e^{-\omega(t-s)} \|y(s) - \overline{y}(s)\| ds$$

is valid. Hence, we attain that

$$e^{\omega t} \|y(t) - \overline{y}(t)\| \leq \frac{2M_0 N}{\omega} e^{\omega a_i} + \frac{N L_2 \gamma \varepsilon}{\omega} \left(e^{\omega t} - e^{\omega a_i}\right)$$
$$+ \int_{a_i}^{t} N L_3 e^{\omega s} \|y(s) - \overline{y}(s)\| ds.$$

Implementing Lemma 2.2 [20] to the last inequality, we obtain

$$e^{\omega t} \|y(t) - \overline{y}(t)\| \leq \frac{2M_0 N}{\omega} e^{\omega a_i} + \frac{N L_2 \gamma \varepsilon}{\omega} \left(e^{\omega t} - e^{\omega a_i}\right)$$
$$+ \int_{a}^{t} N L_3 \left[\frac{2M_0 N}{\omega} e^{\omega a_i} + \frac{N L_2 \gamma \varepsilon}{\omega} \left(e^{\omega s} - e^{\omega a_i}\right)\right] e^{N L_3 (t-s)} ds$$
$$= \frac{N L_2 \gamma \varepsilon}{\omega} e^{\omega t} + \left(\frac{2M_0 N}{\omega} - \frac{N L_2 \gamma \varepsilon}{\omega}\right) e^{\omega a_i} e^{N L_3 (t-a_i)}$$

$$+\frac{N^2L_2L_3\gamma\varepsilon}{\omega(\omega-NL_3)}e^{NL_3t}\left(e^{(\omega-NL_3)t}-e^{(\omega-NL_3)a_i}\right).$$

Multiplying both sides by the term $e^{-\omega t}$, one can verify that

$$\begin{aligned}\|y(t)-\overline{y}(t)\| &\le \frac{2M_0N}{\omega}e^{(NL_3-\omega)(t-a_i)}\\ &+\left(\frac{NL_2\gamma\varepsilon}{\omega}+\frac{N^2L_2L_3\gamma\varepsilon}{\omega(\omega-NL_3)}\right)\left(1-e^{(NL_3-\omega)(t-a_i)}\right)\\ &=\frac{2M_0N}{\omega}e^{(NL_3-\omega)(t-a_i)}+\frac{NL_2\gamma\varepsilon}{\omega-NL_3}\left(1-e^{(NL_3-\omega)(t-a_i)}\right).\end{aligned}$$

If E is sufficiently large such that $E>\dfrac{2}{\omega-NL_3}\ln\left(\dfrac{1}{\gamma\varepsilon}\right)$, then one has

$$e^{(NL_3-\omega)(t-a_i)}<e^{(NL_3-\omega)E_i/2}\le e^{(NL_3-\omega)E/2}<\gamma\varepsilon,$$

for $t\in[a_i+E_i/2,a_i+E_i]$.

Since the natural number i was arbitrarily chosen, for each $i\in\mathbb{N}$, we have that

$$\|y(t)-\overline{y}(t)\|<\left(\frac{2M_0N}{\omega}+\frac{NL_2}{\omega-NL_3}\right)\gamma\varepsilon=\varepsilon,$$

for each $t\in\widetilde{J}_i=\left[\widetilde{a}_i,\widetilde{a}_i+\widetilde{E}_i\right]$, where $\widetilde{a}_i=a_i+E_i/2$ and $\widetilde{E}_i=E_i/2$. Note that for each i the interval $\widetilde{J}_i\subset\mathbb{R}$ has a length no less than $\widetilde{E}=E/2$. As a consequence, the couple of functions $\left(\phi_{x(t)}(t),\phi_{\overline{x}(t)}(t)\right)\in\mathscr{A}_y\times\mathscr{A}_y$ is proximal according to Definition 2.6.

The proof is completed. □

The following lemma indicates the replication of the next characteristic feature of Li–Yorke chaos.

Lemma 2.7 *If a couple $(x(t),\overline{x}(t))\in\mathscr{A}_x\times\mathscr{A}_x$ is frequently (ε_0,Δ)-separated for some positive numbers ε_0 and Δ, then the couple $\left(\phi_{x(t)}(t),\phi_{\overline{x}(t)}(t)\right)\in\mathscr{A}_y\times\mathscr{A}_y$ is frequently $(\varepsilon_1,\overline{\Delta})$-separated for some positive numbers ε_1 and $\overline{\Delta}$.*

Proof Since a given couple of functions $(x(t),\overline{x}(t))\in\mathscr{A}_x\times\mathscr{A}_x$ is frequently (ε_0,Δ)-separated for some $\varepsilon_0>0$ and $\Delta>0$, there exist infinitely many disjoint intervals, each with a length no less than Δ, such that $\|x(t)-\overline{x}(t)\|>\varepsilon_0$ for each t from these intervals. Without loss of generality, assume that these intervals are all closed subsets of $\mathbb{R}$. In that case, one can find a sequence $\{\Delta_i\}$ satisfying $\Delta_i\ge\Delta$, $i\in\mathbb{N}$, and a sequence $\{d_i\}$, $d_i\to\infty$ as $i\to\infty$, such that for each $i\in\mathbb{N}$ the inequality $\|x(t)-\overline{x}(t)\|>\varepsilon_0$ holds for $t\in J_i=[d_i,d_i+\Delta_i]$, and $J_i\cap J_j=\emptyset$ whenever $i\ne j$. Throughout the proof, let us denote $y(t)=\phi_{x(t)}(t)\in\mathscr{A}_y$ and $\overline{y}(t)=\phi_{\overline{x}(t)}(t)\in\mathscr{A}_y$.

Our aim is to show the existence of positive numbers $\varepsilon_1, \overline{\Delta}$ and infinitely many disjoint intervals $\overline{J}_i \subset J_i, i \in \mathbb{N}$, each with length $\overline{\Delta}$, such that the inequality

$$\|y(t) - \overline{y}(t)\| > \varepsilon_1$$

holds for each t from the intervals $\overline{J}_i, i \in \mathbb{N}$.

As in Sect. 2.5, we again suppose that $g(x, y) = \begin{pmatrix} g_1(x, y) \\ g_2(x, y) \\ \vdots \\ g_n(x, y) \end{pmatrix}$, where each $g_j(x, y)$, $1 \leq j \leq n$, is a real valued function. Using the equicontinuity on $\mathbb{R}$ of the family $\mathscr{F}$, which is mentioned in Lemma 2.2, one can find a positive number $\tau < \Delta$, independent of the functions $x(t), \overline{x}(t) \in \mathscr{A}_x$, $y(t), \overline{y}(t) \in \mathscr{A}_y$, such that for any $t_1, t_2 \in \mathbb{R}$ with $|t_1 - t_2| < \tau$ the inequality

$$\begin{aligned} &\Big| \big(g_j\left(x(t_1), y(t_1)\right) - g_j\left(\overline{x}(t_1), y(t_1)\right)\big) \\ &- \big(g_j\left(x(t_2), y(t_2)\right) - g_j\left(\overline{x}(t_2), y(t_2)\right)\big) \Big| \\ &< \frac{L_1 \varepsilon_0}{2n} \end{aligned} \tag{2.6.23}$$

holds for all $1 \leq j \leq n$.

Suppose that the sequence $\{s_i\}$ denotes the midpoints of the intervals J_i, that is, $s_i = d_i + \Delta_i/2$ for each $i \in \mathbb{N}$. Let us define a sequence $\{\theta_i\}$ through the equation $\theta_i = s_i - \tau/2$.

Let us fix an arbitrary natural number i. In a similar way to the method specified in the proof of Lemma 2.3, one can show the existence of an integer $j_i = j_i(s_i)$, $1 \leq j_i \leq n$, such that

$$\left| g_{j_i}(x(s_i), y(s_i)) - g_{j_i}(\overline{x}(s_i), y(s_i)) \right| \geq \frac{L_1}{n} \|x(s_i) - \overline{x}(s_i)\| > \frac{L_1 \varepsilon_0}{n}. \tag{2.6.24}$$

On the other hand, making use of the inequality (2.6.23), it is easy to verify that

$$\begin{aligned} &\left| g_{j_i}\left(x(s_i), y(s_i)\right) - g_{j_i}\left(\overline{x}(s_i), y(s_i)\right) \right| - \left| g_{j_i}\left(x(t), y(t)\right) - g_{j_i}\left(\overline{x}(t), y(t)\right) \right| \\ &\leq \Big| \big(g_{j_i}\left(x(t), y(t)\right) - g_{j_i}\left(\overline{x}(t), y(t)\right)\big) - \big(g_{j_i}\left(x(s_i), y(s_i)\right) - g_{j_i}\left(\overline{x}(s_i), y(s_i)\right)\big) \Big| \\ &< \frac{L_1 \varepsilon_0}{2n}, \end{aligned}$$

for all $t \in [\theta_i, \theta_i + \tau]$. Therefore, by favor of (2.6.24), we obtain that the inequality

$$\begin{aligned}
&\left|g_{j_i}\left(x(t), y(t)\right) - g_{j_i}\left(\overline{x}(t), y(t)\right)\right| \\
&> \left|g_{j_i}\left(x(s_i), y(s_i)\right) - g_{j_i}\left(\overline{x}(s_i), y(s_i)\right)\right| - \frac{L_1\varepsilon_0}{2n} \\
&> \frac{L_1\varepsilon_0}{2n}
\end{aligned} \tag{2.6.25}$$

is valid on the same interval.

Using the mean value theorem for integrals, it is possible to find real numbers $s_1^i, s_2^i, \ldots, s_n^i \in [\theta_i, \theta_i + \tau]$ such that

$$\begin{aligned}
&\left\| \int_{\theta_i}^{\theta_i+\tau} [g(x(s), y(s)) - g(\overline{x}(s), y(s))]\, ds \right\| \\
&= \left\| \begin{pmatrix} \int_{\theta_i}^{\theta_i+\tau} [g_1(x(s), y(s)) - g_1(\overline{x}(s), y(s))]\, ds \\ \int_{\theta_i}^{\theta_i+\tau} [g_2(x(s), y(s)) - g_2(\overline{x}(s), y(s))]\, ds \\ \vdots \\ \int_{\theta_i}^{\theta_i+\tau} [g_n(x(s), y(s)) - g_n(\overline{x}(s), y(s))]\, ds \end{pmatrix} \right\| \\
&= \left\| \begin{pmatrix} \tau\left[g_1(x(s_1^i), y(s_1^i)) - g_1(\overline{x}(s_1^i), y(s_1^i))\right] \\ \tau\left[g_2(x(s_2^i), y(s_2^i)) - g_2(\overline{x}(s_2^i), y(s_2^i))\right] \\ \vdots \\ \tau\left[g_n(x(s_n^i), y(s_n^i)) - g_n(\overline{x}(s_n^i), y(s_n^i))\right] \end{pmatrix} \right\|.
\end{aligned}$$

Hence, the inequality (2.6.25) yields that

$$\begin{aligned}
&\left\| \int_{\theta_i}^{\theta_i+\tau} [g(x(s), y(s)) - g(\overline{x}(s), y(s))]\, ds \right\| \\
&\geq \tau \left| g_{j_i}(x(s_{j_i}^i), y(s_{j_i}^i)) - g_{j_i}(\overline{x}(s_{j_i}^i), y(s_{j_i}^i)) \right| \\
&> \frac{\tau L_1 \varepsilon_0}{2n}.
\end{aligned}$$

For $t \in [\theta_i, \theta_i + \tau]$, the functions $y(t) \in \mathscr{A}_y$ and $\overline{y}(t) \in \mathscr{A}_y$ satisfy the relations

$$y(t) = y(\theta_i) + \int_{\theta_i}^{t} Ay(s)ds + \int_{\theta_i}^{t} g(x(s), y(s))ds,$$

and

$$\overline{y}(t) = \overline{y}(\theta_i) + \int_{\theta_i}^{t} A\overline{y}(s)ds + \int_{\theta_i}^{t} g(\overline{x}(s), \overline{y}(s))ds,$$

respectively, and herewith the equation

$$y(t)-\overline{y}(t)=(y(\theta_i)-\overline{y}(\theta_i))+\int_{\theta_i}^{t}A(y(s)-\overline{y}(s))ds$$
$$+\int_{\theta_i}^{t}[g(x(s),y(s))-g(\overline{x}(s),y(s))]ds$$
$$+\int_{\theta_i}^{t}[g(\overline{x}(s),y(s))-g(\overline{x}(s),\overline{y}(s))]ds$$

is achieved. Taking $t=\theta_i+\tau$ in the last equation, we attain the inequality

$$\begin{aligned}\|y(\theta_i+\tau)-\overline{y}(\theta_i+\tau)\| &\geq \left\|\int_{\theta}^{\theta+\tau}[g(x(s),y(s))-g(\overline{x}(s),y(s))]ds\right\| \\ -\|y(\theta_i)-\overline{y}(\theta_i)\| &- \int_{\theta_i}^{\theta_i+\tau}(\|A\|+L_3)\,\|y(s)-\overline{y}(s)\|\,ds\end{aligned} \tag{2.6.26}$$

Now, assume that $\max_{t\in[\theta_i,\theta_i+\tau]}\|y(t)-\overline{y}(t)\|\leq\dfrac{\tau L_1\varepsilon_0}{2n[2+\tau(L_3+\|A\|)]}$. In this case, one arrives at a contradiction since, by means of the inequalities (2.6.25) and (2.6.26), we have

$$\begin{aligned}&\max_{t\in[\theta_i,\theta_i+\tau]}\|y(t)-\overline{y}(t)\|\geq\|y(\theta_i+\tau)-\overline{y}(\theta_i+\tau)\|\\ &>\frac{\tau L_1\varepsilon_0}{2n}-[1+\tau(L_3+\|A\|)]\max_{t\in[\theta_i,\theta_i+\tau]}\|y(t)-\overline{y}(t)\|\\ &\geq\frac{\tau L_1\varepsilon_0}{2n}-[1+\tau(L_3+\|A\|)]\frac{\tau L_1\varepsilon_0}{2n[2+\tau(L_3+\|A\|)]}\\ &=\frac{\tau L_1\varepsilon_0}{2n}\left(1-\frac{1+\tau(L_3+\|A\|)}{2+\tau(L_3+\|A\|)}\right)\\ &=\frac{\tau L_1\varepsilon_0}{2n[2+\tau(L_3+\|A\|)]}.\end{aligned}$$

Therefore, it is true that $\max_{t\in[\theta_i,\theta_i+\tau]}\|y(t)-\overline{y}(t)\|>\dfrac{\tau L_1\varepsilon_0}{2n[2+\tau(L_3+\|A\|)]}$.

Suppose that the real valued function $\|y(t)-\overline{y}(t)\|$ takes its maximum value for $t\in[\theta_i,\theta_i+\tau]$ at a point η_i. In other words, for some $\eta_i\in[\theta_i,\theta_i+\tau]$, we have that

$$\max_{t\in[\theta_i,\theta_i+\tau]}\|y(t)-\overline{y}(t)\|=\|y(\eta_i)-\overline{y}(\eta_i)\|.$$

Making use of the integral equations

$$y(t) = y(\eta_i) + \int_{\eta_i}^{t} Ay(s)ds + \int_{\eta_i}^{t} g(x(s), y(s))ds,$$

and

$$\overline{y}(t) = \overline{y}(\eta_i) + \int_{\eta_i}^{t} A\overline{y}(s)ds + \int_{\eta_i}^{t} g(\overline{x}(s), \overline{y}(s))ds,$$

on the time interval $[\theta_i, \theta_i + \tau]$, one can obtain that

$$y(t) - \overline{y}(t) = (y(\eta_i) - \overline{y}(\eta_i)) + \int_{\eta_i}^{t} A(y(s) - \overline{y}(s))ds$$
$$+ \int_{\eta_i}^{t} [g(x(s), y(s)) - g(\overline{x}(s), \overline{y}(s))]ds.$$

Define the numbers

$$\overline{\Delta} = \min\left\{\frac{\tau}{2}, \frac{\tau L_1 \varepsilon_0}{8n(M\|A\| + M_0)[2 + \tau(L_3 + \|A\|)]}\right\}$$

and

$$\theta_i^1 = \begin{cases} \eta_i, & \text{if } \eta \le \theta_i + \tau/2 \\ \eta_i - \tau^1, & \text{if } \eta_i > \theta_i + \tau/2 \end{cases}.$$

For each $t \in [\theta_i^1, \theta_i^1 + \overline{\Delta}]$, we have that

$$\begin{aligned} \|y(t) - \overline{y}(t)\| &\ge \|y(\eta_i) - \overline{y}(\eta_i)\| - \left|\int_{\eta_i}^{t} \|A\| \, \|y(s) - \overline{y}(s)\| \, ds\right| \\ &- \left|\int_{\eta_i}^{t} \|g(x(s), y(s)) - g(\overline{x}(s), \overline{y}(s))\| \, ds\right| \\ &> \frac{\tau L_1 \varepsilon_0}{2n[2 + \tau(L_3 + \|A\|)]} - 2M\|A\|\tau^1 - 2M_0\tau^1 \\ &= \frac{\tau L_1 \varepsilon_0}{2n[2 + \tau(L_3 + \|A\|)]} - 2\tau^1(M\|A\| + M_0) \\ &\ge \frac{\tau L_1 \varepsilon_0}{4n[2 + \tau(L_3 + \|A\|)]}. \end{aligned}$$

The information mentioned above is true for an arbitrarily chosen natural number i. Therefore, for each $i \in \mathbb{N}$, the interval $\overline{J}_i = [\theta_i^1, \theta_i^1 + \overline{\Delta}]$ is a subset of $[\theta_i, \theta_i + \tau]$, and hence of J_i. Moreover, for any $i \in \mathbb{N}$, we have $\|y(t) - \overline{y}(t)\| > \varepsilon_1$, $t \in \overline{J}_i$, where $\varepsilon_1 = \dfrac{\tau L_1 \varepsilon_0}{4n[2 + \tau(L_3 + \|A\|)]}$.

Consequently, according to Definition 2.7, the pair $\left(\phi_{x(t)}(t), \phi_{\overline{x}(t)}(t)\right) \in \mathscr{A}_y \times \mathscr{A}_y$ is frequently $(\varepsilon_1, \overline{\Delta})$-separated.

The proof of the lemma is finalized. □

Now, we state and prove the main theorem of the present section. In the proof, we suppose that $\mathscr{G}_x \subset \mathscr{A}_x$ denotes the set of periodic functions inside $\mathscr{A}_x$ and the set $\mathscr{G}_y \subset \mathscr{A}_y$, defined through Eq. (2.5.22), denotes the set of periodic functions inside $\mathscr{A}_y$.

Theorem 2.3 *If the set $\mathscr{A}_x$ is Li–Yorke chaotic, then the same is true for the set $\mathscr{A}_y$.*

Proof It can be easily verified that for any natural number k, $x(t) \in \mathscr{G}_x$ is a kT-periodic function if and only if $\phi_{x(t)}(t) \in \mathscr{G}_y$ is kT-periodic, where $\mathscr{G}_x$ and $\mathscr{G}_y$ denote the sets of all periodic functions inside $\mathscr{A}_x$ and $\mathscr{A}_y$,, respectively. Therefore, the set $\mathscr{A}_y$ admits a kT-periodic function for any $k \in \mathbb{N}$.

Next, suppose that the set $\mathscr{C}_x$ is a scrambled set inside $\mathscr{A}_x$ and define the set

$$\mathscr{C}_y = \left\{\phi_{x(t)}(t) \mid x(t) \in \mathscr{C}_x\right\}. \tag{2.6.27}$$

Condition $(A4)$ implies that there is a one-to-one correspondence between the sets $\mathscr{C}_x$ and $\mathscr{C}_y$. Since the scrambled set $\mathscr{C}_x$ is uncountable, it is clear that the set $\mathscr{C}_y$ is also uncountable. Moreover, using the same condition one can show that no periodic functions exist inside $\mathscr{C}_y$, since no such functions take place inside the set $\mathscr{C}_x$. That is, $\mathscr{C}_y \cap \mathscr{G}_y = \emptyset$.

Since each couple of functions inside $\mathscr{C}_x \times \mathscr{C}_x$ is proximal, Lemma 2.6 implies the same feature for each couple of functions inside $\mathscr{C}_y \times \mathscr{C}_y$.

Similarly, Lemma 2.7 implies that if each couple of functions $(x(t), \overline{x}(t)) \in \mathscr{C}_x \times \mathscr{C}_x$ $(\mathscr{C}_x \times \mathscr{G}_x)$ is frequently (ε_0, Δ)-separated for some positive numbers ε_0 and Δ, then each couple of functions $(y(t), \overline{y}(t)) \in \mathscr{C}_y \times \mathscr{C}_y$ $\left(\mathscr{C}_y \times \mathscr{G}_y\right)$ is frequently $(\varepsilon_1, \overline{\Delta})$-separated for some positive numbers ε_1 and $\overline{\Delta}$. Consequently, the set $\mathscr{C}_y$ is a scrambled set inside $\mathscr{A}_y$, and according to Definition 2.10, $\mathscr{A}_y$ is Li–Yorke chaotic.

The proof of the theorem is accomplished. □

An immediate corollary of Theorem 2.3 is the following:

Corollary 2.5 *If the set $\mathscr{A}_x$ is Li–Yorke chaotic, then the set $\mathscr{A}$ is chaotic in the same way.*

2.7 Morphogenesis of Chaos

Two different mechanisms of chaos extension (morphogenesis) through applying replication are considered in this chapter. The first one is illustrated schematically in Fig. 2.4. The figure represents consecutively connected systems as boxes and the blue arrows symbolize unidirectional couplings between two systems. In the first coupling, we take into account a generator system, the leftmost box in the figure, which is connected with a second system considered as a replicator in the couple. In the next coupling, the second system is considered as a generator with respect to the third one. That is, it changes its role in the extension process. In the third coupling,

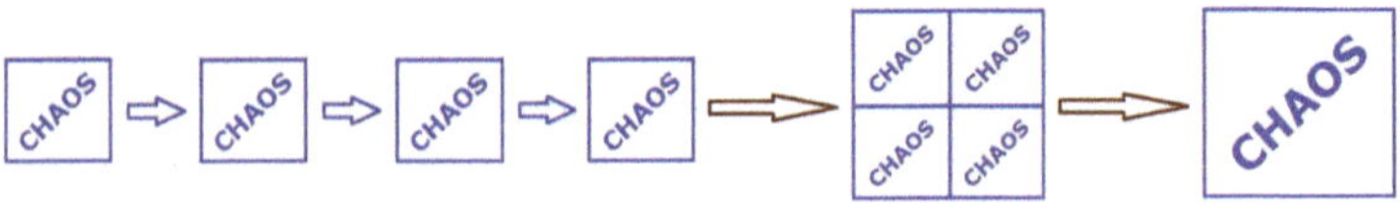

Fig. 2.4 Morphogenesis of chaos through consecutive replications

Fig. 2.5 Morphogenesis of chaos from a prior chaos as a core

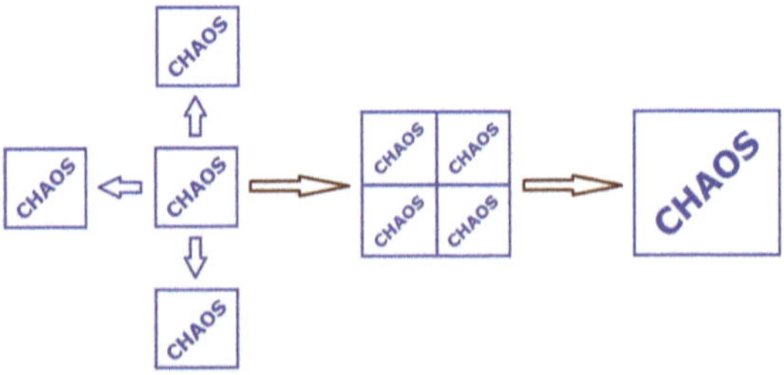

the third system is considered as a generator and the forth one as a replicator. In summary, the mechanism proceeds as follows. We take into account consecutive unidirectionally coupled systems such that the initial one is a generator and at each next coupling the role of the previously chaotified replicator changes and we start to use it as a generator. As a result of the mechanism all individual subsystems are chaotic as well as the system which consists of all subsystems. Moreover, the type of the chaos is saved under this procedure.

In Fig. 2.5 we show another mechanism of chaos extension. Here, the generator is surrounded by three replicators and the blue arrows symbolize, again, unidirectional couplings between two systems. Distinctively from the former mechanism, the replicators do not change their role with respect to each other according to the special topology of connection. The generator is coupled with all other replicators such that it is rather a core than a beginning element. The result of the mechanism is similar to the former such that all replicators as well as the system consisting of all subsystems become chaotic, saving the chaos type of the generator.

We call the two ways as *the chain* and *the core* mechanisms, respectively, and the system which unites the generator and several replicators, of type (2.1.2), in either extension mechanism as *the result-system*. Theoretically, we do not discuss constraints on the dimension of the result-system, but under certain conditions it seems that the dimension is not restricted for both mechanisms. However, this is definitely true for the core mechanism even with infinite dimensions. We will discuss and simulate the chain mechanism in the chapter, mainly, since the core mechanism can be discussed very similarly. One can invent other mechanisms, for example, by considering "composition" of the two mechanisms proposed presently. As an example, one can consider the network pictured in Fig. 1.2.

Next, to exemplify the chaos extension procedure, according to the chain mechanism shown in Fig. 2.4 we set up the following 8-dimensional result-system

$$
\begin{aligned}
x_1' &= x_2 \\
x_2' &= -0.05x_2 - x_1^3 + 7.5\cos t \\
x_3' &= x_4 + x_1 \\
x_4' &= -3x_3 - 2x_4 - 0.008x_3^3 + x_2 \\
x_5' &= x_6 + x_3 \\
x_6' &= -3x_5 - 2.1x_6 - 0.007x_5^3 + x_4 \\
x_7' &= x_8 + x_5 \\
x_8' &= -3.1x_7 - 2.2x_8 - 0.006x_7^3 + x_6.
\end{aligned}
\tag{2.7.28}
$$

We note that system (2.7.28) consists of four subsystems with the coordinates (x_1, x_2), (x_3, x_4), (x_5, x_6), and (x_7, x_8) such that the subsystem (x_1, x_2) is exactly the generator used in system (2.1.5) + (2.1.6), while the subsystem (x_3, x_4) is the replicator of (2.1.5)+(2.1.6).

According to the theoretical results of the present chapter, system (2.7.28) possesses a chaotic attractor in the 8-dimensional phase space. By marking the trajectory of this system with the initial data $x_1(0) = 2$, $x_2(0) = 3$, $x_3(0) = x_5(0) = x_7(0) = -1$, $x_4(0) = x_6(0) = x_8(0) = 1$ stroboscopically at times that are integer multiples of 2π, we obtain the Poincaré section inside the 8-dimensional space. In Fig. 2.6, which informs us about morphogenesis, the 3-dimensional projections of the whole Poincaré section on the $x_2 - x_4 - x_6$ and $x_3 - x_5 - x_7$ spaces are shown. One can see in Fig. 2.6a, b the additional *foldings* which are not possible to observe in the classical strange attractor shown in Fig. 2.2a.

Despite we are restricted to make illustrations at most in 3-dimensional spaces, taking inspiration from Figs. 2.2 and 2.6, one can imagine that the structure of the original Poincaré section in the 8-dimensional space will be similar through its fractal

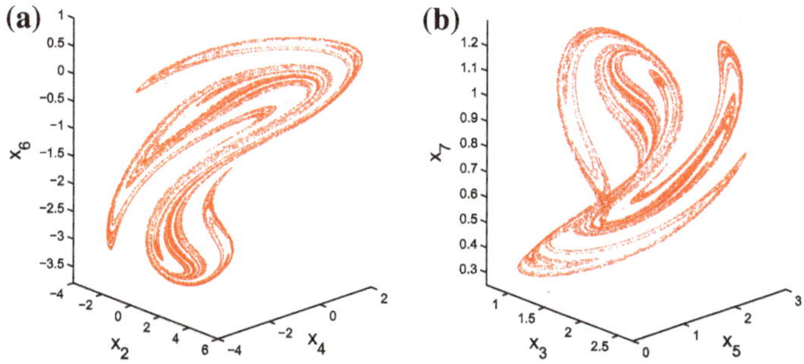

Fig. 2.6 In **a** and **b** projections of the result chaotic attractor on the $x_2 - x_4 - x_6$ and $x_3 - x_5 - x_7$ spaces are respectively presented. One can see in (**a**) and (**b**) the additional *foldings* which are not possible to observe in the 2-dimensional picture of the prior classical chaos shown in Fig. 2.2a. In the same time, the shape of the original attractor is seen in the resulting chaos. The illustrations in (**a**) and (**b**) repeat the structure of the attractor of the generator and the similarity between these pictures is a manifestation of the morphogenesis of chaos

structure, but more beautiful and impressive than its projections. From this point of view, we are not surprised since these facts have been proved theoretically.

Next, we shall handle the problem that whether the chaos extension procedure works for all existing systems in the mechanisms presented above, from the theoretical point of view. Since the core mechanism does not need any additional theoretical discussions, we will consider the chain mechanism.

In addition to the system (2.1.1)+(2.1.2), we take into account the system

$$z' = Bz + h(y(t), z), \tag{2.7.29}$$

where $h : \mathbb{R}^n \times \mathbb{R}^l \to \mathbb{R}^l$ is a continuous function in all of its arguments, the constant $l \times l$ real-valued matrix B has real parts of eigenvalues all negative and $y(t)$ is a solution system (2.1.2).

It is easy to verify the existence of positive numbers $\widetilde{N}$ and $\widetilde{\omega}$ such that $\left\|e^{Bt}\right\| \leq \widetilde{N}e^{-\widetilde{\omega}t}$, for all $t \geq 0$.

In our next theoretical discussions, the system (2.7.29) will serve as the third system in the chain mechanism presented by Fig. 2.4, and we need the following assumptions which are counterparts of the conditions $(A4)$–$(A7)$ presented in Sect. 2.2.

(A8) There exists a positive number $\widetilde{L}_1$ such that

$$\|h(y_1, z) - h(y_2, z)\| \geq \widetilde{L}_1 \|y_1 - y_2\|,$$

for all $y_1, y_2 \in \mathbb{R}^n$, $z \in \mathbb{R}^l$;

(A9) There exist positive numbers $\widetilde{L}_2$ and $\widetilde{L}_3$ such that

$$\|h(y_1, z) - h(y_2, z)\| \leq \widetilde{L}_2 \|y_1 - y_2\|,$$

for all $y_1, y_2 \in \mathbb{R}^n$, $z \in \mathbb{R}^l$, and

$$\|h(y, z_1) - h(y, z_2)\| \leq \widetilde{L}_3 \|z_1 - z_2\|,$$

for all $y \in \mathbb{R}^n$, $z_1, z_2 \in \mathbb{R}^l$;

(A10) There exists a positive number $K_0 < \infty$ such that

$$\sup_{y \in \mathbb{R}^n, z \in \mathbb{R}^l} \|h(y, z)\| \leq K_0;$$

(A11) $\widetilde{N}\widetilde{L}_3 - \widetilde{\omega} < 0.$

Likewise the definition for the set of functions $\mathscr{A}_y$, given by (2.2.10), let us denote by $\mathscr{A}_z$ the set of all bounded on $\mathbb{R}$ solutions of system $z' = Bz + h(y(t), z)$, for any $y(t) \in \mathscr{A}_y$.

In a similar way to Lemma 2.1, one can show that the set $\mathscr{U}_z$ which consists of the solutions of system $z' = Az + g(y(t), z)$ for some $y(t) \in \mathscr{U}_y$ is a basin of $\mathscr{A}_z$.

Furthermore, a similar result of Theorem 2.1 introduced in Sect. 2.4, hold also for the set $\mathscr{A}_z$.

We state in the next theorem that similar results of the Theorems 2.2 and 2.3 presented in Sects. 2.5 and 2.6, respectively, hold also for the set $\mathscr{A}_z$.

We note that, in the case of the presence of arbitrary finite number of systems, which obey conditions that are counterparts of $(A4)$–$(A7)$, one can prove that a similar result of the next theorem holds for the chain mechanism.

Theorem 2.4 *If the set $\mathscr{A}_x$ is Devaney chaotic or Li–Yorke chaotic, then the set $\mathscr{A}_z$ is chaotic in the same way as both $\mathscr{A}_x$ and $\mathscr{A}_y$.*

Proof In the proof, we will show that for each $z(t) \in \mathscr{A}_z$ and arbitrary $\delta > 0$, there exist $\overline{z}(t) \in \mathscr{A}_z$ and $t_0 \in \mathbb{R}$ such that $\|z(t_0) - \overline{z}(t_0)\| < \delta$, which is needed to show sensitivity of $\mathscr{A}_z$. The remaining parts of the proof can be performed in a similar way to the proofs presented in Sects. 2.5 and 2.6, and therefore are omitted.

Suppose that the set $\mathscr{A}_x$ is sensitive. Fix an arbitrary $\delta > 0$ and let $z(t) \in \mathscr{A}_z$ be a given solution of system (2.7.29). In this case, there exists $y(t) = \phi_{x(t)}(t) \in \mathscr{A}_y$, where $x(t) \in \mathscr{A}_x$, such that $z(t)$ is the unique bounded on $\mathbb{R}$ solution of the system $z' = Bz + h(y(t), z)$.

Let us choose a sufficiently small positive number $\overline{\varepsilon} = \overline{\varepsilon}(\delta)$ which satisfies the inequality

$$\left(1 + \frac{\widetilde{N}\widetilde{L}_2}{\widetilde{\omega} - \widetilde{N}\widetilde{L}_3}\right)\left(1 + \frac{NL_2}{\omega - NL_3}\right)\overline{\varepsilon} < \delta$$

and denote $\varepsilon_1 = \left(1 + \dfrac{NL_2}{\omega - NL_3}\right)\overline{\varepsilon}$. Now, take $R = R(\overline{\varepsilon}) < 0$ sufficiently large in absolute value such that both of the inequalities $\dfrac{2M_0N}{\omega}e^{-(NL_3-\omega)R/2} \leq \overline{\varepsilon}$ and $\dfrac{2\widetilde{M}_0\widetilde{N}}{\widetilde{\omega}}e^{-(\widetilde{N}\widetilde{L}_3-\widetilde{\omega})R/2} \leq \varepsilon_1$ are valid, and let $\delta_1 = \delta_1(\overline{\varepsilon}, R) = \overline{\varepsilon}e^{L_0R}$. Since the set $\mathscr{A}_x$ is sensitive, one can find $\overline{x}(t) \in \mathscr{A}_x$ and $t_0 \in \mathbb{R}$ such that the inequality $\|x(t_0) - \overline{x}(t_0)\| < \delta_1$ holds.

As in the case of the proof of Lemma 2.3, for $t \in [t_0 + R, t_0]$, one can verify that $\|x(t) - \overline{x}(t)\| < \overline{\varepsilon}$, and

$$\|y(t) - \overline{y}(t)\| \leq \frac{NL_2\overline{\varepsilon}}{\omega - NL_3} + \frac{2M_0N}{\omega}e^{(NL_3-\omega)(t-t_0-R)}.$$

According to the last inequality, we have $\|y(t) - \overline{y}(t)\| \leq \varepsilon_1$, for $t \in [t_0 + R/2, t_0]$.

Suppose that $\overline{z}(t)$ is the unique bounded on $\mathbb{R}$ solution of the system $z' = Bz + h(\overline{y}(t), z)$. One can see that the relations

$$z(t) = \int_{-\infty}^{t} e^{B(t-s)}h(y(s), z(s))ds$$

and

$$\overline{z}(t) = \int_{-\infty}^{t} e^{B(t-s)} h(\overline{y}(s), \overline{z}(s)) ds,$$

are valid. Using these equations, it can be verified that

$$\|z(t) - \overline{z}(t)\| \le \int_{t_0+\frac{R}{2}}^{t} \widetilde{N} e^{-\widetilde{\omega}(t-s)} \|h(y(s), z(s)) - h(y(s), \overline{z}(s))\| ds$$
$$+ \int_{t_0+\frac{R}{2}}^{t} \widetilde{N} e^{-\widetilde{\omega}(t-s)} \|h(y(s), \overline{z}(s)) - h(\overline{y}(s), \overline{z}(s))\| ds$$
$$+ \int_{-\infty}^{t_0+\frac{R}{2}} \widetilde{N} e^{-\widetilde{\omega}(t-s)} \|h(y(s), z(s)) - h(\overline{y}(s), \overline{z}(s))\| ds.$$

Since $\|y(t) - \overline{y}(t)\| < \varepsilon_1$ for $t \in [t_0 + R/2, t_0]$, one has

$$\|z(t) - \overline{z}(t)\| \le \widetilde{N}\widetilde{L}_3 \int_{t_0+\frac{R}{2}}^{t} e^{-\widetilde{\omega}(t-s)} \|z(s) - \overline{z}(s)\| ds$$
$$+ \widetilde{N}\widetilde{L}_2\varepsilon_1 \int_{t_0+\frac{R}{2}}^{t} e^{-\widetilde{\omega}(t-s)} ds + 2\widetilde{M}_0\widetilde{N} \int_{-\infty}^{t_0+\frac{R}{2}} e^{-\widetilde{\omega}(t-s)} ds$$
$$\le \widetilde{N}\widetilde{L}_3 \int_{t_0+\frac{R}{2}}^{t} e^{-\widetilde{\omega}(t-s)} \|z(s) - \overline{z}(s)\| ds$$
$$+ \frac{\widetilde{N}\widetilde{L}_2\varepsilon_1}{\widetilde{\omega}} e^{-\widetilde{\omega}t} (e^{\widetilde{\omega}t} - e^{\widetilde{\omega}(t_0+R/2)}) + \frac{2\widetilde{M}_0\widetilde{N}}{\widetilde{\omega}} e^{-\widetilde{\omega}(t-t_0-R/2)}.$$

Now, let us introduce the functions $u(t) = e^{\widetilde{\omega}t} \|z(t) - \overline{z}(t)\|$, $k(t) = \frac{\widetilde{N}\widetilde{L}_2\varepsilon_1}{\widetilde{\omega}} e^{\widetilde{\omega}t}$, and $v(t) = c + k(t)$ where $c = \left(\frac{2\widetilde{M}_0\widetilde{N}}{\widetilde{\omega}} - \frac{\widetilde{N}\widetilde{L}_2\varepsilon_1}{\widetilde{\omega}}\right) e^{\widetilde{\omega}(t_0+R/2)}$.

These definitions imply that $u(t) \le v(t) + \int_{t_0+\frac{R}{2}}^{t} \widetilde{N}\widetilde{L}_3 u(s) ds$ and applying Lemma 2.2 [20] leads to

$$u(t) \le v(t) + \widetilde{N}\widetilde{L}_3 \int_{t_0+\frac{R}{2}}^{t} e^{\widetilde{N}\widetilde{L}_3(t-s)} h(s) ds.$$

Therefore, for $t \in [t_0 + R/2, t_0]$ we have

$$u(t) \le c + k(t) + c\left(e^{\widetilde{N}\widetilde{L}_3(t-t_0-R/2)} - 1\right) + \frac{N^2\widetilde{L}_2\widetilde{L}_3\varepsilon_1}{\widetilde{\omega}} e^{\widetilde{N}\widetilde{L}_3 t} \int_{t_0+\frac{R}{2}}^{t} e^{(\widetilde{\omega}-\widetilde{N}\widetilde{L}_3)s} ds$$

$$= \frac{\widetilde{N}\widetilde{L}_2\varepsilon_1}{\widetilde{\omega}} e^{\widetilde{\omega}t} + \left(\frac{2\widetilde{M}_0 N}{\widetilde{\omega}} - \frac{\widetilde{N}\widetilde{L}_2\varepsilon_1}{\widetilde{\omega}}\right) e^{\widetilde{\omega}T} e^{\widetilde{N}\widetilde{L}_3(t-t_0-R/2)}$$

$$+ \frac{\widetilde{N}^2\widetilde{L}_2\widetilde{L}_3\varepsilon_1}{\widetilde{\omega}(\widetilde{\omega}-\widetilde{N}\widetilde{L}_3)} e^{\widetilde{\omega}t}\left[1 - e^{(\widetilde{N}\widetilde{L}_3-\widetilde{\omega})(t-t_0-R/2)}\right],$$

and hence

$$\|z(t) - \overline{z}(t)\| \le \frac{\widetilde{N}\widetilde{L}_2\varepsilon_1}{\widetilde{\omega} - \widetilde{N}\widetilde{L}_3}\left[1 - e^{(\widetilde{N}\widetilde{L}_3-\widetilde{\omega})(t-t_0-R/2)}\right] + \frac{2\widetilde{M}_0 N}{\widetilde{\omega}} e^{(\widetilde{N}\widetilde{L}_3-\widetilde{\omega})(t-t_0-R/2)}.$$

Consequently, the inequality

$$\begin{aligned} \|z(t_0) - \overline{z}(t_0)\| &\le \frac{\widetilde{N}\widetilde{L}_2\varepsilon_1}{\widetilde{\omega} - \widetilde{N}\widetilde{L}_3} + \frac{2\widetilde{M}_0\widetilde{N}}{\widetilde{\omega}} e^{(\widetilde{\omega}-\widetilde{N}\widetilde{L}_3)R/2} \\ &< \left(1 + \frac{\widetilde{N}\widetilde{L}_2}{\widetilde{\omega} - \widetilde{N}\widetilde{L}_3}\right)\varepsilon_1 \\ &< \delta \end{aligned}$$

is valid.

The theorem is proved. □

2.8 Period-Doubling Cascade

We start this section by describing the chaos through period-doubling cascade [22–24] for the set of functions $\mathscr{A}_x$, and deal with its replication by the set of functions $\mathscr{A}_y$, which is defined by Eq. (2.2.10).

Suppose that there exists a function $G : \mathbb{R} \times \mathbb{R}^m \times \mathbb{R} \to \mathbb{R}^m$ which is continuous in all of its arguments such that $F(t, x) = G(t, x, \mu_\infty)$ for some finite number μ_∞, which will be explained below.

To discuss chaos through period-doubling cascade, let us consider the system

$$x' = G(t, x, \mu), \tag{2.8.30}$$

where μ is a parameter.

We say that the set $\mathscr{A}_x$ is chaotic through period-doubling cascade if there exist a natural number k and a sequence of period-doubling bifurcation values $\{\mu_m\}$, $\mu_m \to \mu_\infty$ as $m \to \infty$, such that for each $m \in \mathbb{N}$ as the parameter μ increases (or decreases) through μ_m, system (2.8.30) undergoes a period-doubling bifurcation and a periodic solution with period $k2^m T$ appears. As a consequence, at $\mu = \mu_\infty$, there exist infinitely many unstable periodic solutions of system (2.8.30), and hence of system (2.1.1), all lying in a bounded region. In this case, the set $\mathscr{A}_x$ admits periodic functions of periods $kT, 2kT, 4kT, 8kT, \ldots$.

Now, making use of the Eq. (2.2.9), one can show that for any natural number p, if $x(t) \in \mathscr{A}_x$ is a pT-periodic function then $\phi_{x(t)}(t) \in \mathscr{A}_y$ is also pT-periodic. Moreover, condition $(A4)$ implies that the converse is also true. Consequently, if the set $\mathscr{A}_x$ admits periodic functions of periods $kT, 2kT, 4kT, \ldots$, then the same is valid for $\mathscr{A}_y$, with no additional periodic functions of any other period. Furthermore, the technique indicated in the proof of Lemma 2.3 can be used to show that these periodic solutions are all unstable and this provides us an opportunity to state the following theorem.

Theorem 2.5 *If the set $\mathscr{A}_x$ is chaotic through period-doubling cascade, then the same is true for $\mathscr{A}_y$.*

The following corollary of Theorem 2.5 states that the result-system (2.1.1) + (2.1.2) is chaotic through the period-doubling cascade, provided the system (2.1.1) is.

Corollary 2.6 *If the set $\mathscr{A}_x$ is chaotic through period-doubling cascade, then the same is true for $\mathscr{A}$.*

Our theoretical results show that the replicator system (2.1.2), likewise the generator counterpart, undergoes period-doubling bifurcations as the parameter μ increases or decreases through the values μ_m, $m \in \mathbb{N}$. That is, the sequence $\{\mu_m\}$ of bifurcation parameters is exactly the same for both generator and replicator systems. In this case, if the generator system obeys the Feigenbaum universality [5, 25–27] then one can conclude that the same is true also for the replicator. In other words, when $\lim\limits_{m\to\infty} \dfrac{\mu_m - \mu_{m+1}}{\mu_{m+1} - \mu_{m+2}}$ is evaluated, the universal constant known as the Feigenbaum number 4.6692016... is achieved and this universal number is the same for both generator and replicator.

It is worth saying that the results about replication of period-doubling cascade as well as the Feigenbaum's universal behavior, which can be perceived as another aspect of morphogenesis of chaos, are true also for chaos extension mechanisms shown in Figs. 2.4 and 2.5. In our next example, using the chain mechanism, we will illustrate through simulations the morphogenesis of period-doubling cascade.

In paper [28], it is indicated that the Duffing's equation

$$x'' + 0.3x' + x^3 = \mu \cos t \tag{2.8.31}$$

admits the chaos through period-doubling cascade at the parameter value $\mu = \mu_\infty \equiv 40$. Defining the new variables $x_1 = x$ and $x_2 = x'$, Eq. (2.8.31) can be rewritten as a system in the following form:

$$\begin{aligned} x_1' &= x_2 \\ x_2' &= -0.3x_2 - x_1^3 + \mu \cos t. \end{aligned} \tag{2.8.32}$$

Making use of system (2.8.32) as the generator, let us constitute the following 8-dimensional result-system

$$\begin{aligned}
x_1' &= x_2 \\
x_2' &= -0.3x_2 - x_1^3 + \mu \cos t \\
x_3' &= 2x_3 - x_4 + 0.4 \tan\left((x_1 + x_3)/10\right) \\
x_4' &= 17x_3 - 6x_4 + x_2 \\
x_5' &= -2x_5 + 0.5 \sin x_6 - 4x_4 \\
x_6' &= -x_5 - 4x_6 - \tan\left(x_3/2\right) \\
x_7' &= 2x_7 + 5x_8 - 0.0003(x_7 - x_8)^3 - 1.5x_6 \\
x_8' &= -5x_7 - 8x_8 + 4x_5.
\end{aligned} \tag{2.8.33}$$

System (2.8.33) is designed according to the chain mechanism indicated in Fig. 2.4. In the coupling between the subsystems with coordinates (x_1, x_2) and (x_3, x_4) the former is the generator and the latter is the replicator. In the second coupling between the subsystems with coordinates (x_3, x_4) and (x_5, x_6), this time the former is used as the generator although it was the replicator in the previous coupling. The final coupling between the subsystems with coordinates (x_5, x_6) and (x_7, x_8) is constructed in a similar way. In this exemplification we will refer to subsystems with coordinates (x_1, x_2), (x_3, x_4), (x_5, x_6) and (x_7, x_8) as the first, second, third and the fourth subsystems, respectively.

According to our theoretical discussions, the result-system (2.8.33) with the parameter value $\mu = \mu_\infty \equiv 40$ admits a chaotic attractor in the 8-dimensional phase space, which is obtained through period-doubling cascade. For the parameter value μ between 30 and 40, the bifurcation diagrams corresponding to the x_2, x_4, x_6 and x_8 coordinates of system (2.8.33) are illustrated in Fig. 2.7. The picture shown in Fig. 2.7a is the bifurcation diagram of the system (2.8.32), while the pictures presented in Fig. 2.7b–d correspond to the second, third and the fourth subsystems, respectively. For the parameter values where stable periodic solutions exist, the one-to-one correspondence between the periodic solutions of the subsystems is observable in the figure. Moreover, it is seen in Fig. 2.7b–d that, likewise the first subsystem, all other subsystems undergo period-doubling bifurcations at the same parameter values such that for $\mu = \mu_\infty$ all of them are chaotic. One should recognize that the similarities between the presented bifurcation diagrams indicate morphogenesis of period-doubling cascade.

In Fig. 2.8a–d, we illustrate the 2-dimensional projections of the trajectory of system (2.8.33), with the initial data $x_1(0) = 2.16$, $x_2(0) = -9.28$, $x_3(0) = -0.21$, $x_4(0) = -2.03$, $x_5(0) = 3.36$, $x_6(0) = -0.52$, $x_7(0) = 3.07$, $x_8(0) = -0.32$, on the planes $x_1 - x_2$, $x_3 - x_4$, $x_5 - x_6$, and $x_7 - x_8$, respectively. The picture in Fig. 2.8a shows in fact the attractor of the prior chaos produced by the generator system (2.8.32) and similarly the illustrations in Fig. 2.8b–d correspond to the chaotic attractors of the second, third and the fourth subsystems, respectively. The resemblance between the shapes of the attractors of the subsystems reflect the morphogenesis of chaos in the result-system (2.8.33).

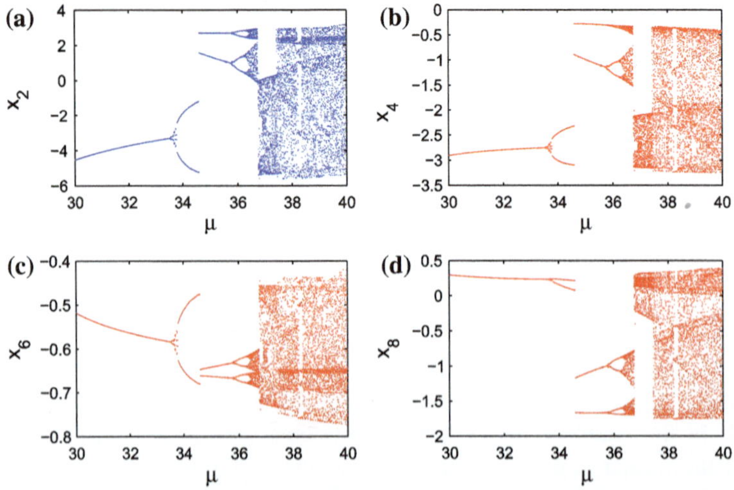

Fig. 2.7 The bifurcation diagrams of system (2.8.33) according to coordinates. **a** The bifurcation diagram corresponding to x_2-coordinate. **b** The bifurcation diagram corresponding to x_4-coordinate. **c** The bifurcation diagram corresponding to x_6-coordinate. **d** The bifurcation diagram corresponding to x_8-coordinate. The picture in (**a**) is the bifurcation diagram of the generator system (2.8.32) and the pictures shown in (**b**), (**c**) and (**d**) correspond to the second, third and fourth replicator systems, respectively. It is observable that all replicators, likewise the generator, undergo period-doubling bifurcations at the same values of the parameter and all of them are chaotic for $\mu = \mu_\infty \equiv 40$. The one-to-one correspondence between the stable periodic solutions of the generator and replicators are also seen in the figure. The resemblances between the bifurcation diagrams corresponding to the coordinates x_2, x_4, x_6, and x_8 reveal morphogenesis of chaos

To obtain a better impression about the chaotic attractor of system (2.8.33), in Fig. 2.9 we demonstrate the 3-dimensional projections of the trajectory with the same initial data as above, on the $x_3 - x_5 - x_7$ and $x_4 - x_6 - x_8$ spaces. Although we are restricted to make illustrations at most in 3-dimensional spaces and not able to provide a picture of the whole chaotic attractor, the results shown both in Figs. 2.8 and 2.9 give us an idea about the spectacular chaotic attractor of system (2.8.33).

We note that system (2.8.33) exhibits a symmetry under the transformation which maps x_i to $-x_i$, $i = 1, 2, \ldots, 8$ and t to $t + \pi$, and the presented attractors are symmetric around the origin due to the symmetry of the result-system (2.8.33) under this transformation.

Now, let us show that the first replicator system which is included inside (2.8.33) satisfies the condition $(A7)$.

When the system

$$\begin{aligned} x_3' &= 2x_3 - x_4 + 0.4 \tan((x_1 + x_3)/10) \\ x_4' &= 17x_3 - 6x_4 + x_2 \end{aligned} \tag{2.8.34}$$

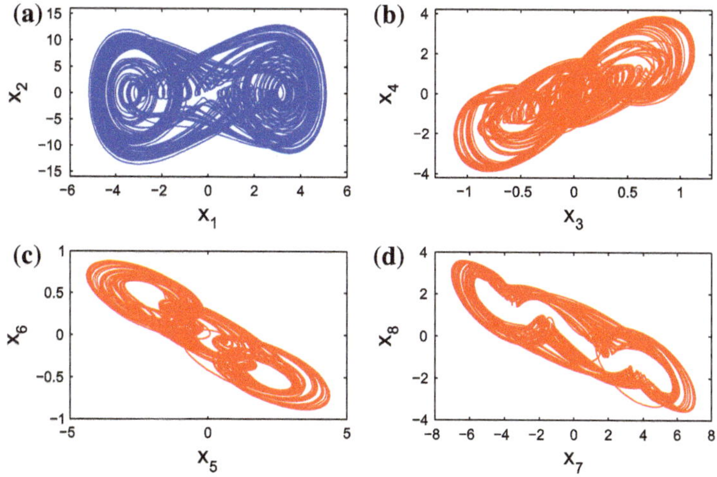

Fig. 2.8 2-dimensional projections of the chaotic attractor of the result-system (2.8.33). **a** Projection on the $x_1 - x_2$ plane. **b** Projection on the $x_3 - x_4$ plane. **c** Projection on the $x_5 - x_6$ plane. **d** Projection on the $x_7 - x_8$ plane. The picture in (**a**) shows the attractor of the prior chaos produced by the generator system (2.8.32) and in (**b**)–(**d**) the chaotic attractors of the remaining subsystems are observable. The illustrations in (**b**)–(**d**) repeated the structure of the attractor shown in (**a**), and these pictures are indicators of the chaos extension

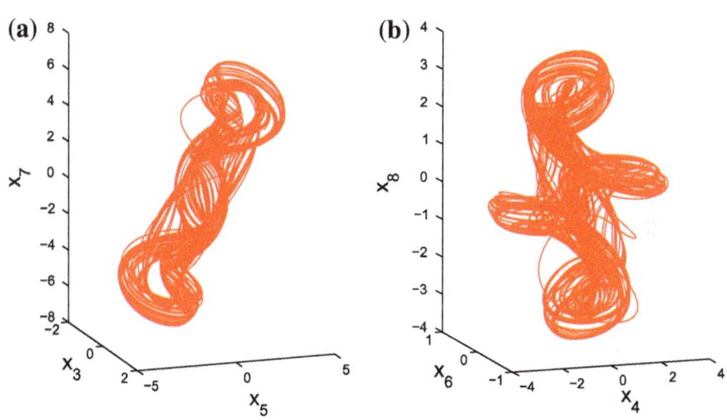

Fig. 2.9 3-dimensional projections of the chaotic attractor of the result-system (2.8.33). **a** Projection on the $x_3 - x_5 - x_7$ space. **b** Projection on the $x_4 - x_6 - x_8$ space. The illustrations presented in (**a**) and (**b**) give information about the impressive chaotic attractor in the 8-dimensional space

is considered in the form of system (2.1.2), one can see that the matrix A can be written as $A = \begin{pmatrix} 2 & -1 \\ 17 & -6 \end{pmatrix}$, which admits the complex conjugate eigenvalues $-2 \mp i$.

The real Jordan form of the matrix A is given by $J = \begin{pmatrix} -2 & -1 \\ 1 & -2 \end{pmatrix}$ and the identity $P^{-1}AP = J$ is satisfied where $P = \begin{pmatrix} 0 & 1 \\ -1 & 4 \end{pmatrix}$. Evaluating the exponential matrix e^{At} we obtain that

$$e^{At} = e^{-2t} P \begin{pmatrix} \cos t & -\sin t \\ \sin t & \cos t \end{pmatrix} P^{-1}. \tag{2.8.35}$$

Taking $N = \|P\| \left\|P^{-1}\right\| < 18$ and $\omega = 2$, one can see that the inequality $\left\|e^{At}\right\| \leq Ne^{-\omega t}$ holds for all $t \geq 0$. The function $g : \mathbb{R}^2 \times \mathbb{R}^2 \to \mathbb{R}^2$ defined as

$$g(x_1, x_2, x_3, x_4) = \left(0.4 \tan\left(\frac{x_1 + x_3}{10}\right), x_2\right)$$

satisfies the conditions $(A4)$ and $(A5)$ with constants $L_1 = \sqrt{2}/50$, $L_2 = \sqrt{2}$ and $L_3 = 0.08$ since the chaotic attractor of system (2.8.33) satisfies the inequalities $|x_1| \leq 6$, $|x_3| \leq 3/2$, and consequently $\left|\dfrac{x_1 + x_3}{10}\right| \leq 3/4$. Therefore, the condition $(A7)$ is satisfied.

In a similar way, for the second replicator system, making use of $|x_3| \leq 3/2$ once again, one can show that the function $h : \mathbb{R}^2 \times \mathbb{R}^2 \to \mathbb{R}^2$ defined as

$$h(x_3, x_4, x_5, x_6) = \left(0.5 \sin x_6 - 4x_4, -\tan\left(\frac{x_3}{2}\right)\right)$$

satisfies the counterparts of the conditions $(A4)$ and $(A5)$ with constants $L_1 = \sqrt{2}/4$, $L_2 = 4\sqrt{2}$ and $L_3 = 1/2$.

Now, we shall focus on the third replicator system

$$\begin{aligned} x_7' &= 2x_7 + 5x_8 - 0.00004(x_7 - x_8)^3 - \tfrac{3}{2}x_6 \\ x_8' &= -5x_7 - 8x_8 + 4x_5. \end{aligned} \tag{2.8.36}$$

The matrix of coefficients of the system (2.8.36) with the assumed coefficients is

$$A = \begin{pmatrix} 2 & 5 \\ -5 & -8 \end{pmatrix}.$$

It can be easily seen that -3 is an eigenvalue of the matrix A with multiplicity 2. The real Jordan form of the matrix A is $J = \begin{pmatrix} -3 & 1 \\ 0 & -3 \end{pmatrix}$ and the identity $J = P^{-1}AP$ is satisfied where $P = \begin{pmatrix} 1 & 0 \\ -1 & 1/5 \end{pmatrix}$. Evaluating the exponential matrix e^{At} we have

$$e^{At} = e^{-3t} P \begin{pmatrix} 1 & t \\ 0 & 1 \end{pmatrix} P^{-1}. \tag{2.8.37}$$

If we denote by I the 2×2 identity matrix, then using Eq. (2.8.37), one can conclude for $t \geq 0$ that

$$\begin{aligned} \left\| e^{At} \right\| &\leq e^{-3t} \left\| P \right\| \left\| P^{-1} \right\| \left\| I + \begin{pmatrix} 0 & t \\ 0 & 0 \end{pmatrix} \right\| \\ &\leq e^{-3t} \left\| P \right\| \left\| P^{-1} \right\| (1+t) \\ &= e^{-2t} \left\| P \right\| \left\| P^{-1} \right\| \frac{1+t}{e^t} \\ &\leq e^{-2t} \left\| P \right\| \left\| P^{-1} \right\| \end{aligned}$$

since $1 + t \leq e^t$ for all $t \geq 0$.

Thus, taking $N = \|P\| \left\| P^{-1} \right\| < 10.2$ and $\omega = 2$, one can see that the inequality $\left\| e^{At} \right\| \leq N e^{-\omega t}$ holds for all $t \geq 0$. Furthermore, the function $k : \mathbb{R}^2 \times \mathbb{R}^2 \to \mathbb{R}^2$ defined by the formula

$$k(x_5, x_6, x_7, x_8) = \left(0.0003(x_7 - x_8)^3 - \frac{3}{2} x_6, 4x_5 \right)$$

satisfies the conditions $(A4)$ and $(A5)$ with constants $L_1 = 3\sqrt{2}/4$, $L_2 = 4\sqrt{2}$ and $L_3 = 0.19$, since the chaotic attractor of system (2.8.36) satisfies the inequalities $|x_7| \leq 8$, $|x_8| \leq 4$. Therefore, $NL_3 - \omega < 0$ and condition $(A7)$ is satisfied.

Remark 2.3 We have proved that the replicator system (2.1.2) exhibits chaos in the sense of Devaney, Li–Yorke, and the one obtained through period-doubling cascade, provided that the generator system (2.1.1) or (2.2.8) exhibits the same types of chaos. Since Lemma 2.1 implies the presence of the criterion (1.1.9) for the unidirectionally coupled system (2.2.8) + (2.1.2), in which an autonomous generator is used, we can say that generalized synchronization takes place in the dynamics of system (2.2.8) + (2.1.2).

The next section is devoted to the results about controlling the replicated chaos.

2.9 Control by Replication

In the previous sections, we have theoretically proved replication of chaos for specific types and controlling the extended chaos is another interesting problem. The next theorem and its corollary indicate a method to control the chaos of the replicator system (2.1.2) and the result-system (2.1.1)+(2.1.2), respectively, and reveal that controlling the chaos of system (2.1.1) is sufficient for this.

Theorem 2.6 *Assume that for arbitrary* $\varepsilon > 0$, *a periodic solution* $x_p(t) \in \mathscr{A}_x$ *is stabilized such that for any solution* $x(t)$ *of system (2.1.1) there exist real numbers* a *and* $E > 0$ *such that the inequality* $\|x(t) - x_p(t)\| < \varepsilon$ *holds for* $t \in [a, a+E]$.

Then, the periodic solution $\phi_{x_p(t)}(t) \in \mathscr{A}_y$ *is stabilized such that for any solution* $y(t)$ *of system (2.1.2) there exists a number* $b \geq a$ *such that the inequality* $\|y(t) - \phi_{x_p(t)}(t)\| < \left(1 + \dfrac{NL_2}{\omega - NL_3}\right)\varepsilon$ *holds for* $t \in [b, a+E]$, *provided that the number* E *is sufficiently large.*

Proof Fix an arbitrary solution $y(t)$ of system $y' = Ay + g(x(t), y)$ for some solution $x(t)$ of system (2.1.1). According to our assumption, there exist numbers a and $E > 0$ such that the inequality $\|x(t) - x_p(t)\| < \varepsilon$ holds for $t \in [a, a+E]$. Let us denote $y_p(t) = \phi_{x_p(t)}(t) \in \mathscr{A}_y$. It is clear that the function $y_p(t)$ is periodic with the same period as $x_p(t)$. Since $y(t)$ and $y_p(t)$ satisfy the integral equations

$$y(t) = e^{A(t-a)}y(a) + \int_a^t e^{A(t-s)}g(x(s), y(s))ds,$$

and

$$y_p(t) = e^{A(t-a)}y_p(a) + \int_a^t e^{A(t-s)}g(x_p(s), y_p(s))ds,$$

respectively, one has

$$\begin{aligned} y(t) - y_p(t) &= e^{A(t-a)}(y(a) - y_p(a)) \\ &+ \int_a^t e^{A(t-s)}\left[g(x(s), y(s)) - g(x(s), y_p(s))\right]ds \\ &+ \int_a^t e^{A(t-s)}\left[g(x(s), y_p(s)) - g(x_p(s), y_p(s))\right]ds. \end{aligned}$$

By the help of the last equation, we have

$$\|y(t) - y_p(t)\| \leq Ne^{-\omega(t-a)}\|y(a) - y_p(a)\| + \frac{NL_2\varepsilon}{\omega}e^{-\omega t}\left(e^{\omega t} - e^{\omega a}\right)$$
$$+NL_3\int_a^t e^{-\omega(t-s)}\|y(s) - y_p(s)\|ds.$$

Let $u : [a, a+E] \to [0, \infty)$ be a function defined as $u(t) = e^{\omega t} \left\| y(t) - y_p(t) \right\|$. In this case, we reach the inequality

$$u(t) \leq N e^{\omega a} \left\| y(a) - y_p(a) \right\| + \frac{N L_2 \varepsilon}{\omega} \left(e^{\omega t} - e^{\omega a} \right) + N L_3 \int_a^t u(s) ds.$$

Implementation of Lemma 2.2 [20] to the last inequality, where $t \in [a, a+E]$, provides us

$$u(t) \leq \frac{N L_2 \varepsilon}{\omega} e^{\omega t} + N \left\| y(a) - y_p(a) \right\| e^{\omega a} e^{N L_3 (t-a)}$$
$$- \frac{N L_2 \varepsilon}{\omega} e^{\omega a} e^{N L_3 (t-a)} + \frac{N^2 L_2 L_3 \varepsilon}{\omega(\omega - N L_3)} e^{\omega t} \left(1 - e^{(N L_3 - \omega)(t-a)} \right).$$

and consequently,

$$\left\| y(t) - y_p(t) \right\| \leq \frac{N L_2 \varepsilon}{\omega} + N \left\| y(a) - y_p(a) \right\| e^{(N L_3 - \omega)(t-a)}$$
$$- \frac{N L_2 \varepsilon}{\omega} e^{(N L_3 - \omega)(t-a)} + \frac{N^2 L_2 L_3 \varepsilon}{\omega(\omega - N L_3)} \left(1 - e^{(N L_3 - \omega)(t-a)} \right)$$
$$< N \left\| y(a) - y_p(a) \right\| e^{(N L_3 - \omega)(t-a)} + \frac{N L_2 \varepsilon}{\omega - N L_3}.$$

If $y(a) = y_p(a)$, then clearly $\left\| y_p(t) - y(t) \right\| < \left(1 + \frac{N L_2}{\omega - N L_3} \right) \varepsilon$, $t \in [a, a+E]$. Suppose that $y(a) \neq y_p(a)$. For $t \geq a + \frac{1}{N L_3 - \omega} \ln \left(\frac{\varepsilon}{N \left\| y(a) - y_p(a) \right\|} \right)$, the inequality $e^{(N L_3 - \omega)(t-a)} \leq \frac{\varepsilon}{N \left\| y(a) - y_p(a) \right\|}$ is satisfied. Assume that the number E is sufficiently large so that $E > \frac{1}{N L_3 - \omega} \ln \left(\frac{\varepsilon}{N \left\| y(a) - y_p(a) \right\|} \right)$. Thus, taking

$$b = \max \left\{ a, a + \frac{1}{N L_3 - \omega} \ln \left(\frac{\varepsilon}{N \left\| y(a) - y_p(a) \right\|} \right) \right\}$$

and

$$\widetilde{E} = \min \left\{ E, E - \frac{1}{N L_3 - \omega} \ln \left(\frac{\varepsilon}{N \left\| y(a) - y_p(a) \right\|} \right) \right\}$$

one attains $\left\|y(t) - y_p(t)\right\| < \left(\dfrac{\omega - NL_3 + NL_2}{\omega - NL_3}\right)\varepsilon$, for $t \in [b, b + \widetilde{E}]$. Here the number $\widetilde{E}$ stands for the duration of control for system (2.1.2). We note that $b \geq a$, $0 < \widetilde{E} \leq E$ and $b + \widetilde{E} = a + E$.

Hence $\left\|y(t) - y_p(t)\right\| < \left(1 + \dfrac{NL_2}{\omega - NL_3}\right)\varepsilon$, for $t \in [b, a + E]$.

The proof of the theorem is finalized. □

An immediate corollary of Theorem 2.6 is the following.

Corollary 2.7 *Assume that the conditions of Theorem 2.6 hold. In this case, the periodic solution* $z_p(t) = \left(x_p(t), \phi_{x_p(t)}(t)\right) \in \mathscr{A}$ *is stabilized such that for any solution* $z(t)$ *of system (2.1.1)+(2.1.2) there exists a number* $b \geq a$ *such that the inequality* $\left\|z_p(t) - z(t)\right\| < \left(2 + \dfrac{NL_2}{\omega - NL_3}\right)\varepsilon$ *holds for* $t \in [b, a + E]$, *provided that the number* E *is sufficiently large.*

Proof Making use of the inequality

$$\left\|z(t) - z_p(t)\right\| \leq \left\|x(t) - x_p(t)\right\| + \left\|y(t) - \phi_{x_p(t)}(t)\right\|,$$

and the conclusion of Theorem 2.6, one can show that the inequality

$$\left\|z_p(t) - z(t)\right\| < \left(2 + \frac{NL_2}{\omega - NL_3}\right)\varepsilon$$

holds for $t \in [b, a + E]$ and for some $b \geq a$. The proof is completed. □

Remark 2.4 As a conclusion of Theorem 2.6, the transient time for control to take effect may increase and the duration of control may decrease as the number of consecutive replicator systems increase.

In the remaining part of this section, our aim is to present an illustration which confirms the results of Theorem 2.6, and for our purposes, we will make use of the Pyragas control method [29]. Therefore, primarily, we continue with a brief explanation of this method.

A delayed feedback control method for the stabilization of unstable periodic orbits of a chaotic system was proposed by Pyragas [29]. In this method, one considers a system of the form

$$x' = H(x, q), \tag{2.9.38}$$

where $q = q(t)$ is an externally controllable parameter and for $q = 0$ it is assumed that the system (2.9.38) is in the chaotic state of interest, whose periodic orbits are to be stabilized [27, 29–31]. According to Pyragas method, an unstable ξ-periodic solution of the system (2.9.38) with $q = 0$, can be stabilized by the control law

$q(t) = C\left[s\left(t-\xi\right) - s(t)\right]$, where the parameter C represents the strength of the perturbation and $s(t) = \sigma\left[x(t)\right]$ is a scalar signal given by some function of the state of the system.

It is indicated in [31] that in order to apply the Pyragas control method to the chaotic Duffing oscillator given by the system

$$\begin{aligned} x_1' &= x_2 \\ x_2' &= -0.10x_2 + 0.5x_1\left(1 - x_1^2\right) + 0.24\sin t, \end{aligned} \tag{2.9.39}$$

one can construct the corresponding control system

$$\begin{aligned} v_1' &= v_2 \\ v_2' &= -0.10v_2 + 0.5v_1\left(1 - v_1^2\right) + 0.24\sin(v_3) + C\left[v_2(t-2\pi) - v_2(t)\right] \\ v_3' &= 1, \end{aligned} \tag{2.9.40}$$

where $q(t) = C\left[v_2(t-2\pi) - v_2(t)\right]$ is the control law, and an unstable 2π-periodic solution can be stabilized by choosing an appropriate value for the parameter C.

Now, let us combine system (2.9.39) with two consecutive replicator systems and set up the following 6-dimensional result-system

$$\begin{aligned} x_1' &= x_2 \\ x_2' &= -0.10x_2 + 0.5x_1\left(1 - x_1^2\right) + 0.24\sin t \\ x_3' &= x_4 - 0.1x_1 \\ x_4' &= -3x_3 - 2x_4 - 0.008x_3^3 + 1.6x_2 \\ x_5' &= x_6 + 0.6x_3 \\ x_6' &= -3.1x_5 - 2.1x_6 - 0.007x_5^3 + 2.5x_4. \end{aligned} \tag{2.9.41}$$

In system (2.9.41) the subsystems with coordinates (x_3, x_4) and (x_5, x_6) correspond to the first and the second replicator systems, respectively. Since our procedure of morphogenesis is valid for specific types of chaos such as in Devaney's and Li–Yorke sense and through period-doubling cascade, we expect that our procedure is also applicable to any other chaotic system with an unspecified type of chaos. Accordingly, system (2.9.41) is chaotic since the generator system (2.9.39) is chaotic.

Theorem 2.6 specifies that in order to control the chaos of system (2.9.41) one should control the chaos of the generator system, which is the subsystem of (2.9.41) with coordinates (x_1, x_2). In accordance with this purpose, we will use the Pyragas control method by means of the system

$$\begin{aligned} v_1' &= v_2 \\ v_2' &= -0.10v_2 + 0.5v_1\left(1 - v_1^2\right) + 0.24\sin(v_3) + C\left[v_2(t-2\pi) - v_2(t)\right] \\ v_3' &= 1 \\ v_4' &= v_5 - 0.1v_1 \\ v_5' &= -3v_4 - 2v_5 - 0.008v_4^3 + 1.6v_2 \\ v_6' &= v_7 + 0.6v_4 \\ v_7' &= -3.1v_6 - 2.1v_7 - 0.007v_6^3 + 2.5v_5, \end{aligned} \tag{2.9.42}$$

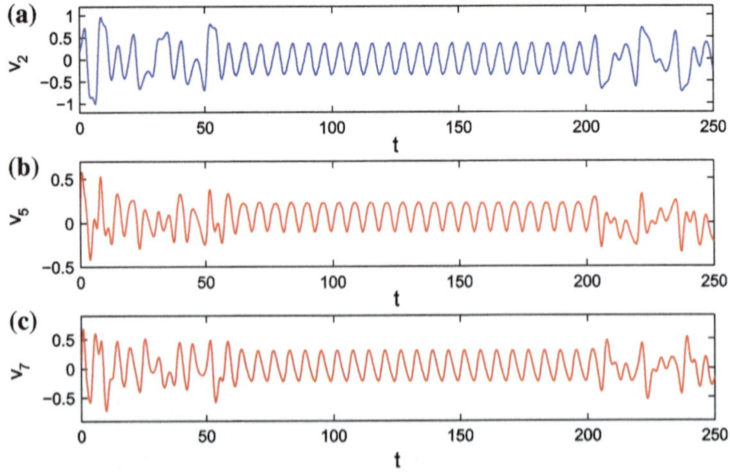

Fig. 2.10 Pyragas control method applied to the result-system (2.9.41) with the aid of the corresponding control system (2.9.42). **a** The graph of v_2-coordinate. **b** The graph of v_5-coordinate. **c** The graph of v_7-coordinate. The result of Pyragas control method applied to the generator system (2.9.39) is seen in (**a**). Through this method, the 2π-periodic solution of the generator and accordingly the 2π-periodic solutions of the first and the second replicator systems are stabilized. In other words, the chaos of the result-system (2.9.41) is controlled. For the coordinates v_1, v_4 and v_6 we have similar results which are not just pictured here. The control starts at $t = 60$ and ends at $t = 200$, after which emergence of the chaos is observable again

which is the control system corresponding to (2.9.41).

Let us consider a solution of system (2.9.42) with the initial data $v_1(0) = 0.2$, $v_2(0) = 0.2$, $v_3(0) = 0$, $v_4(0) = -0.5$, $v_5(0) = 0.1$, $v_6 = -0.2$, and $v_7(0) = 0.1$. We let the system evolve freely taking $C = 0$ until $t = 60$, and at that moment we switch on the control by taking $C = 0.84$. At $t = 200$, we switch off the control and start to use the value of the parameter $C = 0$ again. In Fig. 2.10 one can see the graphs of the v_2, v_5, v_7 coordinates of the solution. Supporting the result of Theorem 2.6, it is observable in Fig. 2.10 that stabilizing a 2π-periodic solution of the generator system provides the stabilization of the corresponding 2π-periodic solutions of the replicator systems. After switching off the control, the 2π-periodic solutions of the generator and replicators lose their stability and chaos emerges again.

2.10 Miscellany

In this part of the chapter, we intend to consider not rigorously proved, but interesting phenomena which can be considered in the framework of our results. So, we shall give some additional light on the results obtained above and say about the possibility for the replication of intermittency, Shilnikov orbits and relay systems. We also demonstrate the possibility of quasiperiodic motions as an infinite basis of chaos.

We start our discussions with replication of intermittency.

2.10.1 Intermittency

In the previous sections, we have rigorously proved replication of specific types of chaos such as period-doubling cascade, Devaney's, and Li–Yorke chaos. Consequently, one can expect that the same procedure also works for the intermittency route.

Pomeau and Manneville [32] observed chaos through intermittency in the Lorenz system (2.5.20), with the coefficients $\sigma = 10$, $b = 8/3$ and values of r slightly larger than the critical value $r_c \approx 166.06$. To observe intermittent behavior in the Lorenz system, let us consider a solution of system (2.5.20) together with the coefficients $\sigma = 10$, $b = 8/3$, $r = 166.25$ using the initial data $x_1(0) = -23.3$, $x_2(0) = 38.3$ and $x_3(0) = 193.4$. The time-series for the x_1, x_2 and x_3 coordinates of the solution are indicated in Fig. 2.11, where one can see that regular oscillations are interrupted by irregular ones.

To perform the replication of intermittency, let us consider the Lorenz system (2.5.20) as the generator and set up the 6-dimensional result-system

$$
\begin{aligned}
x_1' &= \sigma(-x_1 + x_2) \\
x_2' &= -x_2 + rx_1 - x_1x_3 \\
x_3' &= -bx_3 + x_1x_2 \\
x_4' &= -x_4 + 4x_1 \\
x_5' &= x_6 + 2x_2 \\
x_6' &= -3x_5 - 2x_6 - 0.00005x_5^3 + 0.5x_4,
\end{aligned}
\qquad (2.10.43)
$$

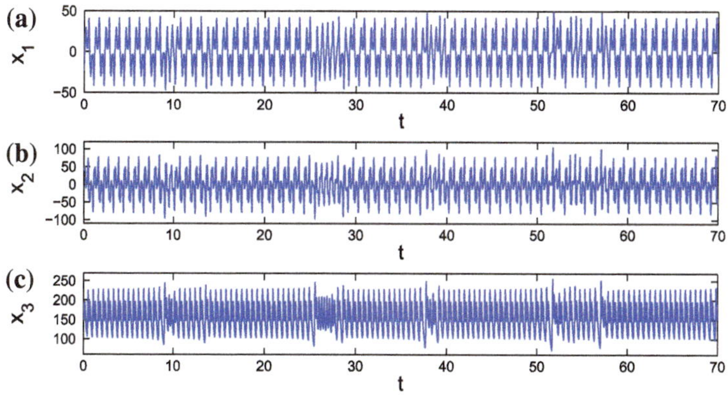

Fig. 2.11 Intermittency in the Lorenz system (2.5.20), where $\sigma = 10$, $b = 8/3$ and $r = 166.25$. **a** The graph of the x_1-coordinate. **b** The graph of the x_2-coordinate. **c** The graph of the x_3-coordinate

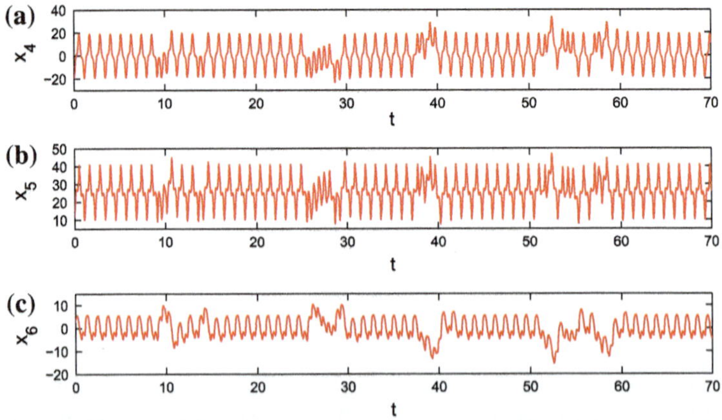

Fig. 2.12 Intermittency in the replicator system. **a** The graph of the x_4-coordinate. **b** The graph of the x_5-coordinate. **c** The graph of the x_6-coordinate. The analogy between the time series of the generator and the replicator systems indicates the morphogenesis of intermittency

again with the coefficients $\sigma = 10$, $b = 8/3$ and $r = 166.25$. It can be easily verified that condition $(A7)$ is valid for system (2.10.43). We consider the trajectory of system (2.10.43) corresponding to the initial data $x_1(0) = -23.3$, $x_2(0) = 38.3$, $x_3(0) = 193.4$, $x_4(0) = -17.7$, $x_5(0) = 11.4$, and $x_6(0) = 2.5$, and represent the graphs for the x_4, x_5 and x_6 coordinates in Fig. 2.12 such that the intermittent behavior in the replicator system is observable. The similarity between the graphs of the coordinates corresponding to the generator and the replicator counterpart reveals the replication of intermittency.

2.10.2 Shilnikov Orbits

To illustrate that by our method it may also be possible to replicate strange attractors [33–35], let us provide simulations of homoclinic and complicated Shilnikov orbits (Figs. 2.13 and 2.14 correspondingly).

As a model for Shilnikov's orbits, the paper [36] considers the system

$$\begin{aligned} x_1' &= x_2 \\ x_2' &= x_3 \\ x_3' &= -x_2 - \beta x_3 + f_\mu(x_1), \end{aligned} \tag{2.10.44}$$

where

$$f_\mu(x) = \begin{cases} 1 - \mu x, & \text{if } x > 0 \\ 1 + \alpha x, & \text{if } x \leq 0. \end{cases} \tag{2.10.45}$$

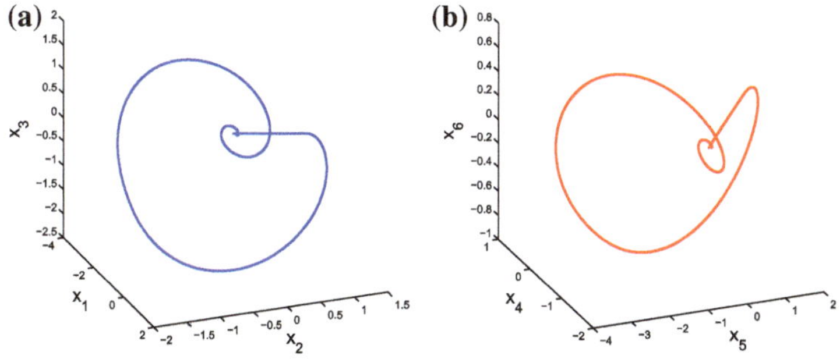

Fig. 2.13 Replication of a Shilnikov type homoclinic orbit. In picture **a**, one can see the projection on the $x_1 - x_2 - x_3$ space of the trajectory of system (2.10.46) corresponding to the initial data $x_1(0) - 1.57590$, $x_2(0) = 0$, $x_3(0) = 0$, $x_4(0) = -0.78795$, $x_5(0) = 0$ and $x_6(0) = 0$. The picture in **b** shows the projection on the $x_4 - x_5 - x_6$ space of the same trajectory. The parameter values $\alpha = 0.633625$, $\beta = 0.3375$ and $\mu = 2.16$ are used in the simulation. The picture in (**a**) represents a Shilnikov type homoclinic orbit corresponding to the generator system (2.10.44), while the picture in (**b**) shows its replication through the system (2.10.46)

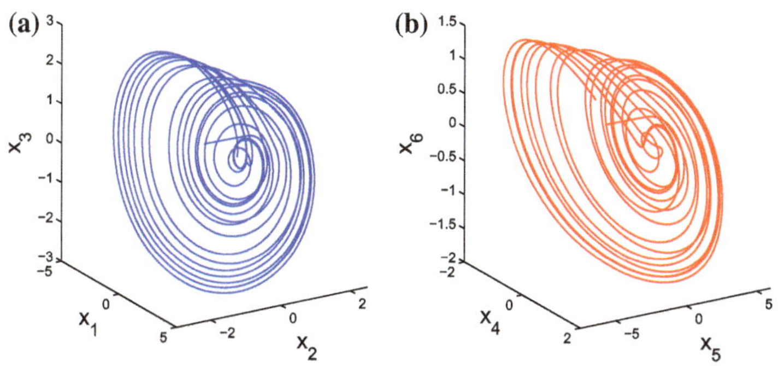

Fig. 2.14 Projections of a complicated orbit of system (2.10.46) with $\alpha = 0.633625$, $\beta = 0.3375$ and $\mu = 0.83$. **a** Projection on the $x_1 - x_2 - x_3$ space. **b** Projection on the $x_4 - x_5 - x_6$ space. The initial data $x_1(0) - 1.57590$, $x_2(0) = 0$, $x_3(0) = 0$, $x_4(0) = -0.78795$, $x_5(0) = 0$, $x_6(0) = 0$ is used for the illustration. The picture in (**a**) represents the behavior of the trajectory corresponding to the generator (2.10.44), while the picture in (**b**) illustrates its replication

The values $\alpha = 0.633625$, $\beta = 0.3375$ and the parameter μ used in system (2.10.44) are taken from [23]. The point $e_0 = (-1/\alpha, 0, 0)$ is an equilibrium point of system (2.10.44), and the eigenvalues of the matrix of linearization at e_0 are 0.4625, $-0.4 \pm 1.1i$ such that the condition of the Shilnikov's theorem about eigenvalues [37] is satisfied. For values of the parameter μ near 2.16, system (2.10.44) possesses a special type homoclinic orbit−*Shilnikov orbit*, and its presence implies chaotic dynamics [23]. In this case, Shilnikov's theorem asserts that every neighborhood of the homoclinic orbit contains a countably infinite number of unstable periodic orbits [36, 37].

To demonstrate numerically the replication of a Shilnikov orbit, let us consider the following system

$$\begin{aligned}
x_1' &= x_2\\
x_2' &= x_3\\
x_3' &= -x_2 - \beta x_3 + f_\mu(x_1)\\
x_4' &= -2x_4 + x_1\\
x_5' &= -0.6x_5 + 2x_2 + 0.1x_2^3\\
x_6' &= -1.2x_6 + 0.001\sin(x_6) + x_3,
\end{aligned} \tag{2.10.46}$$

where, again, the function $f_\mu(x)$ is given by formula (2.10.45).

System (2.10.44) is used as a generator in system (2.10.45), where the last three coordinates are of a replicator. Let us consider system (2.10.46) with the values $\alpha = 0.633625$, $\beta = 0.3375$ and $\mu = 2.16$ once again. In Fig. 2.13 we show the trajectory of this system with initial data $x_1(0) - 1.5759$, $x_2(0) = 0$, $x_3(0) = 0$, $x_4(0) = -0.78795$, $x_5(0) = 0$ and $x_6(0) = 0$. The picture in Fig. 2.13a, where we illustrate the projection of the trajectory on the $x_1 - x_2 - x_3$ space represents, in fact, the Shilnikov orbit corresponding to the generator system (2.10.44). On the other hand, the picture in Fig. 2.13b, shows the projection of the trajectory on the $x_4 - x_5 - x_6$ space and in this picture the replication of the Shilnikov orbit is observable.

Next, we consider system (2.10.46) with the values $\alpha = 0.633625$, $\beta = 0.3375$, $\mu = 0.83$ and take the trajectory of this system with the same initial data as above. In Fig. 2.14a, b, we represent the projections of this trajectory on the $x_1 - x_2 - x_3$ and $x_4 - x_5 - x_6$ spaces, respectively. The picture in (a) represents the complicated behavior of the generator system (2.10.44) and one can see in picture (b) the replication of this behavior.

We suppose that theoretical affirmation of our simulation results can be done if one considers interpretation of Shilnikov's theorem [37] for the multidimensional replicator. That is, we are still questioning whether our approach can be somehow combined with methods indicating chaos through Shilnikov type strange attractors [33, 35]. At least, it is easy to see that a homoclinic trajectory exists for a replicator as well as a denumerable set of unstable periodic solutions.

In next our discussion, we will emphasize by means of simulations the morphogenesis of the double-scroll Chua's attractor in a unidirectionally coupled open chain of Chua circuits. Approaches for the generation of hyperchaotic systems have already been discussed making use of Chua circuits which are all chaotic [38, 39]. It deserves to remark that to create hyperchaotic attractors in previous papers, others consider both involved interacting systems chaotic, but in our case only the first link of the chain is chaotic and other consecutive Chua systems are all non-chaotic.

2.10.3 *Morphogenesis of the Double-Scroll Chua's Attractor*

The type of chaos for the double-scroll Chua circuit is proposed in paper [40]. It is an interesting problem to prove that this type of chaos can be replicated through the method discussed in this chapter. Nevertheless, we will show by simulations that the regular behavior in Chua circuits placed in the extension mechanism can also be seen. This means that next special investigation has to be done. Moreover, this will show how one can use morphogenesis not only for chaos, but also for Chua circuits by uniting them in complexes in electrical (physical) sense, and observing the same properties as a unique separated Chua circuit admits. This is an interesting problem which can give a light for the complex behavior of huge electrical circuits.

There is a well-known result of the chaoticity based on the double-scroll Chua's attractor [41]. It was proven first in the paper [40] rigorously, and the proof is based on the Shilnikov's theorem [37]. Since the Chua circuit and its chaotic behavior is of extreme importance from the theoretical point of view and its usage area in electrical circuits by radio physicists and nonlinear scientists from other disciplines, one can suppose that morphogenesis of the chaos will also be of a practical and a theoretical interest.

We just take into account a simulation result which supports that morphogenesis idea can be developed also from this point of view.

Let us consider the dimensionless form of the Chua's oscillator given by the system

$$\begin{aligned} x_1' &= k\alpha[x_2 - x_1 - f(x_1)] \\ x_2' &= k(x_1 - x_2 + x_3) \\ x_3' &= k(-\beta x_2 - \gamma x_3) \\ f(x) &= bx + 0.5(a - b)\left(|x + 1| + |x - 1|\right), \end{aligned} \tag{2.10.47}$$

where $\alpha, \beta, \gamma, a, b$ and k are constants.

In paper [42], it is indicated that system (2.10.47) with the coefficients $\alpha = 21.32/5.75$, $\beta = 7.8351$, $\gamma = 1.38166392/12$, $a = -1.8459$, $b = -0.86604$ and $k = 1$ admits a stable equilibrium.

In what follows, as the generator, we make use of system (2.10.47) together with the coefficients $\alpha = 15.6$, $\beta = 25.58$, $\gamma = 0$, $a = -8/7$, $b = -5/7$ and $k = 1$ such that a double-scroll Chua's attractor takes place [22], and consider the following 12-dimensional result-system:

$$
\begin{aligned}
x_1' &= 15.6\,[x_2 - (2/7)x_1 + (3/14)\,(|x_1+1| + |x_1-1|)] \\
x_2' &= x_1 - x_2 + x_3 \\
x_3' &= -25.58x_2 \\
x_4' &= (21.32/5.75)[x_5 - 0.13396x_4 \\
&\quad +0.48993\,(|x_4+1| + |x_4-1|)] + 2x_1 \\
x_5' &= x_4 - x_5 + x_6 + 5x_2 \\
x_6' &= -7.8351x_5 - (1.38166392/12)x_6 + 2x_3 \\
x_7' &= (21.32/5.75)[x_8 - 0.13396x_7 \\
&\quad +0.48993\,(|x_7+1| + |x_7-1|)] + 2x_4 \\
x_8' &= x_7 - x_8 + x_9 + 3x_5 \\
x_9' &= -7.8351x_8 - (1.38166392/12)x_9 - 0.001x_6 \\
x_{10}' &= (21.32/5.75)[x_{11} - 0.13396x_{10} \\
&\quad +0.48993\,(|x_{10}+1| + |x_{10}-1|)] + 4x_7 \\
x_{11}' &= x_{10} - x_{11} + x_{12} - 0.1x_8 \\
x_{12}' &= -7.8351x_{11} - (1.38166392/12)x_{12} + 2x_9.
\end{aligned}
\tag{2.10.48}
$$

System (2.10.48) consists of four unidirectionally coupled Chua circuits such that the subsystems with coordinates (x_1, x_2, x_3), (x_4, x_5, x_6), (x_7, x_8, x_9) and (x_{10}, x_{11}, x_{12}) correspond to the first, second, third, and the fourth links of the open chain of circuits.

In Fig. 2.15, we simulate the 3-dimensional projections on the $x_1 - x_2 - x_3$ and $x_4 - x_5 - x_6$ spaces of the trajectory of the result-system (2.10.48) with the initial data $x_1(0) = 0.634$, $x_2(0) = -0.093$, $x_3(0) = -0.921$, $x_4(0) = -8.013$, $x_5(0) = 0.221$, $x_6(0) = 6.239$, $x_7(0) = -50.044$, $x_8(0) = -0.984$, $x_9(0) = 48.513$, $x_{10}(0) = -256.325$, $x_{11}(0) = 7.837$, $x_{12}(0) = 264.331$. The projection on the $x_1 - x_2 - x_3$ space shows the double-scroll Chua's attractor produced by the generator system (2.10.47), and projection on the $x_4 - x_5 - x_6$ space represents the chaotic attractor of the first replicator.

In a similar way, we display the projections of the same trajectory on the $x_7 - x_8 - x_9$ and $x_{10} - x_{11} - x_{12}$ spaces, which correspond to the attractors of the second and the third replicator systems, in Fig. 2.16. The illustrations shown in Figs. 2.15 and 2.16 indicate the extension of chaos in system (2.10.48). Possibly the result-system (2.10.48) produces a double-scroll Chua's attractor with hyperchaos, where the number of positive Lyapunov exponents are more than one and even four.

2.10.4 Quasiperiodicity in Chaos

Now, let us indicate that if there are more than one generator system, then the chaos extension mechanism will lead to some new forms such as periodicity gives birth to quasiperiodicity.

In paper [43], it is mentioned that the Duffing equation

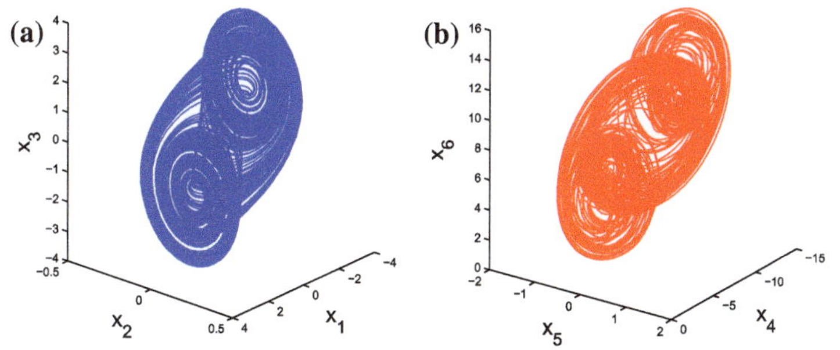

Fig. 2.15 3-dimensional projections of the chaotic attractor of the result-system (2.10.48). **a** Projection on the $x_1 - x_2 - x_3$ space. **b** Projection on the $x_4 - x_5 - x_6$ space. The picture in (**a**) shows the attractor of the original prior chaos of the generator system (2.10.47) and (**b**) represents the attractor of the first replicator. The resemblance between shapes of the attractors of the generator and the replicator systems makes the extension of chaos apparent

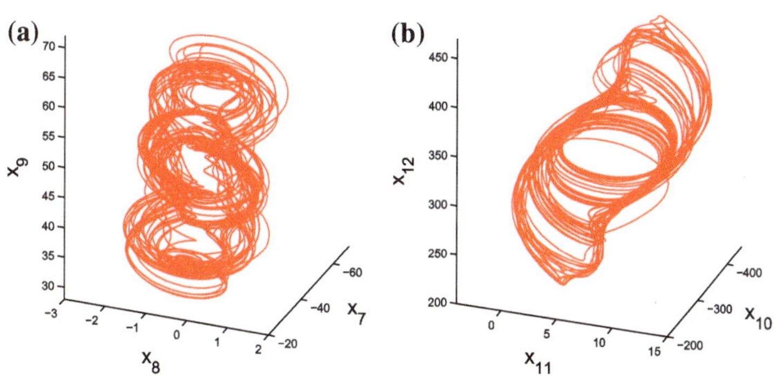

Fig. 2.16 3-dimensional projections of the chaotic attractor of the result-system (2.10.48). **a** Projection on the $x_7 - x_8 - x_9$ space. **b** Projection on the $x_{10} - x_{11} - x_{12}$ space. (**a**) and (**b**) demonstrates the attractors generated by the second and the third replicator systems, respectively

$$x'' + 0.168x' - 0.5x\left(1 - x^2\right) = \mu \sin t, \tag{2.10.49}$$

where μ is a parameter, admits the chaos through period-doubling cascade at the parameter value $\mu = \mu_\infty \equiv 0.21$. That is, at the parameter value $\mu = \mu_\infty$, for each natural number k the Eq. (2.10.49) admits infinitely many periodic solutions with periods $2k\pi$. Using the change of variables $t = 2\pi s$ and $x(t) = y(s)$, and relabeling s as t, one attains the following equation:

$$y'' + 0.168\pi y' - 0.5\pi^2 y\left(1 - y^2\right) = \pi^2 \mu \sin(\pi t). \tag{2.10.50}$$

Likewise Eq. (2.10.49), it is clear that Eq. (2.10.50), when considered with $\mu = \mu_\infty$, also admits the chaos through period-doubling cascade and has infinitely many periodic solutions with periods $2, 4, 8, \ldots$.

Using the new variables $x_1 = x$, $x_2 = x'$ and $x_3 = y$, $x_4 = y'$, one can convert the Eqs. (2.10.49) and (2.10.50) to the systems

$$\begin{aligned} x_1' &= x_2 \\ x_2' &= -0.168x_2 + 0.5x_1\left(1 - x_1^2\right) + \mu \sin t \end{aligned} \tag{2.10.51}$$

and

$$\begin{aligned} x_3' &= x_4 \\ x_4' &= -0.168\pi x_4 + 0.5\pi^2 x_3(1 - x_3^2) + \pi^2\mu \sin(\pi t), \end{aligned} \tag{2.10.52}$$

respectively. Now, we shall make use both of the systems (2.10.51) and (2.10.52), with $\mu = \mu_\infty$, as generators to obtain a chaotic system with infinitely many quasiperiodic solutions. We mean that the two systems admit incommensurate periods and consequently their influence on the replicator will be quasiperiodic. In this case, one can expect that replicator will expose infinitely many quasiperiodic solutions. For that purpose, let us consider the 6-dimensional result-system

$$\begin{aligned} x_1' &= x_2 \\ x_2' &= -0.168x_2 + 0.5x_1(1 - x_1^2) + 0.21 \sin t \\ x_3' &= x_4 \\ x_4' &= -0.168\pi x_4 + 0.5\pi^2 x_3(1 - x_3^2) + 0.21\pi^2 \sin(\pi t) \\ x_5' &= x_6 + x_1 + x_3 \\ x_6' &= -3x_5 - 2x_6 - 0.008x_5^3 + x_2 + x_4, \end{aligned} \tag{2.10.53}$$

where the last two equations are of a replicator.

To reveal existence of quasiperiodic solutions embedded in the chaotic attractor of system (2.10.53) we control the chaos of system (2.10.53) by the Pyragas method through the following control system

$$\begin{aligned} v_1' &= v_2 \\ v_2' &= -0.168v_2 + 0.5v_1(1 - v_1^2) + 0.21 \sin v_3 \\ &+C_1(v_2(t - 2\pi) - v_2(t)) \\ v_3' &= 1 \\ v_4' &= v_5 \\ v_5' &= -0.168\pi v_5 + 0.5\pi^2 v_4(1 - v_4^2) + 0.21\pi^2 \sin(\pi v_6) \\ &+C_2(v_5(t - 2) - v_5(t)) \\ v_6' &= 1 \\ v_7' &= v_8 + v_1 + v_4 \\ v_8' &= -3v_7 - 2v_8 - 0.008v_7^3 + v_2 + v_5. \end{aligned} \tag{2.10.54}$$

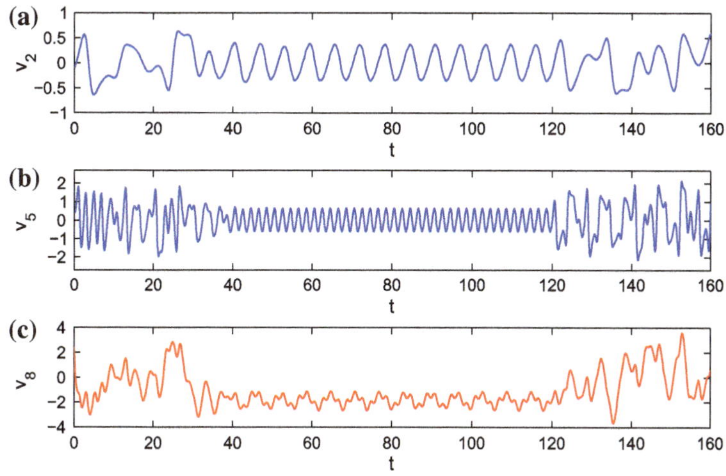

Fig. 2.17 Pyragas control method applied to the result-system (2.10.53) by means of the corresponding control system (2.10.54). **a** The graph of the v_2-coordinate. **b** The graph of the v_5-coordinate. **c** The graph of the v_8-coordinate. The simulation for the result-system (2.10.53) is provided such that in (**a**) and (**b**) periodic solutions with incommensurate periods 2 and 2π are controlled by Pyragas method and in (**c**) a quasiperiodic solution of the replicator system is pictured. The control starts at $t = 35$ and ends at $t = 120$. After switching off the control, chaos emerges again and irregular behavior reappears. For the coordinates v_1, v_4 and v_7 we have similar illustrations which are not indicated here

We take into account the solution of the result-system (2.10.53) with the initial data $v_1(0) = 0.4$, $v_2(0) = -0.1$, $v_3(0) = 0$, $v_4(0) = -0.2$, $v_5(0) = 0.5$, $v_6(0) = 0$, $v_7(0) = 1.1$ and $v_8(0) = 2.5$. The simulation results are shown in Fig. 2.17. The control mechanism starts at $t = 35$ and ends at $t = 120$. The chaos not only in the generator systems, but also in the replicator counterpart is observable before the control is switched on. During the control, we make use of the values of $C_1 = 0.62$ and $C_2 = 2.58$ to stabilize the periodic solutions corresponding to the generator systems (2.10.51) and (2.10.52) with periods 2π and 2, respectively. Up to $t = 35$ and after $t = 120$ the values $C_1 = C_2 = 0$ are used. Between $t = 35$ and $t = 120$, the quasiperiodic solution of the replicator is stabilized and after $t = 120$ chaos in the system (2.10.53) develops again.

Possibly the obtained simulation result and previous theoretical discussions can give a support to the idea of *quasiperiodical cascade* for the appearance of chaos which can be considered as a development of the popular period-doubling route to chaos.

In paper [44], it has been mentioned that, in general, in the place of countable set of periodic solutions to form chaos, one can take an uncountable collection of Poisson stable motions which are dense in a quasi-minimal set. This can be also observed in Horseshoe attractor [45]. These emphasize that our simulation of quasiperiodic solutions can be considered as another evidence for the theoretical results.

2.10.5 Replicators with Nonnegative Eigenvalues

We recall that in our theoretical discussions, all eigenvalues of the real-valued constant matrix A, used in system (2.1.2), are assumed to have negative real parts. Now, as open problems from the theoretical point of view, we shall discuss through simulations the problem of chaos replication in the case when the matrix A possesses an eigenvalue with positive or zero real part.

First, we are going to concentrate on the case of the existence of an eigenvalue with positive real part. Let us make use of the Lorenz system (2.5.20) together with the coefficients $\sigma = 10$, $r = 28$ and $b = 8/3$ as the generator, which is known to be chaotic [21, 46], and set up the 6-dimensional result-system

$$\begin{aligned} x_1' &= -10x_1 + 10x_2 \\ x_2' &= -x_2 + 28x_1 - x_1x_3 \\ x_3' &= -(8/3)x_3 + x_1x_2 \\ x_4' &= -2x_4 + x_1 \\ x_5' &= -3x_5 + x_2 \\ x_6' &= 4x_6 - x_6^3 + x_3. \end{aligned} \tag{2.10.55}$$

It is crucial to note that system (2.10.55) is of the form of system (2.1.1)+(2.1.2), where the matrix A admits the eigenvalues -2, -3 and 4, such that one of them is positive. We take into account the solution of system (2.10.55) with the initial data $x_1(0) = -12.7$, $x_2(0) = -8.5$, $x_3(0) = 36.5$, $x_4(0) = -3.4$, $x_5(0) = -3.2$, $x_6(0) = 3.7$ and visualize in Fig. 2.18 the projections of the corresponding trajectory on the $x_1 - x_2 - x_3$ and $x_4 - x_5 - x_6$ spaces. It is seen that the replicator system admits the chaos and the input–output analysis works for system (2.10.55).

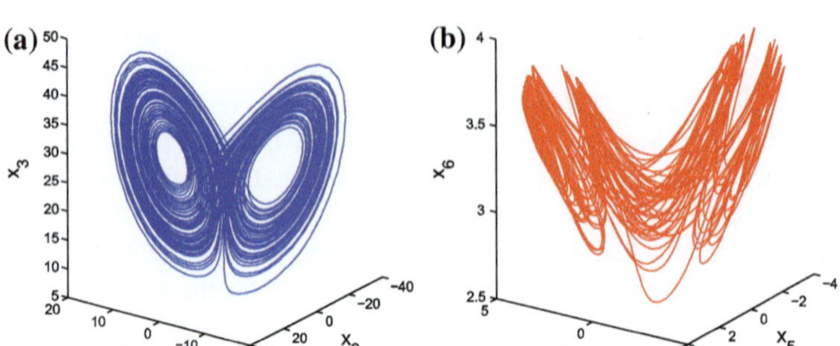

Fig. 2.18 3-dimensional projections of the chaotic attractor of the result-system (2.10.55). **a** Projection on the $x_1 - x_2 - x_3$ space. **b** Projection on the $x_4 - x_5 - x_6$ space. In (**a**), the famous Lorenz attractor produced by the generator system (2.5.20) with coefficients $\sigma = 10$, $r = 28$ and $b = 8/3$ is shown. In (**b**), as in usual way, the projection of the chaotic attractor of the result-system (2.10.55), which can separately be considered as a chaotic attractor, is presented. Possibly one can call the attractor of the result-system as $6D$ Lorenz attractor

Next, we continue to our discussion with the case of the existence of an eigenvalue with a zero real part. This time we consider the chaotic Rössler system [46, 47] described by

$$\begin{aligned} x_1' &= -(x_2 + x_3) \\ x_2' &= x_1 + 0.2x_2 \\ x_3' &= 0.2 + x_3(x_1 - 5.7) \end{aligned} \tag{2.10.56}$$

as the generator and constitute the result-system

$$\begin{aligned} x_1' &= -(x_2 + x_3) \\ x_2' &= x_1 + 0.2x_2 \\ x_3' &= 0.2 + x_3(x_1 - 5.7) \\ x_4' &= -4x_4 + x_1 \\ x_5' &= -x_5 + x_2 \\ x_6' &= -0.2x_6^3 + x_3. \end{aligned} \tag{2.10.57}$$

In this case, one can consider system (2.10.57) as in the form of (2.1.1)+(2.1.2) where the matrix A is a diagonal matrix with entries $-4, -1, 0$ on the diagonal and admits the number 0 as an eigenvalue. We simulate the solution of system (2.10.57) with the initial data $x_1(0) = 4.6, x_2(0) = -3.3, x_3(0) = 0, x_4(0) = 1, x_5(0) = -3.7$ and $x_6(0) = 0.8$. The projections of the trajectory on the $x_1 - x_2 - x_3$ and $x_4 - x_5 - x_6$ spaces are seen in Fig. 2.19. The simulation results confirm that the replicator mimics the complex behavior of the generator system.

These results of the simulations request more detailed investigation which concern not only the theoretical existence of chaos, but also its resistance and stability.

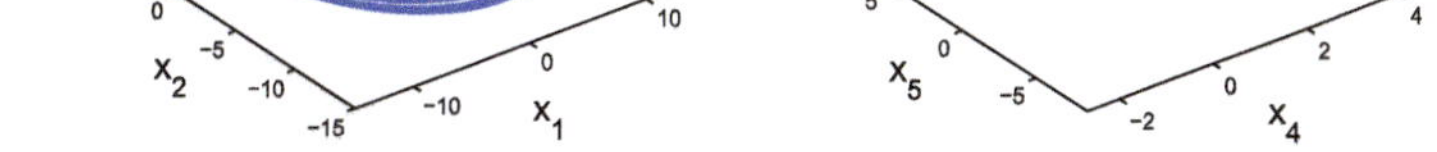

Fig. 2.19 3-dimensional projections of the chaotic attractor of the result-system (2.10.57). **a** Projection on the $x_1 - x_2 - x_3$ space. **b** Projection on the $x_4 - x_5 - x_6$ space. The picture in (**a**) indicates the famous Rössler attractor produced by the generator system (2.10.56). The similarity between the illustrations presented in (**a**) and (**b**) supports the morphogenesis of chaos. The attractor of the result-system (2.10.57) can be possibly called as $6D$ Rössler attractor

2.11 Notes

In this chapter, we show that a known type of chaos, such as the one obtained through period-doubling cascade and in the sense of Devaney or Li–Yorke, can be extended to systems with arbitrary large dimensions. More precisely, we provide the replication of chaos between unidirectionally coupled systems such that a result-system admitting the same type of chaos is obtained. The definitions of chaotic sets as well as the hyperbolic sets of continuous functions are introduced, and the replication of the chaos is proved rigorously. The considered morphogenesis mechanism is based on a chaos generating element inserted in a network of systems. Replication of intermittency as well as Shilnikov orbits are discussed. Morphogenesis of the double-scroll Chua's attractor and quasiperiodical motions as a possible skeleton of a chaotic attractor are demonstrated numerically. The presented technique is useful for creating chaos in systems that are encountered in mechanics, electrical systems, economic theory, meteorology, neural networks theory, and communication systems.

The concept of self-replicating machines, in the abstract sense, starts with the ideas of von Neumann [48], and these ideas are supposed to be the origins of cellular automata theory [49]. Morphogenesis was deeply involved in mathematical discussions through Turing's investigations [50] as well as in the concept of structural stability [51]. In this chapter, the term "morphogenesis" is used in the meaning of "processes creating forms" where we accept the *form* not only as a type of chaos, but also accompanying concepts as the structure of the chaotic attractor, its fractal dimension, form of the bifurcation diagram, the spectra of Lyapunov exponents, inheritance of intermittency, etc. This is similar to the idea such that morphogenesis is used in fields such as urban studies [52], architecture [53], mechanics [54], computer science [55], linguistics [56], and sociology [57, 58]. The results of this chapter were published in the paper [59].

References

1. C. Corduneanu, *Almost Periodic Functions* (Interscience Publishers, New York, 1968)
2. C. Corduneanu, *Integral Equations and Applications* (Cambridge University Press, New York, 2008)
3. C. Corduneanu, *Almost Periodic Oscillations and Waves* (Springer, New York, 2009)
4. A.M. Fink, *Almost Periodic Differential Equations*, Lecture Notes in Mathematics (Springer, Berlin, 1974)
5. J.M.T. Thompson, H.B. Stewart, *Nonlinear Dynamics and Chaos* (Wiley, Chichester, 2002)
6. R.A. Horn, C.R. Johnson, *Matrix Analysis* (Cambridge University Press, Cambridge, 1992)
7. J.K. Hale, *Ordinary Differential Equations* (Krieger Publishing Company, Florida, 1980)
8. C. Corduneanu, *Principles of Differential and Integral Equations* (Chelsea Publishing Company, The Bronx, 1977)
9. R.L. Devaney, *An Introduction to Chaotic Dynamical Systems* (Addison-Wesley, Menlo Park, 1989)
10. E. Akin, S. Kolyada, Li-Yorke sensitivity. Nonlinearity **16**, 1421–1433 (2003)
11. M. Čiklová, Li-Yorke sensitive minimal maps. Nonlinearity **19**, 517–529 (2006)

12. P. Kloeden, Z. Li, Li-Yorke chaos in higher dimensions: a review. J. Differ. Equ. Appl. **12**, 247–269 (2006)
13. T.Y. Li, J.A. Yorke, Period three implies chaos. Am. Math. Mon. **82**, 985–992 (1975)
14. K. Palmer, *Shadowing in Dynamical Systems: Theory and Applications* (Kluwer Academic Publishers, Dordrecht, 2000)
15. L.M. Lerman, L.P. Shil'nikov, Homoclinical structures in nonautonomous systems: nonautonomous chaos. Chaos **2**, 447–454 (1992)
16. K.R. Meyer, G.R. Sell, Homoclinic orbits and Bernoulli bundles in almost periodic systems. Can. Math. Soc. Conf. Proc. **8**, 527 (1987)
17. K.R. Palmer, D. Stoffer, Chaos in almost periodic systems. J. Appl. Math. Phys. (ZAMP) **40**, 592–602 (1989)
18. J. Scheurle, Chaotic solutions of systems with almost periodic forcing. J. Appl. Math. Phys. (ZAMP) **37**, 12–26 (1986)
19. K.V. Zadiraka, Investigation of irregularly perturbed differential equations (Russian), in *Questions of the Theory and History of Differential Equations* (Nauk. Dumka, Kiev, 1968), pp. 81–108
20. E.A. Barbashin, *Introduction to the Theory of Stability* (Wolters-Noordhoff Publishing, Groningen, 1970)
21. E.N. Lorenz, Deterministic nonperiodic flow. J. Atmos. Sci. **20**, 130–141 (1963)
22. K.T. Alligood, T.D. Sauer, J.A. Yorke, *Chaos: An Introduction to Dynamical Systems* (Springer, New York, 1996)
23. J. Hale, H. Koçak, *Dynamics and Bifurcations* (Springer, New York, 1991)
24. E. Sander, J.A. Yorke, Period-doubling cascades galore. Ergod. Theory Dyn. Syst. **31**, 1249–1267 (2011)
25. M.J. Feigenbaum, Universal behavior in nonlinear systems. Los Alamos Sci./Summer 4–27 (1980)
26. H.G. Schuster, W. Just, *Deterministic Chaos, An Introduction* (Wiley-VCH, Weinheim, 2005)
27. I. Zelinka, S. Celikovsky, H. Richter, G. Chen (eds.), *Evolutionary Algorithms and Chaotic Systems* (Springer, Berlin, 2010)
28. S. Sato, M. Sano, Y. Sawada, Universal scaling property in bifurcation structure of Duffing's and of generalized Duffing's equations. Phys. Rev. A **28**, 1654–1658 (1983)
29. K. Pyragas, Continuous control of chaos by self-controlling feedback. Phys. Rev. A **170**, 421–428 (1992)
30. A.L. Fradkov, *Cybernetical Physics: From Control of Chaos to Quantum Control* (Springer, Berlin, 2007)
31. J.M. Gonzáles-Miranda, *Synchronization and Control of Chaos* (Imperial College Press, London, 2004)
32. Y. Pomeau, P. Manneville, Intermittent transition to turbulence in dissipative dynamical systems. Commun. Math. Phys. **74**, 189–197 (1980)
33. F. Drubi, S. Ibáñez, J.A. Rodriguez, Coupling leads to chaos. J. Differ. Equ. **239**(2), 371–385 (2007)
34. J. Guckenheimer, P.J. Holmes, *Nonlinear Oscillations, Dynamical Systems and Bifurcations of Vector Fields* (Springer, New York, 1997)
35. S. Wiggins, *Global Bifurcations and Chaos: Analytical Methods* (Springer, New York, 1988)
36. A. Arneodo, P. Coullet, C. Tresser, Oscillators with chaotic behavior, an illustration of a theorem by Shil'nikov. J. Stat. Phys. **27**, 171–182 (1982)
37. L.P. Shilnikov, A case of the existence of a denumerable set of periodic motions. Sow. Math. Dokl. **6**, 163–166 (1965)
38. V.S. Anishchenko, T. Kapitaniak, M.A. Safonova, O.V. Sosnovzeva, Birth of double-double scroll attractor in coupled Chua circuits. Phys. Lett. A **192**, 207–214 (1994)
39. T. Kapitaniak, L.O. Chua, G. Zhong, Experimental hyperchaos in coupled Chua's circuits. IEEE Trans. Circuits Syst.-I: Fundam. Theory Appl. **41**, 499–503 (1994)
40. L.O. Chua, M. Komuro, T. Matsumoto, The double scroll family, parts I and II. IEEE Trans. Circuit Syst. **CAS-33**, 1072–1118 (1986)

41. T. Matsumoto, L.O. Chua, M. Komuro, The double scroll. IEEE Trans. Circuit Syst. **CAS-32**, 797–818 (1985)
42. L.O. Chua, C.W. Wu, A. Huang, G. Zhong, A universal circuit for studying and generating chaos-part I: routes to chaos. IEEE Trans. Circuits Syst.-I: Fundam. Theory Appl. **40**, 732–744 (1993)
43. E.H. Dowell, C. Pazeshki, On the understanding of chaos in Duffing's equation including a comparison with experiment. J. Appl. Mech. **53**, 5–9 (1986)
44. L. Shilnikov, Bifurcations and strange attractors, in *Proceedings of the International Congress of Mathematicians*, vol. III (Higher Education Press, Beijing 2002), pp. 349–372
45. S. Smale, Differentiable dynamical systems. Bull. Am. Math. Soc. **73**, 747–817 (1967)
46. J.C. Sprott, *Elegant Chaos: Algebraically Simple Chaotic Flows* (World Scientific Publishing, Singapore, 2010)
47. O.E. Rössler, An equation for continuous chaos. Phys. Lett. **57A**, 397–398 (1976)
48. J. Von Neumann, A.W. Burks (eds.), *The Theory of Self-Reproducing Automata* (University of Illinois Press, Urbana, 1966)
49. J.L. Schiff, *Cellular Automata: A Discrete View of the World* (Wiley, Hoboken, 2008)
50. A.M. Turing, The chemical basis of morphogenesis. Philos. Trans. R. Soc. Lond., Ser. B, Biol. Sci. **237**, 37–72 (1952)
51. R. Thom, *Stabilité Structurelle et Morphogénèse* (W.A. Benjamin, New York, 1972)
52. T. Courtat, C. Gloaguen, S. Douady, Mathematics and morphogenesis of cities: a geometrical approach. Phys. Rev. E **83**, 1–12 (2011)
53. S. Roudavski, Towards morphogenesis in architecture. Int. J. Archit. Comput. **7**, 345–374 (2009)
54. L.A. Taber, Towards a unified theory for morphomechanics. Philos. Trans. R. Soc. A **367**, 3555–3583 (2009)
55. P. Bourgine, A. Lesne, *Morphogenesis: Origins of Patterns and Shapes* (Springer, Berlin, 2011)
56. C. Hagége, *The Language Builder: An Essay on the Human Signature in Linguistic Morphogenesis* (John Benjamins Publishing Co., Amsterdam, 1993)
57. M.S. Archer, *Realistic Social Theory: The Morphogenetic Approach* (Cambridge University Press, Cambridge, 1995)
58. W. Buckley, *Sociology and Modern Systems Theory* (Prentice Hall, New Jersey, 1967)
59. M.U. Akhmet, M.O. Fen, Replication of chaos. Commun. Nonlinear Sci. Numer. Simul. **18**, 2626–2666 (2013)

Chapter 3
Chaos Extension in Hyperbolic Systems

This chapter is devoted to the investigation of chaos in the dynamics of chaotically perturbed hyperbolic systems. Extension of chaos in the sense of Devaney and Li–Yorke is taken into account for unidirectionally coupled systems. The rigorously proved results are supported by simulations. A method for controlling the extended chaos is also presented.

3.1 Introduction

In this chapter, we will consider non-chaotic systems and perturb them in a unidirectional way through exogenous chaotic forcing terms to achieve propagation of the chaotic behavior. In other words, the influence of the chaos of a system on another one will be mentioned in the chapter such that as a result the latter behaves also chaotically. Chaotification of systems of differential equations through different techniques can be found in our studies [1–5, 7–10]. Endogenously generated chaotic behavior of systems are well investigated in the literature. The systems of Lorenz [11], Rössler [12], and Chua [13, 14] as well as the Van der Pol [15–17] and Duffing [18–20] oscillators can be considered as systems which are capable of generating chaos endogenously. We will make use of Devaney and Li–Yorke chaos in the extension mechanism through exogenous perturbations, and essentially indicate in the present chapter that not only endogenous structure of systems, but also exogenous chaotic perturbations can give rise to observation of chaotic behavior.

To explain the extension procedure of the present chapter, let us give the following information. It is known that if one considers the evolution equation $u' = L[u]+I(t)$, where $L[u]$ is a linear operator with spectra placed out of the imaginary axis of the complex plane, then a function $I(t)$ being considered as an input with a certain property (boundedness, periodicity, almost periodicity) produces through the equation the output, a solution with a similar property, boundedness/periodicity/almost periodicity. In particular, in this chapter, we solved a similar problem when the linear system

M. Akhmet and M.O. Fen, *Replication of Chaos in Neural Networks, Economics and Physics*, Nonlinear Physical Science,
DOI 10.1007/978-3-662-47500-3_3

has eigenvalues with negative real parts and input is considered as a chaotic set of functions with a known type. Our results are different in the sense that the input and the output are not single functions, but *a collection of functions*. In other words, we prove that both the input and the output are *chaos* of the same type for the discussed equation. The way of our investigation is arranged in the well-accepted traditional mathematical fashion, but with a new and a more complex way of arrangement of the connections between the input and the output. The same is true for the control results discussed in the chapter. If one considers an element of the chaotic set as the chaotic function, then we may consider our results through the analysis where input and output are of the same nature, that is, they are chaotic functions.

Throughout the chapter, we will denote by $\mathbb{R}$ and $\mathbb{N}$ the sets of real numbers and natural numbers, respectively. Moreover, the usual Euclidean norm for vectors and the norm induced by the Euclidean norm for square matrices [21] will be used.

The paper [8], where we discuss an extension mechanism of chaos, is about the replication of specific types of chaos, such as Devaney, Li–Yorke chaos, and chaos obtained through period-doubling cascade. In this process, we consider the generator-replicator systems such that the generator is considered as a system of the form

$$x' = F(t, x), \tag{3.1.1}$$

where $F : \mathbb{R} \times \mathbb{R}^m \to \mathbb{R}^m$ is a continuous function in all its arguments and the replicator is assumed to have the form

$$y' = Ay + g(x, y), \tag{3.1.2}$$

where $g : \mathbb{R}^m \times \mathbb{R}^n \to \mathbb{R}^n$ is a continuous function in all its arguments and the constant $n \times n$ real-valued matrix A has real parts of eigenvalues all negative.

The rigorous results of the extension mechanism emphasize that system (3.1.2) is chaotic in the same way as system (3.1.1). Replication of chaos through intermittency is also shown through simulations in paper [8], where one can find new definitions for chaotic sets of functions, and precise descriptions for the ingredients of Devaney and Li–Yorke chaos in continuous-time dynamics, which are used as tools for the extension procedure.

In the case that the matrix A in (3.1.2) is hyperbolic, we have not been able to find a way to insert a term nonlinear in y in the system to preserve the results of paper [8] and, instead, we are forced to handle system (3.2.6) to achieve success in the theoretical results. In other words, we could not achieve the extension of Devaney and/or Li–Yorke chaos when nonlinearity with respect to y is included in the system and the eigenvalues of the matrix A are allowed to possess positive real parts as well as negative real parts. However, in the present chapter, the chaos extension problem is considered for a hyperbolic matrix A and the nonlinear term of the initially non-perturbed system is removed for a theoretically supported chaotification process, and this is the main difference compared to [8]. The restrictive conditions of the chapters have their own priority in such a way that the present chapter considers hyperbolic linear vector fields in the form of equation (3.2.5) to be forced through chaotic perturbations, while the paper [8] allows us to include nonlinear terms for the existing replicators such that the real-valued matrix A has eigenvalues with negative real parts. An example with a nonlinear term is demonstrated numerically in the last part of the chapter.

It is also worth noting that, at each step of the extension mechanism, one should take into account replicator systems with smaller Lipschitz constants and this may violate the validity of the procedure for infinitely many consecutive systems [8]. Distinctively, in the present chapter, such a condition is not requested and therefore the procedure can be performed for infinite number of consecutive systems and decay for the strength of chaoticity does not occur. We also note that likewise the Smale horseshoe map or the Cantor set of the logistic map $h_\mu(x) = \mu x(1-x)$ with $\mu > 4$, the bounded solutions of system (3.2.6), in general, cannot be considered as an attractor. Consequently, they cannot be simulated, and present results have rather theoretical sense.

Let us denote by

$$\mathscr{B} = \{\psi(t) \mid \psi : \mathbb{R} \to K \text{ is continuous}\} \tag{3.1.3}$$

a collection of functions, where $K \subset \mathbb{R}^p$ is a bounded region. Since the concept of *chaotic set of functions* is used in the theoretical discussions, let us explain briefly the ingredients of Devaney and Li–Yorke chaos for the set $\mathscr{B}$, which are introduced in paper [8]. The proofs indicated in Sect. 3.3 are predicated on the definitions of these ingredients. For more information about Devaney and Li–Yorke chaos, one can see [22–27].

Let us start with introducing the following ingredients of Devaney chaos for the set $\mathscr{B}$.

(D1) $\mathscr{B}$ is called sensitive if there exist positive numbers ε and Δ such that for every $\psi(t) \in \mathscr{B}$ and for arbitrary $\delta > 0$ there exist $\overline{\psi}(t) \in \mathscr{B}$, $t_0 \in \mathbb{R}$ and an interval $J \subset [t_0, \infty)$, with length not less than Δ, such that $\left\|\psi(t_0) - \overline{\psi}(t_0)\right\| < \delta$ and $\left\|\psi(t) - \overline{\psi}(t)\right\| > \varepsilon$, for all $t \in J$;

(D2) $\mathscr{B}$ is called transitive if there exists a function $\psi^*(t) \in \mathscr{B}$ such that for every function $\psi(t) \in \mathscr{B}$, arbitrary small $\varepsilon > 0$ and arbitrary large $E > 0$, there exist a number $\xi > 0$ and an interval $J \subset \mathbb{R}$, with length E, such that $\|\psi(t) - \psi^*(t+\xi)\| < \varepsilon$, for all $t \in J$;

(D3) $\mathscr{B}$ possesses a dense collection $\mathscr{G} \subset \mathscr{B}$ of periodic functions if for every function $\psi(t) \in \mathscr{B}$, arbitrary small $\varepsilon > 0$ and arbitrary large $E > 0$, there exist $\widetilde{\psi}(t) \in \mathscr{G}$ and an interval $J \subset \mathbb{R}$, with length E, such that $\left\|\psi(t) - \widetilde{\psi}(t)\right\| < \varepsilon$, for all $t \in J$.

We say that the set $\mathscr{B}$ of functions is a Devaney's chaotic set if it is sensitive, transitive, and possesses a dense collection of periodic functions.

Now, let us continue with the ingredients in the sense of Li–Yorke.

(LY1) A couple of functions $\big(\psi(t), \overline{\psi}(t)\big) \in \mathscr{B} \times \mathscr{B}$ is called proximal if for arbitrary small $\varepsilon > 0$ and arbitrary large $E > 0$, there exist infinitely many disjoint intervals of length not less than E such that $\big\|\psi(t) - \overline{\psi}(t)\big\| < \varepsilon$, for each t from these intervals;

(LY2) A couple of functions $\big(\psi(t), \overline{\psi}(t)\big) \in \mathscr{B} \times \mathscr{B}$ is frequently $(\varepsilon_0, \Delta)-$separated if there exist positive real numbers ε_0, Δ and infinitely many disjoint intervals of length not less than Δ, such that $\big\|\psi(t) - \overline{\psi}(t)\big\| > \varepsilon_0$, for each t from these intervals.

A couple of functions $\big(\psi(t), \overline{\psi}(t)\big) \in \mathscr{B} \times \mathscr{B}$ is a Li–Yorke pair if they are proximal and frequently $(\varepsilon_0, \Delta)-$separated for some positive numbers ε_0 and Δ. On the other hand, a set $\mathscr{C} \subset \mathscr{B}$ is called a scrambled set if $\mathscr{C}$ does not contain any periodic functions and each couple of different functions inside $\mathscr{C} \times \mathscr{C}$ is a Li–Yorke pair.

$\mathscr{B}$ is called a Li–Yorke chaotic set if there exists a positive real number T_0 such that $\mathscr{B}$ admits a periodic function of period kT_0, for any $k \in \mathbb{N}$, $\mathscr{B}$ possesses an uncountable scrambled set $\mathscr{C}$, and for any function $\psi(t) \in \mathscr{C}$ and any periodic function $\overline{\psi}(t) \in \mathscr{B}$, the couple $\big(\psi(t), \overline{\psi}(t)\big)$ is frequently $(\varepsilon_0, \Delta)-$separated for some positive real numbers ε_0 and Δ.

In the next section, we introduce the systems which will be under investigation and give information about the properties of these systems under some conditions.

3.2 Preliminaries

Let us consider the systems

$$x' = F(t, x) \tag{3.2.4}$$

and

$$u' = Au, \tag{3.2.5}$$

where the function $F : \mathbb{R} \times \mathbb{R}^m \to \mathbb{R}^m$ is continuous in all its arguments, and the eigenvalues of the $n \times n$ constant real-valued matrix A has nonzero real parts.

To extend chaos generated by equation (3.2.4), we perturb system (3.2.5) through the solutions of (3.2.4) to achieve the system

$$y' = Ay + h(x), \tag{3.2.6}$$

where $h : \mathbb{R}^m \to \mathbb{R}^n$ is a continuous function.

The following assumptions are needed throughout the chapter:

(A1) There exists a positive real number L_0 such that

$$\|F(t, x_1) - F(t, x_2)\| \le L_0 \|x_1 - x_2\|,$$

for all $t \in \mathbb{R}$, $x_1, x_2 \in \mathbb{R}^m$;

(A2) There exists a positive real number H_0 such that

$$\sup_{t\in\mathbb{R}, x\in\mathbb{R}^m} \|F(t, x)\| \le H_0;$$

(A3) There exist positive real numbers L_1 and L_2 such that

$$L_1 \|x_1 - x_2\| \le \|h(x_1) - h(x_2)\| \le L_2 \|x_1 - x_2\|,$$

for all $x_1, x_2 \in \mathbb{R}^m$, $y \in \mathbb{R}^n$;

(A4) There exists a positive real number M_0 such that

$$\sup_{x\in\mathbb{R}^m, y\in\mathbb{R}^n} \|h(x)\| \le M_0.$$

It is worth saying that the results of the present chapter are also true even if we replace the nonautonomous system (3.2.4) by the autonomous equation

$$x' = \overline{F}(x), \tag{3.2.7}$$

where the function $\overline{F} : \mathbb{R}^m \to \mathbb{R}^m$ is continuous with conditions which are counterparts of $(A1)$ and $(A2)$.

If we denote by λ_j, $j = 1, \ldots, n$, the eigenvalues of the matrix A, then we suppose that there exists a nonnegative integer k such that $\max\limits_{j=1,\ldots,k} \mathscr{R}\lambda_j < 0$, and $\min\limits_{j=k+1,\ldots,n} \mathscr{R}\lambda_j > 0$, where $\mathscr{R}\lambda_j$ denotes the real part of the eigenvalue λ_j of the matrix A. Without loss of generality, one can assume that

$$A = \begin{pmatrix} A_+ & 0 \\ 0 & A_- \end{pmatrix}, \tag{3.2.8}$$

where the square matrices A_+ and A_- are of dimensions k and $n-k$, respectively, λ_j, $j = 1, \ldots, k$ are the eigenvalues of the matrix A_+ and λ_j, $j = k+1, \ldots, n$ are the eigenvalues of the matrix A_-. Under the circumstances, there exist positive real numbers N and ω such that $\left\|e^{A_+t}\right\| \le Ne^{-\omega t}$, $t \ge 0$ and $\left\|e^{A_-t}\right\| \le Ne^{\omega t}$, $t \le 0$.

We can write equation (3.2.6) in the following form:

$$\begin{aligned} y_+' &= A_+ y_+ + h_+(x), \\ y_-' &= A_- y_- + h_-(x), \end{aligned}$$

where $y = \begin{pmatrix} y_+ \\ y_- \end{pmatrix}$, $y_+ \in \mathbb{R}^k$, $y_- \in \mathbb{R}^{n-k}$ and $h = \begin{pmatrix} h_+ \\ h_- \end{pmatrix}$.

Using the theory of quasilinear equations [28], one can verify that for a given solution $x(t)$ of system (3.2.4), there exists a unique bounded on $\mathbb{R}$ solution $\phi_{x(t)}(t) = \begin{pmatrix} y_+(t) \\ y_-(t) \end{pmatrix}$, of the system $y' = Ay + h(x(t))$, which satisfies the couple of integral equations

$$y_+(t) = \int_{-\infty}^{t} e^{A_+(t-s)} h_+(x(s))ds,$$
$$y_-(t) = -\int_{t}^{\infty} e^{A_-(t-s)} h_-(x(s))ds.$$

Our main assumption is the existence of a nonempty set $\mathscr{A}_x$ of all solutions of system (3.2.4), uniformly bounded on $\mathbb{R}$. That is, there exists a positive real number H such that $\sup_{t\in\mathbb{R}} \|x(t)\| \leq H$, for all $x(t) \in \mathscr{A}_x$.

Let us introduce the sets of functions [8]

$$\mathscr{A}_y = \left\{\phi_{x(t)}(t) \mid x(t) \in \mathscr{A}_x\right\},$$

and

$$\mathscr{A} = \left\{(x(t), \phi_{x(t)}(t)) \mid x(t) \in \mathscr{A}_x\right\}.$$

We note that for each $y(t) \in \mathscr{A}_y$, one has $\sup_{t\in\mathbb{R}} \|y(t)\| \leq M$, where $M = \dfrac{2NM_0}{\omega}$.

In the last part of the chapter, we will demonstrate numerically the possibility of chaos extension in a hyperbolic system involving a nonlinear term. For this purpose, we will make use of a system in the following form:

$$y' = Ay + f(y) + h(x), \tag{3.2.9}$$

where the matrix A and the function h have the same properties as in system (3.2.6) and $f : \mathbb{R}^n \to \mathbb{R}^n$ is a continuous function. Here, equation (3.2.9) is attained through the perturbation of

$$u' = Au + f(u), \tag{3.2.10}$$

by the solutions of the processor system (3.2.4). In the exemplification, a non-chaotic Lorenz system will be perturbed through the solutions of a chaotic Rössler system.

The next section is devoted for the clarification of the theoretical results for the chaos extension in systems of the form (3.2.4)+(3.2.6).

3.3 Extension of Chaos

The present section is devoted for the rigorous proofs for the extension of chaos in the sense of Devaney and Li–Yorke. We start our discussions with the first ingredient, sensitivity, of Devaney chaos.

Lemma 3.1 *If $\mathscr{A}_x$ sensitive then the same is true for the set $\mathscr{A}_y$.*

Proof Fix an arbitrary function $y(t) = \begin{pmatrix} y_+(t) \\ y_-(t) \end{pmatrix} \in \mathscr{A}_y$ and a positive real number δ. Owing to the description of the set $\mathscr{A}_y$, there exists a function $x(t) \in \mathscr{A}_x$ such that $y(t) = \phi_{x(t)}(t)$.

Choose a sufficiently small positive number $\overline{\varepsilon} = \overline{\varepsilon}(\delta)$ such that

$$\left(1 + \frac{2NL_2}{\omega}\right)\overline{\varepsilon} < \delta$$

and take a sufficiently large positive real number $R = R(\overline{\varepsilon})$ which satisfies the inequality $\frac{4M_0N}{\omega}e^{-\omega R} < \overline{\varepsilon}$. Let $\delta_1 = \delta_1(\overline{\varepsilon}, R) = \overline{\varepsilon}e^{-L_0R}$. Since the set $\mathscr{A}_x$ is sensitive, there exist positive real numbers ε_0 and Δ such that the inequalities $\|x(t_0) - \overline{x}(t_0)\| < \delta_1$ and $\|x(t) - \overline{x}(t)\| \geq \varepsilon_0$, $t \in J$, hold for some $\overline{x}(t) \in \mathscr{A}_x$, $t_0 \in \mathbb{R}$ and for some interval $J \subset [t_0, \infty)$ with length not less than Δ.

Suppose that $\overline{y}(t) = \phi_{\overline{x}(t)}(t) \in \mathscr{A}_y$. In the proof our aim is first to show that

$$\|y(t_0) - \overline{y}(t_0)\| < \delta$$

and then to prove the existence of a positive real number ε_1 and an interval $J^1 \subset J$ with a fixed length, which is independent of $y(t), \overline{y}(t) \in \mathscr{A}_y$, such that the inequality $\|y(t) - \overline{y}(t)\| \geq \varepsilon_1$ holds, for all $t \in J^1$.

Making use of the relations

$$x(t) = x(t_0) + \int_{t_0}^{t} F(s, x(s))ds$$

and

$$\overline{x}(t) = \overline{x}(t_0) + \int_{t_0}^{t} F(s, \overline{x}(s))ds$$

we obtain the inequality

$$\|x(t) - \overline{x}(t)\| \leq \|x(t_0) - \overline{x}(t_0)\| + \left|\int_{t_0}^{t} L_0 \|x(s) - \overline{x}(s)\| ds\right|.$$

Applying Gronwall–Bellman Lemma to the last inequality for $t \in [t_0 - R, t_0 + R]$, we obtain that

$$\|x(t) - \overline{x}(t)\| \le \|x(t_0) - \overline{x}(t_0)\| \, e^{L_0|t-t_0|}.$$

Now, since the inequality $\|x(t_0) - \overline{x}(t_0)\| < \delta_1$ holds true, we get $\|x(t) - \overline{x}(t)\| < \overline{\varepsilon}$ for all $t \in [t_0 - R, t_0 + R]$.

Using the relations

$$y_+(t) = \int_{-\infty}^{t} e^{A_+(t-s)} h_+(x(s))ds$$

and

$$\overline{y}_+(t) = \int_{-\infty}^{t} e^{A_+(t-s)} h_+(\overline{x}(s))ds,$$

for $t \in [t_0 - R, t_0 + R]$, we achieve that

$$\begin{aligned} y_+(t) - \overline{y}_+(t) &= \int_{-\infty}^{t} e^{A_+(t-s)} \left[h_+(x(s)) - h_+(\overline{x}(s))\right] \\ &= \int_{-\infty}^{t_0-R} e^{A_+(t-s)} \left[h_+(x(s)) - h_+(\overline{x}(s))\right] \\ &+ \int_{t_0-R}^{t} e^{A_+(t-s)} \left[h_+(x(s)) - h_+(\overline{x}(s))\right]. \end{aligned}$$

Therefore, the inequality

$$\begin{aligned} \left\|y_+(t) - \overline{y}_+(t)\right\| &\le \int_{-\infty}^{t_0-R} N e^{-\omega(t-s)} \left\|h_+(x(s)) - h_+(\overline{x}(s))\right\| ds \\ &+ \int_{t_0-R}^{t} N e^{-\omega(t-s)} \left\|h_+(x(s)) - h_+(\overline{x}(s))\right\| ds \\ &\le \int_{-\infty}^{t_0-R} 2NM_0 e^{-\omega(t-s)} ds + \int_{t_0-R}^{t} N L_2 \overline{\varepsilon} e^{-\omega(t-s)} ds \\ &= \frac{2NM_0}{\omega} e^{-\omega(t-t_0+R)} + \frac{NL_2\overline{\varepsilon}}{\omega}\left(1 - e^{-\omega(t-t_0+R)}\right) \end{aligned}$$

holds for all $t \in [t_0 - R, t_0 + R]$. In particular, the last inequality is true for $t = t_0$ and hence one obtains $\left\|y_+(t_0) - \overline{y}_+(t_0)\right\| \le \dfrac{2NM_0}{\omega} e^{-\omega R} + \dfrac{NL_2\overline{\varepsilon}}{\omega}$.

Similarly, by means of the integral equations

$$y_-(t) = -\int_{t}^{\infty} e^{A_-(t-s)} h_-(x(s))ds$$

and

$$\overline{y}_-(t) = -\int_t^\infty e^{A_-(t-s)} h_-(\overline{x}(s))ds,$$

we have

$$\begin{aligned}
y_-(t) - \overline{y}_-(t) &= -\int_t^\infty e^{A_-(t-s)} \left[h_-(x(s)) - h_-(\overline{x}(s))\right] \\
&= -\int_t^{t_0+R} e^{A_-(t-s)} \left[h_-(x(s)) - h_-(\overline{x}(s))\right] \\
&- \int_{t_0+R}^\infty e^{A_-(t-s)} \left[h_-(x(s)) - h_-(\overline{x}(s))\right].
\end{aligned}$$

Thus, for all $t \in [t_0 - R, t_0 + R]$, we have

$$\begin{aligned}
\left\|y_-(t) - \overline{y}_-(t)\right\| &\le \int_t^{t_0+R} N e^{\omega(t-s)} \left\|h_-(x(s)) - h_-(\overline{x}(s))\right\| ds \\
&+ \int_{t_0+R}^\infty N e^{\omega(t-s)} \left\|h_-(x(s)) - h_-(\overline{x}(s))\right\| ds \\
&\le \int_{t_0+R}^\infty 2NM_0 e^{\omega(t-s)} ds + \int_t^{t_0+R} NL_2\overline{\varepsilon} e^{\omega(t-s)} ds \\
&= \frac{2NM_0}{\omega} e^{-\omega(t_0+R-t)} + \frac{NL_2\overline{\varepsilon}}{\omega}\left(1 - e^{-\omega(t_0+R-t)}\right)
\end{aligned}$$

and as a consequence the inequality

$$\left\|y_-(t_0) - \overline{y}_-(t_0)\right\| \le \frac{2NM_0}{\omega} e^{-\omega R} + \frac{NL_2\overline{\varepsilon}}{\omega}$$

is valid.

Now, it is easy to see that

$$\begin{aligned}
\|y(t_0) - \overline{y}(t_0)\| &\le \left\|y_+(t_0) - \overline{y}_+(t_0)\right\| + \left\|y_-(t_0) - \overline{y}_-(t_0)\right\| \\
&\le \frac{4NM_0}{\omega} e^{-\omega R} + \frac{2NL_2\overline{\varepsilon}}{\omega} \\
&\le \left(1 + \frac{2NL_2}{\omega}\right)\overline{\varepsilon} \\
&< \delta.
\end{aligned}$$

In the remaining part of the proof, we will show the existence of a positive number ε_1 and an interval $J^1 \subset J$, with a fixed length which is independent of $y(t), \overline{y}(t) \in \mathscr{A}_y$, such that the inequality $\|y(t) - \overline{y}(t)\| > \varepsilon_1$ holds for all $t \in J^1$.

Suppose that $h(x) = \begin{pmatrix} h_1(x) \\ h_2(x) \\ \vdots \\ h_n(x) \end{pmatrix}$, where each h_j, $1 \le j \le n$, is a real-valued function.

Since for each $x(t) \in \mathscr{A}_x$, the function $x'(t)$ is inside the tube with radius H_0, one can conclude that the set $\mathscr{A}_x$ is an equicontinuous family on $\mathbb{R}$. Making use of the uniform continuity of the function $\overline{h} : \mathbb{R}^m \times \mathbb{R}^m \to \mathbb{R}^n$ defined as $\overline{h}(x_1, x_2) = h(x_1) - h(x_2)$ on the compact region

$$\mathscr{D} = \left\{(x_1, x_2) \in \mathbb{R}^m \times \mathbb{R}^m \mid \|x_1\| \le H, \|x_2\| \le H\right\},$$

together with the equicontinuity of $\mathscr{A}_x$, one can easily show that the set

$$\mathscr{F} = \left\{h_j(x(t)) - h_j(\overline{x}(t)) \mid 1 \le j \le n,\ x(t) \in \mathscr{A}_x,\ \overline{x}(t) \in \mathscr{A}_x\right\}$$

is an equicontinuous family on $\mathbb{R}$.

Therefore, there exists a positive real number $\tau < \Delta$, independent of the functions $x(t), \overline{x}(t) \in \mathscr{A}_x$, such that for any $t_1, t_2 \in \mathbb{R}_+$ with $|t_1 - t_2| < \tau$ the inequality

$$\left|\left(h_j\left(x(t_1)\right) - h_j\left(\overline{x}(t_1)\right)\right) - \left(h_j\left(x(t_2)\right) - h_j\left(\overline{x}(t_2)\right)\right)\right| < \frac{L_1 \varepsilon_0}{2n} \tag{3.3.11}$$

holds for all $1 \le j \le n$.

Condition $(A3)$ implies that for each $t \in J$ the inequality

$$\|h(x(t)) - h(\overline{x}(t))\| \ge L_1 \|x(t) - \overline{x}(t)\|$$

holds. Therefore, for all $t \in J$, there exists an integer j_0, $1 \le j_0 \le n$, which possibly depends on t, such that

$$\left|h_{j_0}(x(t)) - h_{j_0}(\overline{x}(t))\right| \ge \frac{L_1}{n} \|x(t) - \overline{x}(t)\|.$$

Otherwise, if there exists $s \in J$ such that for all $1 \le j \le n$, the inequality

$$\left|h_j\left(x\left(s\right)\right) - h_j(\overline{x}(s))\right| < \frac{L_1}{n} \|x(s) - \overline{x}(s)\|$$

holds, then one encounters with a contradiction since

$$\|h(x(s)) - h(\overline{x}(s))\| \le \sum_{j=1}^{n} \left|h_j(x(s)) - h_j(\overline{x}(s))\right| < L_1 \|x(s) - \overline{x}(s)\|.$$

Now, let s_0 be the midpoint of the interval J and $\theta = s_0 - \dfrac{\tau}{2}$. One can find an integer $j_0 = j_0(s_0)$, $1 \le j_0 \le n$, such that

$$\left|h_{j_0}(x(s_0)) - h_{j_0}(\overline{x}(s_0))\right| \ge \frac{L_1}{n}\left\|x(s_0) - \overline{x}(s_0)\right\| \ge \frac{L_1\varepsilon_0}{n}. \qquad (3.3.12)$$

On the other hand, making use of the inequality (3.3.11), for all $t \in [\theta, \theta+\tau]$ we have

$$\begin{aligned}&\left|h_{j_0}(x(s_0)) - h_{j_0}(\overline{x}(s_0))\right| - \left|h_{j_0}(x(t)) - h_{j_0}(\overline{x}(t))\right| \\ &\le \left|\left(h_{j_0}(x(t)) - h_{j_0}(\overline{x}(t))\right) - \left(h_{j_0}(x(s_0)) - h_{j_0}(\overline{x}(s_0))\right)\right| \\ &< \frac{L_1\varepsilon_0}{2n}\end{aligned}$$

and therefore, by means of (3.3.12), we obtain that the inequality

$$\begin{aligned}&\left|h_{j_0}(x(t)) - h_{j_0}(\overline{x}(t))\right| > \left|h_{j_0}(x(s_0)) - h_{j_0}(\overline{x}(s_0))\right| - \frac{L_1\varepsilon_0}{2n} \\ &\ge \frac{L_1\varepsilon_0}{2n}\end{aligned} \qquad (3.3.13)$$

holds for all $t \in [\theta, \theta+\tau]$.

One can verify the existence of numbers $s_1, s_2, \ldots, s_n \in [\theta, \theta+\tau]$ such that

$$\begin{aligned}&\left\|\int_\theta^{\theta+\tau}[h(x(s)) - h(\overline{x}(s))]\,ds\right\| \\ &= \left\|\begin{pmatrix}\int_\theta^{\theta+\tau}[h_1(x(s)) - h_1(\overline{x}(s))]\,ds \\ \int_\theta^{\theta+\tau}[h_2(x(s)) - h_2(\overline{x}(s))]\,ds \\ \vdots \\ \int_\theta^{\theta+\tau}[h_n(x(s)) - h_n(\overline{x}(s))]\,ds\end{pmatrix}\right\| \\ &= \left\|\begin{pmatrix}\tau\,[h_1(x(s_1)) - h_1(\overline{x}(s_1))] \\ \tau\,[h_2(x(s_2)) - h_2(\overline{x}(s_2))] \\ \vdots \\ \tau\,[h_n(x(s_n)) - h_n(\overline{x}(s_n))]\end{pmatrix}\right\|.\end{aligned}$$

Thus, using (3.3.13), one can obtain that

$$\begin{aligned}&\left\|\int_\theta^{\theta+\tau}[h(x(s)) - h(\overline{x}(s))]\,ds\right\| \ge \tau\left|h_{j_0}(x(s_{j_0})) - h_{j_0}(\overline{x}(s_{j_0}))\right| \\ &> \frac{\tau L_1\varepsilon_0}{2n}.\end{aligned} \qquad (3.3.14)$$

It is clear that, for $t \in [\theta, \theta + \tau]$, $y(t)$ and $\overline{y}(t)$ satisfy the integral equations

$$y(t) = y(\theta) + \int_{\theta}^{t} Ay(s)ds + \int_{\theta}^{t} h(x(s))ds,$$

and

$$\overline{y}(t) = \overline{y}(\theta) + \int_{\theta}^{t} A\overline{y}(s)ds + \int_{\theta}^{t} h(\overline{x}(s))ds,$$

respectively, and herewith the equation

$$\begin{aligned} y(t) - \overline{y}(t) &= (y(\theta) - \overline{y}(\theta)) + \int_{\theta}^{t} A(y(s) - \overline{y}(s))ds \\ &+ \int_{\theta}^{t} [h(x(s)) - h(\overline{x}(s))]ds \end{aligned}$$

is achieved. Hence, we have the inequality

$$\begin{aligned} \|y(\theta + \tau) - \overline{y}(\theta + \tau)\| &\geq \left\| \int_{\theta}^{\theta+\tau} [h(x(s)) - h(\overline{x}(s))]ds \right\| \\ &- \|y(\theta) - \overline{y}(\theta)\| - \int_{\theta}^{\theta+\tau} \|A\| \, \|y(s) - \overline{y}(s)\| \, ds. \end{aligned} \tag{3.3.15}$$

Now, assume that $\max_{t \in [\theta, \theta+\tau]} \|y(t) - \overline{y}(t)\| \leq \dfrac{\tau L_1 \varepsilon_0}{2n(2 + \tau \|A\|)}$. In this case, one arrives at a contradiction since, by means of the inequalities (3.3.14) and (3.3.15), we have

$$\begin{aligned} \max_{t \in [\theta, \theta+\tau]} \|y(t) - \overline{y}(t)\| &\geq \|y(\theta + \tau) - \overline{y}(\theta + \tau)\| \\ &> \frac{\tau L_1 \varepsilon_0}{2n} - (1 + \tau \|A\|) \max_{t \in [\theta, \theta+\tau]} \|y(t) - \overline{y}(t)\| \\ &\geq \frac{\tau L_1 \varepsilon_0}{2n} - (1 + \tau \|A\|) \frac{\tau L_1 \varepsilon_0}{2n(2 + \tau \|A\|)} \\ &= \frac{\tau L_1 \varepsilon_0}{2n} \left(1 - \frac{1 + \tau \|A\|}{2 + \tau \|A\|}\right) \\ &= \frac{\tau L_1 \varepsilon_0}{2n(2 + \tau \|A\|)}. \end{aligned}$$

Therefore, we have $\max_{t \in [\theta, \theta+\tau]} \|y(t) - \overline{y}(t)\| > \dfrac{\tau L_1 \varepsilon_0}{2n(2 + \tau \|A\|)}$.

Suppose that the real-valued function $\|y(t) - \overline{y}(t)\|$ takes its maximum on the interval $[\theta, \theta + \tau]$ at η, that is,

$$\max_{t \in [\theta, \theta+\tau]} \|y(t) - \overline{y}(t)\| = \|y(\eta) - \overline{y}(\eta)\|$$

for some $\theta \leq \eta \leq \theta + \tau$.

For $t \in [\theta, \theta + \tau]$, by favor of the integral equations

$$y(t) = y(\eta) + \int_{\eta}^{t} Ay(s)ds + \int_{\eta}^{t} h(x(s))ds,$$

and

$$\overline{y}(t) = \overline{y}(\eta) + \int_{\eta}^{t} A\overline{y}(s)ds + \int_{\eta}^{t} h(\overline{x}(s))ds,$$

we obtain

$$y(t) - \overline{y}(t) = (y(\eta) - \overline{y}(\eta)) + \int_{\eta}^{t} A(y(s) - \overline{y}(s))ds$$
$$+ \int_{\eta}^{t} [h(x(s)) - h(\overline{x}(s))]ds.$$

Define

$$\tau^1 = \min\left\{\frac{\tau}{2}, \frac{\tau L_1 \varepsilon_0}{8n(M\|A\| + M_0)(2 + \tau\|A\|)}\right\}$$

and let

$$\theta^1 = \begin{cases} \eta, & \text{if } \eta \leq \theta + \frac{\tau}{2} \\ \eta - \tau^1, & \text{if } \eta > \theta + \frac{\tau}{2} \end{cases}.$$

We note that the interval $J^1 = [\theta^1, \theta^1 + \tau^1]$ is a subset of $[\theta, \theta + \tau]$ and hence a subset of J.

For $t \in J^1$, we have

$$\begin{aligned}
\|y(t) - \overline{y}(t)\| &\geq \|y(\eta) - \overline{y}(\eta)\| - \left|\int_{\eta}^{t} \|A\| \|y(s) - \overline{y}(s)\| ds\right| \\
&- \left|\int_{\eta}^{t} \|h(x(s)) - h(\overline{x}(s))\| ds\right| \\
&> \frac{\tau L_1 \varepsilon_0}{2n(2 + \tau\|A\|)} - 2M\|A\|\tau^1 - 2M_0\tau^1 \\
&= \frac{\tau L_1 \varepsilon_0}{2n(2 + \tau\|A\|)} - 2\tau^1(M\|A\| + M_0) \\
&\geq \frac{\tau L_1 \varepsilon_0}{4n(2 + \tau\|A\|)}.
\end{aligned}$$

Consequently, we get $\|y(t) - \overline{y}(t)\| > \varepsilon_1$, $t \in J^1$, where $\varepsilon_1 = \dfrac{\tau L_1 \varepsilon_0}{4n(2 + \tau \|A\|)}$ and the length τ^1 of the interval J^1 does not depend on the functions $x(t), \overline{x}(t) \in \mathscr{A}_x$.

This finalizes the proof of the lemma.□

We shall continue in the next lemma by the extension of transitivity feature.

Lemma 3.2 *Transitivity of* $\mathscr{A}_x$ *implies the same feature for the set* $\mathscr{A}_y$.

Proof Fix arbitrary $\varepsilon > 0$, $E > 0$ and $y(t) = \begin{pmatrix} y_+(t) \\ y_-(t) \end{pmatrix} \in \mathscr{A}_y$. In this case, there exists a function $x(t) \in \mathscr{A}_x$ such that $y(t) = \phi_{x(t)}(t)$. Let $\gamma = \dfrac{\omega}{4NM_0 + 2NL_2}$. Since there exists a solution $x^*(t) \in \mathscr{A}_x$, which is dense in $\mathscr{A}_x$, one can find $\xi > 0$ and an interval $J \subset \mathbb{R}$ with length E such that $\|x(t) - x^*(t + \xi)\| < \gamma\varepsilon$, for all $t \in J$. Without loss of generality, assume that J is a closed interval, that is, $J = [a, a + E]$ for some real number a. Let $y^*(t) = \phi_{x^*(t)}(t) = \begin{pmatrix} y_+^*(t) \\ y_-^*(t) \end{pmatrix}$.

The functions $y_+(t)$ and $y_+^*(t)$ satisfy the relations

$$y_+(t) = \int_{-\infty}^{t} e^{A_+(t-s)} h_+(x(s)) ds$$

and

$$y_+^*(t) = \int_{-\infty}^{t} e^{A_+(t-s)} h_+(x^*(s)) ds,$$

respectively.

Making use of the latter, one can show that

$$y_+^*(t + \xi) = \int_{-\infty}^{t} e^{A_+(t-s)} h_+(x^*(s + \xi)) ds.$$

Therefore we have

$$\begin{aligned} y_+(t) - y^*(t + \xi) &= \int_{-\infty}^{t} e^{A_+(t-s)} \left[h_+(x(s)) - h_+(x^*(s + \xi)) \right] ds \\ &= \int_{-\infty}^{a} e^{A_+(t-s)} \left[h_+(x(s)) - h_+(x^*(s + \xi)) \right] ds \\ &+ \int_{a}^{t} e^{A_+(t-s)} \left[h_+(x(s)) - h_+(x^*(s + \xi)) \right] ds. \end{aligned}$$

The last equation implies that

$$\|y_+(t) - y_+^*(t+\xi)\| \le \int_{-\infty}^{a} \left\|e^{A_+(t-s)}\right\| \|h_+(x(s)) - h_+(x^*(s+\xi))\| \, ds$$
$$+ \int_a^t \left\|e^{A_+(t-s)}\right\| \|h_+(x(s)) - h_+(x^*(s+\xi))\| \, ds$$
$$\le \int_{-\infty}^{a} Ne^{-\omega(t-s)} 2M_0 ds + \int_a^t Ne^{-\omega(t-s)} L_2 \gamma \varepsilon ds$$
$$= \frac{2NM_0}{\omega} e^{-\omega(t-a)} + \frac{NL_2\gamma\varepsilon}{\omega} \left(1 - e^{-\omega(t-a)}\right).$$

On the other hand, for $t \in [a, a+E]$, $y_-(t)$ and $y_-^*(t)$ satisfy the integral equations

$$y_-(t) = -\int_t^{\infty} e^{A_-(t-s)} h_-(x(s)) ds$$

and

$$y_-^*(t) = -\int_t^{\infty} e^{A_-(t-s)} h_-(x^*(s)) ds,$$

respectively.

It is easy to verify that the equation

$$y_-^*(t+\xi) = -\int_t^{\infty} e^{A_-(t-s)} h_-(x^*(s+\xi)) ds$$

holds for all $t \in [a, a+E]$.

Thus, the equation

$$y_-(t) - y_-^*(t+\xi) = -\int_t^{\infty} e^{A_-(t-s)} \left[h_-(x(s)) - h_-(x^*(s+\xi))\right] ds$$
$$= -\int_{a+E}^{\infty} e^{A_-(t-s)} \left[h_-(x(s)) - h_-(x^*(s+\xi))\right] ds$$
$$- \int_t^{a+E} e^{A_-(t-s)} \left[h_-(x(s)) - h_-(x^*(s+\xi))\right] ds$$

is valid and as a consequence the inequality

$$\|y_-(t) - y_-^*(t+\xi)\| \le \int_{a+E}^{\infty} \left\|e^{A_-(t-s)}\right\| \|h_-(x(s)) - h_-(x^*(s+\xi))\| \, ds$$
$$+ \int_t^{a+E} \left\|e^{A_-(t-s)}\right\| \|h_-(x(s)) - h_-(x^*(s+\xi))\| \, ds$$

$$\leq \int_{a+E}^{\infty} 2NM_0 e^{\omega(t-s)} ds + \int_{t}^{a+E} NL_2\gamma\varepsilon e^{\omega(t-s)} ds$$
$$= \frac{2NM_0}{\omega} e^{-\omega(a+E-t)} + \frac{NL_2\gamma\varepsilon}{\omega}\left(1 - e^{-\omega(a+E-t)}\right)$$

is satisfied for $t \in [a, a+E]$.

Now, making use of the inequality

$$\left\|y(t) - y^*(t+\xi)\right\| \leq \left\|y_+(t) - y_+^*(t+\xi)\right\| + \left\|y_-(t) - y_-^*(t+\xi)\right\|,$$

we obtain that

$$\left\|y_-(t) - y_-^*(t+\xi)\right\| \leq \frac{2NM_0}{\omega}\left[e^{-\omega(t-a)} + e^{-\omega(a+E-t)}\right]$$
$$+ \frac{NL_2\gamma\varepsilon}{\omega}\left[\left(1 - e^{-\omega(t-a)}\right) + \left(1 - e^{-\omega(a+E-t)}\right)\right]$$
$$\leq \frac{2NM_0}{\omega}\left[e^{-\omega(t-a)} + e^{-\omega(a+E-t)}\right] + \frac{2NL_2\gamma\varepsilon}{\omega}.$$

Suppose that $E > \dfrac{3}{\omega}\ln\left(\dfrac{1}{\gamma\varepsilon}\right)$. In this case, for $t \in \left[a + \frac{E}{3}, a + \frac{2E}{3}\right]$ one can see that both of the inequalities $e^{-\omega(t-a)} \leq e^{-\omega E/3}$ and $e^{-\omega(a+E-t)} \leq e^{-\omega E/3}$ are valid, and consequently

$$\left\|y_-(t) - y_-^*(t+\xi)\right\| < \left(\frac{4NM_0 + 2NL_2}{\omega}\right)\gamma\varepsilon = \varepsilon.$$

Lemma is proved. □

We handle the extension of the last ingredient of Devaney chaos in the next lemma.

Lemma 3.3 *If $\mathscr{A}_x$ admits a dense countable collection of periodic functions, then the same is true for $\mathscr{A}_y$.*

Proof Fix arbitrary numbers $\varepsilon > 0$, $E > 0$ and $y(t) = \begin{pmatrix} y_+(t) \\ y_-(t) \end{pmatrix} \in \mathscr{A}_y$. Owing to the construction of the set $\mathscr{A}_y$, one can find a function $x(t) \in \mathscr{A}_x$ such that $y(t) = \phi_{x(t)}(t)$. Assume that the set $\mathscr{A}_x$ admits a dense countable collection $\mathscr{G}_x$ of periodic functions. Let $\gamma = \dfrac{\omega}{4NM_0 + 2NL_2}$.

By density of $\mathscr{G}_x \subset \mathscr{A}_x$, there exist $\widetilde{x}(t) \in \mathscr{G}_x$ and an interval $J \subset \mathbb{R}$ with length E such that $\|x(t) - \widetilde{x}(t)\| < \gamma\varepsilon$, for all $t \in J$. Without loss of generality, assume that J is a closed interval, that is, $J = [a, a+E]$ for some real number a.

We note that by condition $(A3)$ there is a one-to-one correspondence between the sets $\mathscr{G}_x$ and $\mathscr{G}_y = \left\{\phi_{x(t)}(t) \mid x(t) \in \mathscr{G}_x\right\}$. Moreover, $x(t) \in \mathscr{G}_x$ and $\phi_{x(t)}(t) \in \mathscr{G}_y$ admit the same periods. Therefore, $\mathscr{G}_y \subset \mathscr{A}_y$ is a countable collection of periodic functions and our aim is to show that the set $\mathscr{G}_y$ is dense in $\mathscr{A}_y$.

Let $\widetilde{y}(t) = \phi_{\widetilde{x}(t)}(t) = \begin{pmatrix} \widetilde{y}_+(t) \\ \widetilde{y}_-(t) \end{pmatrix} \in \mathscr{G}_y$. The functions $y_+(t)$ and $\widetilde{y}_+(t)$ satisfy the relations

$$y_+(t) = \int_{-\infty}^{t} e^{A_+(t-s)} h_+(x(s)) ds$$

and

$$\widetilde{y}_+(t) = \int_{-\infty}^{t} e^{A_+(t-s)} h_+(\widetilde{x}(s)) ds,$$

respectively.

Therefore we have

$$y_+(t) - \widetilde{y}_+(t) = \int_{-\infty}^{t} e^{A_+(t-s)} \left[h_+(x(s)) - h_+(\widetilde{x}(s)) \right] ds$$
$$= \int_{-\infty}^{a} e^{A_+(t-s)} \left[h_+(x(s)) - h_+(x^*(s)) \right] ds$$
$$+ \int_{a}^{t} e^{A_+(t-s)} \left[h_+(x(s)) - h_+(\widetilde{x}(s)) \right] ds.$$

The last equation implies that

$$\|y_+(t) - \widetilde{y}_+(t)\| \leq \int_{-\infty}^{a} \left\| e^{A_+(t-s)} \right\| \|h_+(x(s)) - h_+(\widetilde{x}(s))\| ds$$
$$+ \int_{a}^{t} \left\| e^{A_+(t-s)} \right\| \|h_+(x(s)) - h_+(\widetilde{x}(s))\| ds$$
$$\leq \int_{-\infty}^{a} N e^{-\omega(t-s)} 2M_0 ds + \int_{a}^{t} N e^{-\omega(t-s)} L_2 \gamma \varepsilon ds$$
$$= \frac{2NM_0}{\omega} e^{-\omega(t-a)} + \frac{NL_2\gamma\varepsilon}{\omega} \left(1 - e^{-\omega(t-a)}\right).$$

On the other hand, for $t \in [a, a+E]$, $y_-(t)$ and $\widetilde{y}_-(t)$ satisfy the integral equations

$$y_-(t) = -\int_{t}^{\infty} e^{A_-(t-s)} h_-(x(s)) ds$$

and

$$\widetilde{y}_-(t) = -\int_{t}^{\infty} e^{A_-(t-s)} h_-(\widetilde{x}(s)) ds,$$

respectively.

Thus, the equation

$$
\begin{aligned}
y_-(t) - \widetilde{y}_-(t) &= -\int_t^{\infty} e^{A_-(t-s)} \left[h_-(x(s)) - h_-(\widetilde{x}(s))\right] ds \\
&= -\int_{a+E}^{\infty} e^{A_-(t-s)} \left[h_-(x(s)) - h_-(\widetilde{x}(s))\right] ds \\
&\quad -\int_t^{a+E} e^{A_-(t-s)} \left[h_-(x(s)) - h_-(\widetilde{x}(s))\right] ds
\end{aligned}
$$

is valid and as a consequence the inequality

$$
\begin{aligned}
\|y_-(t) - \widetilde{y}_-(t)\| &\le \int_{a+E}^{\infty} \left\|e^{A_-(t-s)}\right\| \|h_-(x(s)) - h_-(\widetilde{x}(s))\| \, ds \\
&+ \int_t^{a+E} \left\|e^{A_-(t-s)}\right\| \|h_-(x(s)) - h_-(\widetilde{x}(s))\| \, ds \\
&\le \int_{a+E}^{\infty} 2NM_0 e^{\omega(t-s)} ds + \int_t^{a+E} NL_2\gamma\varepsilon e^{\omega(t-s)} ds \\
&= \frac{2NM_0}{\omega} e^{-\omega(a+E-t)} + \frac{NL_2\gamma\varepsilon}{\omega}\left(1 - e^{-\omega(a+E-t)}\right)
\end{aligned}
$$

is satisfied for $t \in [a, a+E]$.

Now, making use of the inequality

$$
\|y(t) - \widetilde{y}(t)\| \le \|y_+(t) - \widetilde{y}_+(t)\| + \|y_-(t) - \widetilde{y}_-(t)\|,
$$

we obtain that

$$
\begin{aligned}
\|y(t) - \widetilde{y}(t)\| &\le \frac{2NM_0}{\omega}\left[e^{-\omega(t-a)} + e^{-\omega(a+E-t)}\right] \\
&+ \frac{NL_2\gamma\varepsilon}{\omega}\left[\left(1 - e^{-\omega(t-a)}\right) + \left(1 - e^{-\omega(a+E-t)}\right)\right] \\
&\le \frac{2NM_0}{\omega}\left[e^{-\omega(t-a)} + e^{-\omega(a+E-t)}\right] + \frac{2NL_2\gamma\varepsilon}{\omega}.
\end{aligned}
$$

Suppose that $E > \dfrac{3}{\omega} \ln\left(\dfrac{1}{\gamma\varepsilon}\right)$. In this case, for $t \in \left[a + \frac{E}{3}, a + \frac{2E}{3}\right]$ one can see that both of the inequalities $e^{-\omega(t-a)} \le e^{-\omega E/3}$ and $e^{-\omega(a+E-t)} \le e^{-\omega E/3}$ are valid, and consequently

$$
\|y(t) - \widetilde{y}(t)\| < \left(\frac{4NM_0 + 2NL_2}{\omega}\right)\gamma\varepsilon = \varepsilon.
$$

Proof of the lemma is accomplished. □

Next, we state the main theorem about Devaney chaos, whose proof follows from Lemmas 3.1–3.3.

Theorem 3.1 *If $\mathscr{A}_x$ is a Devaney chaotic set, then the same is true for $\mathscr{A}_y$.*

A corollary of Theorem 3.1 is the following one.

Corollary 3.1 *Under the condition of Theorem 3.1, the collection $\mathscr{A}$ is a Devaney chaotic set.*

Now, we continue by indicating the extension of chaos in the sense of Li–Yorke for the system (3.2.4) + (3.2.6). The proof of the next theorem is given briefly, since the technique is similar to the previous lemmas.

Theorem 3.2 *If $\mathscr{A}_x$ is a Li–Yorke chaotic set, then the same is true for $\mathscr{A}_y$.*

Proof Assume that the set $\mathscr{A}_x$ is Li–Yorke chaotic. According to the one-to-one correspondence between the periodic solutions of (3.2.4) and (3.2.6), if for some $T > 0$ the set $\mathscr{A}_x$ admits a $kT-$periodic function for any natural number k, then the same is true for the set $\mathscr{A}_y$.

Now, suppose that $\mathscr{C}_x$ is an uncountable scrambled set inside $\mathscr{A}_x$. Let us introduce

$$\mathscr{C}_y = \left\{\phi_{x(t)}(t) \, : \, x(t) \in \mathscr{C}_x\right\}. \tag{3.3.16}$$

Condition $(A3)$ implies that there is a one-to-one correspondence between the sets $\mathscr{C}_x$ and $\mathscr{C}_y$. Therefore, $\mathscr{C}_y$ is also uncountable. Under same condition, it is easy verify that there does not exist any periodic function inside the set $\mathscr{C}_y$.

Since the collection $\mathscr{A}_x$ is assumed to be chaotic in the sense of Li–Yorke, each couple of functions inside $\mathscr{C}_x \times \mathscr{C}_x$ is proximal. Under the circumstances, one can use the method of the proof of Lemma 3.3 to show that the same is valid for each couple inside $\mathscr{C}_y \times \mathscr{C}_y$.

On the other hand, one can follow the technique used in the proof of Lemma 3.1 to verify the existence of positive real numbers ε_1 and Δ_1 such that each couple of functions $(y(t), \overline{y}(t)) \in \mathscr{C}_y \times \mathscr{C}_y$ are frequently $(\varepsilon_1, \Delta_1)-$separated. The same property is true also for each couple of sequences inside $\left(\mathscr{C}_y \times \mathscr{G}_y\right)$, where $\mathscr{G}_y$ represents the set of periodic functions inside $\mathscr{A}_y$. Consequently, the set of functions $\mathscr{A}_y$ is a Li–Yorke chaotic set.

The proof of the theorem is finalized. □

We end up the present section by stating the following corollary of Theorem 3.2.

Corollary 3.2 *Under the condition of Theorem 3.2, the collection $\mathscr{A}$ is a Li–Yorke chaotic set.*

3.4 Simulations

Let us consider the Duffing's chaotic oscillator [20]

$$x'' + 0.05x' + x^3 = 7.5 \cos t \tag{3.4.17}$$

and the equation

$$u'' + 2u' + 3u = 0. \tag{3.4.18}$$

Let us combine the equations (3.4.17) and (3.4.18) in a unidirectional way to build the following couple of differential equations

$$\begin{aligned} &x'' + 0.05x' + x^3 = 7.5 \cos t \\ &y'' + 2y' + 3y = 2(x + x^3). \end{aligned} \tag{3.4.19}$$

One can see that the coupled system (3.4.19) is constructed in such a way that equation (3.4.18) is forced by the exogenous term $h(x) = 2(x + x^3)$, where the solutions of the Duffing's oscillator (3.4.17) are used.

Making use of the new variables $x_1 = x, y_1 = y, x_2 = x'$, and $y_2 = y'$, one can rewrite the couple of differential equations (3.4.19) as a 4-dimensional system as follows:

$$\begin{aligned} &x_1 = x_2 \\ &x_2' = -0.05x_2 - x_1^3 + 7.5 \cos t \\ &y_1' = y_2 \\ &y_2' = -2y_2 - 3y_1 + 2(x_1 + x_1^3). \end{aligned} \tag{3.4.20}$$

We note that system (3.4.20) is in the form of (3.2.4)+(3.2.6), where the matrix $A = \begin{pmatrix} 0 & 1 \\ -3 & -2 \end{pmatrix}$ has eigenvalues $-1 \pm i\sqrt{2}$. According to the theoretical results of the chapter, we say that system (3.4.20) is chaotic. For the visualization of the chaotic behavior, let us consider the trajectory of system (3.4.20) with the initial data $x_1(0) = 3.07, x_2(0) = 4.18, y_1(0) = 0.15$, and $y_2 = 0.24$. Figure 3.1 represents the 2-dimensional projections of the trajectory on the $x_1 - x_2$ and $y_1 - y_2$ planes. The picture shown in Fig. 3.1a is, in fact, the chaotic attractor of the nonlinear system

$$\begin{aligned} &x_1 = x_2 \\ &x_2' = -0.05x_2 - x_1^3 + 7.5 \cos t, \end{aligned} \tag{3.4.21}$$

which is the subsystem of (3.4.20) corresponding to the first two coordinates, and the chaotic attractor shown in Fig. 3.1b signifies the extension of chaos in system (3.4.20) and supports our rigorously approved results.

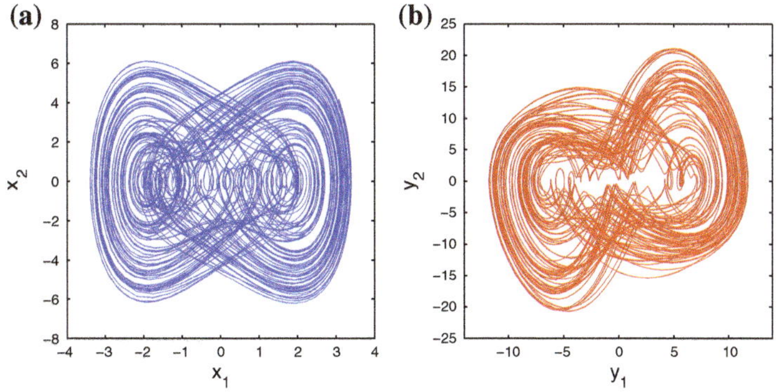

Fig. 3.1 The 2-dimensional projections of the chaotic attractor of system (3.4.20). **a** The projection on the $x_1 - x_2$ plane. **b** The projection on the $y_1 - y_2$ plane. The picture in (**a**) represents, in fact, the chaotic attractor of system (3.4.21). Extension of chaos in system (3.4.20) is apparent in picture (**b**) which supports our theoretical results

Next, let us demonstrate numerically that the chaos achieved in system (3.4.20) is controllable and we shall make use of the Pyragas method [29] for stabilizing the unstable periodic solutions of (3.4.20). In the Pyragas control procedure, one considers a system in the form

$$v' = H(v, q), \tag{3.4.22}$$

where $q = q(t)$ is an externally controllable parameter and for $q = 0$ it is assumed that the system (3.4.22) is in the chaotic state of interest, whose periodic orbits are to be stabilized [29–32]. According to Pyragas method, by means of the control law $q(t) = C\,[s\,(t - p_0) - s(t)]$, an unstable periodic solution with period p_0 of system (3.4.22) with $q = 0$ can be stabilized. Here, the parameter C represents the strength of the perturbation and $s(t) = \sigma\,[x(t)]$ is a scalar signal given by some function of the state of the system.

Through the Pyragas control method, one can verify numerically that it is possible to stabilize the unstable 2π-periodic solution of system (3.4.21) by means of the following control system

$$\begin{aligned} v_1' &= v_2 \\ v_2' &= -0.05v_2 - v_1^3 + 7.5\cos(v_3) + C\,[v_2(t - 2\pi) - v_2(t)] \\ v_3' &= 1, \end{aligned}$$

by taking $C = 0.36$, where $q(t) = C\,[v_2(t - 2\pi) - v_2(t)]$ is the control law.

Now, let us show how it is possible to control chaos of system (3.4.20). We propose that if a periodic solution of the system (3.4.21), which is used as the source of chaotic perturbation, is stabilized, then the same is true for the corresponding unstable periodic solution of (3.4.20).

To control chaos of system (3.4.20), we set up the system

$$\begin{aligned} v_1' &= v_2 \\ v_2' &= -0.05v_2 - v_1^3 + 7.5\cos(v_3) + 0.36\,[v_2(t-2\pi) - v_2(t)] \\ v_3' &= 1 \\ y_1' &= y_2 \\ y_2' &= -2y_2 - 3y_1 + 2(v_1 + v_1^3), \end{aligned} \tag{3.4.23}$$

which we call as the control system corresponding to (3.4.20).

We consider the solution of system (3.4.23) with the initial data $v_1(0) = 3.07$, $v_2(0) = 4.18$, $v_3(0) = 0$, $y_4(0) = 0.15$, and $y_5(0) = 0.24$. The system is allowed to evolve freely by taking $C = 0$ until the time $t = 70$, and at that moment the control is switched on using the value $C = 0.36$. At $t = 180$, the control mechanism is switched off and after this moment we start to use the value of the parameter $C = 0$ once again. Figure 3.2, which reveals the chaos control for system (3.4.20), represents the time-series for the v_2 and y_2 coordinates of the solution. After the control is switched off, the stabilized 2π-periodic solution loses its stability and chaos emerges again. Similar pictures can be obtained for the other coordinates of system (3.4.23), which are not just pictured here. Figure 3.2 supports the idea to control the chaos of system (3.4.20), it is sufficient to control the chaos of (3.4.21) and for this purpose the Pyragas control method is suitable.

Our next example is about the numerical demonstration of the chaos extension for systems in the form of (3.2.4)+(3.2.10).

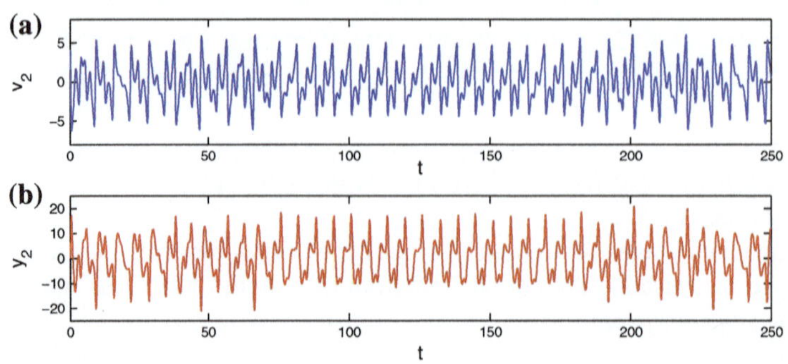

Fig. 3.2 Application of the Pyragas control method to system (3.4.20). **a** The time-series for the v_2 coordinate. **b** The time-series for the y_2 coordinate. The chaos control for system (3.4.20) is achieved by means of the corresponding control system (3.4.23). The picture in (**a**) shows that by means of the control system (3.4.23), it is possible to stabilize the 2π-periodic solution of system (3.4.21), and correspondingly the picture in (**b**) reveals that the corresponding periodic solution of system (3.4.20) is controlled. The control mechanism starts at the time $t = 70$ and ends at $t = 180$. After switching off the control mechanism, the stabilized 2π-periodic solution of (3.4.20) loses its stability and irregular behavior develops again

Let us consider the following Lorenz system [11]

$$\begin{aligned} u_1' &= -10u_1 + 10u_2 \\ u_2' &= 7u_1 - u_2 - u_1u_3 \\ u_3' &= u_1u_2 - (8/3)u_3. \end{aligned} \tag{3.4.24}$$

System (3.4.24) admits two fixed points $C^+ = (4, 4, 6)$ and $C^- = (-4, -4, 6)$, which are both stable, and it can be written in the form of equation (3.2.10), where the matrix

$$A = \begin{pmatrix} -10 & 10 & 0 \\ 7 & -1 & 0 \\ 0 & 0 & -8/3 \end{pmatrix}$$

has eigenvalues -15, $-8/3$ and 4.

We perturb system (3.4.24) through the solutions of the chaotic Rössler system [12]

$$\begin{aligned} x_1' &= -x_2 - x_3 \\ x_2' &= x_1 + 0.2x_2 \\ x_3' &= 0.2 - 5.7x_3 + x_1x_3 \end{aligned} \tag{3.4.25}$$

and set up the following 6-dimensional system

$$\begin{aligned} x_1' &= -x_2 - x_3 \\ x_2' &= x_1 + 0.2x_2 \\ x_3' &= 0.2 - 5.7x_3 + x_1x_3 \\ y_1' &= -10y_1 + 10y_2 + x_1 \\ y_2' &= 7y_1 - y_2 - y_1y_3 + 2x_2 \\ y_3' &= y_1y_2 - (8/3)y_3 + x_3. \end{aligned} \tag{3.4.26}$$

Figure 3.3 shows the 3-dimensional projections of the trajectory of system (3.4.26) with the initial data $x_1(0) = 10.563$, $x_2(0) = -1.594$, $x_3(0) = 4.024$ $y_1(0) = -3.441$, $y_2(0) = -4.365$, and $y_3(0) = 7.398$. The picture in Fig. 3.3a corresponds to the chaotic attractor of the Rössler system (3.4.25), while Fig. 3.3b indicates the extension of chaos in system (3.4.26). Although the initially non-perturbed system (3.4.24) is not chaotic, the perturbation acts in such a way that the system starts to behave chaotically.

The result of the last example shows that it is also possible to make use of systems of the form (3.2.4)+(3.2.10) to extend chaos. The present chapter does not give a theoretical background for this behavior since the nonlinearity is included in the system. One can try to linearize system (3.4.24) around one of the fixed points C^+ or C^- to give a theoretical support for the numerical results through the discussions of the paper [8]. Such a discussion is also not possible, since mathematical results of the paper [8] request the smallness for the Lipschitz constants of the nonlinear

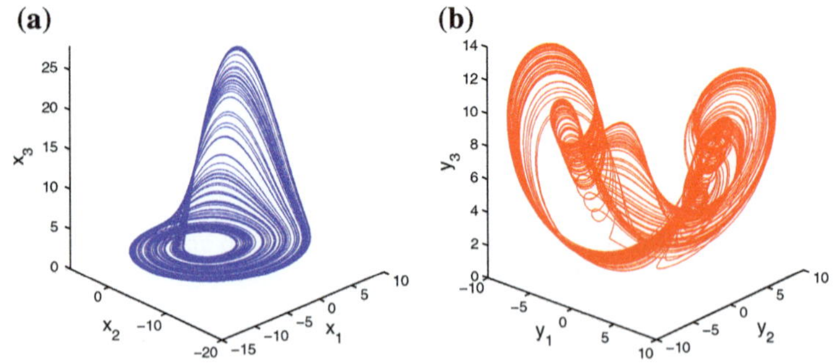

Fig. 3.3 The 3-dimensional projections of the chaotic attractor of system (3.4.26). **a** Projection on the $x_1 - x_2 - x_3$ space. **b** Projection on the $y_1 - y_2 - y_3$ space. The picture in (**a**) shows, in fact, the chaotic attractor of the Rössler system (3.4.25). The picture presented in (**b**) supports the idea that it is also possible to make use of systems of the form (3.2.4)+(3.2.10) to extend chaos

terms, which is not satisfied by the system (3.4.26). Accordingly, we should say that the theoretical discussions for the Lorenz system have to be handled by a different and a new method, compared to our present and previous results, and we will discuss the problem in next investigations.

3.5 Notes

In this chapter, we consider the extension of chaos in hyperbolic systems with arbitrary large dimensions. Our investigations comprise chaos in the sense of both Devaney and Li–Yorke. We provide a mechanism for unidirectionally coupled systems through the insertion of chaos from one system to another, where the latter is initially non-chaotic. In the chaos extension procedure, we take advantage of chaotic sets of functions in order to provide mathematically approved results. The theoretical results are supported through simulations by means of a Duffing oscillator. A control technique for the extended chaos is demonstrated numerically. The results of Chap. 3 were published in the paper [6].

References

1. M.U. Akhmet, Creating a chaos in a system with relay. Int. J. Qual. Theory Differ. Equ. Appl. **3**, 3–7 (2009)
2. M.U. Akhmet, Devaney's chaos of a relay system. Commun. Nonlinear Sci. Numer. Simul. **14**, 1486–1493 (2009)
3. M.U. Akhmet, Li–Yorke chaos in the system with impacts. J. Math. Anal. Appl. **351**, 804–810 (2009)

4. M.U. Akhmet, M.O. Fen, Chaotic period-doubling and OGY control for the forced Duffing equation. Commun. Nonlinear Sci. Numer. Simul. **17**, 1929–1946 (2012)
5. M.U. Akhmet, M.O. Fen, The period-doubling route to chaos in the relay system, in *Proceedings of Dynamic Systems and Applications*, vol. 6, ed. by G.S. Ladde, N.G. Medhin, C. Peng, M. Sambandham (Dynamic Publisher Inc., Atlanta, 2012), pp. 22–26
6. M.U. Akhmet, M.O. Fen, Chaos generation in hyperbolic systems. Interdiscip. J. Discontin. Nonlinearity Complex. **1**, 367–386 (2012)
7. M. Akhmet, *Principles of Discontinuous Dynamical Systems* (Springer, New York, 2010)
8. M.U. Akhmet, M.O. Fen, Replication of chaos. Commun. Nonlinear Sci. Numer. Simul. **18**, 2626–2666 (2013)
9. M.U. Akhmet, M.O. Fen, Shunting inhibitory cellular neural networks with chaotic external inputs. Chaos **23**, 023112 (2013)
10. M.U. Akhmet, The complex dynamics of the cardiovascular system. Nonlinear Anal.: Theory Methods Appl. **71**, e1922–e1931 (2009)
11. E.N. Lorenz, Deterministic nonperiodic flow. J. Atmos. Sci. **20**, 130–141 (1963)
12. O.E. Rössler, An equation for continuous chaos. Phys. Lett. A **57**, 397–398 (1976)
13. L.O. Chua, M. Komuro, T. Matsumoto, The double scroll family, parts I and II. IEEE Trans. Circuit Syst. **CAS-33**, 1072–1118 (1986)
14. L.O. Chua, C.W. Wu, A. Huang, G. Zhong, A universal circuit for studying and generating chaos-part I: routes to chaos. IEEE Trans. Circuits Syst.-I: Fundam. Theory Appl. **40**, 732–744 (1993)
15. M. Cartwright, J. Littlewood, On nonlinear differential equations of the second order I: the equation $\ddot{y} - k(1 - y^2)'y + y = bkcos(\lambda t + a)$, k large. J. Lond. Math. Soc. **20**, 180–189 (1945)
16. M. Levi, *Qualitative Analysis of the Periodically Forced Relaxation Oscillations* (Memoirs of the American Mathematical Society, Providence, 1981)
17. N. Levinson, A second order differential equation with singular solutions. Ann. Math. **50**, 127–153 (1949)
18. A.C.J. Luo, *Regularity and Complexity in Dynamical Systems* (Springer, New York, 2012)
19. F.C. Moon, *Chaotic Vibrations: An Introduction for Applied Scientists and Engineers* (Wiley, Hoboken, 2004)
20. J.M.T. Thompson, H.B. Stewart, *Nonlinear Dynamics and Chaos* (Wiley, Chichester, 2002)
21. R.A. Horn, C.R. Johnson, *Matrix Analysis* (Cambridge University Press, New York, 1992)
22. E. Akin, S. Kolyada, Li–Yorke sensitivity. Nonlinearity **16**, 1421–1433 (2003)
23. M. Čiklová, Li–Yorke sensitive minimal maps. Nonlinearity **19**, 517–529 (2006)
24. R.L. Devaney, *An Introduction to Chaotic Dynamical Systems* (Addison, Menlo Park, 1989)
25. P. Kloeden, Z. Li, Li–Yorke chaos in higher dimensions: a review. J. Differ. Equ. Appl. **12**, 247–269 (2006)
26. T.Y. Li, J.A. Yorke, Period three implies chaos. Am. Math. Mon. **82**, 985–992 (1975)
27. K. Palmer, *Shadowing in Dynamical Systems: Theory and Applications* (Kluwer Academic Publishers, Dordrecht, 2000)
28. J.K. Hale, *Ordinary Differential Equations* (Krieger Publishing Company, Malabar, 1980)
29. K. Pyragas, Continuous control of chaos by self-controlling feedback. Phys. Rev. A **170**, 421–428 (1992)
30. A.L. Fradkov, *Cybernetical Physics: From Control of Chaos to Quantum Control* (Springer, Berlin, 2007)
31. J.M. Gonzáles-Miranda, *Synchronization and Control of Chaos* (Imperial College Press, London, 2004)
32. E. Schöll, H.G. Schuster, *Handbook of Chaos Control* (Wiley, Weinheim, 2008)

Chapter 4
Entrainment by Chaos

In this chapter, a new phenomenon, the entrainment of limit cycles by chaos, which results as the appearance of cyclic irregular behavior, is discussed. Sensitivity is considered as the main ingredient of chaos to be captured, and the period-doubling cascade is chosen for extension. Theoretical results are supported by simulations and discussions regarding Chua's oscillators, entrainment of toroidal attractors by chaos, synchronization, and controlling problems. It is demonstrated that the entrainment cannot be considered as generalized synchronization of chaotic systems.

4.1 Introduction

Christiaan Huygens was the first to introduce the concept of entrainment when he observed that two pendulum clocks mounted next to each other on the same support often become synchronized [1]. One can also mention the practice of fine-tuning brainwaves to a desired frequency, that is, brainwave entrainment [2, 3], or the idea of entrainment in biomusicology, which is understood to be the synchronization of organisms to an external rhythm [4]. The entrainment phenomenon is also known in hydrodynamics as the movement of one fluid induced by another [5]. In the present chapter, we discuss the entrainment of limit cycles by chaos and demonstrate that entrainment, in mathematical theory, is not confined to the notions of frequency, period, or phase [6–8] but extends to the concept of chaos, as well. Unidirectional coupling, which has been extensively studied in physics [8–12], is investigated in this work. The results presented can be used to generate the entrainment by chaos in business cycle models [13], chaotic cycles in electrical circuits, such as those obtained via the Van der Pol equations [14], and chaotic oscillations in Belousov–Zhabotinsky reactions [15]. Indeed, the results can be applied and developed in any field in which limit cycles are observed.

M. Akhmet and M.O. Fen, *Replication of Chaos in Neural Networks, Economics and Physics*, Nonlinear Physical Science,
DOI 10.1007/978-3-662-47500-3_4

The entrainment by chaos is understood in this chapter as the deformation of limit cycles to chaotic cycles. The presence of chaotic behavior is confirmed by checking for the existence of infinitely many unstable periodic solutions and sensitivity, which is the main ingredient of chaos [16–19].

In studies [20–25], we considered systems with asymptotically stable and hyperbolic equilibria and perturbed them chaotically. It was found that the solutions admit the same type of chaos that perturbations do. Unlike the mechanism discussed in [25], where we considered chaos near fixed points, in the present chapter, we take into account systems with orbitally stable limit cycles and perturb them with chaos. As a result, we obtain *chaotic cycles*, that is, motions that behave cyclically and chaotically at the same time.

To the best of our knowledge, the generation of chaos is considered in synchronization theory [26–33]. However, in the literature, it is required that the chaos of the response system be asymptotically close to that of the driver, and this property is used for verification of chaos. Since we do not use this proximity, our method is of different type in generation of chaos than that used in synchronization theory. Moreover, we have shown that our results are not reducible to the synchronization, in general.

In Sect. 4.7, we present the results of simulations of a chaotic torus and a Chua oscillator, though these entrainment phenomena must be investigated further. It is also of great interest to prove the entrainment by chaos around hyperbolic periodic solutions [34–36].

To illustrate the main idea of this chapter, we present the example of an oscillating chemical reaction. Paper [37] considers the chlorine dioxide–iodine–malonic acid ($ClO_2{-}I_2{-}MA$) chemical reaction, which arises from the following three component reactions:

$$\begin{aligned} &MA + I_2 \longrightarrow IMA + I^- + H^+,\\ &ClO_2 + I^- \longrightarrow ClO_2^- + \frac{1}{2}I_2,\\ &ClO_2^- + 4I^- + 4H^+ \longrightarrow Cl^- + 2I_2 + 2H_2O. \end{aligned} \tag{4.1.1}$$

After making reasonable simplifications and nondimensionalizations, Lengyel et al. [37] reduced the rate equations to the system

$$\begin{aligned} u_1' &= a - u_1 - \frac{4u_1u_2}{1+u_1^2},\\ u_2' &= bu_1\left(1 - \frac{u_2}{1+u_1^2}\right), \end{aligned} \tag{4.1.2}$$

where u_1 and u_2 represent the dimensionless concentrations of I^- and ClO_2^- ions, respectively, and the parameters $a > 0$ and $b > 0$ depend on the empirical rate constants and the concentrations of the slow reactants.

For a given value of parameter a, system (4.1.2) undergoes a Hopf bifurcation at the parameter value $b = b_0 \equiv 3a/5 - 25/a$, such that when $b > b_0$ all trajectories spiral into the stable fixed point $(u_1^*, u_2^*) = (a/5, 1 + a^2/25)$, whereas for $b < b_0$, trajectories are attracted to an orbitally stable limit cycle. If we consider system (4.1.2) with the coefficient $a = 11$, Hopf bifurcation occurs for $b_0 = 238/55$, and an orbitally stable limit cycle takes place for $b = 2.1$ [38].

Next, we take into account the Birkhoff-Shaw chaotic attractor [39, 40] which is generated by the following system of differential equations:

$$\begin{aligned} x_1' &= 0.7x_2 + 10x_1(0.1 - x_2^2), \\ x_2' &= -x_1 + 0.25\sin(1.57t). \end{aligned} \tag{4.1.3}$$

The following system is obtained by perturbing system (4.1.2) using solutions of (4.1.3):

$$\begin{aligned} y_1' &= 11 - y_1 - \frac{4y_1y_2}{1+y_1^2} + 0.5\tan\left(\frac{x_1(t)}{2}\right), \\ y_2' &= 2.1y_1\left(1 - \frac{y_2}{1+y_1^2}\right) + 0.4x_2(t). \end{aligned} \tag{4.1.4}$$

The present chapter rigorously demonstrates that system (4.1.4) displays chaotic motions around the orbitally stable limit cycle of system (4.1.2). Figure 4.1a depicts the chaotic trajectory, $x(t)$, of (4.1.3), with $x_1(0) = 0.2$, $x_2(0) = 0.3$. If one substitutes $x(t)$ into (4.1.4), then the system admits a chaotic trajectory, $y(t)$, with $y_1(0) = 0.75$, $y_2(0) = 4.82$, as shown in Fig. 4.1b. That is, an entrainment by chaos is observed. Moreover, the irregular behavior of the y_2 coordinate over time is illustrated in Fig. 4.2.

Now, let us use the auxiliary system approach [26, 27] to investigate the couple (4.1.3) + (4.1.4) for generalized synchronization.

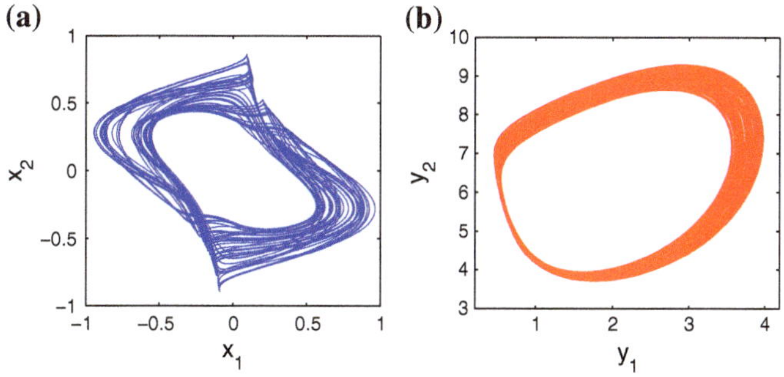

Fig. 4.1 The chaotic trajectory in **a** corresponds to system (4.1.3) and the irregular structure around the limit cycle in **b** is a manifestation of the entrainment by chaos

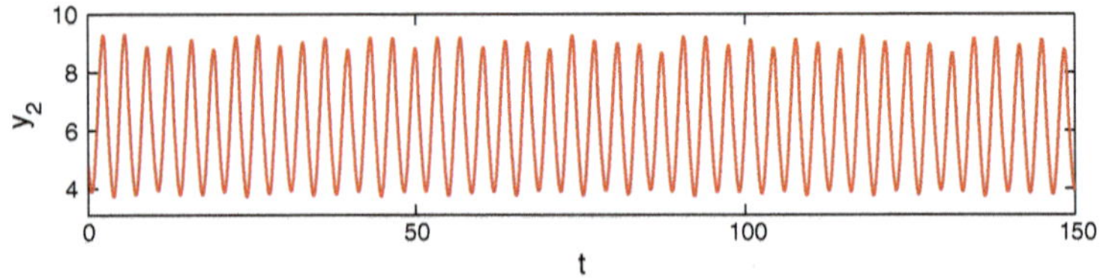

Fig. 4.2 The graph presents the irregular behavior of the y_2 coordinate and supports the existence of the entrainment by chaos

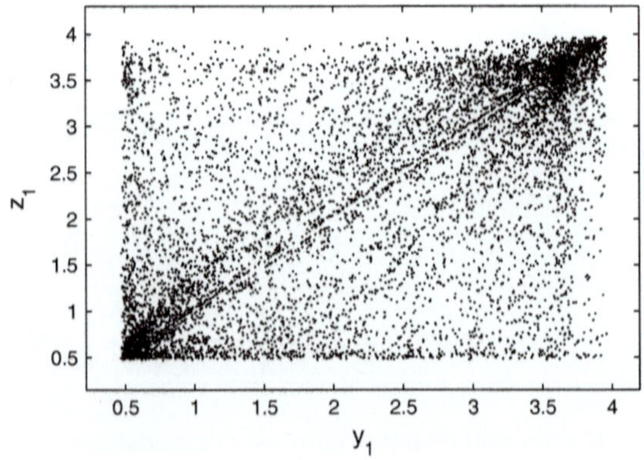

Fig. 4.3 The auxiliary system approach shows that the systems (4.1.3) and (4.1.4) are unsynchronized

Consider the auxiliary system

$$\begin{aligned} z_1' &= 11 - z_1 - \frac{4z_1z_2}{1+z_1^2} + 0.5\tan\left(\frac{x_1(t)}{2}\right), \\ z_2' &= 2.1z_1\left(1 - \frac{z_2}{1+z_1^2}\right) + 0.4x_2(t). \end{aligned} \tag{4.1.5}$$

By marking the trajectory of system (4.1.3)+(4.1.4)+(4.1.5) with initial data $x_1(0) = 0.2$, $x_2(0) = 0.3$, $y_1(0) = 0.75$, $y_2(0) = 4.82$, $z_1(0) = 2.92$, $z_2(0) = 8.78$ at times t that are integer multiples of $2\pi/1.57$ and omitting the first 200 iterations, we obtain the stroboscopic plot whose projection on the $y_1 - z_1$ plane is shown in Fig. 4.3. Since the plot is not placed on the line $z_1 = y_1$, we conclude that generalized synchronization does not occur in the couple (4.1.3)+(4.1.4).

4.2 Preliminaries

Throughout the chapter, $\mathbb{R}$ and $\mathbb{R}_+$ will denote the set of real numbers and the interval $[0, \infty)$, respectively. We will make use of the usual Euclidean norm for vectors and the norm induced by the Euclidean norm for square matrices [41].

Let us introduce the system

$$x' = F(t, x), \tag{4.2.6}$$

where $F : \mathbb{R}_+ \times \mathbb{R}^m \to \mathbb{R}^m$ is a continuous function in all its arguments. We treat system (4.2.6) as a source of chaos and thus call it the *generator* system.

Now, consider the system

$$u' = f(u), \tag{4.2.7}$$

where $f : \mathbb{R}^n \to \mathbb{R}^n$ is a continuously differentiable function. We subject system (4.2.7) to the entrainment mechanism in the following way:

$$y' = f(y) + \mu g(x), \tag{4.2.8}$$

such that we now consider the system

$$y' = f(y) + \mu g(x(t)), \tag{4.2.9}$$

where $x(t)$ are solutions of system (4.2.6), μ is a nonzero constant and $g : \mathbb{R}^m \to \mathbb{R}^n$ is a continuous function. Here, the couple (4.2.6)+(4.2.8) is a system with a skew product structure.

Remark 4.1 The results presented in the remaining sections are valid even if we replace the nonautonomous system (4.2.6) with the autonomous equation

$$x' = \overline{F}(x), \tag{4.2.10}$$

where $\overline{F} : \mathbb{R}^m \to \mathbb{R}^m$ is a continuous function.

The following conditions are required:

(A1) There exists a positive number L_f such that

$$\|f(y_1) - f(y_2)\| \le L_f \|y_1 - y_2\|,$$

for all $y_1, y_2 \in \mathbb{R}^n$;

(A2) There exists a positive number L_g such that

$$\|g(x_1) - g(x_2)\| \ge L_g \|x_1 - x_2\|,$$

for all $x_1, x_2 \in \mathbb{R}^m$;

(A3) There exist positive numbers M_F, M_f and M_g such that

$$\sup_{t \in \mathbb{R}_+, x \in \mathbb{R}^m} \|F(t, x)\| \le M_F, \ \sup_{y \in \mathbb{R}^n} \|f(y)\| \le M_f, \ \sup_{x \in \mathbb{R}^m} \|g(x)\| \le M_g.$$

For a given $x(t)$, the existence of solutions of system (4.2.9) as well as their continuation to $+\infty$, follow from the Lipschitz condition for the function f because the perturbation $\mu g(x(t))$ depends only on t and from the fact that the domain of the equation is the entire space $\mathbb{R}^n$ [42].

We mainly assume that system (4.2.6) ((4.2.10)) admits a chaotic attractor, let us say a set in $\mathbb{R}^m$ for (4.2.10). Fix x_0 from the attractor and take a solution $x(t)$ of (4.2.10) with $x(0) = x_0$. Since we use the solution $x(t)$ as a perturbation in system (4.2.8), we call it as *chaotic function*. The chaotic functions may be irregular as well as regular (periodic and unstable) [43–49].

We also assume that the nonlinear autonomous system (4.2.7) possesses a non-constant T-periodic solution $p(t)$ for some positive number T and consider system (4.2.8) in a neighborhood of the orbit

$$\gamma = \{\sigma \in \mathbb{R}^n : \sigma = p(t),\ t \in [0, T]\}. \tag{4.2.11}$$

It is clear that $p'(t)$ is a nontrivial T-periodic solution of the variational system

$$v' = A(t)v, \tag{4.2.12}$$

where $A(t) = \dfrac{\partial f(p(t))}{\partial u}$ is an $n \times n$ real, continuous, T-periodic matrix function, and consequently, 1 is a characteristic multiplier of system (4.2.12).

In what follows, we assume that 1 is a simple characteristic multiplier of the variational system (4.2.12) and the remaining $n-1$ characteristic multipliers are less than 1 in modulus. Under this assumption, according to the Andronov–Witt Theorem [50], the periodic solution $p(t)$ of system (4.2.7) is asymptotically orbitally stable, with the asymptotic phase property.

In the following, we will understand chaos in terms of sensitivity and the existence of infinitely many unstable periodic solutions in a bounded region.

4.3 Sensitivity

In this section, the sensitivity of system (4.2.9) will be extended. We begin by defining the sensitivity of Eq. (4.2.6) and continue with its replication in (4.2.8).

In what follows, for a given chaotic solution $x(t)$ of system (4.2.6), the function $\eta_{x(t)}(t, \eta_0)$ will represent the solution of (4.2.9) with $\eta_{x(t)}(0, \eta_0) = \eta_0$.

System (4.2.6) is called sensitive if there exist positive numbers ε_0 and Δ such that for an arbitrary positive number δ_0 and for each chaotic solution $x(t)$ of system (4.2.6), there exist a chaotic solution $\overline{x}(t)$ of the same system and an interval $J \subset \mathbb{R}_+$, with a length no less than Δ, such that $\|x(0) - \overline{x}(0)\| < \delta_0$ and $\|x(t) - \overline{x}(t)\| > \varepsilon_0$ for all $t \in J$.

We say that system (4.2.8) replicates the sensitivity of (4.2.6) if there exist positive numbers ε_1 and $\overline{\Delta}$ such that for an arbitrary positive number δ_1 and for each solution $\eta_{x(t)}(t, \eta_0)$, there exist an interval $J^1 \subset \mathbb{R}_+$, with a length no less than $\overline{\Delta}$, and a solution $\eta_{\overline{x}(t)}(t, \eta_1)$ such that

$$\|\eta_0 - \eta_1\| < \delta_1$$

and

$$\left\|\eta_{x(t)}(t, \eta_0) - \eta_{\overline{x}(t)}(t, \eta_1)\right\| > \varepsilon_1$$

for all $t \in J^1$.

Theorem 4.1 *If conditions* $(A1)-(A3)$ *hold, then system (4.2.8) replicates the sensitivity of system (4.2.6).*

Proof Fix an arbitrary positive number δ_1 and a solution $\eta_{x(t)}(t, \eta_0)$ of (4.2.9). Since system (4.2.6) is sensitive, there exist positive numbers ε_0 and Δ such that for arbitrary $\delta_0 > 0$ the inequalities $\|x(0) - \overline{x}(0)\| < \delta_0$ and $\|x(t) - \overline{x}(t)\| > \varepsilon_0$, $t \in J$, hold for some chaotic solution $\overline{x}(t)$ of (4.2.6) and for some interval $J \subset \mathbb{R}_+$ with length not less than Δ.

Now, let us fix arbitrary $\eta_1 \in \mathbb{R}^n$ such that $\|\eta_0 - \eta_1\| < \delta_1$. Our aim is to determine positive numbers ε_1, $\overline{\Delta}$ and an interval J^1 with length $\overline{\Delta}$, such that $\left\|\eta_{x(t)}(t, \eta_0) - \eta_{\overline{x}(t)}(t, \eta_1)\right\| > \varepsilon_1$ for all $t \in J^1$.

Suppose that $g(x) = (g_1(x), g_2(x), \ldots, g_n(x))$, where each g_j, $1 \le j \le n$, is a real valued function.

Since for each chaotic solution $x(t)$ of (4.2.6) the inequality $\sup_{t \in \mathbb{R}_+} \left\|x'(t)\right\| \le M_F$ holds, one can conclude that the collection of chaotic solutions of Eq. (4.2.6) constitutes an equicontinuous family on $\mathbb{R}_+$. According to our assumption that system (4.2.6) possesses a chaotic attractor, there exists a positive number M such that $\sup_{t \in \mathbb{R}_+} \|x(t)\| \le M$ for each chaotic solution of (4.2.6). Making use of the uniform continuity of the function $\overline{g} : \mathbb{R}^m \times \mathbb{R}^m \to \mathbb{R}^n$, defined as $\overline{g}(x_1, x_2) = g(x_1) - g(x_2)$, on the compact region $\{(x_1, x_2) \in \mathbb{R}^m \times \mathbb{R}^m : \|x_1\| \le M, \|x_2\| \le M\}$, together with the equicontinuity of the collection of chaotic solutions of system (4.2.6), one can verify that the set consisting of the elements of the form $g_j(x(t)) - g_j(\overline{x}(t))$, $1 \le j \le n$, where $x(t)$ and $\overline{x}(t)$ are chaotic solutions of (4.2.6), is an equicontinuous family on $\mathbb{R}_+$. Therefore, there exists a positive number $\tau < \Delta$, independent of the functions $x(t)$ and $\overline{x}(t)$, such that for any $t_1, t_2 \in \mathbb{R}_+$ with $|t_1 - t_2| < \tau$, the inequality

$$\left|\left(g_j\left(x(t_1)\right) - g_j\left(\overline{x}(t_1)\right)\right) - \left(g_j\left(x(t_2)\right) - g_j\left(\overline{x}(t_2)\right)\right)\right| < \frac{L_g \varepsilon_0}{2n}$$

holds for all $1 \le j \le n$.

Condition $(A2)$ implies for all $t \in J$ that $\|g(x(t)) - g(\overline{x}(t))\| \ge L_g \|x(t) - \overline{x}(t)\|$. Thus, for each $t \in J$, there exists an integer j_0, $1 \le j_0 \le n$, which possibly depends on t, such that

$$\left|g_{j_0}(x(t)) - g_{j_0}(\overline{x}(t))\right| \geq \frac{L_g}{n} \|x(t) - \overline{x}(t)\| \, .$$

Let s_0 be the midpoint of the interval J and $\theta = s_0 - \tau/2$. One can find an integer $j_0 = j_0(s_0)$, $1 \leq j_0 \leq n$, such that

$$\left|g_{j_0}(x(s_0)) - g_{j_0}(\overline{x}(s_0))\right| > \frac{L_g \varepsilon_0}{n}. \tag{4.3.13}$$

On the other hand, for all $t \in [\theta, \theta + \tau]$ we have

$$\left|g_{j_0}\,(x(s_0)) - g_{j_0}\,(\overline{x}(s_0))\right| - \left|g_{j_0}\,(x(t)) - g_{j_0}\,(\overline{x}(t))\right| < \frac{L_g \varepsilon_0}{2n}$$

and by means of (4.3.13) we obtain

$$\left|g_{j_0}\,(x(t)) - g_{j_0}\,(\overline{x}(t))\right| > \frac{L_g \varepsilon_0}{2n}, \; t \in [\theta, \theta + \tau] \, .$$

The last inequality implies that

$$\left\| \int_{\theta}^{\theta+\tau} [g(x(s)) - g(\overline{x}(s))] \, ds \right\| > \frac{\tau L_g \varepsilon_0}{2n}. \tag{4.3.14}$$

Using the inequality

$$\begin{aligned} \left\|\eta_{x(t)}(\theta+\tau, \eta_0) - \eta_{\overline{x}(t)}(\theta+\tau, \eta_1)\right\| &\geq |\mu| \left\| \int_{\theta}^{\theta+\tau} [g(x(s)) - g(\overline{x}(s))] ds \right\| \\ - \left\|\eta_{x(t)}(\theta, \eta_0) - \eta_{\overline{x}(t)}(\theta, \eta_1)\right\| &- \int_{\theta}^{\theta+\tau} L_f \left\|\eta_{x(t)}(s, \eta_0) - \eta_{\overline{x}(t)}(s, \eta_1)\right\| ds \end{aligned}$$

together with (4.3.14), one can verify that

$$\max_{t \in [\theta, \theta+\tau]} \left\|\eta_{x(t)}(t, \eta_0) - \eta_{\overline{x}(t)}(t, \eta_1)\right\| > \frac{|\mu| \, \tau L_g \varepsilon_0}{2n(2 + \tau L_f)}.$$

Suppose that the function $\left\|\eta_{x(t)}(t, \eta_0) - \eta_{\overline{x}(t)}(t, \eta_1)\right\|$ takes its maximum on the interval $[\theta, \theta + \tau]$ at the point ξ.

Let us define

$$\overline{\Delta} = \min \left\{ \frac{\tau}{2}, \frac{|\mu| \, \tau L_g \varepsilon_0}{8n(M_f + M_g \, |\mu|)(2 + \tau L_f)} \right\}$$

and

$$\theta^1 = \begin{cases} \xi, & \text{if } \xi \leq \theta + \tau/2 \\ \xi - \overline{\Delta}, & \text{if } \xi > \theta + \tau/2 \end{cases}.$$

We note that the interval $J^1 = [\theta^1, \theta^1 + \overline{\Delta}]$ is a subset of J. For $t \in J^1$, it can be verified that

$$\left\|\eta_{x(t)}(t, \eta_0) - \eta_{\overline{x}(t)}(t, \eta_1)\right\| > \varepsilon_1, \tag{4.3.15}$$

where $\varepsilon_1 = \dfrac{|\mu|\,\tau L_g \varepsilon_0}{4n(2+\tau L_f)}$, and the length $\overline{\Delta}$ of the interval J^1 does not depend on $x(t)$ and $\overline{x}(t)$. Consequently, system (4.2.8) replicates the sensitivity of system (4.2.6). □

4.4 Unstable Periodic Solutions

We begin this section by describing period-doubling cascade for system (4.2.6) and continue with its extension to system (4.2.9) through system (4.2.8).

Assume, in this section, that system (4.2.6) admits a period-doubling cascade. That is, there exists an equation

$$x' = G(t, x, \lambda), \tag{4.4.16}$$

where λ is a parameter and the function $G : \mathbb{R}_+ \times \mathbb{R}^m \times \mathbb{R} \to \mathbb{R}^m$ is such that for some finite number λ_∞, $G(t, x, \lambda_\infty)$ is equal to the function $F(t, x)$ in the right-hand side of system (4.2.6).

The following condition is required:

(A4) There exists a positive number ω such that the periodicity property $G(t + \omega, x, \lambda) = G(t, x, \lambda)$ holds for all $t \in \mathbb{R}_+$, $x \in \mathbb{R}^m$ and $\lambda \in \mathbb{R}$.

System (4.2.6) is said to admit a period-doubling cascade [43, 45, 46, 48, 49] if there exist a natural number k_0 and a sequence of period-doubling bifurcation values $\{\lambda_j\}$, $\lambda_j \to \lambda_\infty$ as $j \to \infty$, such that for each natural number j, a periodic solution with period $k_0 2^j \omega$ appears, and as the parameter λ increases or decreases through λ_j, system (4.4.16) undergoes a period-doubling bifurcation. As a consequence, at the parameter value $\lambda = \lambda_\infty$, there exist infinitely many unstable periodic solutions of system (4.4.16), and hence of system (4.2.6), all lying in a bounded region.

Now, let us introduce the following definition [51]. We say that the solutions of the nonautonomous system (4.2.9), with a fixed $x(t)$, are ultimately bounded if there exists a number $B > 0$ such that for every solution $y(t)$, $y(t_0) = y_0$, of system (4.2.9), there exists a positive number R such that the inequality $\|y(t)\| < B$ holds for all $t \geq t_0 + R$.

The following condition is required in the next theorem, which can be verified using Theorem 15.8 [51].

(A5) Solutions of system (4.2.9) are ultimately bounded by a bound common for all $x(t)$.

We say that system (4.2.8) replicates the period-doubling cascade of system (4.2.6) if for each periodic solution $x(t)$ of (4.2.6), system (4.2.9) admits a periodic solution with the same period.

Theorem 4.2 *If conditions* $(A1)-(A5)$ *hold, then system (4.2.8) replicates the period-doubling cascade of system (4.2.6).*

We emphasize that the instability of the infinite number of periodic solutions of system (4.2.8) is ensured by Theorem 4.1. Condition $(A5)$ can be verified directly, for example, by using Lyapunov functions, as in the case of system (4.6.26) presented in Sect. 4.6.

4.5 Main Result

Let H be the set of all solutions $\eta_{x(t)}(t, \eta_0)$ of (4.2.9). Based on the previous results, one can say that solutions in H are sensitive and there are infinitely many unstable periodic solutions in the set; that is, H is chaotic.

We say that the entrainment of the limit cycle by chaos is observed in system (4.2.8) if there exists a neighborhood N of γ where the chaos is developed. Moreover, there exists an open ball in $\mathbb{R}^n$ centered at $p(0)$ such that each chaotic solution $\eta_{x(t)}(t, \eta_0)$ that starts inside the ball remains in N for all $t \geq 0$.

Theorem 4.3 *Suppose that conditions* $(A1)-(A5)$ *hold. If* $|\mu|$ *is sufficiently small, then there is an entrainment of system (4.2.8) by the chaos.*

Proof Assume, without loss of generality, that $p(0) = 0$ and

$$p'(0) = (\bar{p}_1, 0, 0, \ldots, 0)$$

for some positive number $\bar{p}_1$. At first, we are going to show that for sufficiently small $|\mu|$, the solutions of system (4.2.9) remain and rotate in a neighborhood of the limit cycle. That is, $\eta_{x(t)}(\theta_i, \eta_0)$ belongs to a neighborhood of the origin, if $\|\eta_0\|$ is sufficiently small, for a sequence $\theta_i \to \infty$ as $i \to \infty$ with uniformly bounded $\theta_{i+1} - \theta_i$.

Let us denote by $\zeta(t, \zeta_0)$ the solution of Eq. (4.2.7) with $\zeta(0, \zeta_0) = \zeta_0$. There exists a hypersurface S such that the orbit γ of the periodic solution $p(t)$ intersects this surface transversally, as shown in Sect. 4.8. Therefore, there exists a number $\varepsilon_1 > 0$ such that if $\|\zeta(t, \zeta_0) - p(t)\| < \varepsilon_1$ for each $t \in [0, 2T]$, then $\zeta(t, \zeta_0)$ intersects S at some moment $t_1 \in [0, 2T]$.

Suppose that a positive number $\delta = \delta(\varepsilon_1)$ is chosen such that $\delta \leq \varepsilon_1 e^{-2L_f T}$. Throughout the proof, B_δ will stand for the open ball in $\mathbb{R}^n$ centered at the origin with radius δ. Let an arbitrary $\zeta_0 \in B_\delta$ be given.

The solutions $\zeta(t, \zeta_0)$ and $p(t) = \zeta(t, 0)$ satisfy the relation

$$\|\zeta(t, \zeta_0) - p(t)\| \leq \|\zeta_0\| + \int_0^t L_f \, \|\zeta(s, \zeta_0) - p(s)\| \, ds.$$

Using $\|\zeta_0\| < \delta$, one can verify that if $0 \leq t \leq 2T$ then $\|\zeta(t, \zeta_0) - p(t)\| < \delta e^{2L_f T}$. Therefore,

$$\|\zeta(t, \zeta_0) - p(t)\| < \varepsilon_1$$

for $t \in [0, 2T]$, and $\zeta_1 = \zeta(t_1(\zeta_0), \zeta_0)$ belongs to S for some $t_1(\zeta_0) \in [0, 2T]$. It is clear that $\|\zeta_1\| < R$, where $R = \varepsilon_1 + \rho$ and $\rho = \max_{t \in [0,T]} \|p(t)\|$.

Now, let us fix an arbitrary number $l \in (0, 1)$. In accordance with inequality (4.8.48), presented in Sect. 4.8, there exists a natural number $n_0 = n_0\left(\frac{l\delta}{R}\right)$, independent of ζ_0, such that

$$\|\zeta(n_0 T + t_1(\zeta_0), \zeta_0)\| < l\delta. \tag{4.5.17}$$

Let $\varepsilon = \delta\left(\dfrac{1-l}{2}\right)$ and suppose that the nonzero number $|\mu|$ is sufficiently small so that $|\mu| < \dfrac{\varepsilon L_f}{M_g\left[e^{L_f(n_0+2)T} - 1\right]}$.

Take a solution $\eta_{x(t)}(t, \eta_0)$, with $\eta_0 \in B_\delta$. By the previous discussions, there exists a number $t_1(\eta_0) \in [0, 2T]$ such that $\zeta(t_1(\eta_0), \eta_0)$ belongs to S.

It can be verified that

$$\left\|\eta_{x(t)}(t, \eta_0) - \zeta(t, \eta_0)\right\| \leq |\mu| \, M_g t + \int_0^t L_f \left\|\eta_{x(t)}(s, \eta_0) - \zeta(s, \eta_0)\right\| ds,$$

and therefore, we have

$$\left\|\eta_{x(t)}(t, \eta_0) - \zeta(t, \eta_0)\right\| \leq \frac{|\mu| \, M_g}{L_f}\left(e^{L_f t} - 1\right), \; t \leq (n_0 + 2)T.$$

In this case, we obtain the inequality

$$\left\|\eta_{x(t)}(n_0 T + t_1(\eta_0), \eta_0) - \zeta(n_0 T + t_1(\eta_0), \eta_0)\right\| < \varepsilon,$$

and by means of (4.5.17) we have $\eta_1 = \eta_{x(t)}(\theta_1, \eta_0) \in B_\delta$, where $\theta_1 = n_0 T + t_1(\eta_0)$. We note that the point η_1 depends on both η_0 and $x(t)$.

Similarly to the above, one can find that the inequality

$$\left\|\eta_{x(t)}(t, \eta_0) - \zeta(t - \theta_1, \eta_1)\right\| \leq \frac{|\mu| \, M_g}{L_f}\left(e^{L_f(n_0+2)T} - 1\right)$$

holds for all $t \in [\theta_1, \theta_1 + (n_0 + 2)T]$. Additionally, the existence of a number $t_2(\eta_1) \in [0, 2T]$ such that $\zeta(t_2(\eta_1), \eta_1) \in S$ can be verified. Therefore, we have

$$\left\|\eta_{x(t)}(2n_0T + t_1(\eta_0) + t_2(\eta_1), \eta_0) - \zeta(n_0T + t_2(\eta_1), \eta_1)\right\| < \varepsilon,$$

and hence $\eta_2 = \eta_{x(t)}(\theta_2, \eta_0) \in B_\delta$, where $\theta_2 = 2n_0T + t_1(\eta_0) + t_2(\eta_1)$.

One can continue in the same manner to construct a sequence $\{t_j\}$, which satisfies $0 \le t_j \le 2T$, $j \ge 1$, and

$$\left\|\eta_{x(t)}(t, \eta_0) - \zeta(t - \theta_i, \eta_i)\right\| \le \frac{|\mu|\, M_g}{L_f}\left(e^{L_f(n_0+2)T} - 1\right), \qquad (4.5.18)$$

for $t \in [\theta_i, \theta_i + (n_0 + 2)T]$, where $\eta_i = \eta_{x(t)}(\theta_i, \eta_0) \in B_\delta$, $i \ge 0$, $\theta_0 = 0$ and

$$\theta_i = in_0T + \sum_{j=1}^{i} t_j, \ i \ge 1. \qquad (4.5.19)$$

We emphasize that for any $i \ge 1$, it is true that θ_i belongs to $[in_0T, i(n_0 + 2)T]$ and $\theta_i - \theta_{i-1} = n_0T + t_i \le (n_0 + 2)T$. The procedure of the proof for $t \in [\theta_i, \theta_{i+1}]$ is illustrated in Fig. 4.4.

In the remaining part of the proof, we will demonstrate the boundedness of $\eta_{x(t)}(t, \eta_0) - p(t)$, which implies the boundedness of $\eta_{x(t)}(t, \eta_0)$.

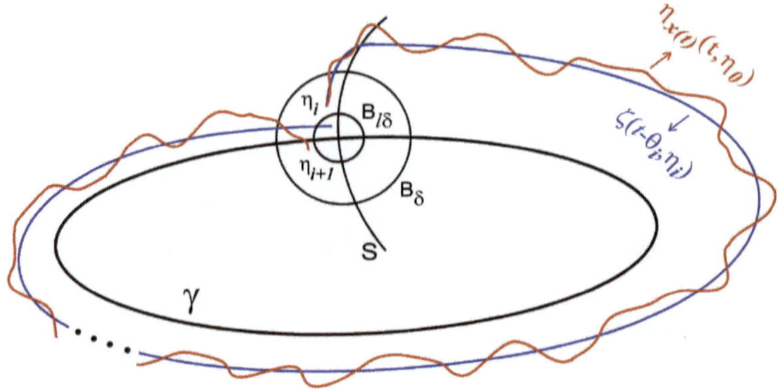

Fig. 4.4 The schematic representation of the proof of Theorem 4.3. The trajectory in *red* shows the function $\eta_{x(t)}(t, \eta_0)$, while the trajectory in *blue* represents $\zeta(t - \theta_i, \eta_i)$, where the sequence $\{\theta_i\}$ is defined in (4.5.19) and $\eta_i = \eta_{x(t)}(\theta_i, \eta_0)$. The presented illustration covers the way of the 2-dimensional case of the proof on the time interval $[\theta_i, \theta_{i+1}]$, for an arbitrary $i \ge 0$. In the figure $B_{l\delta}$ and B_δ denote the open balls centered at the origin with radii $l\delta$ and δ, respectively. At the moment $t = \theta_{i+1}$, the solution $\zeta(t - \theta_i, \eta_i)$ belongs to $B_{l\delta}$ and η_{i+1} is inside B_δ

For a fixed i, using the couple of relations $\zeta(t-\theta_i, \eta_i) = \eta_i + \int_{\theta_i}^{t} f(\zeta(s-\theta_i, \eta_i))ds$ and $p(t) = p(\theta_i) + \int_{\theta_i}^{t} f(p(s))ds$ one can obtain the following inequality

$$\|\zeta(t-\theta_i, \eta_i) - p(t)\| \le (\delta + \rho)e^{2L_f T}, \ \theta_i \le t \le \theta_i + 2T.$$

Hence, we have $\|\zeta(t_{i+1}(\eta_i), \eta_i)\| \le \rho + (\delta + \rho)e^{2L_f T}$.

Since the point $\zeta(t_{i+1}(\eta_i), \eta_i)$ is on the surface S, according to (4.8.50), it is true for $t \in \mathbb{R}_+$ that $\|\zeta(t + t_{i+1}(\eta_i), \eta_i) - p(t)\| \le 4K_1 \left\|P^{-1}(0)\right\| \left[\rho + (\delta + \rho)\, e^{2L_f T}\right]$. Thus, we find for $t \in \mathbb{R}_+$ that

$$\left\|\eta_{x(t)}(t, \eta_0) - p(t)\right\| \le \frac{|\mu|\, M_g}{L_f}\left(e^{L_f(n_0+2)T} - 1\right) + H_0(\delta, \rho), \quad (4.5.20)$$

where

$$H_0(\delta, \rho) = \max\left\{(\delta + \rho)e^{2L_f T}, 4K_1 \left\|P^{-1}(0)\right\| \left[\rho + (\delta + \rho)\, e^{2L_f T}\right]\right\}.$$

It is worth noting that $H_0(\delta, \rho) \to 0$ as $\delta \to 0$ and $\rho \to 0$, and $\left\|\eta_{x(t)}(t, \eta_0) - p(t)\right\|$ can be made arbitrarily small by suitable choices of μ, δ and ρ.

Consequently, any solution $\eta_{x(t)}(t, \eta_0)$, where $\eta_0 \in B_\delta$, is bounded on $\mathbb{R}_+$, and remains near the limit cycle in accordance with formula (4.5.20).

In compliance with the results of Theorems 4.1 and 4.2, the set H exhibits sensitivity and contains infinitely many unstable periodic solutions. For each chaotic $x(t)$, the trajectories of (4.2.9) starting inside the ball B_δ constitute a subfamily of H and behave chaotically around the limit cycle γ. Therefore, the entrainment of the limit cycle by chaos takes place in system (4.2.8). □

Given the presence of chaos in (4.2.6), we have obtained chaos for the couple $(x(t), y(t))$, so that one can talk not only of the entrainment by chaos, but also of the extension of chaos to a higher dimensional system.

4.6 Examples

We consider the system

$$\begin{aligned} u_1' &= \alpha u_1 - u_2 - u_1(u_1^2 + u_2^2), \\ u_2' &= u_1 + \alpha u_2 - u_2(u_1^2 + u_2^2), \end{aligned} \quad (4.6.21)$$

which is in the form of (4.2.7), where α is a positive number and

$$f(u_1, u_2) = \begin{pmatrix} \alpha u_1 - u_2 - u_1(u_1^2 + u_2^2) \\ u_1 + \alpha u_2 - u_2(u_1^2 + u_2^2) \end{pmatrix}.$$

One can verify that $p(t) = \left(\sqrt{\alpha}\cos t, \sqrt{\alpha}\sin t\right)$ is a periodic solution of (4.6.21). Evaluating $A(t) = \dfrac{\partial f(p(t))}{\partial u}$ gives us

$$A(t) = \begin{pmatrix} -2\alpha\cos^2 t & -1-\alpha\sin(2t) \\ 1-\alpha\sin(2t) & -2\alpha\sin^2 t \end{pmatrix}. \tag{4.6.22}$$

Evidently, the first multiplier of the corresponding variational system is $\rho_1 = 1$, and according to Lemma 7.3 [52] $\rho_2 = \exp\left(\int_0^{2\pi} trA(s)ds\right) = e^{-4\pi\alpha}$. Thus, the periodic solution $p(t)$ is asymptotically orbitally stable according to Andronov–Witt Theorem.

As the generator we will make use of Duffing equations in the form

$$x'' + D_1x' + D_2x^3 = \lambda\cos t, \tag{4.6.23}$$

where D_1, D_2 and λ are constants. Defining the variables $x_1 = x$ and $x_2 = x'$, Eq. (4.6.23) can be rewritten as

$$\begin{aligned} x_1' &= x_2, \\ x_2' &= -D_1x_2 - D_2x_1^3 + \lambda\cos t. \end{aligned} \tag{4.6.24}$$

Example 4.1 Consider system (4.6.24) with $D_1 = 0.05$, $D_2 = 1$ and $\lambda = 7.5$ such that the system possesses chaotic motions seen through simulations [40]. Perturbing system (4.6.21) with solutions of (4.6.24) and setting $\alpha = 9$, we obtain the following system

$$\begin{aligned} y_1' &= 9y_1 - y_2 - y_1(y_1^2 + y_2^2) + 0.5x_1(t), \\ y_2' &= y_1 + 9y_2 - y_2(y_1^2 + y_2^2) + 3.6x_2(t). \end{aligned} \tag{4.6.25}$$

In Fig. 4.5a, b, we depict the chaotic trajectories of systems (4.6.24) and (4.6.25), respectively. The initial data $x_1(0) = 3.05$, $x_2(0) = 4.153$, $y_1(0) = 2.8$, $y_2(0) = 0.5$ are used. Figure 4.5b shows the chaotic motion in a neighborhood of the limit cycle of (4.6.21). The pictures support the results of the present chapter predicting the entrainment by chaos.

Figure 4.6 depicts the Poincaré sections, which are obtained by marking the trajectories of systems (4.6.24) and (4.6.25) with $x_1(0) = 2$, $x_2(0) = 3$, $y_1(0) = 3$, $y_2(0) = 0$ stroboscopically at times t that are integer multiples of 2π. Figure 4.6a presents the strange attractor of the first system, and Fig. 4.6b demonstrates the entrained chaotic behavior.

Now, to show through simulations the replication of sensitivity, we consider two initially close solutions of system (4.6.24) + (4.6.25), one with the initial data $x_1(0) = 3.07$, $x_2(0) = 4.18$, $y_1(0) = 1.57$, $y_2(0) = -0.25$, which is presented in blue, and another with the initial data $x_1(0) = 3.22$, $x_2(0) = 4.14$, $y_1(0) = 1.35$, $y_2(0) = -0.22$, which is pictured in red. In Fig. 4.7, we present these trajectories. Figure 4.7a

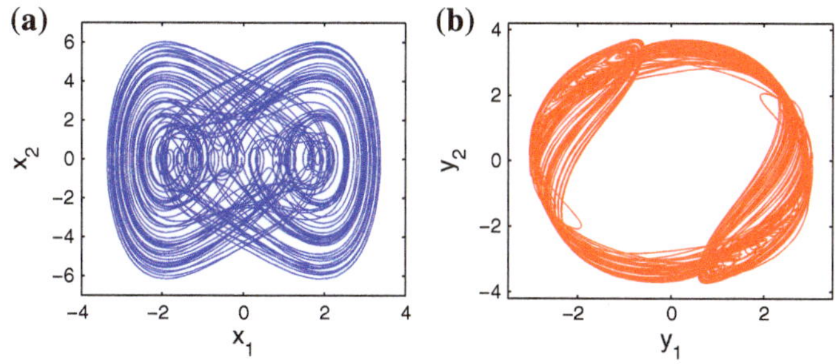

Fig. 4.5 The chaotic behavior of system (4.6.24) is pictured in (**a**), and the chaotic motion generated around the limit cycle is shown in (**b**)

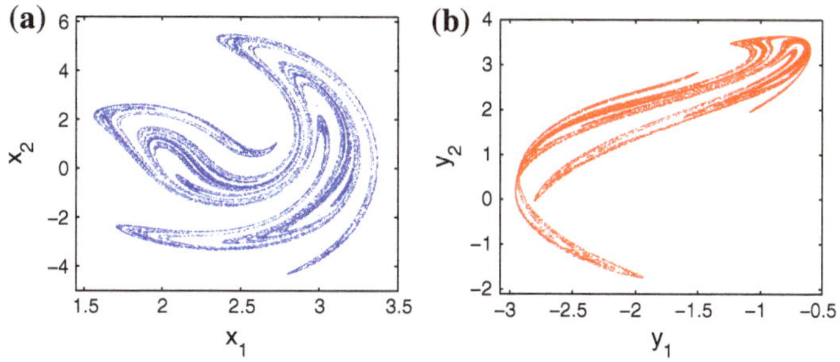

Fig. 4.6 Poincaré sections of systems (4.6.24) and (4.6.25)

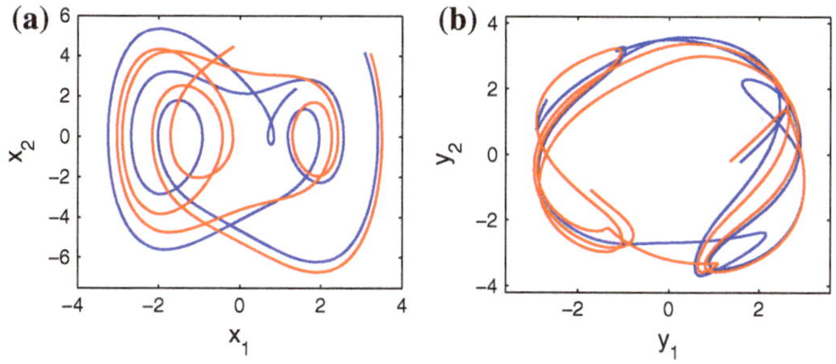

Fig. 4.7 The picture in **a** shows two initially close trajectories corresponding to system (4.6.24) that eventually diverge. The replication of sensitivity is observed in (**b**), where the *blue* and *red* trajectories are initially close to each other and are then separated

shows the existence of sensitivity in system (4.6.24), while Fig. 4.7b illustrates the replication of this feature.

Formula (4.3.15) implies that the strength of sensitivity of system (4.2.8) is proportional to the strength of the chaotic perturbation, $\mu g(x)$, used in the system. Therefore, despite the fact that the extension of sensitivity is guaranteed by Theorem 4.1, if one considers (4.2.8) with weak perturbations, it may not be visible in simulation results. On the other hand, according to formula (4.5.18), strong perturbations may diminish the cyclical behavior of the chaotic solutions. For that reason, given the strength of the perturbation used in system (4.6.25), Fig. 4.7b displays the extension of sensitivity, but does not indicate cyclical behavior.

We will continue with an example that demonstrates the extension of period-doubling cascade.

Example 4.2 In paper [53], it is mentioned that system (4.6.24), in which λ is considered as a parameter, displays period-doubling bifurcations for the coefficients $D_1 = 0.3$, $D_2 = 1$, and the sequence of bifurcation parameter values accumulates at $\lambda = \lambda_\infty \equiv 40$ such that the system admits infinitely many unstable periodic orbits.

To illustrate the entrainment by chaos, system (4.6.24) with the specified coefficients and $\lambda = \lambda_\infty$ will be utilized as the generator. Let us use the solutions of (4.6.24) to perturb (4.6.21) and build the system

$$\begin{aligned} y_1' &= \alpha y_1 - y_2 - y_1\left(y_1^2 + y_2^2\right) + \mu x_1(t), \\ y_2' &= y_1 + \alpha y_2 - y_2\left(y_1^2 + y_2^2\right) + \mu x_2(t), \end{aligned} \tag{4.6.26}$$

where μ is a nonzero constant.

We will make use of the Lyapunov function $V(y_1, y_2) = y_1^2 + y_2^2$ to show the validity of condition $(A5)$ for system (4.6.26). One can verify that

$$\begin{aligned} V'_{(4.6.26)}(y_1, y_2) = -2\sqrt{y_1^2+y_2^2}\Big[& \left(y_1^2 + y_2^2 - \alpha\right)\sqrt{y_1^2+y_2^2} \\ & - \frac{\mu}{\sqrt{y_1^2+y_2^2}}(x_1 y_1 + x_2 y_2)\Big]. \end{aligned}$$

Let us fix a positive number r_0 and suppose that $\sqrt{y_1^2+y_2^2} > \sqrt{\alpha} + r_0$. Under this condition we have $\left(y_1^2 + y_2^2 - \alpha\right)\sqrt{y_1^2+y_2^2} > r_0^3 + 3r_0^2\sqrt{\alpha} + 2r_1\alpha$. Since the chaotic attractor of system (4.6.24) satisfies $|x_1| < 6$ and $|x_2| < 15$, we obtain that $\left|\frac{\mu}{\sqrt{y_1^2+y_2^2}}(x_1 y_1 + x_2 y_2)\right| \le 21\,|\mu|$. Therefore, if $|\mu|$ is sufficiently small so that $|\mu| \le r_0^3/21$, then $V'_{(4.6.26)}(y_1, y_2) < 0$ for $\sqrt{y_1^2+y_2^2} > \sqrt{\alpha} + r_1$, and condition $(A5)$ holds for system (4.6.26).

In conformity with the discussion above, one can identify a bounded region G in $\mathbb{R}^2$ such that for sufficiently small $|\mu|$, Massera's Theorem [51, 54] implies the existence of a periodic solution of the planar system (4.6.26) inside the region G for each periodic $(x_1(t), x_2(t))$. Moreover, all these periodic solutions are unstable.

In Fig. 4.8, the trajectories of systems (4.6.24) and (4.6.26), where $\alpha = 0.002$ and $\mu = 0.008$, with $x_1(0) = 3.5$, $x_2(0) = -2$, $y_1(0) = 0.02$, $y_2(0) = 0.038$ are seen. Figure 4.8a illustrates the chaotic behavior of system (4.6.24) and Fig. 4.8b shows the irregular motion around the limit cycle.

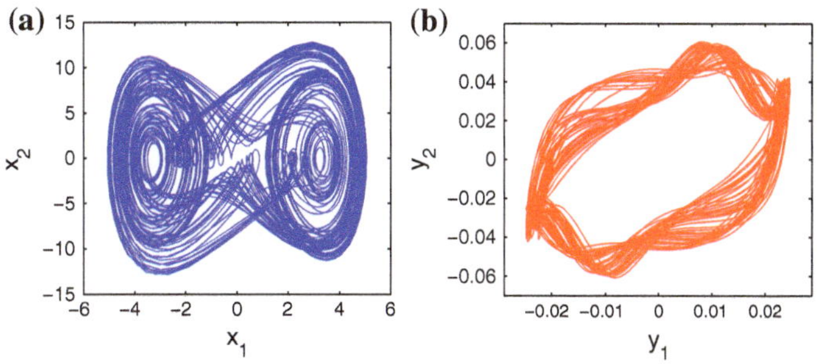

Fig. 4.8 The chaotic trajectories of the unidirectionally coupled systems (4.6.24) and (4.6.26). The coefficients $D_1 = 0.3$, $D_2 = 1$, $\lambda = 40$, $\alpha = 0.002$ and $\mu = 0.008$ are used in the simulation

4.7 Miscellany

This section is devoted to discussions and simulations of the entrainment of toroidal attractors by chaos and entrainment in Chua's oscillators, as well as controlling and synchronization problems. We start with the demonstration of chaos generation around tori.

4.7.1 Chaotic Tori

In previous parts of the chapter, we have discussed the entrainment of limit cycles by chaos. Now, the question is whether a similar approach is possible around tori. In this part, we will investigate numerically the problem of capture of chaos by toroidal attractors.

Let us consider the following system [55, 56]:

$$\begin{aligned} u_1' &= (\lambda - 3)u_1 - 0.25u_2 + u_1\left(u_3 + 0.2(1 - u_3^2)\right), \\ u_2' &= 0.25u_1 + (\lambda - 3)u_2 + u_2\left(u_3 + 0.2(1 - u_3^2)\right), \\ u_3' &= \lambda u_3 - (u_1^2 + u_2^2 + u_3^2), \end{aligned} \tag{4.7.27}$$

where λ is a parameter.

For small and positive values of the parameter λ, system (4.7.27) admits an asymptotically stable equilibrium point with a positive u_3 coordinate close to the origin. At $\lambda \approx 1.68$, the equilibrium point loses its stability and an hyperbolic, asymptotically orbitally stable limit cycle emerges. At the parameter value $\lambda = 2$, the periodic orbit is still asymptotically orbitally stable, but not hyperbolic. For $\lambda > 2$, the limit cycle is no longer stable and an attracting invariant torus is formed near the periodic orbit. With the increasing values of λ, the invariant torus grows rapidly [55].

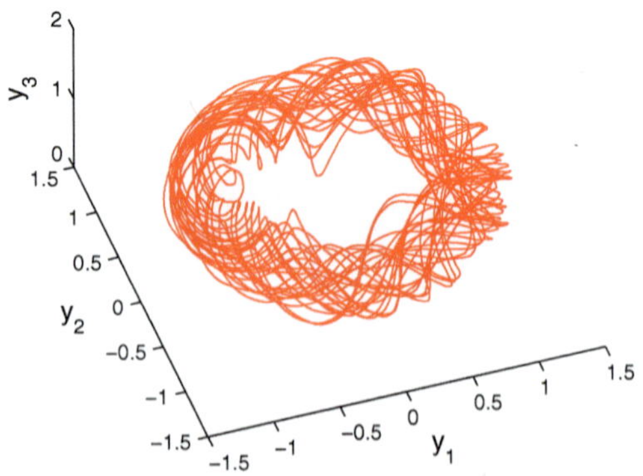

Fig. 4.9 The emergence of chaotic motion around a torus demonstrates the entrainment by chaos

To produce chaotic motions around the torus, we use the chaotic Lorenz system [17]

$$
\begin{aligned}
x_1' &= -10x_1 + 10x_2,\\
x_2' &= -x_1x_3 + 28x_1 - x_2,\\
x_3' &= x_1x_2 - (8/3)x_3,
\end{aligned} \tag{4.7.28}
$$

as the generator and set up the following system:

$$
\begin{aligned}
y_1' &= (\lambda - 3)y_1 - 0.25y_2 + y_1\left(y_3 + 0.2(1 - y_3^2)\right) + 0.003x_1(t),\\
y_2' &= 0.25y_1 + (\lambda - 3)y_2 + y_2\left(y_3 + 0.2(1 - y_3^2)\right) + 0.004x_2(t),\\
y_3' &= \lambda y_3 - (y_1^2 + y_2^2 + y_3^2) + 0.002x_3(t),
\end{aligned} \tag{4.7.29}
$$

where $\lambda = 2.003$.

Figure 4.9 shows the trajectory of system (4.7.29) with $y_1(0) = 0.0793$, $y_2(0) = -1.1761$, $y_3(0) = 0.9449$, where $x(t)$ is a solution of (4.7.28) with $x_1(0) = -6.7453$, $x_2(0) = 0.3435$, $x_3(0) = 32.7629$. One can see that the motion is chaotic and surrounds the torus.

4.7.2 Entrainment in Chua's Oscillators

We continue the discussion by presenting a simulation result for the entrainment by chaos in a Chua's oscillator.

Using system (4.6.24) with $D_1 = 0.3$, $D_2 = 1$ and $\lambda = 40$ as the generator of chaos, we demonstrated in Sect. 4.6 that system (4.6.26) with $\alpha = 0.002$ and $\mu = 0.008$ exhibits motions which behave chaotically and cyclically, so that the entrainment by chaos is present. Now, we consider a Chua's oscillator which admits an asymptotically stable equilibrium in dimensionless form [57], and perturb it with the solutions of (4.6.26):

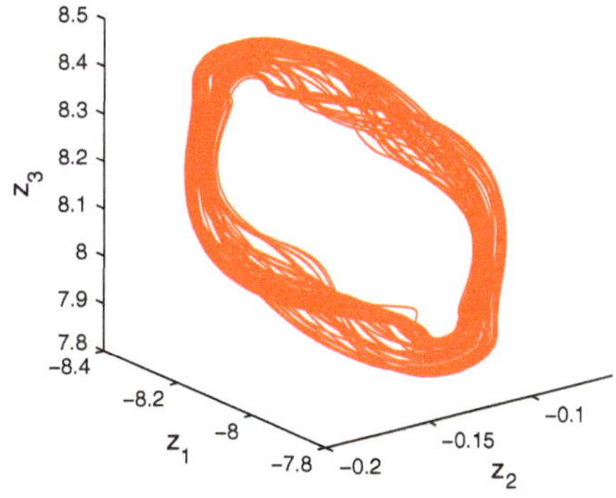

Fig. 4.10 The chaotic and cyclic motion generated by the perturbed Chua system (4.7.30)

$$\begin{aligned} z_1' &= (21.32/5.75)[z_2 - 0.13396 z_1 + 0.48993(|z_1+1| + |z_1-1|)] \\ &\quad +0.5 y_1(t), \\ z_2' &= z_1 - z_2 + z_3 + 2y_2(t), \\ z_3' &= -7.8351 z_2 - (1.38166392/12) z_3 + 3y_2(t). \end{aligned} \tag{4.7.30}$$

Note that system (4.6.26), which itself is a perturbed system, is the generator. According to [25], we have to observe chaotic behavior in the oscillator. We consider a trajectory of system (4.6.24)+(4.6.26) with initial data $x_1(0) = 3.5$, $x_2(0) = -2$, $y_1(0) = 0.02$, $y_2(0) = 0.038$, and plot the corresponding trajectory of (4.7.30) with $z_1(0) = -8.016$, $z_2(0) = -0.084$, $z_3(0) = 7.792$ in Fig. 4.10. It confirms that chaotic motion around a cycle emerges in the perturbed Chua system, which is a manifestation of the entrainment by chaos.

The obtained result highlights the possibility of employing existing cyclic chaos to generate a new one in systems with stable equilibria, and particularly in Chua's oscillators. Furthermore, it is seen in Fig. 4.10 that the resulting motion resembles the spiral Chua's attractor, which occurs in the case of a period-doubling cascade [57, 58].

4.7.3 Controlling Chaos

The Pyragas control method [59–62] is an effective instrument for stabilizing the unstable periodic orbits of chaotic systems. It is also very useful for visually discerning the periodic solutions, which are otherwise indistinguishable in the set of irregular motions.

As an example, we will describe the procedure for stabilizing unstable periodic solutions of systems of the form (4.2.6)+(4.2.8).

It is demonstrated in [27] that to apply the Pyragas control method to the chaotic Duffing oscillator given by the system

$$\begin{aligned} x_1' &= x_2, \\ x_2' &= -0.10x_2 + 0.5x_1\left(1 - x_1^2\right) + 0.24\sin t, \end{aligned} \tag{4.7.31}$$

one can construct the corresponding control system

$$\begin{aligned} z_1' &= z_2, \\ z_2' &= -0.10z_2 + 0.5z_1\left(1 - z_1^2\right) + 0.24\sin(z_3) \\ &\quad + C\left[z_2(t-\tau_0) - z_2(t)\right], \\ z_3' &= 1, \end{aligned} \tag{4.7.32}$$

where $q(t) = C\left[z_2(t-\tau_0) - z_2(t)\right]$ is the control law and the parameter C represents the strength of the perturbation. An unstable 2π-periodic solution can be stabilized by choosing the value $\tau_0 = 2\pi$.

Using system (4.7.31) as the generator, we set up the following system:

$$\begin{aligned} y_1' &= 7y_1 - y_2 - y_1(y_1^2 + y_2^2) + 5x_1(t), \\ y_2' &= y_1 + 7y_2 - y_2(y_1^2 + y_2^2) + 4(x_2(t) + x_2^3(t)). \end{aligned} \tag{4.7.33}$$

According to the theoretical discussions, system (4.7.31)+(4.7.33) is chaotic, and there is entrainment by chaos such that (4.7.33) exhibits chaotic motions around the limit cycle of system (4.6.21) with $\alpha = 7$.

Our current objective is to show numerically how to control the chaos of system (4.7.31)+(4.7.33). We suggest that if a periodic solution of the generator system (4.7.31) is stabilized, then the chaos of system (4.7.31)+(4.7.33) is controlled.

To apply the Pyragas method to control the chaos of (4.7.31)+(4.7.33), we set up the system

$$\begin{aligned} v_1' &= 7v_1 - v_2 - v_1(v_1^2 + v_2^2) + 5z_1(t), \\ v_2' &= v_1 + 7v_2 - v_2(v_1^2 + v_2^2) + 4(z_2(t) + z_2^3(t)), \end{aligned} \tag{4.7.34}$$

where $z_1(t)$ and $z_2(t)$ refer to the first and second coordinates of the solutions of the control system (4.7.32).

Let us consider the solution of (4.7.32)+(4.7.34) with initial data $z_1(0) = 0.2$, $z_2(0) = 0.4$, $z_3(0) = 0$, $v_1(0) = -2.5$ and $v_2(0) = 0.8$. We allow system (4.7.32)+(4.7.34) to evolve freely by taking $C = 0$ until $t = 70$, and at that moment, we switch on the control and use $C = 0.84$. When $t = 210$, the control mechanism is switched off, and henceforth, the value $C = 0$ is utilized. Figure 4.11, which depicts chaos control, shows the z_2 and v_2 coordinates of the solution. It can be observed that after switching off the control mechanism, the stabilized 2π-periodic solution of system (4.7.31)+(4.7.33) loses its stability, and chaos emerges again. We note that one can obtain similar graphs for the other coordinates of (4.7.32)+(4.7.34).

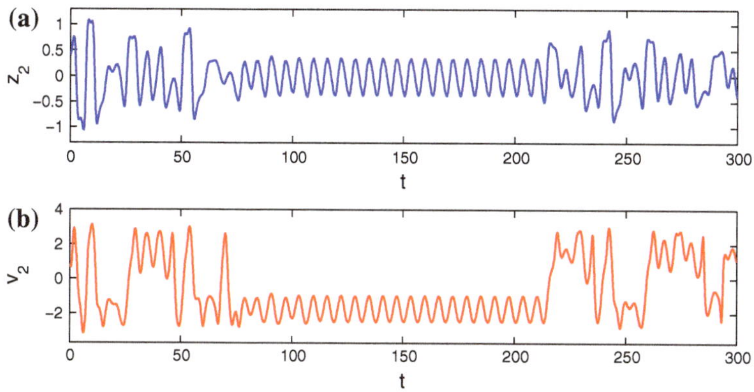

Fig. 4.11 Application of the Pyragas control method to system (4.7.31)+(4.7.33) by means of system (4.7.32)+(4.7.34). **a** Graph of the z_2 coordinate, **b** Graph of the v_2 coordinate

4.7.4 Entrainment and Synchronization

In this subsection, we will show that our results cannot be considered as generalized synchronization (GS) results [33].

GS characterizes the dynamics of a response system that is driven by the output of a chaotic driving system [26–28, 30, 33]. Suppose that the dynamics of the drive and response are governed by the following systems with a skew product structure

$$x' = D(x) \tag{4.7.35}$$

and

$$y' = R(y, K(x)), \tag{4.7.36}$$

respectively, where $x \in \mathbb{R}^p$, $y \in \mathbb{R}^q$. Synchronization [33] is said to occur if there exist sets I_x, I_y of initial conditions and a transformation ϕ, defined on the chaotic attractor of (4.7.35), such that for all $x(0) \in I_x$, $y(0) \in I_y$ the relation

$$\lim_{t\to\infty} \|y(t) - \phi(x(t))\| = 0$$

holds. In this case, a motion that starts on $I_x \times I_y$ collapses onto a manifold $M \subset I_x \times I_y$ of synchronized motions. The transformation ϕ is not required to exist for the transient trajectories. When ϕ is the identity, the identical synchronization takes place [27, 32]. The case of differentiable ϕ is considered in [28].

It is formulated in paper [30] that GS occurs if and only if for all $x_0 \in I_x$, $y_{10}, y_{20} \in I_y$, the following asymptotic stability criterion holds:

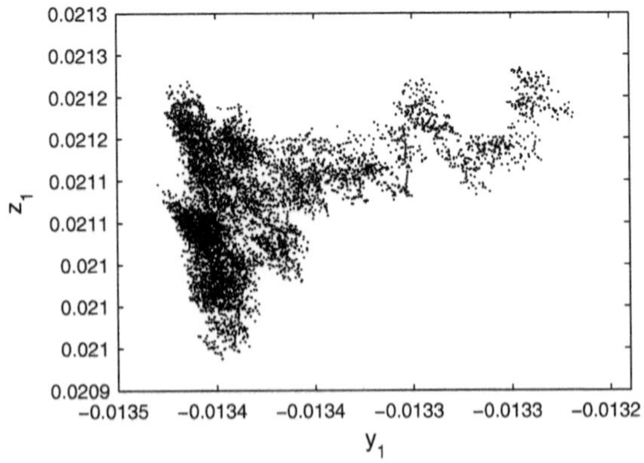

Fig. 4.12 The projection of the stroboscopic plot of the system (4.6.24) + (4.6.26) + (4.7.37) on the $y_1 - z_1$ plane

$$\lim_{t\to\infty} \|y(t, x_0, y_{10}) - y(t, x_0, y_{20})\| = 0,$$

where $y(t, x_0, y_{10})$, $y(t, x_0, y_{20})$ denote the solutions of (4.7.36) with the initial data $y(0, x_0, y_{10}) = y_{10}$, $y(0, x_0, y_{20}) = y_{20}$ and the same $x(t)$, $x(0) = x_0$.

To make comparison of our approach with that of GS, let us apply the auxiliary system method [26, 27] to indicate the presence or absence of GS in the couple (4.2.6) + (4.2.8) ((4.2.10) + (4.2.8)), considered this time as drive-response systems (as it is accepted in the synchronization theory).

Let us start the procedure by the couple (4.6.24) + (4.6.26) with $D_1 = 0.3$, $D_2 = 1$, $\lambda = 40$, $\alpha = 0.002$ and $\mu = 0.008$ such that the entrainment by chaos takes place as demonstrated in Sect. 4.6. The corresponding auxiliary system is

$$\begin{aligned} z_1' &= 0.002z_1 - z_2 - z_1\left(z_1^2 + z_2^2\right) + 0.008x_1(t),\\ z_2' &= z_1 + 0.002z_2 - z_2\left(z_1^2 + z_2^2\right) + 0.008x_2(t), \end{aligned} \tag{4.7.37}$$

which is an identical copy of system (4.6.26).

The projection of the stroboscopic plot of system (4.6.24) + (4.6.26) + (4.7.37) on the $y_1 - z_1$ plane is depicted in Fig. 4.12. The figure is obtained by marking the trajectory with the initial data $x_1(0) = 3.5$, $x_2(0) = -2$, $y_1(0) = -0.01$, $y_2(0) = -0.03$, $z_1(0) = 0.02$, $z_2(0) = 0.038$ at times t that are integer multiples of 2π and by omitting the first 4000 iterations. It is observable in Fig. 4.12 that the stroboscopic plot is not on the line $z_1 = y_1$, and therefore GS does not take place in the system (4.6.24) + (4.6.26).

To have a more detailed comparison of the present results with GS, let us consider a Rössler–Lorenz couple. GS was observed in [26, 27] with specific values of coefficients and perturbations. Let us take into account the couple with our particular data which issues from present investigations.

Consider the Lorenz system

$$\begin{aligned} u_1' &= -10u_1 + 10u_2, \\ u_2' &= -u_1u_3 + 350u_1 - u_2, \\ u_3' &= u_1u_2 - (8/3)u_3. \end{aligned} \tag{4.7.38}$$

According to Sparrow [63], system (4.7.38) possesses a globally attracting limit cycle. We perturb system (4.7.38) with the solutions of the chaotic Rössler system [64]

$$\begin{aligned} x_1' &= -(x_2 + x_3), \\ x_2' &= x_1 + 0.2x_2, \\ x_3' &= 0.2 + x_3(x_1 - 5.7), \end{aligned} \tag{4.7.39}$$

and set up the system

$$\begin{aligned} y_1' &= -10y_1 + 10y_2 + 2.8x_1(t), \\ y_2' &= -y_1y_3 + 350y_1 - y_2 + 7x_2(t), \\ y_3' &= y_1y_2 - (8/3)y_3 + 4.5x_3(t). \end{aligned} \tag{4.7.40}$$

Using the solution of system (4.7.39) with $x_1(0) = 2.1$, $x_2(0) = -7.7$, $x_3(0) = 0.1$, we represent the trajectory of system (4.7.40) corresponding to the initial data $y_1(0) = -19.2$, $y_2(0) = -63.9$, $y_3(0) = 296.1$ in Fig. 4.13a. The projection of the same trajectory on the $y_1 - y_2$ plane is shown in Fig. 4.13b. The simulation results show that chaotic behavior appears near the limit cycle. Moreover, the chaotic behavior of the y_3 coordinate is illustrated in Fig. 4.14.

For system (4.7.39) + (4.7.40), we can construct the corresponding auxiliary system of the form

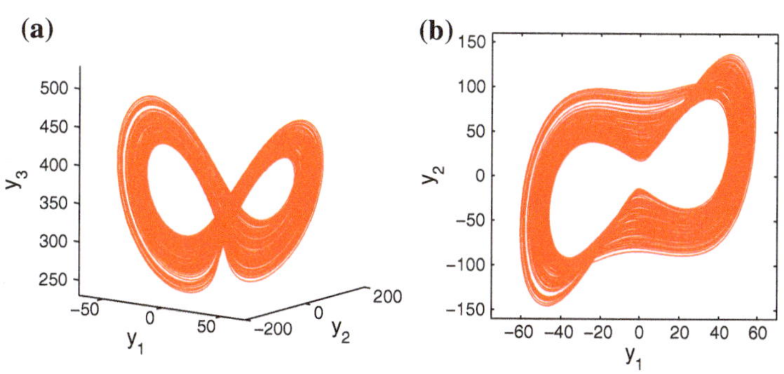

Fig. 4.13 A chaotic trajectory of system (4.7.40) near the limit cycle

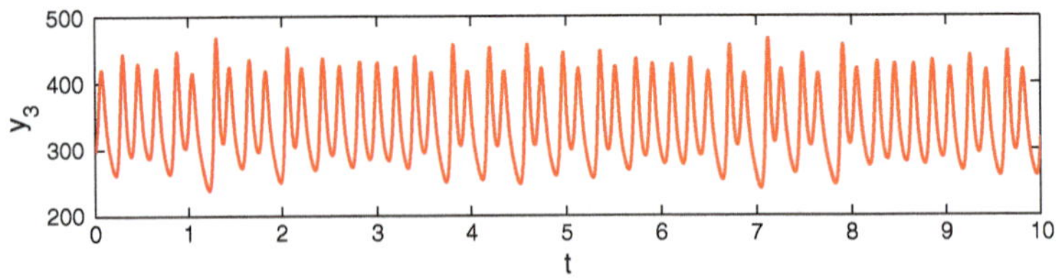

Fig. 4.14 The irregular behavior of the y_3 coordinate of system (4.7.40)

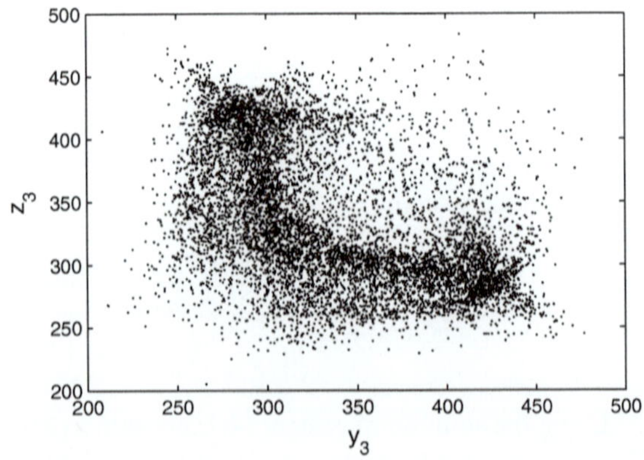

Fig. 4.15 Application of the auxiliary system approach to the system (4.7.39) + (4.7.40) indicates that GS does not exist for the couple

$$\begin{aligned} z_1' &= -10z_1 + 10z_2 + 2.8x_1(t), \\ z_2' &= -z_1z_3 + 350z_1 - z_2 + 7x_2(t), \\ z_3' &= z_1z_2 - (8/3)z_3 + 4.5x_3(t). \end{aligned} \tag{4.7.41}$$

The projection of the stroboscopic plot of system (4.7.39) + (4.7.40) + (4.7.41) on the $y_3 - z_3$ plane is shown in Fig. 4.15. The initial data $x_1(0) = 2.1$, $x_2(0) = -7.7$, $x_3(0) = 0.1$, $y_1(0) = -19.2$, $y_2(0) = -63.9$, $y_3(0) = 296.1$, $z_1(0) = -14.9$, $z_2(0) = -75.6$, $z_3(0) = 325.4$ are used and the first 200 iterations are omitted. To have GS indicated by the auxiliary system approach we need the stroboscopic plot to be placed on the line $z_3 = y_3$. Since this is not the case as seen in Fig. 4.15, we can conclude that the entrainment by chaos is not GS.

Next, let us use the auxiliary system approach to analyze the coupled system (4.7.28) + (4.7.29) with $\lambda = 2.003$ such that the entrainment by chaos takes place as shown in Sect. 4.7.1. The auxiliary system in this case is

$$\begin{aligned} z_1' &= -0.997z_1 - 0.25z_2 + z_1\left(z_3 + 0.2(1 - z_3^2)\right) + 0.003x_1(t), \\ z_2' &= 0.25z_1 - 0.997z_2 + z_2\left(z_3 + 0.2(1 - z_3^2)\right) + 0.004x_2(t), \\ z_3' &= 2.003z_3 - (z_1^2 + z_2^2 + z_3^2) + 0.002x_3(t). \end{aligned} \tag{4.7.42}$$

Making use of the initial data $x_1(0) = -6.74$, $x_2(0) = 0.34$, $x_3(0) = 32.76$, $y_1(0) = 0.07$, $y_2(0) = -1.17$, $y_3(0) = 0.94$, $z_1(0) = 0.85$, $z_2(0) = -0.24$, $z_3(0) =$

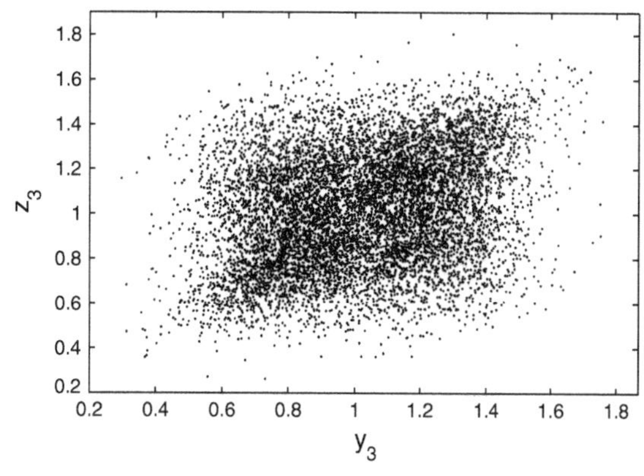

Fig. 4.16 Application of the auxiliary system approach to the system (4.7.28) + (4.7.29) reveals that entrainment of toroidal attractors by chaos is not GS

0.74, and omitting the first 200 iterations, we depict in Fig. 4.16 the projection of the stroboscopic plot of system (4.7.39) + (4.7.40) + (4.7.41) on the $y_3 - z_3$ plane. One can see in Fig. 4.15 that the stroboscopic plot is not placed on the line $z_3 = y_3$. Therefore, we conclude that GS is not achieved in the dynamics of the coupled system (4.7.39) + (4.7.40).

4.8 The Regular Motion Near the Limit Cycle

In this part, we provide the needed information from the proof of Andronov–Witt Theorem [50] and also precise the decay of the solutions regarding the initial value.

Without loss of generality, let us assume that $p(0)=0$ and $p'(0)= (\bar{p}_1, 0, 0, \ldots, 0)$ for some positive number $\bar{p}_1$.

According to our assumption that system (4.2.12) admits the number 1 as a simple characteristic multiplier and the remaining $n-1$ characteristic multipliers are smaller than one in modulus, system (4.2.12) has a real fundamental matrix $\Phi(t)$ of the form $\Phi(t) = P(t)\begin{pmatrix} 1 & 0 \\ 0 & e^{B_1 t} \end{pmatrix}$, where $P(t)$ is a regular, continuously differentiable T-periodic matrix and B_1 is an $(n-1) \times (n-1)$ matrix all of whose eigenvalues have negative real parts.

We emphasize that for an arbitrary solution $u(t)$ of Eq. (4.2.7), the differential equation satisfied by the function $z(t) = u(t) - p(t)$ is

$$z' = A(t)z + \varphi(t, z), \tag{4.8.43}$$

where $A(t) = \dfrac{\partial f(p(t))}{\partial u}$ and $\varphi(t, z) = f(p(t) + z) - f(p(t)) - A(t)z$. It is clear that $\varphi(t + T, z) = \varphi(t, z)$ and $\varphi(t, 0) = \varphi_z(t, 0) = 0$ for all $t \in \mathbb{R}_+$.

Since $\varphi_z(t,z) = o(1)$ as $z \to 0$ uniformly in $t \in \mathbb{R}_+$, there exist numbers $L_\varphi > 0$ and $\widetilde{\delta}(L_\varphi) > 0$ such that if $\|z_1\| < \widetilde{\delta}(L_\varphi)$, $\|z_2\| < \widetilde{\delta}(L_\varphi)$, then the inequality $\|\varphi(t,z_1) - \varphi(t,z_2)\| \le L_\varphi \|z_1 - z_2\|$ holds uniformly on $t \in \mathbb{R}_+$.

Suppose that $a = (0, a_2, a_3, \ldots, a_n)$ is an n-dimensional vector, which is orthogonal to $p'(0)$. There exist positive numbers K_1 and α such that $\|\Phi(t)a\| \le K_1 \|a\| e^{-\alpha t}$ for all $t \in \mathbb{R}_+$. Moreover, if $\|a\| < \widetilde{\delta}(L_\varphi)/(2K_1)$ then a solution $z(t,a)$ of (4.8.43) exists on $[0,\infty)$ and satisfies the following inequality

$$\|z(t,a)\| \le 2K_1 \|a\| e^{-\alpha t/2}, \quad t \ge 0. \tag{4.8.44}$$

A solution $\zeta(t,\zeta_0)$ of (4.2.7) satisfies the relation $\zeta(t,\zeta_0) = z(t,a) + p(t)$, where $z(t,a)$ is a solution of (4.8.43) with $z(0,a) = \zeta_0$. Additionally, the equation

$$\zeta_0 = P(0)a - \widetilde{h}(a) \tag{4.8.45}$$

holds, where $\widetilde{h}(a) = \big(\widetilde{h}_1(a_2,\ldots,a_n), 0, \ldots, 0\big)$, for some continuously differentiable function $\widetilde{h}_1$, and $\widetilde{h}(a) = o(\|a\|)$.

Suppose that $\zeta_0 = (\zeta_1^0, \zeta_2^0, \ldots, \zeta_n^0)$ and p_{ij} are the coordinates of the matrix $P(0)$, where $i, j = 1, 2, \ldots, n$. Equation (4.8.45) is equivalent to

$$\zeta_1^0 + \sum_{i=2}^{n} q_i \zeta_i^0 - h(\eta_2^0, \zeta_3^0, \ldots, \zeta_n^0) = 0, \tag{4.8.46}$$

where q_i, $i = 2, \ldots, n$, are constants and h is a continuously differentiable function such that

$$h(\zeta_2^0, \ldots, \zeta_n^0) = o\left(\left(\sum_{i=2}^{n} (\zeta_i^0)^2\right)^{1/2}\right).$$

Denote by S the $(n-1)$ dimensional, C^1 manifold determined by the equation

$$x_1 + \sum_{i=2}^{n} q_i x_i - h(x_2, x_3, \ldots, x_n) = 0. \tag{4.8.47}$$

The hypersurface S crosses the orbit γ, which is defined by Eq. (4.2.11), transversally so that for any solution $\zeta(t,\zeta_0)$ starting on this initial manifold, we have $\|\zeta(t,\zeta_0) - p(t)\| \to 0$ exponentially as $t \to \infty$.

Now, we are going to prove that for each number $l \in (0,1)$ there exists a natural number $n_0 = n_0(l)$ such that if ζ_0 belongs to S then

$$\|\zeta(n_0 T, \zeta_0)\| \le l \, \|\zeta_0\|. \tag{4.8.48}$$

Let $\overline{\varepsilon} = 1/\left(2\left\|P^{-1}(0)\right\|\right)$. It is possible to find a number $\overline{\delta}\left(\overline{\varepsilon}\right) > 0$ such that if

$$\|a\| < \min\left\{\widetilde{\delta}(L_\varphi)/(2K_1), \overline{\delta}\left(\overline{\varepsilon}\right)\right\},$$

then the inequality

$$\left\|\widetilde{h}(a)\right\| < \overline{\varepsilon}\,\|a\| \tag{4.8.49}$$

is valid.

Let us fix a solution $\zeta(t, \zeta_0)$ such that ζ_0 belongs to S. In the case $\|a\| < \min\left\{\widetilde{\delta}(L_\varphi)/(2K_1), \overline{\delta}\left(\overline{\varepsilon}\right)\right\}$, taking advantage of (4.8.45) and (4.8.49) one can find that $\|a\| \le 2\left\|P^{-1}(0)\right\| \|\zeta_0\|$ and according to (4.8.44), we have

$$\|\zeta(t, \zeta_0) - p(t)\| \le 4K_1\left\|P^{-1}(0)\right\| \|\zeta_0\|\, e^{-\alpha t/2},\ t \ge 0. \tag{4.8.50}$$

Let us fix an arbitrary number $l \in (0, 1)$. There exists a natural number $n_0 = n_0(l)$ such that $4K_1\left\|P^{-1}(0)\right\| e^{-\alpha T n_0/2} < l$. Making use of (4.8.50) we obtain that $\|\zeta(n_0T, \zeta_0) - p(n_0T)\| < l\,\|\zeta_0\|$. Since $p(n_0T) = 0$, inequality (4.8.48) holds.

4.9 Notes

The concept of entrainment is extended to introduce the notion of the entrainment of limit cycles by chaos. Our theoretical results can be effectively adapted to arbitrarily high-dimensional systems that possess asymptotically orbitally stable limit cycles. Examples of such systems can be found in mechanics, electronics, economics, the neural sciences, chemistry, and population dynamics [65–71]. Employing the method presented, one can obtain motions that behave cyclically and chaotically at the same time.

We prove the presence of chaos through the notions of period-doubling cascade and sensitivity. It is known that [17, 18] sensitivity is the main ingredient of chaos. Verifying other ingredients of chaos, namely the transitivity and density of periodic motions, is more difficult.

The entrainment of toroidal attractors by chaos and entrainment in Chua's oscillators are demonstrated numerically. Moreover, the existence of unstable periodic solutions is evidenced through the Pyragas method [60] and simulations.

One of the important peculiarities of our approach is that entrainment by chaos cannot be embedded as a part of synchronization theory [26–28, 30, 33].

Cyclical behavior in chaotic attractors has been widely observed in the literature. We can note the famous Rössler attractor, Chua's spiral attractor and even the classical Lorenz attractor, where one can observe two-center cyclical behavior, as examples. Our results for obtaining cyclical behavior are different from those presented in the literature, because exogenous perturbations are applied in our case. In fact, the mechanism proposed in this chapter could be the unsuspected underlying force that gives rise to some chaotic attractors discussed in the literature.

Some of our results, for example, boundedness of solutions around the limit cycle, can be obtained if one applies results in [72] on the existence of invariant manifolds and their persistence under perturbation or by reduction to discrete equations with respect to both phase and time variables [42].

The conclusions of this chapter can be replicated in cases in which cycles are attracting when time decreases to $-\infty$. Another theoretically challenging problem is to consider hyperbolic cycles, as well as the critical cases [34–36]. Moreover, our results are useful for generating multidimensional chaos, especially if one requires a rigorous proof for the phenomenon [73]. We can formally compare our results with those of Ruelle and Takens [74] on the appearance of turbulence through three successive Hopf bifurcations. Unlike Ruelle and Takens, we observe chaos to emerge after fewer than three bifurcations and we use chaotic perturbations. The results of the present chapter were published in the paper [75].

References

1. C. Huygens, Letter to de Sluse, in *Oeuveres Completes de Christian Huygens* (letters; no. 1333 of 24 February 1665, no. 1335 of 26 February 1665, no. 1345 of 6 March 1665), (Societe Hollandaise Des Sciences, Martinus Nijhoff, La Haye, 1893)
2. G. Oster, Auditory beats in the brain. Sci. Am. **229**, 94–102 (1973)
3. V.J. Walter, W.G. Walter, The central effects of rhythmic sensory stimulation. Electroencephalogr. Clin. Neurophysiol. **1**, 57–86 (1949)
4. M. Clayton, R. Sager, U. Will, In time with the music: the concept of entrainment and its significance for ethnomusicology. ESEM Counterpoint **1**, 1–75 (2004)
5. C. Dombrowski, B. Lewellyn, A.I. Pesci, J.M. Restrepo, J.O. Kessler, R.E. Goldstein, Coiling, entrainment, and hydrodynamic coupling of decelerated fluid jets. Phys. Rev. Lett. **95**(184501), 1–4 (2005)
6. N. Minorsky, *Introduction to Non-linear Mechanics* (J.W. Edwards, Ann Arbor, 1947)
7. A. Pikovsky, M. Rosenblum, J. Kurths, *Synchronization: A Universal Concept in Nonlinear Sciences* (Cambridge University Press, New York, 2001)
8. I. Sendiña-Nadal, I. Leyva, J.M. Buldú, J.A. Almendral, S. Boccaletti, Entraining the topology and the dynamics of a network of phase oscillators. Phys. Rev. E **79**, 1–8 (2009)
9. V.S. Anishchenko, T. Kapitaniak, M.A. Safonova, O.V. Sosnovzeva, Birth of double-double scroll attractor in coupled Chua circuits. Phys. Lett. A **192**, 207–214 (1994)
10. A. Caneco, J.L. Rocha, C. Grácio, Topological entropy in the synchronization of piecewise linear and monotone maps, coupled Duffing oscillators. Int. J. Bifurc. Chaos **19**, 3855–3868 (2009)
11. G. Keller, R. Zweimüller, Unidirectionally coupled interval maps: between dynamics and statistical mechanics. Nonlinearity **15**, 1–24 (2002)
12. J. Wu, L. Jiao, Synchronization in complex delayed dynamical networks with nonsymmetric coupling. Phys. A **386**, 513–530 (2007)
13. E.N. Lorenz, *The Essence of Chaos* (UCL Press, Reading, 1993)
14. B.D. Hassard, N.D. Kazarinoff, Y.H. Wan, *Theory and Applications of Hopf Bifurcation* (Cambridge University Press, Cambridge, 1981)
15. R.J. Field, L. Györgyi, *Chaos in Chemistry and Biochemistry* (World Scientific, Singapore, 1993)

16. J. Guckenheimer, P.J. Holmes, *Nonlinear Oscillations, Dynamical Systems and Bifurcations of Vector Fields* (Springer, New York, 1997)
17. E.N. Lorenz, Deterministic nonperiodic flow. J. Atmos. Sci. **20**, 130–141 (1963)
18. C. Robinson, *Dynamical Systems: Stability, Symbolic Dynamics, and Chaos* (CRC Press, Boca Raton, 1995)
19. S. Wiggins, *Global Bifurcations and Chaos: Analytical Methods* (Springer, New York, 1988)
20. M. Akhmet, *Principles of Discontinuous Dynamical Systems* (Springer, New York, 2010)
21. M.U. Akhmet, Devaney's chaos of a relay system. Commun. Nonlinear Sci. Numer. Simul. **14**, 1486–1493 (2009)
22. M.U. Akhmet, Li-Yorke chaos in the system with impacts. J. Math. Anal. Appl. **351**, 804–810 (2009)
23. M.U. Akhmet, M.O. Fen, Chaos generation in hyperbolic systems. Interdiscip. J. Discontin. Nonlinearity Complex. **1**, 367–386 (2012)
24. M.U. Akhmet, M.O. Fen, Chaotic period-doubling and OGY control for the forced Duffing equation. Commun. Nonlinear. Sci. Numer. Simul. **17**, 1929–1946 (2012)
25. M.U. Akhmet, M.O. Fen, Replication of chaos. Commun. Nonlinear. Sci. Numer. Simul. **18**, 2626–2666 (2013)
26. H.D.I. Abarbanel, N.F. Rulkov, M.M. Sushchik, Generalized synchronization of chaos: the auxiliary system approach. Phys. Rev. E **53**, 4528–4535 (1996)
27. J.M. Gonzáles-Miranda, *Synchronization and Control of Chaos* (Imperial College Press, London, 2004)
28. B.R. Hunt, E. Ott, J.A. Yorke, Differentiable generalized synchronization of chaos. Phys. Rev. E **55**(4), 4029–4034 (1997)
29. T. Kapitaniak, Synchronization of chaos using continuous control. Phys. Rev. E **50**, 1642–1644 (1994)
30. L. Kocarev, U. Parlitz, Generalized synchronization, predictability, and equivalence of unidirectionally coupled dynamical systems. Phys. Rev. Lett. **76**(11), 1816–1819 (1996)
31. E.E.N. Macau, C. Grebogi, Y.-C. Lai, Active synchronization in nonhyperbolic hyperchaotic systems. Phys. Rev. E **65**, 027202 (2002)
32. L.M. Pecora, T.L. Carroll, Synchronization in chaotic systems. Phys. Rev. Lett. **64**, 821–825 (1990)
33. N.F. Rulkov, M.M. Sushchik, L.S. Tsimring, H.D.I. Abarbanel, Generalized synchronization of chaos in directionally coupled chaotic systems. Phys. Rev. E **51**(2), 980–994 (1995)
34. B. Aulbach, Behaviour of solutions near manifolds of periodic solutions. J. Differ. Equ. **39**, 345–377 (1981)
35. J.K. Hale, A.P. Stokes, Behaviour of solutions near integral manifolds. Arch. Ration. Mech. Anal. **6**, 133–170 (1960)
36. Yu.A. Mitropolskij, O.B. Lykova, *Integral Manifolds in Nonlinear Mechanics (Russian)* (Nauka Dumka, Moscow, 1973)
37. I. Lengyel, G. Rábai, I.R. Epstein, Experimental and modeling study of oscillations in the chlorine dioxide-iodine-melanic acid reaction. J. Am. Chem. Soc. **112**, 9104–9110 (1990)
38. S.H. Strogatz, *Nonlinear Dynamics and Chaos with Applications to Physics, Biology, Chemistry, and Engineering* (Perseus Books, New York, 1994)
39. R. Shaw, Strange attractors, chaotic behavior and information flow. Z. Naturf. **36a**, 80–112 (1981)
40. J.M.T. Thompson, H.B. Stewart, *Nonlinear Dynamics and Chaos* (Wiley, Chichester, 2002)
41. R.A. Horn, C.R. Johnson, *Matrix Analysis* (Cambridge University Press, Cambridge, 1992)
42. P. Hartman, *Ordinary Differential Equations* (Wiley, New York, 1964)
43. K.T. Alligood, T.D. Sauer, J.A. Yorke, *Chaos: An Introduction to Dynamical Systems* (Springer, New York, 1996)
44. R.L. Devaney, *An Introduction to Chaotic Dynamical Systems* (Addison-Wesley, Menlo Park, 1989)
45. M.J. Feigenbaum, Universal behavior in nonlinear systems. Los Alamos Sci./Summer 4-27 (1980)

46. I. Kovacic, M.J. Brennan (eds.), *The Duffing Equation: Nonlinear Oscillations and Their Behavior* (Wiley, New York, 2011)
47. K. Palmer, *Shadowing in Dynamical Systems: Theory and Applications* (Kluwer Academic Publishers, Dordrecht, 2000)
48. E. Sander, J.A. Yorke, Period-doubling cascades galore. Ergod. Theory Dyn. Syst. **31**, 1249–1267 (2011)
49. E. Sander, J.A. Yorke, Connecting period-doubling cascades to chaos. Int. J. Bifurc. Chaos **22**, 1–16 (2012)
50. M. Farkas, *Periodic Motions* (Springer, New York, 2010)
51. T. Yoshizawa, *Stability Theory and the Existence of Periodic Solutions and Almost Periodic Solutions* (Springer, New York, 1975)
52. J.K. Hale, *Ordinary Differential Equations* (Krieger Publishing Company, Malabar, 1980)
53. S. Sato, M. Sano, Y. Sawada, Universal scaling property in bifurcation structure of Duffing's and of generalized Duffing's equations. Phys. Rev. A **28**, 1654–1658 (1983)
54. J.L. Massera, The existence of periodic solutions of systems of differential equations. Duke Math. J. **17**, 457–475 (1950)
55. J. Hale, H. Koçak, *Dynamics and Bifurcations* (Springer, New York, 1991)
56. W. Langford, Unfolding of degenerate bifurcations, in *Chaos, Fractals, and Dynamics*, ed. by P. Fisher, W. Smith (Marcel Dekker, New York, 1985), pp. 87–103
57. L.O. Chua, C.W. Wu, A. Huang, G. Zhong, A universal circuit for studying and generating chaos-Part I: routes to chaos. IEEE Trans. Circuits Syst.-I Fundam. Theory Appl. **40**, 732–744 (1993)
58. M. Lakshmanan, S. Rajasekar, *Nonlinear Dynamics: Integrability, Chaos and Patterns* (Springer, Berlin, 2003)
59. A.L. Fradkov, *Cybernetical Physics: From Control of Chaos to Quantum Control* (Springer, Berlin, 2007)
60. K. Pyragas, Continuous control of chaos by self-controlling feedback. Phys. Rev. A **170**, 421–428 (1992)
61. E. Schöll, H.G. Schuster, *Handbook of Chaos Control* (Wiley-VCH, Weinheim, 2008)
62. I. Zelinka, S. Celikovsky, H. Richter, G. Chen (eds.), *Evolutionary Algorithms and Chaotic Systems* (Springer, Berlin, 2010)
63. C. Sparrow, *The Lorenz Equations: Bifurcations, Chaos and Strange Attractors* (Springer, New York, 1982)
64. O.E. Rössler, An equation for continuous chaos. Phys. Lett. **57A**, 397–398 (1976)
65. D. D'Humieres, M.R. Beasley, B.A. Huberman, A. Libchaber, Chaotic states and routes to chaos in the forced pendulum. Phys. Rev. A **26**, 3483–3496 (1982)
66. W. Jiang, K.M. Tsang, Z. Hua, Hopf bifurcation in the Hodgkin-Huxley model exposed to ELF electrical field. Chaos Solitons Fractals **20**, 759–764 (2004)
67. T. Kostova, R. Ravindran, M. Schonbek, Fitzhugh-Nagumo revisited: types of bifurcations, periodical forcing and stability regions by a Lyapunov functional. Int. J. Bifurc. Chaos **14**, 913–925 (2004)
68. S.A. Morton, P.S. Beran, Hopf-bifurcation analysis of airfoil flutter at transonic speeds. J. Aircr. **36**, 421–429 (1999)
69. U. Parlitz, W. Lauterborn, Superstructure in the bifurcation set of the Duffing equations. Phys. Lett. **107A**, 351–355 (1985)
70. M. Wang, Stability and Hopf bifurcation for a prey-predator model with prey-stage structure and diffusion. Math. Biosci. **212**, 149–160 (2008)
71. S. Zhang, D. Tan, L. Chen, Chaotic behavior of a chemostat model with Beddington-DeAngelis functional response and periodically impulsive invasion. Chaos Solitons Fractals **29**, 474–482 (2006)
72. M.W. Hirsch, C.C. Pugh, M. Shub, *Invariant Manifolds* (Springer, Berlin, 1977)
73. F.R. Marotto, Snap-back repellers imply chaos in $\mathbb{R}^n$. J. Math. Anal. Appl. **63**, 199–223 (1978)
74. D. Ruelle, F. Takens, On the nature of turbulence. Commun. Math. Phys. **20**, 167–192 (1971)
75. M.U. Akhmet, M.O. Fen, Entrainment by chaos. J. Nonlinear Sci. **24**, 411–439 (2014)

Chapter 5
Chaotification of Impulsive Systems

In this chapter, we present a new method for chaos generation in non-autonomous impulsive systems. We prove the presence of chaos in the sense of Li–Yorke by implementing chaotic perturbations. An impulsive Duffing oscillator is used to show the effectiveness of our technique, and simulations that support the theoretical results are depicted. Moreover, a procedure to stabilize the unstable periodic solutions is proposed.

5.1 Introduction

It is well known how discrete dynamics is important for the chaos theory [1–4]. Very interesting examples of applications of discrete dynamics to continuous chaos analysis were provided in papers [5–8]. In these studies, the general technique of dynamical synthesis [5] was developed. Besides that, it is of big interest to consider chaotic processes where continuous dynamics is intermingled with discontinuity [9–13].

Impulsive differential equations describe the dynamics of real-world processes in which abrupt changes occur. Such equations play an increasingly important role in various fields such as mechanics, electronics, neural networks, communication systems, and population dynamics [14–20]. In this chapter, we present a rigorous method for chaotification of arbitrary high dimensional impulsive systems.

Throughout the chapter $\mathbb{R}$, $\mathbb{Z}$ and $\mathbb{N}$ will denote the sets of real numbers, integers, and natural numbers, respectively. We will make use of the usual Euclidean norm for vectors and the norm induced by the Euclidean norm for square matrices [21].

The main purpose of our investigation is as follows. Consider the collection of functions

$$\mathscr{A} = \left\{ \varphi(t) : \mathbb{R} \to \mathbb{R}^n \mid \sup_{t\in\mathbb{R}} \|\varphi(t)\| \le H_0 \right\},$$

M. Akhmet and M.O. Fen, *Replication of Chaos in Neural Networks, Economics and Physics*, Nonlinear Physical Science,
DOI 10.1007/978-3-662-47500-3_5

where H_0 is a positive number, and suppose that $\mathscr{A}$ is an equicontinuous family on $\mathbb{R}$. We perturb the impulsive system

$$\begin{aligned} x'(t) &= Ax(t) + f(t, x(t)), \ t \neq \theta_k, \\ \Delta x|_{t=\theta_k} &= Bx(\theta_k) + W(x(\theta_k)), \end{aligned} \tag{5.1.1}$$

by the functions $\varphi(t) \in \mathscr{A}$ to obtain the system

$$\begin{aligned} y'(t) &= Ay(t) + f(t, y(t)) + \varphi(t), \ t \neq \theta_k, \\ \Delta y|_{t=\theta_k} &= By(\theta_k) + W(y(\theta_k)), \end{aligned} \tag{5.1.2}$$

where the functions $f : \mathbb{R} \times \mathbb{R}^n \to \mathbb{R}^n$ and $W : \mathbb{R}^n \to \mathbb{R}^n$ are continuous in all their arguments, A and B are $n \times n$ constant real-valued matrices, the sequence $\{\theta_k\}$, $k \in \mathbb{Z}$, of impulsive moments is strictly increasing, $\Delta y|_{t=\theta_k} = y(\theta_k+) - y(\theta_k)$ and $y(\theta_k+) = \lim_{t\to\theta_k+} y(t)$. The main objective of the present chapter is the verification of chaos in the dynamics of system (5.1.2), provided that the collection $\mathscr{A}$ is chaotic. The description of chaotic collection of functions will be presented in the next section.

The term chaos, as a mathematical notion, has first been used in [2] for one dimensional difference equations. According to Li and Yorke [2], a continuous map $F : J \to J$, where $J \subset \mathbb{R}$ is an interval, exhibits chaos if: (i) For every natural number p, there exists a p-periodic point of F in J; (ii) There is an uncountable set $S \subset J$ containing no periodic points such that for every $s_1, s_2 \in S$ with $s_1 \neq s_2$ we have $\limsup_{j\to\infty} \left|F^j(s_1) - F^j(s_2)\right| > 0$ and $\liminf_{j\to\infty} \left|F^j(s_1) - F^j(s_2)\right| = 0$; (iii) For every $s \in S$ and periodic point $\sigma \in J$ we have $\limsup_{j\to\infty} \left|F^j(s) - F^j(\sigma)\right| > 0$.

The concept of snap-back repellers for high dimensional maps was introduced in [22]. According to Marotto [22], if a multidimensional continuously differentiable map has a snap-back repeller, then it is Li–Yorke chaotic. Marotto's Theorem was used in [23] to prove the existence of Li–Yorke chaos in a spatiotemporal chaotic system. Li–Yorke sensitivity, which links the Li–Yorke chaos with the notion of sensitivity, was studied in [24]. Moreover, generalizations of Li–Yorke chaos to mappings in Banach spaces and complete metric spaces were provided in [25–27]. In the present chapter, we develop the concept of Li–Yorke chaos to piecewise continuous functions, and prove its presence rigorously in impulsive systems of the form (5.1.2) without any restriction on the dimension.

Taking advantage of chaotically changing impulsive moments, which are functionally dependent on the initial moment, the presence of Li–Yorke chaos in a nonautonomous impulsive differential equation was rigorously proved in [9]. On the other hand, the existence of Li–Yorke chaos and its control in an autonomous impulsive differential system were discussed both theoretically and numerically in the paper [12], where the presence of a snap-back repeller was proved based on the qualitative analysis using the Poincaré map and the Lambert W-function. A system of impulsive differential equations with moments of impulses generated by a sensitive map which depends on a parameter was taken into account in [13], and sensitivity was considered as a chaotic property. The existence of chaos in singular impulsive systems was shown in [11] by means of transversal homoclinic points. Moreover, chaos in the sense of Devaney [1] was studied in an impulsive model of the cardiovascular system by means of chaotically changing impulsive moments within the scope of the paper [10]. Distinctively from the papers [9–13], we make use of chaotic perturbations to prove the existence of Li–Yorke chaos, and this is the main novelty of this chapter.

Small perturbations applied to control parameters can be used to stabilize chaos, keeping the parameters in the neighborhood of their nominal values [28, 29], and this idea is first introduced by Ott et al. [30]. Experimental applications of the OGY control method requires a permanent computer analysis of the system's state. Since the method deals with a Poincaré map, the parameter changes are discrete in time. By this method, it is possible to stabilize only those periodic orbits whose maximal Lyapunov exponent is small compared to the reciprocal of the time interval between parameter changes [31].

In the example presented in Sect. 5.4, to obtain a collection of chaotic functions, we will use a Duffing oscillator which is forced by a relay function. On the other hand, to support our new theoretical results, an impulsive Duffing oscillator will be utilized. The presented example shows the effectiveness of our technique. Moreover, making use of the OGY control method [30], we will demonstrate that the chaos of system (5.1.2) is controllable. This method is useful for visually discerning the periodic solutions, which are otherwise indistinguishable in the set of irregular motions.

A concept in which impulsive differential equations are effectively used is the impulsive synchronization of chaotic systems [32–38]. This technique is appropriate for the synchronization of Lorenz systems [39–41], Chua oscillators [19, 42] and Rössler systems [43]. In the framework of impulsive synchronization, one can set up an impulsive error system which admits the synchronization error $e = y - x$ as a solution, where x and y denote the states of the drive and response systems, respectively, and require this system to possess a stable equilibrium point. According to the last equation, the synchronized systems must have the same dimensions. However, this is not requested in our results. Therefore, we significantly extend the chaos generation possibilities. Generally speaking, we consider the chaotification procedure in the most definitive and general form. On the other hand, in our theory, it is not necessary to use a drive-response couple. Instead, one can take into account an impulsive system and perturb it by a previously obtained chaotic data. Moreover, in the theory of impulsive synchronization the chaos type of the drive and response systems is not considered. Contrarily, in our results, we rigorously prove that the impulsive system exhibits the same type of chaos as the chaotic perturbations.

The rest of the chapter is organized as follows. In Sect. 5.2, we introduce the ingredients of Li–Yorke chaos for collections of piecewise continuous functions and give sufficient conditions for the presence of chaotic dynamics in system (5.1.2). Moreover, we verify the attractiveness property of the bounded on $\mathbb{R}$ solutions of system (5.1.2). Section 5.3 is devoted to theoretical results such that the ingredients of Li–Yorke chaos for system (5.1.2) are rigorously proved. Our method of chaotification is applied to an impulsive Duffing oscillator in Sect. 5.4, and a procedure to stabilize the existing unstable periodic solutions is presented. Finally, some concluding remarks are indicated in Sect. 5.5.

5.2 Preliminaries

We say that a function $\psi(t) : \mathbb{R} \to \mathbb{R}^l$, $l \in \mathbb{N}$, belongs to the set $\mathscr{PC}(\mathbb{R})$ if it is left-continuous and continuous except, possibly, at the points where it has discontinuities of the first kind. The definition of a Li–Yorke chaotic set of piecewise continuous functions is as follows.

Suppose that $\mathscr{D}$ is a set of uniformly bounded functions $\psi(t) : \mathbb{R} \to \mathbb{R}^l$ which belong to $\mathscr{PC}(\mathbb{R})$ and have common points of discontinuity.

We say that a couple $\left(\psi(t), \widetilde{\psi}(t)\right) \in \mathscr{D} \times \mathscr{D}$ is proximal if for arbitrary small $\varepsilon > 0$ and arbitrary large $E > 0$, there exists an interval J with a length no less than E such that $\left\|\psi(t) - \widetilde{\psi}(t)\right\| < \varepsilon$ for $t \in J$. On the other hand, a couple $\left(\psi(t), \widetilde{\psi}(t)\right) \in \mathscr{D} \times \mathscr{D}$ is called frequently (ε_0, Δ)-separated if there exist positive numbers ε_0, Δ and infinitely many disjoint intervals, each with a length no less than Δ, such that $\left\|\psi(t) - \widetilde{\psi}(t)\right\| > \varepsilon_0$ for each t from these intervals, and each of these intervals contains at most one discontinuity point of both $\psi(t)$ and $\widetilde{\psi}(t)$. It is worth noting that the numbers ε_0 and Δ depend on the functions $\psi(t)$ and $\widetilde{\psi}(t)$.

A couple $\left(\psi(t), \widetilde{\psi}(t)\right) \in \mathscr{D} \times \mathscr{D}$ is a Li–Yorke pair if it is proximal and frequently (ε_0, Δ)-separated for some positive numbers ε_0 and Δ. Moreover, an uncountable set $\mathscr{C} \subset \mathscr{D}$ is called a scrambled set if $\mathscr{C}$ does not contain any periodic functions and each couple of different functions inside $\mathscr{C} \times \mathscr{C}$ is a Li–Yorke pair.

We say that the collection $\mathscr{D}$ is a Li–Yorke chaotic set if: (i) There exists a positive number T_0 such that $\mathscr{D}$ possesses a periodic function of period mT_0 for each $m \in \mathbb{N}$; (ii) $\mathscr{D}$ possesses a scrambled set $\mathscr{C}$; (iii) For any function $\psi(t) \in \mathscr{C}$ and any periodic function $\widetilde{\psi}(t) \in \mathscr{D}$, the couple $\left(\psi(t), \widetilde{\psi}(t)\right)$ is frequently (ε_0, Δ)-separated for some positive numbers ε_0 and Δ.

One can obtain a new Li–Yorke chaotic set of functions from a given one as follows. Suppose that $h : \mathbb{R}^l \to \mathbb{R}^{\bar{l}}$ is a function which satisfies for all $u_1, u_2 \in \mathbb{R}^l$ that

$$L_1 \|u_1 - u_2\| \le \|h(u_1) - h(u_2)\| \le L_2 \|u_1 - u_2\|, \tag{5.2.3}$$

where L_1 and L_2 are positive numbers. One can verify that if $\mathscr{D}$ is a Li–Yorke chaotic set, then the collection $\mathscr{D}_h = \{h(\psi(t)) \mid \psi(t) \in \mathscr{D}\}$ is also Li–Yorke chaotic.

The following conditions are needed:

(A1) The matrices A and B commute and $\det(I + B) \neq 0$, where I is the $n \times n$ identity matrix;

(A2) There exists a positive number T and a natural number p such that $f(t + T, y) = f(t, y)$ for all $t \in \mathbb{R}$, $y \in \mathbb{R}^n$ and $\theta_{k+p} = \theta_k + T$ for all $k \in \mathbb{Z}$;

(A3) The eigenvalues of the matrix $A + \frac{p}{T}\ln(I + B)$ have negative real parts;

(A4) There exist positive numbers M_f and M_W such that

$$\sup_{t\in\mathbb{R}, y\in\mathbb{R}^n} \|f(t, y)\| \leq M_f \text{ and } \sup_{y\in\mathbb{R}^n} \|W(y)\| \leq M_W;$$

(A5) There exists a positive number L_f such that

$$\|f(t, y_1) - f(t, y_2)\| \leq L_f \|y_1 - y_2\|,$$

for all $t \in \mathbb{R}$ and $y_1, y_2 \in \mathbb{R}^n$;

(A6) There exists a positive number L_W such that

$$\|W(y_1) - W(y_2)\| \leq L_W \|y_1 - y_2\|,$$

for all $y_1, y_2 \in \mathbb{R}^n$.

Let us denote by $U(t, s)$ the transition matrix of the linear homogeneous system

$$\begin{aligned} u'(t) &= Au(t), \ t \neq \theta_k, \\ \Delta u|_{t=\theta_k} &= Bu(\theta_k). \end{aligned}$$

Under the conditions $(A1)-(A3)$, there exist positive numbers N and ω such that $\|U(t, s)\| \leq Ne^{-\omega(t-s)}$ for $t \geq s$ [14, 44].

The following conditions are also required:

(A7) $N\left(\frac{L_f}{\omega} + \frac{pL_W}{1 - e^{-\omega T}}\right) < 1;$

(A8) $-\omega + NL_f + \frac{p}{T}\ln(1 + NL_W) < 0;$

(A9) $L_W \left\|(I + B)^{-1}\right\| < 1.$

We say that a left-continuous function $y(t) : \mathbb{R} \to \mathbb{R}^n$ is a solution of (5.1.2) if: (i) It has discontinuities only at the points θ_k, $k \in \mathbb{Z}$, and these discontinuities are of the first kind; (ii) The derivative $y'(t)$ exists at each point $t \in \mathbb{R}\backslash\{\theta_k\}$, and the left-sided derivative exists at the points θ_k, $k \in \mathbb{Z}$; (iii) The differential equation is satisfied by $y(t)$ on $\mathbb{R}\backslash\{\theta_k\}$, and it holds for the left derivative of $y(t)$ at every point θ_k, $k \in \mathbb{Z}$; (iv) The jump equation is satisfied by $y(t)$ for every $k \in \mathbb{Z}$.

According to the results of [14, 44], for any function $\varphi(t) \in \mathscr{A}$, one can confirm under the conditions $(A1)-(A7)$ that there exists a unique bounded on $\mathbb{R}$ solution $\phi_\varphi(t)$ of system (5.1.2) which satisfies the relation

$$\phi_\varphi(t) = \int_{-\infty}^{t} U(t, s)\left[f(s, \phi_\varphi(s)) + \varphi(s)\right] ds + \sum_{\theta_k < t} U(t, \theta_k) W(\phi_\varphi(\theta_k)).$$

It can be verified for each $\varphi(t) \in \mathscr{A}$ that the inequality $\sup_{t\in\mathbb{R}} \left\|\phi_\varphi(t)\right\| \le K_0$ holds, where

$$K_0 = \frac{N(M_f + H_0)}{\omega} + \frac{pNM_W}{1 - e^{-\omega T}}.$$

By means of the collection $\mathscr{A}$, let us construct the set

$$\mathscr{B} = \left\{\phi_\varphi(t) \mid \varphi(t) \in \mathscr{A}\right\}.$$

For a given function $\varphi(t) \in \mathscr{A}$, let us denote by $y_\varphi(t, y_0)$ the solution of (5.1.2) with $y_\varphi(0, y_0) = y_0$. We say that the collection $\mathscr{B}$ is an attractor if for any $\varphi(t) \in \mathscr{A}$ and $y_0 \in \mathbb{R}^n$, we have $\left\|y_\varphi(t, y_0) - \phi_\varphi(t)\right\| \to 0$ as $t \to \infty$. The attractiveness feature of the collection $\mathscr{B}$ is mentioned in the next assertion.

Lemma 5.1 *If the conditions* $(A1)-(A8)$ *are valid, then the collection* $\mathscr{B}$ *is an attractor.*

Proof Fix an arbitrary function $\varphi(t) \in \mathscr{A}$ and $y_0 \in \mathbb{R}^n$. Taking advantage of the relations

$$\begin{aligned} y_\varphi(t, y_0) &= U(t,0)y_0 + \int_0^t U(t,s)\left[f(s, y_\varphi(s, y_0)) + \varphi(s)\right] ds \\ &+ \sum_{0\le\theta_k<t} U(t,\theta_k) W(y_\varphi(\theta_k, y_0)) \end{aligned}$$

and

$$\phi_\varphi(t) = U(t,0)\phi_\varphi(0) + \int_0^t U(t,s)\left[f(s, \phi_\varphi(s)) + \varphi(s)\right] ds + \sum_{0\le\theta_k<t} U(t,\theta_k)W(\phi_\varphi(\theta_k)),$$

for $t \ge 0$ we obtain the inequality

$$\begin{aligned} e^{\omega t}\left\|y_\varphi(t, y_0) - \phi_\varphi(t)\right\| &\le N\left\|y_0 - \phi_\varphi(0)\right\| + \int_0^t NL_f e^{\omega s}\left\|y_\varphi(s, y_0) - \phi_\varphi(s)\right\| \\ &+ \sum_{0\le\theta_k<t} NL_W e^{\omega\theta_k}\left\|y_\varphi(\theta_k, y_0) - \phi_\varphi(\theta_k)\right\|. \end{aligned}$$

Applying the Gronwall-Bellman Lemma for piecewise continuous functions, one can verify that

$$\left\|y_\varphi(t, y_0) - \phi_\varphi(t)\right\| \le N(1+NL_W)^p\left\|y_0 - \phi_\varphi(0)\right\| e^{[-\omega+NL_f+(p/T)\ln(1+NL_W)]t},\ t \ge 0.$$

Consequently, in accordance with condition $(A8)$, we have that

$$\left\| y_\varphi(t, y_0) - \phi_\varphi(t) \right\| \to 0$$

as $t \to \infty$.

In the next section, we will prove that if the collection $\mathscr{A}$ is chaotic in the sense of Li–Yorke, then the same is true for the collection $\mathscr{B}$.

5.3 Chaotic Dynamics

The ingredients of Li–Yorke chaos for system (5.1.2) will be considered in Lemmas 5.2 and 5.3. The main conclusion of the present chapter will be stated in Theorem 5.1.

In the proof of the next assertion, we will denote by $i((a,b))$ the number of the terms of the sequence $\{\theta_k\}$ which belong to the interval (a,b), where a and b are real numbers such that $a < b$. Clearly, $i((a,b)) \le p + \frac{p}{T}(b-a)$.

Lemma 5.2 *Suppose that the conditions* $(A1)-(A8)$ *hold. If a couple of functions* $(\varphi(t), \overline{\varphi}(t)) \in \mathscr{A} \times \mathscr{A}$ *is proximal, then the same is true for the couple* $\left(\phi_\varphi(t), \phi_{\overline{\varphi}}(t)\right) \in \mathscr{B} \times \mathscr{B}$.

Proof Fix an arbitrary small positive number ε and an arbitrary large positive number E. Let us denote $\alpha = \omega - NL_f - \frac{p}{T}\ln(1 + NL_W)$ and set

$$\gamma = 1 + \frac{N}{\omega} + \frac{N^2 L_f}{\omega\alpha}(1 + NL_W)^p + \left(\frac{N^2 L_W}{\omega}\right)\left(\frac{pe^{2\alpha T}}{e^{\alpha T} - 1}\right)(1 + NL_W)^p.$$

Since the couple $(\varphi(t), \overline{\varphi}(t)) \in \mathscr{A} \times \mathscr{A}$ is proximal, there exists an interval $J = [\sigma, \sigma + E_1]$, where $E_1 \ge E$, such that $\|\varphi(t) - \overline{\varphi}(t)\| < \varepsilon/\gamma$ for $t \in J$.

For the sake of clarity, we will denote $y(t) = \phi_\varphi(t) \in \mathscr{B}$ and $\overline{y}(t) = \phi_{\overline{\varphi}}(t) \in \mathscr{B}$. The functions $y(t)$ and $\overline{y}(t)$ satisfy the relations

$$y(t) = \int_{-\infty}^{t} U(t,s)\left[f(s, y(s)) + \varphi(s)\right] ds + \sum_{\theta_k < t} U(t, \theta_k) W(y(\theta_k))$$

and

$$\overline{y}(t) = \int_{-\infty}^{t} U(t,s)\left[f(s, \overline{y}(s)) + \overline{\varphi}(s)\right] ds + \sum_{\theta_k < t} U(t, \theta_k) W(\overline{y}(\theta_k)),$$

respectively. By means of these relations, we obtain that

$$y(t)-\overline{y}(t)=\int_{-\infty}^{\sigma}U(t,s)\left[f(s,y(s))-f(s,\overline{y}(s))+\varphi(s)-\overline{\varphi}(s)\right]ds$$
$$+\int_{\sigma}^{t}U(t,s)\left[f(s,y(s))-f(s,\overline{y}(s))+\varphi(s)-\overline{\varphi}(s)\right]ds$$
$$+\sum_{\theta_k\le\sigma}U(t,\theta_k)\left[W(y(\theta_k))-W(\overline{y}(\theta_k))\right]$$
$$+\sum_{\sigma<\theta_k<t}U(t,\theta_k)\left[W(y(\theta_k))-W(\overline{y}(\theta_k))\right].$$

Using the inequalities

$$\left\|\sum_{\theta_k\le\sigma}U(t,\theta_k)\left[W(y(\theta_k))-W(\overline{y}(\theta_k))\right]\right\|\le\sum_{\theta_k\le\sigma}2NM_We^{-\omega(t-\theta_k)}$$
$$=2NM_We^{-\omega t}\sum_{l=0}^{\infty}\sum_{\sigma-(l+1)T<\theta_k\le\sigma-lT}e^{\omega\theta_k}$$
$$\le 2NM_We^{-\omega t}\sum_{l=0}^{\infty}pe^{\omega(\sigma-lT)}$$
$$=\frac{2pNM_W}{1-e^{-\omega T}}e^{-\omega(t-\sigma)}$$

and

$$\left\|\int_{-\infty}^{\sigma}U(t,s)\left[f(s,y(s))-f(s,\overline{y}(s))+\varphi(s)-\overline{\varphi}(s)\right]ds\right\|\le\frac{2N(M_f+H_0)}{\omega}e^{-\omega(t-\sigma)},$$

it can be verified for $t\in J$ that

$$\|y(t)-\overline{y}(t)\|\le\left(\frac{2N(M_f+H_0)}{\omega}+\frac{2pNM_W}{1-e^{-\omega T}}\right)e^{-\omega(t-\sigma)}+\frac{N\varepsilon}{\gamma\omega}\left(1-e^{-\omega(t-\sigma)}\right)$$
$$+NL_f\int_{\sigma}^{t}e^{-\omega(t-s)}\|y(s)-\overline{y}(s)\|\,ds+NL_W\sum_{\sigma<\theta_k<t}e^{-\omega(t-\theta_k)}\|y(\theta_k)-\overline{y}(\theta_k)\|.$$

Now, let $u(t)=e^{\omega t}\|y(t)-\overline{y}(t)\|$. Under the circumstances we have that

$$u(t)\le c+\frac{N\varepsilon}{\gamma\omega}e^{\omega t}+NL_f\int_{\sigma}^{t}u(s)ds+NL_W\sum_{\sigma<\theta_k<t}u(\theta_k),\ t\in J,$$

where $c=e^{\omega\sigma}\left(\frac{2N(M_f+H_0)}{\omega}+\frac{2pNM_W}{1-e^{-\omega T}}-\frac{N\varepsilon}{\gamma\omega}\right)$.

Implication of the analogue of Gronwall's Lemma for piecewise continuous functions leads to the inequality

$$u(t) \le \frac{N\varepsilon}{\gamma\omega}e^{\omega t} + c + \int_\sigma^t NL_f\left(\frac{N\varepsilon}{\gamma\omega}e^{\omega s} + c\right)(1+NL_W)^{i((s,t))}e^{NL_f(t-s)}ds$$
$$+ \sum_{\sigma<\theta_k<t} NL_W\left(\frac{N\varepsilon}{\gamma\omega}e^{\omega\theta_k} + c\right)(1+NL_W)^{i((\theta_k,t))}e^{NL_f(t-\theta_k)}.$$

With the aid of the equation

$$(1+NL_W)^{i((\sigma,t))}e^{NL_f(t-\sigma)} = 1 + \int_\sigma^t NL_f(1+NL_W)^{i((s,t))}e^{NL_f(t-s)}ds$$
$$+ \sum_{\sigma<\theta_k<t} NL_W(1+NL_W)^{i((\theta_k,t))}e^{NL_f(t-\theta_k)}$$

one can attain that

$$u(t) \le \frac{N\varepsilon}{\gamma\omega}e^{\omega t} + c(1+NL_W)^{i((\sigma,t))}e^{NL_f(t-\sigma)}$$
$$+\frac{N^2L_f\varepsilon}{\gamma\omega}\int_\sigma^t e^{\omega s}(1+NL_W)^{i((s,t))}e^{NL_f(t-s)}ds$$
$$+\frac{N^2L_W\varepsilon}{\gamma\omega}\sum_{\sigma<\theta_k<t} e^{\theta_k}(1+NL_W)^{i((\theta_k,t))}e^{NL_f(t-\theta_k)}$$
$$\le \frac{N\varepsilon}{\gamma\omega}e^{\omega t} + c(1+NL_W)^p e^{(\omega-\alpha)(t-\sigma)} + \frac{N^2L_f\varepsilon}{\gamma\omega\alpha}(1+NL_W)^p e^{\omega t}\left(1-e^{-\alpha(t-\sigma)}\right)$$
$$+\frac{N^2L_W\varepsilon}{\gamma\omega}(1+NL_W)^p e^{(\omega-\alpha)t}\sum_{\sigma<\theta_k<t} e^{\alpha\theta_k}.$$

Let $q = q(t) = \left\lfloor \frac{t-\sigma}{T} \right\rfloor$, that is, q is the greatest integer which not larger than $\frac{t-\sigma}{T}$. Under the circumstances we have that

$$\sum_{\sigma<\theta_k<t} e^{\alpha\theta_k} \le \sum_{\sigma<\theta_k<\sigma+(q+1)T} e^{\alpha\theta_k} \le \sum_{l=0}^{q}\sum_{\sigma+lT\le\theta_k<\sigma+(l+1)T} e^{\alpha\theta_k} < \sum_{l=0}^{q} pe^{\alpha[\sigma+(l+1)T]}$$
$$= pe^{(\sigma+T)\alpha}\frac{e^{(q+1)\alpha T}-1}{e^{\alpha T}-1} \le \frac{pe^{(\sigma+T)\alpha}}{e^{\alpha T}-1}\left(e^{\alpha(t-\sigma+T)}-1\right) < \frac{pe^{2\alpha T}}{e^{\alpha T}-1}e^{\alpha t}.$$

The last inequality implies that

$$
\begin{aligned}
u(t) &< \frac{N\varepsilon}{\gamma\omega} e^{\omega t}\left(1-(1+NL_W)^p e^{-\alpha(t-\sigma)}\right) \\
&+\frac{N^2 L_f \varepsilon}{\gamma\omega\alpha}(1+NL_W)^p e^{\omega t}\left(1-e^{-\alpha(t-\sigma)}\right) \\
&+\left(\frac{N^2 L_W \varepsilon}{\gamma\omega}\right)\left(\frac{pe^{2\alpha T}}{e^{\alpha T}-1}\right)(1+NL_W)^p e^{\omega t} \\
&+(1+NL_W)^p\left(\frac{2N(M_f+H_0)}{\omega}+\frac{2pNM_W}{1-e^{-\omega T}}\right)e^{\omega t}e^{-\alpha(t-\sigma)},
\end{aligned}
$$

and multiplying both sides by $e^{-\omega t}$ one can obtain the inequality

$$
\begin{aligned}
\|y(t)-\overline{y}(t)\| &< \frac{N\varepsilon}{\gamma\omega}\left(1-(1+NL_W)^p e^{-\alpha(t-\sigma)}\right) \\
&+\frac{N^2 L_f \varepsilon}{\gamma\omega\alpha}(1+NL_W)^p\left(1-e^{-\alpha(t-\sigma)}\right) \\
&+\left(\frac{N^2 L_W \varepsilon}{\gamma\omega}\right)\left(\frac{pe^{2\alpha T}}{e^{\alpha T}-1}\right)(1+NL_W)^p \\
&+(1+NL_W)^p\left(\frac{2N(M_f+H_0)}{\omega}+\frac{2pNM_W}{1-e^{-\omega T}}\right)e^{-\alpha(t-\sigma)} \\
&< \frac{N\varepsilon}{\gamma\omega}+\frac{N^2 L_f \varepsilon}{\gamma\omega\alpha}(1+NL_W)^p+\left(\frac{N^2 L_W \varepsilon}{\gamma\omega}\right)\left(\frac{pe^{2\alpha T}}{e^{\alpha T}-1}\right)(1+NL_W)^p \\
&+(1+NL_W)^p\left(\frac{2N(M_f+H_0)}{\omega}+\frac{2pNM_W}{1-e^{-\omega T}}\right)e^{-\alpha(t-\sigma)}.
\end{aligned}
$$

Set $\beta=(1+NL_W)^p\left(\frac{2N(M_f+H_0)}{\omega}+\frac{2pNM_W}{1-e^{-\omega T}}\right)$, and suppose that the number E is sufficiently large such that $E \geq \frac{2}{\alpha}\ln\left(\frac{\gamma\beta}{\varepsilon}\right)$. In this case, $\beta e^{-\alpha(t-\sigma)} < \varepsilon/\gamma$ for $t \in [\sigma+E/2, \sigma+E_1]$. Thus, the inequality

$$
\begin{aligned}
&\|y(t)-\overline{y}(t)\| \\
&< \frac{\varepsilon}{\gamma}\left[1+\frac{N}{\omega}+\frac{N^2 L_f}{\omega\alpha}(1+NL_W)^p+\left(\frac{N^2 L_W}{\omega}\right)\left(\frac{pe^{2\alpha T}}{e^{\alpha T}-1}\right)(1+NL_W)^p\right] \\
&= \varepsilon
\end{aligned}
$$

holds for $t \in [\sigma+E/2, \sigma+E_1]$. The interval $\widetilde{J}=[\sigma+E/2, \sigma+E_1]$ has a length no less than $E/2$. Consequently, the couple $\left(\phi_{\varphi}(t), \phi_{\overline{\varphi}}(t)\right) \in \mathscr{B}\times\mathscr{B}$ is proximal.

Next, we shall continue with the second ingredient of Li–Yorke chaos in the following lemma.

Lemma 5.3 *Suppose that the conditions* $(A1)-(A7)$, $(A9)$ *are fulfilled. If a couple of functions* $(\varphi(t), \overline{\varphi}(t)) \in \mathscr{A} \times \mathscr{A}$ *are frequently* (ε_0, Δ)*-separated for some positive numbers* ε_0 *and* Δ, *then the couple* $\left(\phi_{\varphi}(t), \phi_{\overline{\varphi}}(t)\right) \in \mathscr{B} \times \mathscr{B}$ *are frequently* $(\varepsilon_1, \overline{\Delta})$*-separated for some positive numbers* ε_1 *and* $\overline{\Delta}$.

Proof Since the couple of functions $(\varphi(t), \overline{\varphi}(t)) \in \mathscr{A} \times \mathscr{A}$ is frequently (ε_0, Δ) separated for some positive numbers ε_0 and Δ, there exists infinitely many disjoint intervals J_i, $i \in \mathbb{N}$, each with a length no less than Δ, such that

$$\|\varphi(t) - \overline{\varphi}(t)\| > \varepsilon_0$$

for each t from these intervals. Without loss of generality, suppose that the intervals $J_i, i \in \mathbb{N}$, are all open subsets of $\mathbb{R}$. In that case, one can find a sequence $\{\Delta_i\}$ satisfying $\Delta_i \geq \Delta$, $i \in \mathbb{N}$, and a sequence $\{\alpha_i\}$, $\alpha_i \to \infty$ as $i \to \infty$, such that $J_i = (\alpha_i, \alpha_i + \Delta_i)$.

In the proof, we will verify the existence of positive numbers ε_1, $\overline{\Delta}$ and infinitely many disjoint intervals J_i^1, $i \in \mathbb{N}$, each with length $\overline{\Delta}$, such that the inequality $\left\|\phi_{\varphi}(t) - \phi_{\overline{\varphi}}(t)\right\| > \varepsilon_1$ holds for each t from these intervals.

Suppose that

$$\varphi(t) = (\varphi_1(t), \varphi_2(t), \ldots, \varphi_n(t))$$

and

$$\overline{\varphi}(t) = \left(\overline{\varphi}_1(t), \overline{\varphi}_2(t), \ldots, \overline{\varphi}_n(t)\right),$$

where φ_j and $\overline{\varphi}_j$, $1 \leq j \leq n$, are real-valued functions. According to the equicontinuity of the collection $\mathscr{A}$, one can find a positive number $\tau < \Delta$, such that for any $t_1, t_2 \in \mathbb{R}$ with $|t_1 - t_2| < \tau$, the inequality

$$\left|\left(\varphi_j(t_1) - \overline{\varphi}_j(t_1)\right) - \left(\varphi_j(t_2) - \overline{\varphi}_j(t_2)\right)\right| < \frac{\varepsilon_0}{2n} \tag{5.3.4}$$

holds for all $1 \leq j \leq n$.

For each i, let $\eta_i = \alpha_i + \dfrac{\Delta_i}{2}$. That is, η_i is the midpoint of the interval J_i. Moreover, define a sequence $\{\zeta_i\}$, $i \in \mathbb{N}$, through the equation $\zeta_i = \eta_i - \dfrac{\tau}{2}$.

Fix a natural number i. For each $t \in J_i$, there exists an integer $j_i = j_i(t)$, $1 \leq j_i \leq n$, such that

$$\left|\varphi_{j_i}(t) - \overline{\varphi}_{j_i}(t)\right| \geq \frac{1}{n} \|\varphi(t) - \overline{\varphi}(t)\|.$$

Otherwise, if there exists $t_0 \in J_i$ such that for all $1 \leq j \leq n$ the inequality

$$\left|\varphi_j(t_0) - \overline{\varphi}_j(t_0)\right| < \frac{1}{n} \|\varphi(t_0) - \overline{\varphi}(t_0)\|$$

holds, then we encounter with a contradiction since

$$\|\varphi(t_0)-\overline{\varphi}(t_0)\| \leq \sum_{j=1}^{n} \left|\varphi_j(t_0)-\overline{\varphi}_j(t_0)\right| < \|\varphi(t_0)-\overline{\varphi}(t_0)\|.$$

For this reason, there exists an integer $j_i = j_i(\eta_i)$, $1 \leq j_i \leq n$, such that

$$\left|\varphi_{j_i}(\eta_i)-\overline{\varphi}_{j_i}(\eta_i)\right| \geq \frac{1}{n}\|\varphi(\eta_i)-\overline{\varphi}(\eta_i)\| > \frac{\varepsilon_0}{n}. \tag{5.3.5}$$

On the other hand, making use of the inequality (5.3.4), it is easy to verify for all $t \in [\zeta_i, \zeta_i+\tau]$ that

$$\begin{aligned} &\left|\varphi_{j_i}(\eta_i)-\overline{\varphi}_{j_i}(\eta_i)\right| - \left|\varphi_{j_i}(t)-\overline{\varphi}_{j_i}(t)\right| \\ &\leq \left|\left(\varphi_{j_i}(t)-\overline{\varphi}_{j_i}(t)\right) - \left(\varphi_{j_i}(\eta_i)-\overline{\varphi}_{j_i}(\eta_i)\right)\right| \\ &< \frac{\varepsilon_0}{2n}. \end{aligned}$$

Therefore, by virtue of (5.3.5), we obtain the inequality

$$\left|\varphi_{j_i}(t)-\overline{\varphi}_{j_i}(t)\right| > \left|\varphi_{j_i}(\eta_i)-\overline{\varphi}_{j_i}(\eta_i)\right| - \frac{\varepsilon_0}{2n} > \frac{\varepsilon_0}{2n},\ t \in [\zeta_i, \zeta_i+\tau]. \tag{5.3.6}$$

It is possible to find numbers $s_1^i, s_2^i, \ldots, s_n^i \in [\zeta_i, \zeta_i+\tau]$ such that

$$\begin{aligned} &\int_{\zeta_i}^{\zeta_i+\tau} (\varphi(s)-\overline{\varphi}(s))\, ds \\ &= \tau\left(\varphi_1(s_1^i)-\overline{\varphi}_1(s_1^i), \varphi_2(s_2^i)-\overline{\varphi}_2(s_2^i), \ldots, \varphi_n(s_n^i)-\overline{\varphi}_n(s_n^i)\right). \end{aligned}$$

Hence, the inequality (5.3.6) implies that

$$\left\|\int_{\zeta_i}^{\zeta_i+\tau} (\varphi(s)-\overline{\varphi}(s))\, ds\right\| \geq \tau\left|\varphi_{j_i}(s_{j_i}^i)-\overline{\varphi}_{j_i}(s_{j_i}^i)\right| > \frac{\tau\varepsilon_0}{2n}. \tag{5.3.7}$$

For the sake of clarity, let us denote $y(t) = \phi_\varphi(t)$ and $\overline{y}(t) = \phi_{\overline{\varphi}}(t)$. For $t \in [\zeta_i, \zeta_i+\tau]$, using the couple of relations

$$y(t) = y(\zeta_i) + \int_{\zeta_i}^{t} [Ay(s)+f(s,y(s))+\varphi(s)]\, ds + \sum_{\zeta_i \leq \theta_k < t} [By(\theta_k)+W(y(\theta_k))]$$

and

$$\overline{y}(t) = \overline{y}(\zeta_i) + \int_{\zeta_i}^{t} [A\overline{y}(s) + f(s, \overline{y}(s)) + \overline{\varphi}(s)]\, ds + \sum_{\zeta_i \le \theta_k < t} \left[B\overline{y}(\theta_k) + W(\overline{y}(\theta_k))\right],$$

one can verify that

$$\|y(\zeta_i + \tau) - \overline{y}(\zeta_i + \tau)\| \ge \left\| \int_{\zeta_i}^{\zeta_i + \tau} (\varphi(s) - \overline{\varphi}(s))\, ds \right\| - \|y(\zeta_i) - \overline{y}(\zeta_i)\|$$
$$- \int_{\zeta_i}^{\zeta_i+\tau} (\|A\| + L_f)\, \|y(s) - \overline{y}(s)\|\, ds - \sum_{\zeta_i \le \theta_k < \zeta_i + \tau} (\|B\| + L_W)\, \|y(\theta_k) - \overline{y}(\theta_k)\|.$$

Making use of the last inequality together with (5.3.7), we obtain that

$$\sup_{t \in [\zeta_i, \zeta_i + \tau]} \|y(t) - \overline{y}(t)\| > \frac{\tau \varepsilon_0}{2n} - \left[1 + \tau(\|A\| + L_f)\right] \sup_{t \in [\zeta_i, \zeta_i + \tau]} \|y(t) - \overline{y}(t)\|$$
$$- \frac{p}{T}(T + \tau)\, (\|B\| + L_W) \sup_{t \in [\zeta_i, \zeta_i + \tau]} \|y(t) - \overline{y}(t)\|,$$

and therefore $\sup_{t \in [\zeta_i, \zeta_i + \tau]} \|y(t) - \overline{y}(t)\| > M_0$, where

$$M_0 = \frac{\tau \varepsilon_0}{2n\left[2 + \tau(\|A\| + L_f) + \frac{p}{T}(T + \tau)\, (\|B\| + L_W)\right]}.$$

Set $\underline{\theta} = \min_{1 \le k \le p} (\theta_{k+1} - \theta_k)$, and define the numbers

$$\varepsilon_1 = \frac{M_0}{2} \min \left\{ \frac{1 - L_W \left\|(I + B)^{-1}\right\|}{\left\|(I + B)^{-1}\right\|}, \frac{1}{1 + \|B\| + L_W} \right\}$$

and

$$\overline{\Delta} = \min \Bigg\{ \underline{\theta}, \frac{M_0}{4\left[(\|A\| + L_f)K_0 + H_0\right](2 + \|B\| + L_W)},$$
$$\frac{M_0 \left(1 - L_W \left\|(I + B)^{-1}\right\|\right)}{4\left[(\|A\| + L_f)K_0 + H_0\right]\left[1 + (1 - L_W)\left\|(I + B)^{-1}\right\|\right]} \Bigg\}.$$

First, suppose that there exists $\xi_i \in [\zeta_i, \zeta_i + \tau]$ such that

$$\sup_{t \in [\zeta_i, \zeta_i + \tau]} \|y(t) - \overline{y}(t)\| = \|y(\xi_i) - \overline{y}(\xi_i)\|.$$

Let $\zeta_i^1 = \begin{cases} \xi_i, & \text{if } \xi_i \le \zeta_i + \tau/2 \\ \xi_i - \overline{\Delta}, & \text{if } \xi_i > \zeta_i + \tau/2 \end{cases}$. Since $\overline{\Delta} \le \underline{\theta}$, there exists at most one impulsive moment on the interval $(\zeta_i^1, \zeta_i^1 + \overline{\Delta})$.

We shall start by considering the case $\xi_i > \zeta_i + \dfrac{\tau}{2}$. Assume that there exists an impulsive moment $\theta_j \in (\zeta_i^1, \zeta_i^1 + \overline{\Delta})$. Under the circumstances, one can verify for $t \in (\theta_j, \zeta_i^1 + \overline{\Delta})$ that

$$\begin{aligned}
\|y(t) - \overline{y}(t)\| &\ge \|y(\xi_i) - \overline{y}(\xi_i)\| - \left\| \int_{\xi_i}^{t} A(y(s) - \overline{y}(s))ds \right\| \\
&- \left\| \int_{\xi_i}^{t} [f(s, y(s)) - f(s, \overline{y}(s))]\, ds \right\| - \left\| \int_{\xi_i}^{t} [\varphi(s) - \overline{\varphi}(s)]\, ds \right\| \\
&> M_0 - 2\overline{\Delta}\left[\left(\|A\| + L_f\right) K_0 + H_0\right] \\
&> \frac{M_0}{2} \\
&> \varepsilon_1.
\end{aligned}$$

In particular, the inequality $\left\|y(\theta_j+) - \overline{y}(\theta_j+)\right\| > M_0 - 2\overline{\Delta}\left[\left(\|A\| + L_f\right) K_0 + H_0\right]$ is valid. Taking advantage of the relations $y(\theta_j+) = (I + B)y(\theta_j) + W(y(\theta_j))$ and $\overline{y}(\theta_j+) = (I + B)\overline{y}(\theta_j) + W\left(\overline{y}(\theta_j)\right)$, we obtain that

$$\left\|y(\theta_j) - \overline{y}(\theta_j)\right\| \ge \frac{\left\|y(\theta_j+) - \overline{y}(\theta_j+)\right\|}{1 + \|B\| + L_W} > \frac{M_0 - 2\overline{\Delta}\left[\left(\|A\| + L_f\right) K_0 + H_0\right]}{1 + \|B\| + L_W}.$$

The last inequality implies for $t \in (\zeta_i^1, \theta_j]$ that

$$\begin{aligned}
\|y(t) - \overline{y}(t)\| &\ge \left\|y(\theta_j) - \overline{y}(\theta_j)\right\| - \left\| \int_{\theta_j}^{t} A(y(s) - \overline{y}(s))ds \right\| \\
&- \left\| \int_{\theta_j}^{t} [f(s, y(s)) - f(s, \overline{y}(s))]\, ds \right\| - \left\| \int_{\theta_j}^{t} [\varphi(s) - \overline{\varphi}(s)]\, ds \right\| \\
&> \frac{M_0 - 2\overline{\Delta}\left[\left(\|A\| + L_f\right) K_0 + H_0\right]}{1 + \|B\| + L_W} - 2\overline{\Delta}\left[\left(\|A\| + L_f\right) K_0 + H_0\right] \\
&= \frac{1}{1 + \|B\| + L_W}\left[M_0 - 2\overline{\Delta}\left(2 + \|B\| + L_W\right)\left(\left(\|A\| + L_f\right) K_0 + H_0\right)\right] \\
&\ge \frac{M_0}{2\left(1 + \|B\| + L_W\right)} \\
&\ge \varepsilon_1.
\end{aligned}$$

On the other hand, if none of the impulsive moments belong to $(\zeta_i^1, \zeta_i^1 + \overline{\Delta})$, then for each t from this interval we have that $\|y(t) - \overline{y}(t)\| > \dfrac{M_0}{2}$. Therefore, in the case of $\xi_i > \zeta_i + \frac{\tau}{2}$, the inequality $\|y(t) - \overline{y}(t)\| > \varepsilon_1$ holds for all $t \in (\zeta_i^1, \zeta_i^1 + \overline{\Delta})$, regardless of the existence of an impulsive moment in this interval.

Next, we continue with the case $\xi_i \leq \zeta_i + \frac{\tau}{2}$. If there exists an impulsive moment $\theta_j \in (\zeta_i^1, \zeta_i^1 + \overline{\Delta})$, then it is easy to show for $t \in (\zeta_i^1, \theta_j]$ that

$$\|y(t) - \overline{y}(t)\| > M_0 - 2\overline{\Delta}\left[\left(\|A\| + L_f\right) K_0 + H_0\right] > \varepsilon_1.$$

Since $\left\|y(\theta_j) - \overline{y}(\theta_j)\right\| > M_0 - 2\overline{\Delta}\left[\left(\|A\| + L_f\right) K_0 + H_0\right]$, the condition $(A9)$ implies that

$$\begin{aligned}
&\left\|y(\theta_j+) - \overline{y}(\theta_j+)\right\| \geq \left\|(I + B)(y(\theta_j) - \overline{y}(\theta_j))\right\| - L_W \left\|y(\theta_j) - \overline{y}(\theta_j)\right\| \\
&\geq \frac{\left\|y(\theta_j) - \overline{y}(\theta_j)\right\|}{\left\|(I + B)^{-1}\right\|} - L_W \left\|y(\theta_j) - \overline{y}(\theta_j)\right\| \\
&> \left(\frac{1 - L_W \left\|(I + B)^{-1}\right\|}{\left\|(I + B)^{-1}\right\|}\right)\left[M_0 - 2\overline{\Delta}\left(\left(\|A\| + L_f\right) K_0 + H_0\right)\right].
\end{aligned}$$

Making use of the last inequality, we attain for all $t \in (\theta_j, \zeta_i^1 + \overline{\Delta})$ that

$$\begin{aligned}
&\|y(t) - \overline{y}(t)\| \geq \left\|y(\theta_j+) - \overline{y}(\theta_j+)\right\| - 2\overline{\Delta}\left[(\|A\| + L_f)K_0 + H_0\right] \\
&> \left(\frac{1 - L_W \left\|(I + B)^{-1}\right\|}{\left\|(I + B)^{-1}\right\|}\right) \\
&\times \left[M_0 - 2\overline{\Delta}\left(\left(\|A\| + L_f\right) K_0 + H_0\right)\right] - 2\overline{\Delta}\left[(\|A\| + L_f)K_0 + H_0\right] \\
&\geq \frac{M_0}{2}\left(\frac{1 - L_W \left\|(I + B)^{-1}\right\|}{\left\|(I + B)^{-1}\right\|}\right).
\end{aligned}$$

Therefore, for all $t \in (\zeta_i^1, \zeta_i^1 + \overline{\Delta})$ it is clear that $\|y(t) - \overline{y}(t)\| > \varepsilon_1$. Besides, the same inequality holds even if the interval $(\zeta_i^1, \zeta_i^1 + \overline{\Delta})$ does not contain an impulsive moment.

Now, suppose that there exists an impulsive moment $\theta_l \in [\zeta_i, \zeta_i + \tau]$ such that

$$\sup_{t \in [\zeta_i, \zeta_i + \tau]} \|y(t) - \overline{y}(t)\| = \|y(\theta_l+) - \overline{y}(\theta_l+)\|.$$

Let us define $\zeta_i^1 = \begin{cases} \theta_l, & \text{if } \theta_l \leq \zeta_i + \tau/2 \\ \theta_l - \overline{\Delta}, & \text{if } \theta_l > \zeta_i + \tau/2 \end{cases}$. In the case that $\theta_l > \zeta_i + \frac{\tau}{2}$, taking advantage of the inequality

$$\|y(\theta_l) - \overline{y}(\theta_l)\| \geq \frac{\|y(\theta_l+) - \overline{y}(\theta_l+)\|}{1 + \|B\| + L_W} > \frac{M_0}{1 + \|B\| + L_W},$$

one can verify that

$$\begin{aligned}\|y(t)-\overline{y}(t)\| &\geq \|y(\theta_l)-\overline{y}(\theta_l)\| - 2\overline{\Delta}\left[(\|A\|+L_f)K_0+H_0\right] \\ &> \frac{M_0}{2(1+\|B\|+L_W)} \\ &\geq \varepsilon_1,\end{aligned}$$

for all $t \in (\zeta_i^1, \zeta_i^1 + \overline{\Delta})$. In a similar way, if $\theta_l \leq \zeta_i + \dfrac{\tau}{2}$, then we have for $t \in (\zeta_i^1, \zeta_i^1 + \overline{\Delta})$ that

$$\|y(t)-\overline{y}(t)\| \geq \|y(\theta_l+)-\overline{y}(\theta_l+)\| - 2\overline{\Delta}\left[(\|A\|+L_f)K_0+H_0\right] > \frac{M_0}{2} > \varepsilon_1.$$

Consequently, on each of the intervals $J_i^1 = (\zeta_i^1, \zeta_i^1 + \overline{\Delta})$, $i \in \mathbb{N}$, the inequality $\|y(t)-\overline{y}(t)\| > \varepsilon_1$ holds. Therefore, the couple of functions $\left(\phi_\varphi(t), \phi_{\overline{\varphi}}(t)\right) \in \mathscr{B} \times \mathscr{B}$ is frequently $(\varepsilon_1, \overline{\Delta})$-separated.

The main theorem of the present chapter is as follows.

Theorem 5.1 *Suppose that the conditions* $(A1)-(A9)$ *are valid. If* $\mathscr{A}$ *is a Li–Yorke chaotic set which possesses an* mT*-periodic function for each natural number* m*, then* $\mathscr{B}$ *is also a Li–Yorke chaotic set.*

Proof By means of conditions $(A1)-(A7)$, one can confirm that if $\varphi(t) \in \mathscr{A}$ is mT-periodic for some natural number m, then $\phi_\varphi(t) \in \mathscr{B}$ is a periodic function with the same period, and vice versa.

Suppose that the set $\mathscr{C}_{\mathscr{A}}$ is a scrambled set inside $\mathscr{A}$, and define the set

$$\mathscr{C}_{\mathscr{B}} = \left\{\phi_{\varphi(t)}(t) \mid \varphi(t) \in \mathscr{C}_{\mathscr{A}}\right\}.$$

It is easy to verify that there is a one-to-one correspondence between the sets $\mathscr{C}_{\mathscr{A}}$ and $\mathscr{C}_{\mathscr{B}}$. Since the set $\mathscr{C}_{\mathscr{A}}$ is uncountable, the same is true for $\mathscr{C}_{\mathscr{B}}$. Moreover, no periodic functions exist inside $\mathscr{C}_{\mathscr{B}}$, since no such functions take place inside the set $\mathscr{C}_{\mathscr{A}}$.

Because each pair of functions that belong to $\mathscr{C}_{\mathscr{A}} \times \mathscr{C}_{\mathscr{A}}$ is proximal, Lemma 5.2 implies the same feature for each pair inside $\mathscr{C}_{\mathscr{B}} \times \mathscr{C}_{\mathscr{B}}$. In connection with Lemma 5.3, there exists positive numbers ε_1 and $\overline{\Delta}$ such that each couple of functions from $\mathscr{C}_{\mathscr{B}} \times \mathscr{C}_{\mathscr{B}}$ are frequently $(\varepsilon_1, \overline{\Delta})$-separated. Hence, $\mathscr{C}_{\mathscr{B}}$ is a scrambled set inside $\mathscr{B}$. If we denote by $\mathscr{P}_{\mathscr{A}}$ and $\mathscr{P}_{\mathscr{B}}$ the sets of periodic functions inside $\mathscr{A}$ and $\mathscr{B}$, respectively, then a similar discussion holds for each pair inside $\mathscr{C}_{\mathscr{B}} \times \mathscr{P}_{\mathscr{B}}$, since the same is true for any pair from the set $\mathscr{C}_{\mathscr{A}} \times \mathscr{P}_{\mathscr{A}}$. Consequently, the collection $\mathscr{B}$ is Li–Yorke chaotic.

In the next section, we will apply our results to an impulsive Duffing oscillator.

5.4 An Example

This section is devoted to an illustrative example. In Theorem 5.1 we have shown that if the perturbation term $\varphi(t)$ in system (5.1.2) belongs to a collection which is chaotic in the sense of Li–Yorke, then system (5.1.2) exhibits chaotic motions. Therefore, to actualize our results, we need a source of Li–Yorke chaotic functions. To construct such a collection, we will consider a Duffing oscillator which is forced with a relay function [45–49]. The switching moments of the relay function are generated through the logistic map

$$F_\mu(s) = \mu s(1-s), \tag{5.4.8}$$

which is chaotic in the sense of Li–Yorke for the parameter μ between 3.84 and 4 [2]. We note that the interval $[0, 1]$ is invariant under the iterations of the map $F_\mu(s)$ if $0 < \mu \leq 4$ [50]. An impulsive Duffing oscillator will be used for the main illustration. Moreover, in the example, a procedure to control the chaos of the impulsive system will be presented. In our evaluations, we will make use of the usual Euclidean norm [21].

Example 5.1 Consider, the forced Duffing oscillator

$$z''(t) + 0.6z'(t) + 5z(t) - 0.02z^3(t) = \nu(t, t_0, \mu), \tag{5.4.9}$$

where $t \in \mathbb{R}$, t_0 belongs to the interval $[0, 1]$ and the relay function $\nu(t, t_0)$ is defined as

$$\nu(t, t_0, \mu) = \begin{cases} 0.6, & \text{if } \zeta_{2j}(t_0, \mu) < t \leq \zeta_{2j+1}(t_0, \mu), \ j \in \mathbb{Z}, \\ 2.5, & \text{if } \zeta_{2j-1}(t_0, \mu) < t \leq \zeta_{2j}(t_0, \mu), \ j \in \mathbb{Z}, \end{cases}. \tag{5.4.10}$$

In Eq. (5.4.10), the switching moments $\zeta_j(t_0, \mu)$, $j \in \mathbb{Z}$, are defined through the equation $\zeta_j(t_0, \mu) = j + \kappa_j(t_0, \mu)$, where the sequence $\{\kappa_j(t_0, \mu)\}$, $\kappa_0(t_0, \mu) = t_0$, is generated by the logistic map (5.4.8), that is, $\kappa_{j+1}(t_0, \mu) = F_\mu(\kappa_j(t_0, \mu))$. More information about the dynamics of relay systems can be found in [45–49].

By means of the variables $z_1 = z$ and $z_2 = z'$, Eq. (5.4.9) can be reduced to the system

$$\begin{aligned} z_1'(t) &= z_2(t), \\ z_2'(t) &= -5z_1(t) - 0.6z_2(t) + 0.02z_1^3(t) + \nu(t, t_0, \mu). \end{aligned} \tag{5.4.11}$$

According to the results of [45], system (5.4.11) with the parameter value $\mu = 3.9$ is Li–Yorke chaotic. Moreover, for each natural number m, the system admits different unstable periodic solutions with periods $2m$.

In system (5.4.11) we set $\mu = 3.9$, and represent in Fig. 5.1 the z_1 and z_2 coordinates of the solution of the system with $z_1(t_0) = 0.492$ and $z_2(t_0) = -0.143$, where $t_0 = 0.385$. It is seen in Fig. 5.1 that system (5.4.11) possesses chaotic motions.

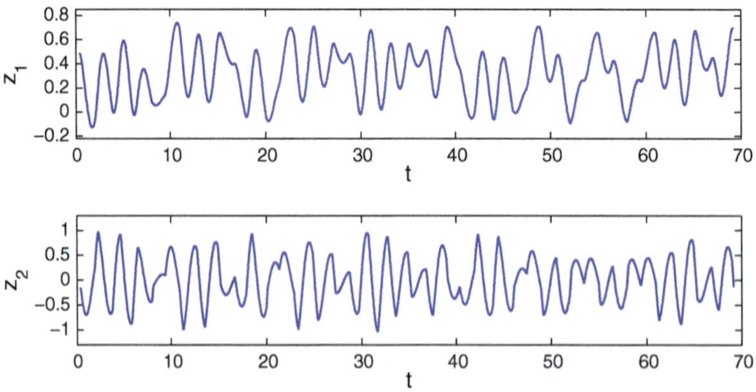

Fig. 5.1 The chaotic behavior of system (5.4.11) with $\mu = 3.9$. The initial data $z_1(t_0) = 0.492$, $z_2(t_0) = -0.143$, where $t_0 = 0.385$, is used in the simulation.

The function $h(z_1, z_2) = (z_1 + 0.5z_1^3, z_2)$ satisfies the inequality (5.2.3) with $L_1 = 1/\sqrt{2}$ and $L_2 = 1.96\sqrt{2}$ on the compact region in which the chaotic attractor of system (5.4.11) with $\mu = 3.9$ takes place. Therefore, the collection consisting of functions of the form $(z_1(t) + 0.5z_1^3(t), z_2(t))$, where $(z_1(t), z_2(t))$ is a bounded on $\mathbb{R}$ solution of system (5.4.11), is Li–Yorke chaotic.

Next, we take into account the impulsive Duffing oscillator

$$\begin{aligned} &x''(t) + 2x'(t) + 3x(t) + 0.025x^3(t) = 0.2\cos(\pi t), \ t \neq \theta_k, \\ &\Delta x|_{t=\theta_k} = -\frac{3}{4}x(\theta_k), \\ &\Delta x'|_{t=\theta_k} = -\frac{3}{4}x'(\theta_k) + 0.05(x'(\theta_k))^2, \end{aligned} \tag{5.4.12}$$

where $t \in \mathbb{R}$ and $\theta_k = 2k$, $k \in \mathbb{Z}$. Clearly, $\theta_{k+p} = \theta_k + T$, where $p = 1$ and $T = 2$.

Defining the new variables $x_1 = x$ and $x_2 = x'$, one can reduce (5.4.12) to the system

$$\begin{aligned} &x_1'(t) = x_2(t), \\ &x_2'(t) = -3x_1(t) - 2x_2(t) - 0.025x_1^3(t) + 0.2\cos(\pi t), \ t \neq \theta_k, \\ &\Delta x_1|_{t=\theta_k} = -\frac{3}{4}x_1(\theta_k), \\ &\Delta x_2|_{t=\theta_k} = -\frac{3}{4}x_2(\theta_k) + 0.05(x_2(\theta_k))^2. \end{aligned} \tag{5.4.13}$$

Let us demonstrate numerically that the system (5.4.13) possesses an asymptotically stable periodic solution. Figure 5.2 shows the graphs of the x_1 and x_2 coordinates of system (5.4.13). The initial data $x_1(1) = -0.019$, $x_2(1) = -0.056$ is used in the simulation. The existence of an asymptotically stable periodic solution is observable in the figure, and therefore, one can conclude that system (5.4.13) is not chaotic.

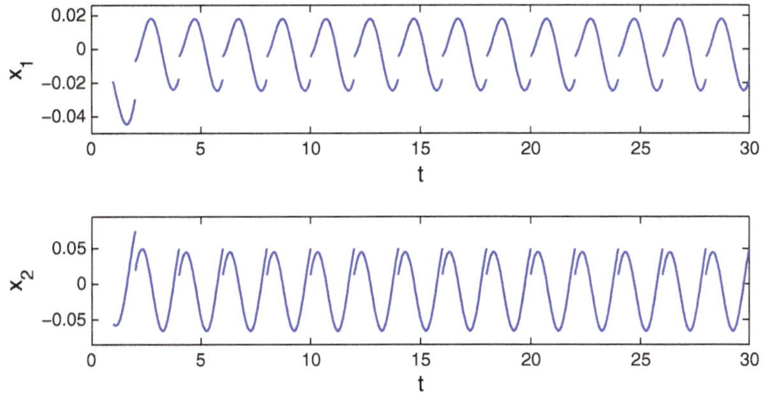

Fig. 5.2 The graphs of the x_1 and x_2 coordinates of system (5.4.13).

We perturb (5.4.13) by the solutions of (5.4.11) to set up the system

$$\begin{aligned} y_1'(t) &= y_2(t) + z_1(t) + 0.5z_1^3(t), \\ y_2'(t) &= -3y_1(t) - 2y_2(t) - 0.025y_1^3(t) + 0.2\cos(\pi t) + z_2(t), \ t \neq \theta_k, \\ \Delta y_1|_{t=\theta_k} &= -\frac{3}{4}y_1(\theta_k), \\ \Delta y_2|_{t=\theta_k} &= -\frac{3}{4}y_2(\theta_k) + 0.05(y_2(\theta_k))^2. \end{aligned} \tag{5.4.14}$$

System (5.4.14) is in the form of (5.1.2), where

$$A = \begin{pmatrix} 0 & 1 \\ -3 & -2 \end{pmatrix}, \ B = \begin{pmatrix} -3/4 & 0 \\ 0 & -3/4 \end{pmatrix},$$

$f(t, y_1, y_2) = (0, -0.025y_1^3 + 0.2\cos(\pi t))$ and $W(y_1, y_2) = (0, 0.05y_2^2)$. The matrices A and B commute, and the matrix $A + \frac{p}{T}\ln(I + B) = \begin{pmatrix} -\ln 2 & 1 \\ -3 & -2-\ln 2 \end{pmatrix}$ has eigenvalues $\lambda_{1,2} = -1 - \ln 2 \pm i\sqrt{2}$.

Let us denote by $U(t, s)$ the transition matrix of the linear homogeneous system

$$\begin{aligned} u_1'(t) &= u_2(t), \\ u_2'(t) &= -3u_1(t) - 2u_2(t), \ t \neq \theta_k, \\ \Delta u_1|_{t=\theta_k} &= -\frac{3}{4}u_1(\theta_k), \\ \Delta u_2|_{t=\theta_k} &= -\frac{3}{4}u_2(\theta_k). \end{aligned} \tag{5.4.15}$$

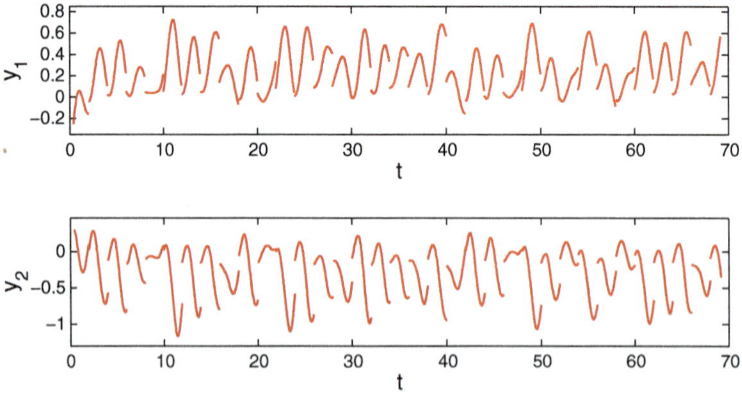

Fig. 5.3 The chaotic behavior of the perturbed impulsive Duffing oscillator (5.4.14). The initial data $y_1(t_0) = -0.254$, $y_2(t_0) = 0.297$ is used, where $t_0 = 0.385$. The irregular behavior observed in the graphs support our theoretical results.

One can verify that

$$U(t,s) = e^{-(t-s)} P \begin{pmatrix} \cos(\sqrt{2}(t-s)) & -\sin(\sqrt{2}(t-s)) \\ \sin(\sqrt{2}(t-s)) & \cos(\sqrt{2}(t-s)) \end{pmatrix} P^{-1} (I+B)^{i([s,t))},\ t > s,$$

where $i([s,t))$ is the number of the terms of the sequence $\{\theta_k\}$ that belong to the interval $[s,t)$ and $P = \begin{pmatrix} 0 & 1 \\ \sqrt{2} & -1 \end{pmatrix}$. The inequality $\|U(t,s)\| \leq N e^{-\omega(t-s)}$ holds for $t \geq s$, where $\omega = 1$ and $N = 2.415$.

The conditions $(A4)-(A9)$ are valid for system (5.4.14) with $M_f = 0.2183$, $M_W = 0.0845$, $L_f = 0.0608$, and $L_W = 0.13$. Thus, according to Theorem 5.1, system (5.4.14) is Li–Yorke chaotic.

Making use of the initial data $y_1(t_0) = -0.254$ and $y_2(t_0) = 0.297$, where $t_0 = 0.385$, we illustrate in Fig. 5.3 the y_1 and y_2 coordinates of the solution of system (5.4.14) with the solution $(z_1(t), z_2(t))$ of system (5.4.11) which is illustrated in Fig. 5.1. On the other hand, Fig. 5.4 depicts the trajectory of the same solution on the $y_1 - y_2$ plane. Even if the system (5.4.13) is not chaotic, the simulation results shown in Figs. 5.3 and 5.4 support our theoretical results such that a chaotic attractor takes place in the dynamics of system (5.4.14).

Now, we shall present a method to control the chaos of system (5.4.14). This procedure depends on the idea that to control the chaos of system (5.4.14) it is sufficient to stabilize an unstable periodic solution of system (5.4.11). For this reason, we will apply the OGY control method for the logistic map [29, 30], since the map gives rise to the chaotic behavior in system (5.4.11). Let us explain the method briefly.

Suppose that the parameter μ in the logistic map (5.4.8) is allowed to vary in the range $[3.9 - \varepsilon, 3.9 + \varepsilon]$, where ε is a given small positive number. Consider, an

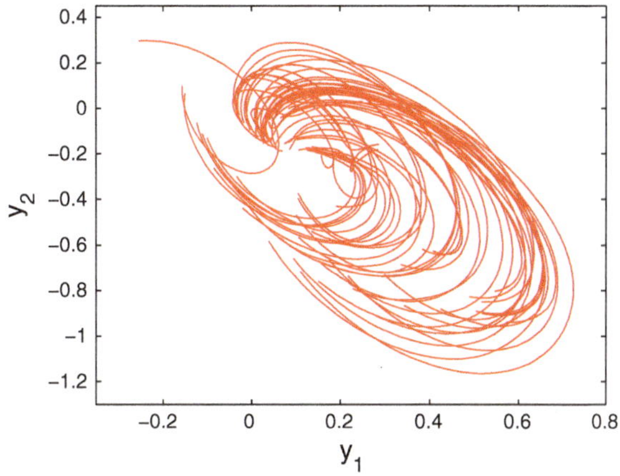

Fig. 5.4 The chaotic trajectory of system (5.4.14)

arbitrary solution $\{\kappa_j\}$, $\kappa_0 \in [0, 1]$, of the map and denote by $\kappa^{(i)}$, $i = 1, 2, \ldots, p_0$, the target unstable p_0-periodic orbit to be stabilized. In the OGY control method [29], at each iteration step j after the control mechanism is switched on, we consider the logistic map with the parameter value $\mu = \bar{\mu}_j$, where

$$\bar{\mu}_j = 3.9\left(1 + \frac{(2\kappa^{(i)} - 1)(\kappa_j - \kappa^{(i)})}{\kappa^{(i)}(1 - \kappa^{(i)})}\right), \tag{5.4.16}$$

provided that the number on the right-hand side of the formula (5.4.16) belongs to the interval $[3.9 - \varepsilon, 3.9 + \varepsilon]$. In other words, formula (5.4.16) is valid if the trajectory $\{\kappa_j\}$ is sufficiently close to the target periodic orbit. Otherwise, we take $\bar{\mu}_j = 3.9$, so that the system evolves at its original parameter value, and wait until the trajectory $\{\kappa_j\}$ enters in a sufficiently small neighborhood of the periodic orbit $\kappa^{(i)}$, $i = 1, 2, \ldots, p_0$, such that the inequality $-\varepsilon \le 3.9\dfrac{(2\kappa^{(i)} - 1)(\kappa_j - \kappa^{(i)})}{\kappa^{(i)}(1 - \kappa^{(i)})} \le \varepsilon$ holds. If this is the case, the control of chaos is not achieved immediately after switching on the control mechanism. Instead, there is a transition time before the desired periodic orbit is stabilized. The transition time increases if the number ε decreases [28].

To apply the OGY method for controlling the chaos of system (5.4.11), we replace the parameter μ in system (5.4.11) with $\bar{\mu}_j$, which is introduced by formula (5.4.16), and set up the system

$$\begin{aligned} z_1'(t) &= z_2(t), \\ z_2'(t) &= -5z_1(t) - 0.6z_2(t) + 0.02z_1^3(t) + \nu(t, t_0, \bar{\mu}_j). \end{aligned} \tag{5.4.17}$$

System (5.4.17) is the control system conjugate to (5.4.11).

We consider the solution of system (5.4.17) with $z_1(t_0) = 0.492$ and $z_2(t_0) = -0.143$, where $t_0 = 0.385$, and apply the OGY control method around the period-1

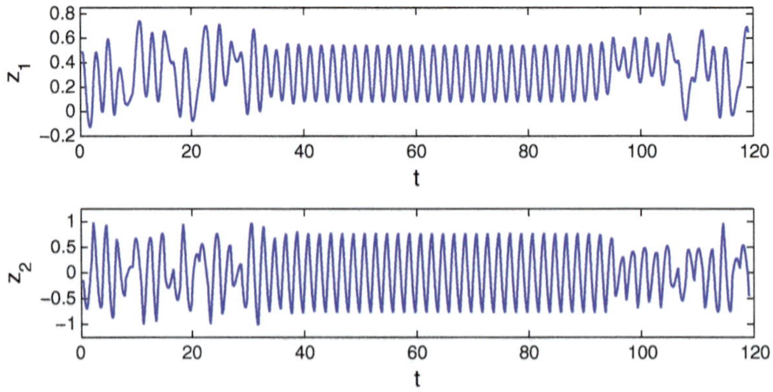

Fig. 5.5 Chaos control of system (5.4.11) by means of the corresponding control system (5.4.17). In the simulation, the value $\varepsilon = 0.05$ is used. The control is switched on at $t = \zeta_{20}$ and switched off at $t = \zeta_{40}$. It is seen in the figure that an unstable 2-periodic solution of system (5.4.11) is stabilized.

orbit, that is the fixed point 2.9/3.9, of the logistic map $F_{3.9}(s)$. Figure 5.5 depicts the simulation results. One can observe in the figure that a 2-periodic solution of system (5.4.17) is stabilized. The value $\varepsilon = 0.05$ is used. The control mechanism is switched on at $t = \zeta_{20}$ and switched off at $t = \zeta_{40}$. The control becomes dominant approximately at $t = 37$ and its effect lasts approximately until $t = 93$, after which the instability becomes dominant and irregular behavior develops again.

In the next simulation, we demonstrate that the chaos of system (5.4.14) can be controlled by stabilizing an unstable periodic solution of system (5.4.11). We consider system (5.4.14) with the solution $(z_1(t), z_2(t))$ of system (5.4.17) which is

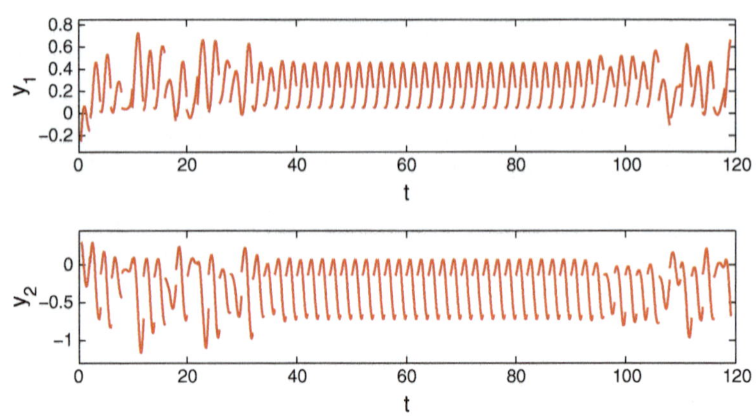

Fig. 5.6 Chaos control of the perturbed impulsive Duffing oscillator (5.4.14). To control the chaos of this system, the OGY control method is applied to system (5.4.11), which gives rise to the presence of chaos in (5.4.14).

illustrated in Fig. 5.5, and simulate in Fig. 5.6 the solution $(y_1(t), y_2(t))$ of system (5.4.14) with $y_1(t_0) = -0.254$ and $y_2(t_0) = 0.297$, where $t_0 = 0.385$. It is seen in Fig. 5.6 that a 2-periodic solution of the system is stabilized. The moments where the control is switched on and switched off and the period of time in which the stabilization becomes dominant are the same with the results presented in Fig. 5.5. The simulations seen in Fig. 5.6 confirm that to control the chaos of system (5.4.14) it is sufficient to stabilize an unstable periodic solution of system (5.4.11).

5.5 Notes

In this chapter, we present a technique to obtain chaotic impulsive systems with the aid of chaotic perturbations. Chaotic collections of piecewise continuous functions are introduced based on the Li–Yorke definition of chaos. Our results are useful for generating multidimensional discontinuous chaos, especially if one requires a rigorous proof for the phenomenon.

We applied our method to an impulsive Duffing oscillator to show the feasibility. According to their instability, the existing periodic solutions of system (5.1.2) are invisible in the simulations. A periodic solution of the perturbed impulsive Duffing oscillator is illustrated by means of the OGY control method [30] applied to the logistic map. Other control procedures, such as the Pyragas method [31], can also be used for this purpose. The results of the present chapter are convenient for construction and stabilization of chaotic mechanical systems and electrical circuits with impulses. Moreover, our approach can be applied to other types of chaos such as the one analyzed through period-doubling cascade [51]. The results of this chapter were published in the paper [52].

References

1. R.L. Devaney, *An Introduction to Chaotic Dynamical Systems* (Addison-Wesley, Menlo Park, California, 1989)
2. T.Y. Li, J.A. Yorke, Period three implies chaos. Am. Math. Mon. **82**, 985–992 (1975)
3. E.N. Lorenz, Deterministic nonperiodic flow. J. Atmos. Sci. **20**, 130–141 (1963)
4. S. Wiggins, *Global Bifurcations and Chaos: Analytical Methods* (Springer, New York, 1988)
5. R. Brown, L. Chua, Dynamical synthesis of Poincaré maps. Int. J. Bifurc. Chaos **3**, 1235–1267 (1993)
6. R. Brown, L. Chua, From almost periodic to chaotic: the fundamental map. Int. J. Bifurc. Chaos **6**, 1111–1125 (1996)
7. R. Brown, L. Chua, Chaos: generating complexity from simplicity. Int. J. Bifurc. Chaos **7**, 2427–2436 (1997)
8. R. Brown, R. Berezdivin, L. Chua, Chaos and complexity. Int. J. Bifurc. Chaos **11**, 19–26 (2001)
9. M.U. Akhmet, Li-Yorke chaos in the system with impacts. J. Math. Anal. Appl. **351**, 804–810 (2009)

10. M.U. Akhmet, The complex dynamics of the cardiovascular system. Nonlinear Anal. Theory Methods Appl. **71**, e1922–e1931 (2009)
11. F. Battelli, M. Fečkan, Chaos in singular impulsive O.D.E. Nonlinear Anal. Theory Methods Appl. **28**, 655–671 (1997)
12. G. Jiang, Q. Lu, L. Qian, Chaos and its control in an impulsive differential system. Chaos Solitons Fractals **34**, 1135–1147 (2007)
13. W. Lin, Description of complex dynamics in a class of impulsive differential equations. Chaos Solitons Fractals **25**, 1007–1017 (2005)
14. M. Akhmet, *Principles of Discontinuous Dynamical Systems* (Springer, New York, 2010)
15. M.U. Akhmet, E. Yılmaz, Impulsive Hopfield-type neural network system with piecewise constant argument. Nonlinear Anal. Real World Appl. **11**, 2584–2593 (2010)
16. A. Khadra, X. Liu, X. Shen, Application of impulsive synchronization to communication security. IEEE Trans. Circuits Syst.-I Fundam. Theory Appl. **50**, 341–351 (2003)
17. X. Liu, Stability results for impulsive differential systems with applications to population growth models. Dyn. Stab. Syst. **9**, 163–174 (1994)
18. A. Ruiz-Herrera, Chaos in predator-prey systems with/without impulsive effect. Nonlinear Anal. Real World Appl. **13**, 977–986 (2012)
19. T. Yang, L.O. Chua, Impulsive stabilization for control and synchronization of chaotic systems: theory and application to secure communication. IEEE Trans. Circuits Syst.-I Fundam. Theory Appl. **44**, 976–988 (1997)
20. Y. Yang, J. Cao, Stability and periodicity in delayed cellular neural networks with impulsive effects. Nonlinear Anal. Real World Appl. **8**, 362–374 (2007)
21. R.A. Horn, C.R. Johnson, *Matrix Analysis* (Cambridge University Press, Cambridge, 1992)
22. F.R. Marotto, Snap-back repellers imply chaos in $\mathbb{R}^n$. J. Math. Anal. Appl. **63**, 199–223 (1978)
23. P. Li, Z. Li, W.A. Halang, G. Chen, Li-Yorke chaos in a spatiotemporal chaotic system. Chaos Solitons Fractals **33**(2), 335–341 (2007)
24. E. Akin, S. Kolyada, Li-Yorke sensitivity. Nonlinearity **16**, 1421–1433 (2003)
25. P. Kloeden, Z. Li, Li-Yorke chaos in higher dimensions: a review. J. Differ. Equ. Appl. **12**, 247–269 (2006)
26. Y. Shi, G. Chen, Chaos of discrete dynamical systems in complete metric spaces. Chaos Solitons Fractals **22**(3), 555–571 (2004)
27. Y. Shi, G. Chen, Discrete chaos in Banach spaces, science in China. Ser. A Math. **48**, 222–238 (2005)
28. J.M. Gonzáles-Miranda, *Synchronization and Control of Chaos* (Imperial College Press, London, 2004)
29. H.G. Schuster, *Handbook of Chaos Control* (Wiley-VCH, Weinheim, 1999)
30. E. Ott, C. Grebogi, J.A. Yorke, Controlling chaos. Phys. Rev. Lett. **64**, 1196–1199 (1990)
31. K. Pyragas, Continuous control of chaos by self-controlling feedback. Phys. Rev. A **170**, 421–428 (1992)
32. M. Haeri, M. Dehghani, Impulsive synchronization of different hyperchaotic (chaotic) systems. Chaos Solitons Fractals **38**, 120–131 (2008)
33. C. Li, X. Liao, R. Zhang, Impulsive synchronization of nonlinear coupled chaotic systems. Phys. Lett. A **328**, 47–50 (2004)
34. Q. Li, X.-S. Yang, F. Yang, Hyperchaos in Hopfield-type neural networks. Neurocomputing **67**, 275–280 (2005)
35. B. Liu, X. Liu, G. Chen, H. Wang, Robust impulsive synchronization of uncertain dynamical networks. IEEE Trans. Circuits Syst.-I Regul. Pap. **52**, 1431–1441 (2005)
36. Q. Ren, J. Zhao, Impulsive synchronization of coupled chaotic systems via adaptive-feedback approach. Phys. Lett. A **355**, 342–347 (2006)
37. M. Sun, C. Zeng, Y. Tao, L. Tian, Adaptive-impulsive synchronization in drive-response networks of continuous systems and its application. Phys. Lett. A **373**, 3041–3046 (2009)
38. X. Wan, J. Sun, Adaptive-impulsive synchronization of chaotic systems. Math. Comput. Simul. **81**, 1609–1617 (2011)

39. J. Sun, Y. Zhang, Q. Wu, Impulsive control for the stabilization and synchronization of Lorenz systems. Phys. Lett. A **298**, 153–160 (2002)
40. J. Wu, L. Jiao, Synchronization in complex delayed dynamical networks with nonsymmetric coupling. Phys. A **386**, 513–530 (2007)
41. T. Yang, L.O. Chua, Impulsive control and synchronization of nonlinear dynamical systems and application to secure communication. Int. J. Bifurc. Chaos **7**, 645–664 (1997)
42. T. Yang, *Impulsive Control Theory* (Springer, Berlin, 2001)
43. R. Zhang, M. Hu, Z. Xu, Impulsive synchronization of Rössler systems with parameter driven by an external signal. Phys. Lett. A **364**, 239–243 (2007)
44. A.M. Samoilenko, N.A. Perestyuk, *Impulsive Differential Equations* (World Scientific, Singapore, 1995)
45. M.U. Akhmet, Creating a chaos in a system with relay. Int. J. Qual. Theory Differ. Equ. Appl. **3**, 3–7 (2009)
46. M.U. Akhmet, Devaney's chaos of a relay system. Commun. Nonlinear Sci. Numer. Simul. **14**, 1486–1493 (2009)
47. M.U. Akhmet, Dynamical synthesis of quasi-minimal sets. Int. J. Bifurc. Chaos **19**, 2423–2427 (2009)
48. M.U. Akhmet, M.O. Fen, Chaotic period-doubling and OGY control for the forced Duffing equation. Commun. Nonlinear Sci. Numer. Simul. **17**, 1929–1946 (2012)
49. M.U. Akhmet, M.O. Fen, Replication of chaos. Commun. Nonlinear Sci. Numer. Simul. **18**, 2626–2666 (2013)
50. J. Hale, H. Koçak, *Dynamics and Bifurcations* (Springer, New York, 1991)
51. E. Sander, J.A. Yorke, Period-doubling cascades galore. Ergod. Theory Dyn. Syst. **31**, 1249–1267 (2011)
52. M.U. Akhmet, M.O. Fen, Chaotification of impulsive systems by perturbations. Int. J. Bifurc. Chaos **24**, 1450078 (2014)

Chapter 6
Chaos Generation in Continuous/Discrete-Time Models

6.1 Devaney's Chaos of a Relay System

We address the differential equation with a pulse function, whose moments of discontinuity depend on the initial moment. The existence of a chaotic attractor and the complex behavior of all solutions are investigated. Appropriate simulations are presented.

6.1.1 Introduction and Preliminaries

The irregular behavior of dynamical systems [1–7] has been discovered and investigated intensively during the last decades. One of the ways to look for complex behavior on the basis of the qualitative theory of differential equations is the approach based on the topological ingredients, which were summarized in [8]. We investigate the non-autonomous differential equation with a pulse function in the right-hand side, using the topological ingredients for guidance. The moments where this function changes its value are dependent on the initial moment. Another issue of relevance to the section is nonlinear dynamics of electric circuits, of mechanical models [9], and of control systems [10] which convert a discrete data to a continuous output. We believe that our results may be applied to models with a pulsating control, which depends on the initial data. Extremely close to our results in this sense is the investigation of relay systems. That is, linear systems which can be analyzed by means of existing linear theory, and where at certain instants the relay releases discontinuous actions in one direction or another. The discontinuities are the results of idealizations used in the representation of nonlinear characteristics. Moreover, one can see that the set of solutions of the initial value problem is not linear, either. Consequently, the system we consider concerns the nonlinear discontinuous dynamics [9–13].

M. Akhmet and M.O. Fen, *Replication of Chaos in Neural Networks, Economics and Physics*, Nonlinear Physical Science,
DOI 10.1007/978-3-662-47500-3_6

In this section, we provide the definitions of chaos and of chaotic attractors of non-autonomous differential equations, and define conditions of their existence.

We begin with the description of the symbolic dynamics [14, 15], which is in the basis of an initial value problem with a pulse function. Consider the sequence space [8, 14]

$$\Sigma_2 = \{s = (s_0 s_1 s_2 \ldots) : s_j = 0 \text{ or } 1\}$$

with the metric

$$d[s, \bar{s}] = \sum_{i=0}^{\infty} \frac{|s_i - \bar{s}_i|}{2^i},$$

where $\bar{s} = (\bar{s}_0 \bar{s}_1 \ldots) \in \Sigma_2$, and the shift map $\sigma : \Sigma_2 \to \Sigma_2$, such that $\sigma(s) = (s_1 s_2 \ldots)$. The semidynamics (Σ_2, σ) is the symbolic dynamics [15].

The map is continuous, *card* $Per_n(\sigma) = 2^n$, $Per(\sigma)$ is dense in Σ_2, and there exists a dense orbit in Σ_2.

Dynamics of the logistic map $h(x, \mu) \equiv \mu x(1 - x)$, $\mu > 0$, is another central instrument. The dynamics has a positively invariant subset $\Lambda \subseteq I = [0, 1]$, such that $\Lambda = I$, if $\mu \le 4$. If $\mu > 4$, then Λ is a Cantor set, and is chaotic on Λ [14]. That is, h has sensitive dependence on the initial conditions; periodic points are dense in Λ, and there exists a solution with every natural period p; and h is topologically transitive, that is there exists a trajectory of h, dense in Λ.

If $\mu > 4$, we denote

$$I_0 = \left[0, \frac{1}{2} - \sqrt{\frac{1}{4} - \frac{1}{\mu}}\right], A_0 = \left(\frac{1}{2} - \sqrt{\frac{1}{4} - \frac{1}{\mu}}, \frac{1}{2} + \sqrt{\frac{1}{4} - \frac{1}{\mu}}\right), I_1 = \left[\frac{1}{2} + \sqrt{\frac{1}{4} - \frac{1}{\mu}}, 1\right],$$

so that $I = I_0 \cup A_0 \cup I_1$, $\Lambda \subset I_0 \cup I_1$, $h(I_0) = h(I_1) = I$, $h(A_0) \cap I = \emptyset$.

Consider the itinerary of x, $S(x) = (s_0 s_1 \ldots)$, where $s_j = 0$, if $h^j(x) \in I_0$, and $s_j = 1$, if $h^j(x) \in I_1$. The function $S(x)$ is a homeomorphism between Λ and Σ_2, and $S \circ h = \sigma \circ S$. That is, h and σ are topologically conjugate.

For every $t_0 \in \Lambda$, one can construct a sequence $\kappa(t_0)$ of real numbers $\kappa_i, i \in \mathbb{Z}$, in the following way. If $i \ge 0$, then $\kappa_{i+1} = h(\kappa_i)$ and $\kappa_0 = t_0$. Let us show, how the sequence is defined for negative i. Denote $s^0 = S(t_0)$, $s^0 = (s_0^0 s_1^0 \ldots)$. Consider elements $\underline{s} = (0 s_0^0 s_0^1 \ldots)$, $\overline{s} = (1 s_0^0 s_0^1 \ldots)$ of Σ_2, such that $\sigma(\underline{s}) = \sigma(\overline{s}) = s^0$ and $\underline{t} = S^{-1}(\underline{s})$, $\overline{t} = S^{-1}(\overline{s})$. The homeomorphism implies that $h(\overline{t}) = h(\underline{t}) = t_0$. Set $h^{-1}(t_0)$ may consist of not more than two elements $\overline{t}, \underline{t} \in \Lambda$. Each of these two values can be chosen as $\kappa_{-1}(t_0)$. Obviously, one can continue the process to $-\infty$, choosing always one element from the set h^{-1}. We have finalized the construction of the sequence, and, moreover, it is proved that $\kappa(t_0) \subset \Lambda$. Fix one of the sequences and introduce a sequence $\zeta(t_0) = \{\zeta_i\}$, $\zeta_i = i + \kappa_i, i \in \mathbb{Z}$. The sequence $\zeta(t_0)$ has the *periodicity property* if there exists $p \in \mathbb{N}$ such that $\zeta_{i+p} = \zeta_i + p$, for all $i \in \mathbb{Z}$. If we denote by Π the set of all such sequences $\{\zeta_i\}, i \in \mathbb{Z}$, then a multivalued functional $w : I \to \Pi$ is defined such that each of the sequences $\zeta(t_0)$ is one of values of $w(t_0)$.

Let $J \subseteq \mathbb{R}$ be an interval. Introduce the following distance $\|\zeta(t_0) - \zeta(t_1)\|_J = \sup_{\zeta_i(t_0),\zeta_i(t_1)\in J} |\zeta_i(t_0) - \zeta_i(t_1)|$. Let us formulate two important, for our discussion, consequences of the topological conjugacy [14] of the symbolical dynamics and the dynamics generated by the logistic map, in the following assertion.

Lemma 6.1 *If* $\mu > 4$, *then*

(a) *for each* $\zeta(t_0) \in \Pi$, *arbitrarily small* $\varepsilon > 0$, *and arbitrarily large positive number* E, *there exists a sequence* $\zeta(t_1) \in \Pi$ *with the periodicity property such that* $|\zeta(t_0) - \zeta(t_1)|_J < \varepsilon$, *where* $J = (0, E)$ *;*
(b) *there exists a sequence* $\zeta(t^*) \in \Pi$ *such that for each* $t_0 \in \Lambda$, *for arbitrarily small* $\varepsilon > 0$, *and arbitrarily large positive number* E, *there exists an integer* m *such that* $|\zeta(t_0) - \zeta(t^*, m)|_J < \varepsilon$, *where* $J = (0, E)$.

Let $\mathbb{Z}$, $\mathbb{N}$, and $\mathbb{R}$ be the sets of all integers, natural and real numbers, respectively. Denote by $\|\cdot\|$ the Euclidean norm in $\mathbb{R}^n$, $n \in \mathbb{N}$.

For every $t_0 \in \Lambda$ one can construct a sequence $\kappa(t_0)$ of real numbers κ_i, $i \in \mathbb{Z}$, such that $\kappa_{i+1} = h(\kappa_i, \mu)$ and $\kappa_0 = t_0$ if $i \geq 0$.

Fix a nonzero vector $m_0 \in \mathbb{R}^n$. For each $\zeta(t_0)$, $t_0 \in \Lambda$, we introduce a pulse function

$$f(t, t_0) = \begin{cases} m_0 & \text{if} \quad \zeta_{2i}(t_0) < t \leq \zeta_{2i+1}(t_0),\ i \in \mathbb{Z}, \\ 0 & \text{if} \quad \zeta_{2i-1}(t_0) < t \leq \zeta_{2i}(t_0),\ i \in \mathbb{Z}. \end{cases}$$

It is worth mentioning that we can consider other types of pulse functions to obtain similar results, for instance, one may discuss,

$$F(t, t_0) = \begin{cases} m_0 & \text{if} \quad \zeta_{2i}(t_0) < t \leq \zeta_{2i+1}(t_0),\ i \in \mathbb{Z}, \\ m_1 & \text{if} \quad \zeta_{2i-1}(t_0) < t \leq \zeta_{2i}(t_0),\ i \in \mathbb{Z}, \end{cases}$$

$m_0, m_1 \in \mathbb{R}^n$.

The main object of our investigation is the following special initial value problem

$$\begin{aligned} z'(t) &= Az(t) + f(t, t_0), \\ z(t_0) &= z_0,\ (t_0, z_0) \in \Lambda \times \mathbb{R}^n, \end{aligned} \tag{6.1.1}$$

where $z \in \mathbb{R}^n$, $t \in \mathbb{R}$. Following [10], we call (6.1.1) the pulse system.

The following assumption is required throughout the section: A is an $n \times n$ constant real-valued matrix such that $\Re\lambda_j < 0$, $j = 1, 2, \ldots, m$, $\Re\lambda_j > 0$, $j = m+1, m+2, \ldots, n$, where m is a natural number, $0 \leq m \leq n$, and $\Re\lambda_j$ denotes the real part of the eigenvalue λ_j of the matrix A. That is the matrix A is hyperbolic. Denote $\alpha = \max_j \Re\lambda_j$, $j = 1, 2, \ldots, m$, and $\beta = \min_j \Re\lambda_j$, $j = m+1, m+2, \ldots, n$.

We shall use the following definitions of solutions of (6.1.1). They coincide with the definitions for differential equations with piecewise constant arguments of generalized type [16], see also [17].

Definition 6.1 A function $z(t), z(t_0) = z_0$, is a solution of (6.1.1) on $\mathbb{R}$ if: (i) $z(t)$ is continuous on $\mathbb{R}$; (ii) the derivative $z'(t)$ exists at each point $t \in \mathbb{R}$ with the possible exception of the points $\zeta_i(t_0), i \in \mathbb{Z}$, where one-sided derivatives exist; (iii) Equation (6.1.1) is satisfied on each interval $(\zeta_i(t_0), \zeta_{i+1}(t_0)], i \in \mathbb{Z}$.

Definition 6.2 A solution $z(t), z(t_0) = z_0$, of (6.1.1) on $[t_0, \infty)$ is a continuous function such that (i) the derivative $z'(t)$ exists at each point $t \in [t_0, \infty)$, with the possible exception of the points $\zeta_j(t_0), j \geq 0$, where left-sided derivatives exist; (ii) Equation (6.1.1) is satisfied by $z(t)$ on each interval $(\zeta_j(t_0), \zeta_{j+1}(t_0)], j \geq 0$.

It can be easily verified that problem (6.1.1) has a unique solution in the sense of Definition 6.1, as well as Definition 6.2, for each $t_0 \in \Lambda, z_0 \in \mathbb{R}^n$,.

In what follows we denote by $z(t, \xi, v), \xi \in \mathbb{R}, v \in \mathbb{R}^n$, a solution of (6.1.1) with $t_0 = \xi, z_0 = v$.

There exists a constant matrix B such that $B^{-1}AB = diag\{C_-, C_+\}$, where C_- and C_+ are $m \times m$ and $(n-m) \times (n-m)$ matrices, respectively, C_+ has eigenvalues with positive real part, and C_- has eigenvalues with negative real part. If we apply the linear transformation $z = Bx$ to system (6.1.1), then one can see that the pulse function f will be transformed to a pulse function. Hence, without any loss of generality, we may assume that matrix A itself has the box-diagonal form so that $A = diag\{A_-, A_+\}$, where all eigenvalues of A_+ have positive real part, and all eigenvalues of A_- have negative real part. Consequently, there exist positive numbers N and ω such that

$$\|e^{A_- t}\| \leq Ne^{-\omega t}, t \geq 0, \quad \|e^{A_+ t}\| \leq Ne^{\omega t}, t \leq 0. \tag{6.1.2}$$

Let us denote $Z(t,s) = diag\{Z_-(t,s), Z_+(t,s)\}, Z_-(t,s) = e^{A_-(t-s)}$, $Z_+(t,s) = e^{A_+(t-s)}, t, s \in \mathbb{R}$, where $Z(t,s)$ is the transition matrix of the linear homogeneous system of differential equations associated with (6.1.1). It can be easily checked (for a detailed explanation see [16]) that the solution $z(t) = z(t, t_0, z_0), t_0 \in \Lambda, z_0 \in \mathbb{R}^n$, of (6.1.1) has the form

$$z(t) = Z(t, t_0)z_0 + \int_{t_0}^{t} Z(t,s) f(s, t_0)\, ds, \tag{6.1.3}$$

and is defined on $\mathbb{R}$.

Moreover, using the standard technique, one can verify that for every $t_0 \in \Lambda$ there exists a unique vector $v_0 \in \mathbb{R}^n$ such that $z(t, t_0, v_0)$ is a bounded on $\mathbb{R}$ solution of (6.1.1). Denote $z(t, t_0) = z(t, t_0, v_0)$, and $z(t, t_0) = (u(t, t_0), v(t, t_0)), u \in \mathbb{R}^m, v \in \mathbb{R}^{n-m}$. One can see that the bounded solution is equal to

$$\begin{aligned} u(t, t_0) &= \int_{-\infty}^{t} Z_-(t,s) f_-(s, t_0) ds, \\ v(t, t_0) &= -\int_{t}^{\infty} Z_+(t,s) f_+(s, t_0) ds, \end{aligned} \tag{6.1.4}$$

if we denote $f(t, t_0) = (f_-(t, t_0), f_+(t, t_0))$. It is easy to show that $\|z(t, t_0)\| < \frac{2N\|m_0\|}{\omega}$, $t \in \mathbb{R}$. Denote $\mathscr{C}\mathscr{B} = \{z(t, t_0) : t_0 \in \Lambda\}$.

Thus, we have that all bounded on $\mathbb{R}$ solutions of (6.1.1) are placed in the tube with the radius $\frac{2N\|m_0\|}{\omega}$. If $\kappa(t_0)$, $t_0 \in \Lambda$, is a p-periodic sequence, then $z(t, t_0)$ is periodic with period p. Denote the periodic solution by $\phi(t, t_0)$.

Remark 6.1 Since sequences $\kappa(t)$ do not intersect for different $t \in \Lambda$, one can see that there exists a p-periodic solution for each $p \in \mathbb{N}$. Consequently, there are infinitely many periodic solutions of (6.1.1).

One can easily verify (see also [16]) that a solution $z(t)$ of (6.1.1) is bounded on $[0, \infty)$ if and only if $z(t) = (u(t), v(t)) = z(t, t_0, z_0)$, $z_0 = (u_0, v_0)$,

$$u(t) = Z_-(t, t_0)u_0 + \int_{t_0}^{t} Z_-(t, s) f_-(s, t_0)\, ds,$$

$$v(t) = -\int_{t}^{\infty} Z_+(t, s) f_+(s, t_0) ds. \tag{6.1.5}$$

We denote the solutions defined by (6.1.5) as $z(t, t_0, u_0)$. Then $\mathscr{C} = \{z(t, t_0, u_0) : t_0 \in \Lambda, u_0 \in \mathbb{R}^m\}$ is the set of all solutions of (6.1.1) bounded on $[0, \infty)$. One can confirm that

$$\|z(t, t_0, u_0) - z(t, t_0)\| < N e^{-\omega(t-t_0)}(\|u_0\| + \|m_0\|/\omega),\ t \geq t_0. \tag{6.1.6}$$

That is, every solution $z(t, t_0, u_0) \in \mathscr{C} \backslash \mathscr{C}\mathscr{B}$ is attracted by a bounded solution $z(t, t_0) \in \mathscr{C}\mathscr{B}$. These solutions have a common set of discontinuity points $\zeta(t_0)$. Thus, $\mathscr{C}\mathscr{B}$ is an attractor with the basin $\mathscr{C}$. Obviously, $\mathscr{C}\mathscr{B} \subset \mathscr{C}$. We intend to address the topological ingredients for $\mathscr{C}\mathscr{B}$ and $\mathscr{C}$.

The section is organized as follows. In Sect. 6.1.2 we consider the main subjects: the ingredients of the chaos, the existence of a chaotic attractor, the period-doubling cascade, and an appropriate example. The conclusion is formulated at the end of the section.

6.1.2 The Chaos

Everywhere in this subsection we assume that $\mu > 4$, with the exception of the part addressing the period-doubling cascade. At first we are going to describe the ingredients for solutions of the initial value problem, which do not necessarily belong to the attractor, but they are attracted by the bounded solutions from this set. Then the chaos on the attractor will be defined. Finally, we will consider the period-doubling cascade for the problem, and an illustrative example will be constructed.

Definition 6.3 We say that (6.1.1) is sensitive on Λ if there exists positive number ε_0 such that for every number $t_0 \in \Lambda$ and for each $\delta > 0$ one can find a number

$t_1 \in \Lambda$, $|t_0 - t_1| < \delta$, such that for each pair of solutions $z(t, t_1, u_1))$, $z(t, t_0, u_0)$, $u_1, u_0 \in \mathbb{R}^n$, there exists a moment $\xi > \max(t_0, t_1)$, which satisfies $||z(\xi, t_1, u_1)) - z(\xi, t_0, u_0)|| > \varepsilon_0$, $||z(\xi, t_1, u_1))\|$, $\|z(\xi, t_0, u_0)|| < \dfrac{2N\|m_0\|}{\omega} + 1$.

Definition 6.4 The set of all periodic solutions is called dense in $\mathscr{C}$ if for every solution $z(t) = z(t, t_1, u_0) \in \mathscr{C}, t_1 \in \Lambda$, and each $\varepsilon > 0, E > 0$, there exists a periodic solution $\phi(t, t_0)$, $t_0 \in \Lambda$, and an interval $J \subset [t_1, \infty)$, with length E, such that $\|\phi(t, t_0) - z(t, t_1, z_0)\| < \varepsilon, t \in J$.

Definition 6.5 A solution $z(t, t^*, u_0) \in \mathscr{C}$ is called dense in $\mathscr{C}$ if for every solution $z(t, t_1, z_1) \in \mathscr{C}$ and each $\varepsilon > 0$, $E > 0$, there exists a number $\xi > 0$ and an interval $J \subset [\max\{t_1, t^*\}, \infty)$ with length E, such that $\|z(t, t_1, z_1) - z(t + \xi, t^*, z_0)\| < \varepsilon$, for all $t \in J$.

Theorem 6.1 *Problem(6.1.1) is sensitive on Λ.*

Proof Fix $t_0 \in \Lambda$, $u_0, u_1 \in \mathbb{R}^n$, and solutions of (6.1.1), $z(t) = z(t, t_0, u_0)$, $z_1(t) = z(t, t_1, u_1)$. Let $S(t_0) = s^0 = (s_0^0, s_1^0, \ldots)$. Take a number $t_1 \in \Lambda$ such that $S(t_1) = s^1 = (s_0^0, s_1^0, \ldots, s_{k-1}^0, s_k^1, s_{k+1}^0, s_{k+2}^0, \ldots)$, $s_k^1 \neq s_k^0$, for some $k > 0$. We have that

$$d[\sigma^i s^0, \sigma^i s^1] = \begin{cases} \frac{1}{2^{k-i}}, & \text{if } 0 \le i \le k, \\ 0, & \text{if } i > k. \end{cases}$$

Assume that k is sufficiently large so that by (6.1.6) $||z(t, t_1, u_1))\|$, $\|z(t, t_0, u_0)||< \frac{2N\|m_0\|}{\omega} + 1$, if $t > \min(\zeta_k(t_0), \zeta_k(t_1)) - 1$.

Since S is a homeomorphism and the set Σ_2 is compact there exists a positive number $\mu_0 < 1$ so that $|\kappa_k(t_0) - \kappa_k(t_1)| > \mu_0$. Without loss of generality, assume that $\kappa_k(t_0) < \kappa_k(t_1)$.

Denote $\overline{m} = \max\limits_{\mu_0 \le u \le 1} \|e^{Au}\|$, $\underline{M} = \min\limits_{\mu_0 \le u \le 1} \|\int_0^u e^{As} m_0 ds\|$.

We shall show that the constant ε_0 and the moment ξ of Definition 6.3 can be taken equal to $\varepsilon_0 = \frac{\underline{M}}{2(1+\overline{m})}$, and ξ and $\zeta_k(t_0)$ or $\zeta_k(t_1)]$, relatively.

If $\|z(\zeta_k(t_0)) - z_1(\zeta_k(t_0))\| \le \varepsilon_0$, then we have that for $t \in [\zeta_k(t_0), \zeta_k(t_1)]$,

$$z(t) = e^{A(t-\zeta_k(t_0))} z((\zeta_k(t_0)) + \int_{\zeta_k(t_0)}^{t} e^{A(t-s)} f(s, t_0) ds,$$

$$z_1(t) = e^{A(t-\zeta_k(t_0))} z_1((\zeta_k(t_0)),$$

and

$$\|z(\zeta_k(t_1)) - z_1(\zeta_k(t_1))\| \ge \underline{M} - \overline{m}\varepsilon_0 > \varepsilon_0.$$

The theorem is proved.

Theorem 6.2 *The set of all periodic solutions $\phi(t, t_0)$, $t_0 \in \Lambda$, of (6.1.1) is dense in $\mathscr{C}$.*

Proof Let us fix $t_1 \in \Lambda$ and $\varepsilon, E > 0$, and denote $z(t) = (u, v) = z(t, t_1, u_0)$. Fix a positive number δ sufficiently small so that $2N\|m_0\|\delta \dfrac{e^{2\omega}}{1 - e^{-2\omega}} < \varepsilon/2$.

By Lemma 6.1 (*a*) for δ and an arbitrarily large number $\tilde{T}$, there exists a periodic sequence $\zeta(t_0) \in \Pi$ such that $\|\zeta(t_1) - \zeta(t_0)\|_Q < \delta$, where $Q = (t_1, t_1 + \tilde{T} + E)$. We shall find numbers δ and $\tilde{T}$ such that $\|z(t) - \phi(t, t_0)\| < \varepsilon$ on $J = (t_1 + \tilde{T}, t_1 + \tilde{T} + E)$. Denote $M_1 = 1 + \frac{2N\|m_0\|}{\omega}$. By (6.1.6) there exists a number $\bar{T} = \bar{T}(z_0, 1) > t_1$ such that $\|z(t)\| < M_1$, if $t \geq \bar{T}$. Denote $\phi(t) = (\phi_-, \phi_+) = \phi(t, t_0)$, the periodic solution. Assuming, without loss of generality, that $\zeta_i(t_0) \leq \zeta_i(t_1)$, for all $i \in \mathbb{Z}$, we have

$$\|z(t) - \phi(t)\| = \|u(t) - \phi_-(t)\| + \|v(t) - \phi_+(t)\| \leq \|u(\bar{T}) - \phi_-(\bar{T})\|\|Z_-(t, \bar{T})\| +$$

$$\int_{\bar{T}}^{t} \|Z_-(t, s)\|\|f_-(s, t_0) - f_-(s, t_1)\|ds + \int_{-\infty}^{\bar{T}} \|Z_-(t, s)\|\|f_-(s, t_0)\|ds \leq$$

$$2Ne^{-\omega(t - \bar{T})}M_1 + \sum_{\bar{T} \leq \zeta_j(t_1) < t} \left[\int_{\zeta_{2j}(t_0))}^{\zeta_{2j}(t_1))} 2Ne^{-\omega(t-s)}\|m_0\|ds + \right.$$

$$\left. \int_{\zeta_{2j+1}(t_0))}^{\zeta_{2j+1}(t_1))} 2Ne^{-\omega(t-s)}\|m_0\|ds \right] + \int_{-\infty}^{\bar{T}} e^{-\omega(t-s)}\|m_0\|ds \leq$$

$$N\Big[2M_1 e^{-\omega(t-\bar{T})} + 4\delta \frac{e^{2\omega}\|m_0\|}{1 - e^{-2\omega}} + \|m_0\|e^{-\omega(t-\bar{T})}1/\omega\Big].$$

Now, if $\tilde{T} \geq \bar{T}$ is sufficiently large so that

$$2NM_1 e^{-\omega(\tilde{T} - \bar{T})} + N\|m_0\|e^{-\omega(\tilde{T} - \bar{T})}\frac{1}{\omega} < \frac{\varepsilon}{2},$$

then $\|z(t) - \phi(t, t_0)\| < \varepsilon$ for all $t \in J$. The theorem is proved.

Theorem 6.3 *There exists a solution of (6.1.1) dense in $\mathscr{C}$.*

Proof Fix positive E, ε. By Lemma 6.1 (*b*), there exists $t^* \in \Lambda$ such that $\zeta(t^*)$ is dense in Π. There exists a unique bounded on $\mathbb{R}$ solution $z_*(t) = (u_*, v_*) = z(t, t^*) = z(t, t^*, z_0)$. Let us prove that $z_*(t)$ is the dense solution.

Consider an arbitrary solution $z(t) = z(t, t_1, u_1)$, $t_1 \in \Lambda$, of (6.1.1). There exists $\bar{T}$, such that $\|z(t)\| < M_1$, if $t > \bar{T}$. Consider an interval $J_1 = (0, \bar{T} + E_1)$, where

E_1 is an arbitrarily large positive number. By Lemma 6.1 (b), there exists a natural m such that

$$\|\zeta(t_1) - \zeta(t^*, m)\|_{J_1} < \delta < \varepsilon, \tag{6.1.7}$$

where δ will be defined more precisely below. We have that

$$\begin{aligned} u_*(t+m) &= \int_{-\infty}^{t+m} Z_-(t+m, s) f_-(s, t^*) ds, \\ v_*(t+m) &= -\int_{t+m}^{\infty} Z_+(t+m, s) f_+(s, t^*) ds, \end{aligned} \tag{6.1.8}$$

and

$$\begin{aligned} u(t) &= Z_-(t, \bar{T}) u(\bar{T}) + \int_{\bar{T}}^{t} Z_-(t, s) f_-(s, t_1)\, ds, \\ v(t) &= -\int_t^{\infty} Z_+(t, s) f_+(s, t_1) ds. \end{aligned} \tag{6.1.9}$$

Now, using the last two formulas and (6.1.7), and emulating the proof of Theorem 6.2, we shall complete the proof. We have for $t \geq \bar{T}$ that

$$\|u_*(t+m) - u(t)\| = \| \int_{-\infty}^{t+m} Z_-(t+m, s) f_-(s, t^*) ds - Z_-(t, \bar{T}) u(\bar{T}) -$$

$$\int_{\bar{T}}^{t} Z_+(t, s) f_+(s, t_1)\, ds\| = \|Z_-(t, \bar{T}) u(\bar{T})\| + \| \int_{-\infty}^{\bar{T}+m} Z_-(t+m, s) f_-(s, t^*) ds\| +$$

$$\| \int_{\bar{T}+m}^{t+m} Z_-(t+m, s) f_-(s, t^*) ds - \int_{\bar{T}}^{t} Z_-(t, s) f_-(s, t_1)\, ds\| = \|Z_-(t, \bar{T}) u(\bar{T})\| +$$

$$\| \int_{-\infty}^{\bar{T}} Z_-(t+m, s+m) f_-(s+m, t^*) ds\| + \int_{\bar{T}}^{t} [\|Z_-(t+m, s+m) - Z_-(t, s)\|$$

$$\times \|f_-(s+m, t^*)\| + |Z_-(t, s)\| \|f_-(s+m, t^*) - f_-(s, t_1)\|]\, ds \leq$$

$$N e^{-\omega(t-\bar{T})}[M_1 + \|m_0\|/\omega] + \delta N \|m_0\|/\omega.$$

Similarly,

$$\|v_*(t+m) - v(t)\| = \| \int_{t+m}^{\infty} Z_+(t+m, s) f_+(s, t^*) ds - \int_t^{\infty} Z_+(t, s) f_+(s, t_1) ds\| \leq$$

$$\int_t^{\infty} [\|Z_+(t+m, s+m) - Z_+(t,s)\| \|f_+(s,t^*)\| + \|Z_+(t,s)\| \|f_+(s,t^*) - f_+(s,t_1)\|] ds \leq$$

$$\delta N \|m_0\|/\omega.$$

Fix $\tilde{T} > \bar{T}$ and δ sufficiently large and small, respectively, so that

$$N e^{-\omega(\tilde{T}-\bar{T})}[M_1 + \|m_0\|/\omega] + 2\delta N \|m_0\|/\omega < \varepsilon.$$

Then the last two inequalities imply that $\|z(t) - z_*(t+m)\| < \varepsilon$, on the interval $J = [\tilde{T}, \tilde{T} + E]$, if $E_1 = \tilde{T} + E - \bar{T}$.

The theorem is proved.

6.1.3 The Chaos on the Attractor

This subsection is devoted to the discussion of the chaotic ingredients of bounded solutions from $\mathscr{C}\mathscr{B}$. The first of these definitions is significantly different from its counterpart for $\mathscr{C}$, as it requires closeness of the initial values.

Definition 6.6 We say that (6.1.1) is sensitive on $\mathscr{C}\mathscr{B}$ if there exist positive real numbers $\varepsilon_0, \varepsilon_1$ such that for each $t_0 \in \Lambda$, and for every $\delta > 0$ one can find $t_1 \in \Lambda, z_1 \in \mathbb{R}^n$, $\|z_1 - z_0\| + |t_0 - t_1| < \delta$, and an interval Q from $[0, \infty)$ with length no less than ε_1 such that $||z(t, t_0) - z(t, t_1)|| \geq \varepsilon_0, t \in Q$, and there are no points of discontinuity of $z(t, t_0), z(t, t_1)$ on Q.

Definition 6.7 The set of all periodic solutions is called dense in $\mathscr{C}\mathscr{B}$ if for every solution $z(t) = z(t, t_1), t_1 \in \Lambda$, and each $\varepsilon > 0, E > 0$, there exists a periodic solution $\phi(t, t_0), t_0 \in \Lambda$, and an interval $J \subset [t_1, \infty)$, with length E, such that $\|\phi(t, t_0) - z(t, t_1, z_0)\| < \varepsilon, t \in J$.

Definition 6.8 A solution $z(t, t^*) \in \mathscr{C}\mathscr{B}$ is called dense in $\mathscr{C}\mathscr{B}$ if for every solution $z(t, t_1) \in \mathscr{C}\mathscr{B}$, and each $\varepsilon > 0$, $E > 0$, there exists a number $\xi > 0$ and an interval $J \subset [\max\{t_1, t^*\}, \infty)$ with length E such that $||z(t, t_1) - z(t + \xi, t^*)|| < \varepsilon$, for all $t \in J$.

We call the attractor *chaotic* if: (i) problem (6.1.1) is sensitive in $\mathscr{C}\mathscr{B}$; (ii) there are infinitely many periodic solutions $\phi(t, t_0), t_0 \in \Lambda$, and they are dense in $\mathscr{C}\mathscr{B}$; (iii) there exists a solution $z(t, t_0), t_0 \in \Lambda$, which is dense in $\mathscr{C}\mathscr{B}$.

Theorem 6.4 *The manifold $\mathscr{C}\mathscr{B}$ is a chaotic attractor.*

Proof Let us start with sensitivity in $\mathscr{C}\mathscr{B}$. Fix a solution $z(t, t_0) = (u, v) = z(t, t_0, z_0), z_0 = (u_0, v_0)$, in $\mathscr{C}\mathscr{B}$.

If we take into account the proof of Theorem 6.1 applied to $z(t, t_0) \in \mathscr{C}\mathscr{B}$, we need only to show that for an arbitrarily small $\delta > 0$ there exist t_1, z_1, which are considered in the proof, such that $|t_0 - t_1|$, $\|z_0 - z_1\| < \delta/2$, and $z_1 = (u_1, v_1) = z(t_1, t_1)$. In other words,

$$u_1 = \int_{-\infty}^{t_1} Z_-(t_1, s) f_-(s, t_1) ds, \quad v_1 = -\int_{t_1}^{\infty} Z_+(t_1, s) f_+(s, t_1) ds.$$

Let $(\dots s^1_{-k} s^1_{-(k-1)} \dots s^1_0 s^1_1 \dots)$ be a bi-infinite sequence such that $s^1_i = s^0_i, i < n, s^1_n \neq s^0_n$, where n is the number discussed in the proof of Theorem 6.1. Denote $s^1 = (s^1_0 s^1_1 \dots)$. Fix a positive $\delta_1 < \delta/2$, which will be defined more precisely below, and choose a number n sufficiently large so that $\|\zeta(t_0) - \zeta(t_1)\|_{[0,n/2]} < \delta_1$. Obviously, $|\zeta_{-k}(t_0) - \zeta_{-k}(t_1)| < \delta_1, k \geq 1$.

Now, assuming without any loss of generality that $t_0 < t_1$, we have that

$$\|u_0 - u_1\| = \| \int_{-\infty}^{t_0} Z_-(t_0, s) f_-(s, t_0) ds - \int_{-\infty}^{t_1} Z_-(t_1, s) f_-(s, t_1) ds \| \leq$$

$$\| \int_{t_0}^{t_1} Z_+(t_1, s) f_+(s, t_1) ds \| + \int_{-\infty}^{t_0} [\|Z_+(t_0, s) - Z_+(t_1, s)\| \|f_+(s, t_0)\| +$$

$$\|Z_+(t_1, s)\| \|f_+(s, t_1) - f_+(s, t_0)\|] ds \leq \|m_0\| \left\{ \frac{N}{\omega} (\kappa(\delta_1) + \delta_1) + \delta_1 m^- \right\},$$

where κ is a continuous function, such that $\|I - e^{A_- u}\| \leq \kappa(\delta_1)$ if $0 \leq |u| < \delta_1$, $m^- = \max_{0 \leq |u| \leq 1} \|e^{A_- u}\|$.

Similarly, we have that

$$\|v_0 - v_1\| = \| \int_{t_0}^{\infty} Z_+(t_0, s) f_+(s, t_0) ds - \int_{t_1}^{\infty} Z_+(t_1, s) f_+(s, t_1) ds \| \leq$$

$$\| \int_{t_0}^{t_1} Z_+(t_1, s) f_+(s, t_0) ds \| + \int_{t_0}^{n/2} [\|Z_+(t_0, s) - Z_+(t_1, s)\| \|f_+(s, t_0)\| +$$

$$\|Z_+(t_1, s)\| \|f_+(s, t_1) - f_+(s, t_0)\|] ds + \int_{n/2}^{\infty} [\|Z_+(t_0, s) - Z_+(t_1, s)\| \|f_+(s, t_0)\| +$$

$$\|Z_+(t_1, s)\| \|f_+(s, t_1) - f_+(s, t_0)\|] ds \leq$$
$$\|m_0\| \left\{ \frac{N}{\omega} e^{\omega} [\kappa_1(\delta_1) + \delta_1 + e^{\omega(t_0 - \frac{n}{2})})(1 - \delta_1) + \delta_1 m^+ \right\},$$

where κ_1 is a continuous function, such that $\|I - e^{A_+ u}\| \leq \kappa_1(\delta_1)$ if $0 \leq |u| < \delta_1$, $m^+ = \max_{0 \leq |u| \leq 1} \|e^{A_+ u}\|$.

If we suppose that n and δ_1 are sufficiently large and small, respectively, so that

$$\|m_0\| \left\{ \frac{N}{\omega} [e^{\omega}(\kappa_1(\delta_1) + \delta_1 + e^{\omega(t_0 - \frac{n}{2})})(1 - \delta_1)) + \kappa(\delta_1) + \delta_1] + \delta_1(m^+ + m^-) \right\} < \frac{\delta}{2},$$

then from the last two inequalities $\|z_1 - z_0\| < \delta/2$. Sensitivity is proved.

The existence of infinitely many periodic solutions is considered in the Sect. 6.1.1. The density of periodic solutions in $\mathscr{C}\mathscr{B}$ follows immediately from Theorem 6.2. The existence of a dense solution in $\mathscr{C}\mathscr{B}$ can be proved in exactly the same way as Theorem 6.3.

The theorem is proved.

6.1.4 The Period-Doubling Cascade and Intermittency: An Example

The logistic map has been used to shape the chaos in the multidimensional system. Consequently, one can expect to observe the period-doubling cascade and intermittency.

Let us consider $\mu > 0$, μ being the parameter for the logistic map $h(t, \mu) \equiv \mu t(1 - t)$. It is known [5], that there exists an infinite sequence $3 < \mu_1 < \mu_2 < \cdots < \mu_k \cdots < 3.8284\cdots$ such that $h(t, \mu_i), i \geq 1$, has an asymptotically stable prime period-2^i point t_i^* with a region of attraction $(t_i^* - \delta_i, t_i^* + \delta_i)$. And beyond the value $3.8284\ldots$, there are cycles with every integer period [3].

One can easily see that there is a 2^i-periodic solution $\phi(t, t_i^*, \mu_i)$ of (6.1.1) for each i, and for different i these periodic solutions do not coincide. The periodic solutions are in the bounded region $\|x\| < \dfrac{2N\|m_0\|}{\omega}$ of the space $\mathbb{R}^n$. The chaotic attractor is also placed in the region. Finally, the cascade generates infinitely many periodic solutions.

The numerical simulation of the chaos is not an easy task since even the verification of sensitivity requires two close values of the initial moment in the Cantor set Λ, which cannot be found easily. Hammel et al. [18] have given a computer-assisted proof that an approximate trajectory of the logistic map can be shadowed by a true trajectory for a long time. This result and the continuous dependence of the solutions on the sequence of discontinuity points make possible the following appropriate simulations.

To illustrate the chaotic nature of the discussed system, let us show that the chaotic attractor and intermittency can be observed in the next example.

Example 6.1 Consider the sequence $\zeta_i = i + \kappa_i, \kappa_i = 3.8282\kappa_{i-1}(1 - \kappa_{i-1}), \kappa_0 = t_0, t_0 \in [0, 1], i \geq 0$. The coefficient's value of 3.8282 is such that the logistic map admits intermittency [19]. Let the following pendulum equation be given

$$x'' + 2x' + 1.5x = f_2(t, t_0), \tag{6.1.10}$$

where $f_2(t, t_0)$ is a scalar pulse function with $m_0 = 1$. Using new variables $x_1 = x, x_2 = x'$, one can reduce (6.1.10) to the system

$$\begin{aligned} x_1' &= x_2, \\ x_2' &= -1.5x_1 - 2x_2 + f_2(t, t_0). \end{aligned} \tag{6.1.11}$$

One can easily verify that both eigenvalues of the matrix of coefficients have negative real parts. Fix $t_0 = 0.5$ and take a solution $(x_1(t), x_2(t))$ of the last system with the initial condition $x_1(t_0) = 0.02, x_2(t_0) = -0.025$. The result of simulation can be seen in Fig. 6.1. It demonstrates the intermittency phenomenon for the pulse mechanical model.

Next, consider Eq. (6.1.10) with $\mu = 4$, and the solution $(x_1(t), x_2(t))$ that has been chosen for the intermittency observation. In Fig. 6.2, the chaotic attractor is shown using the points $(x_1(n), x_2(n)), n = 1, 2, 3, \ldots, 75{,}000$, in the x_1, x_2-plane.

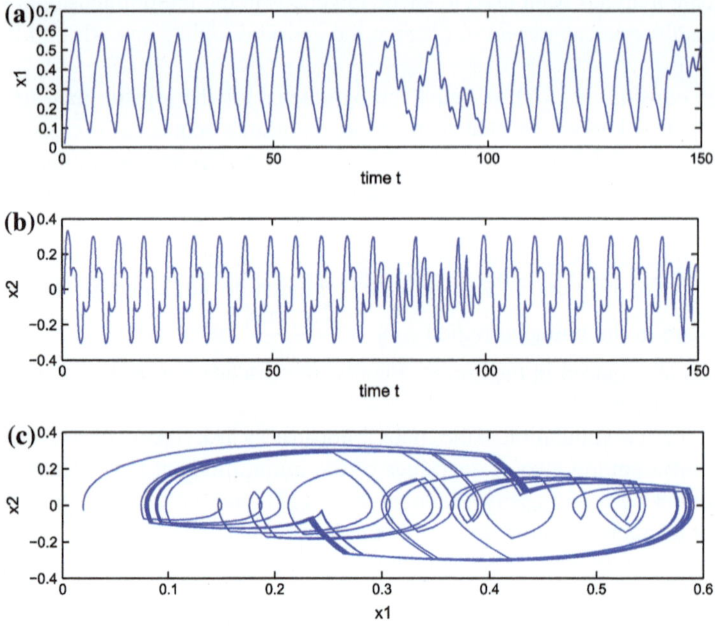

Fig. 6.1 Simulation results. **a** The graph of the x_1 coordinate. **b** The graph of the x_2 coordinate. **c** The trajectory of the solution $(x_1(t), x_2(t))$

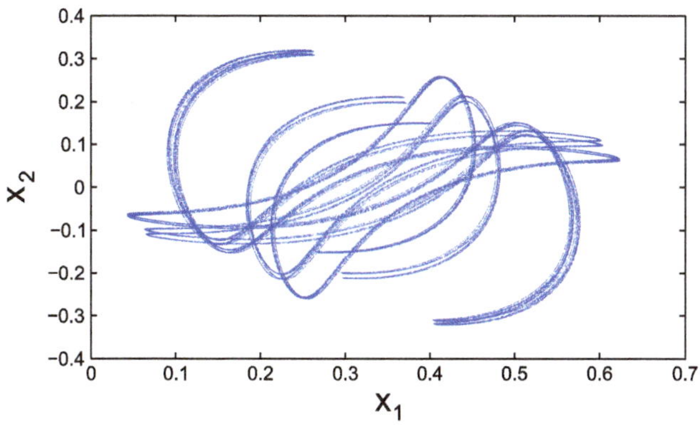

Fig. 6.2 The chaotic attractor by a stroboscopic sequence $(x_1(n), x_2(n)), 1 \leq n \leq 75{,}000$, is observable

6.2 Li–Yorke Chaos in Systems with Impacts

The analogue of Li–Yorke chaos [3] for a special initial value problem of a non-autonomous impulsive differential equation is developed in this section. It is proved that Li–Yorke sensitivity [20] is also proper for the problem.

6.2.1 Introduction and Preliminaries

The first mathematical definition of chaos was introduced by Li and Yorke [3], and it became one of the most discussed topics for the last several decades. Li and Yorke proved that if a map on an interval has a point of period three, then it has points of all periods. Moreover, there exists an uncountable *scrambled* subset of the interval. While the existence of periodic solutions, as it was discovered later, is a particular case of Sharkovsy's theorem [21], the scrambled set remains the feature, which distinguishes Li–Yorke chaos from other definitions [22]. Another fact which makes the chaos attractive for applications is that it can be developed for a multidimensional case [23–26].

The following special initial value problem for the impulsive differential equation is the main subject of this section

$$\begin{aligned} z'(t) &= Az(t) + f(z), \\ \Delta z|_{t=\zeta_i(t_0)} &= Bz(\zeta_i(t_0)) + W(z(\zeta_i(t_0))), \\ z(t_0) &= z_0, (t_0, z_0) \in \Lambda \times \mathbb{R}^n, \end{aligned} \tag{6.2.12}$$

where $z \in \mathbb{R}^n, t \in \mathbb{R}_+ = [0, \infty), i \geq 0$. Cantor set $\Lambda \subset I = [0, 1]$, and the strictly increasing and unbounded sequence of impulsive moments $\zeta_i(t_0)$ will be fully described later. We shall need the following assumptions throughout the section:

(C1) A, B are $n \times n$ constant real-valued matrices, $\det(\mathcal{I} + B) \neq 0$, where $\mathcal{I}$ is $n \times n$ identity matrix;

(C2) the functions $f : \mathbb{R}^n \to \mathbb{R}^n$, $W : \mathbb{R}^n \to \mathbb{R}^n$, satisfy

$$||f(x_1) - f(x_2)|| + \|W(x_1) - W(x_2\| \leq L||x_1 - x_2||, \quad (6.2.13)$$

for all $x_1, x_2 \in \mathbb{R}^n$, where $L > 0$ is a constant;

(C3)

$$Bx + W(x) \neq 0, \forall x \in \mathbb{R}^n; \quad (6.2.14)$$

(C4) the functions f and W are uniformly bounded so that

$$\sup_{x \in \mathbb{R}^n} ||f(x)|| + \sup_{x \in \mathbb{R}^n} ||W(x)|| = M_0 < \infty. \quad (6.2.15)$$

(C5) the matrices A and B commute, and the real parts of all eigenvalues of the matrix $A + \ln(\mathcal{I} + B)$ are negative.

A left continuous function $z(t) : [t_0, \infty) \to \mathbb{R}^n$ belongs to the set of functions $\mathcal{PC}^1(t_0)$, where $t_0 \in \Lambda$ is fixed, if:

(i) it has discontinuities only at points $\zeta_i(t_0)$, $i \geq 0$, and these discontinuities are of the first kind;
(ii) the derivative $z'(t)$ exists at each point $t \in [t_0, \infty) \backslash \{\zeta_i(t_0)\}$, and the left-sided derivative exists at points $\zeta_i(t_0)$, $i > 0$.

A solution $z(t)$ of (6.2.12) on $[t_0, \infty)$ is a function $z(t) \in \mathcal{PC}^1(t_0)$ such that:

(1) the differential equation is satisfied by $z(t)$ on $[t_0, \infty) \backslash \{\zeta_i(t_0)\}$, and it holds for the left derivative of $z(t)$ at every point $\zeta_i(t_0), i > 0$.
(2) the jumps equation is satisfied by $z(t)$ for every $i \geq 0$.

In what follows we denote by $z(t, \xi, v), \xi \in \mathbb{R}_+, v \in \mathbb{R}^n$, a solution of (6.2.12) with $t_0 = \xi, z_0 = v$.

Conditions $(C1)$, $(C2)$ imply that for every $(t_0, z_0) \in \Lambda \times \mathbb{R}^n$ there exists a unique solution $z(t, t_0, z_0)$ of (6.2.12) from $\mathcal{PC}^1(t_0)$ [27].

We attempt to shape the Li–Yorke chaos for system (6.2.12) by implementing a special initial value problem, where the moments of impulsive action are functionally dependent on the initial moment, and using the results of the theory of impulsive differential equations [27–32].

The description of the main subject of this section should begin with the discussion of the moments of impulses, as their generation is most important for the emergence of chaos.

Let us recall the definition of the chaos for maps. Consider an infinite nonvoid compact metric space (X, ρ) with metric ρ and $T : X \to X$, a continuous map.

Definition 6.9 The map $T : X \to X$ is Li–Yorke chaotic, if:

(i) it has points with all periods $p \in \mathbb{N}$;
(ii) there exists an uncountable subset $X' \subseteq X$, the scrambled set, that does not contain periodic points and

$$\limsup_{i\to\infty} \rho(T^i(x), T^i(\tilde{x})) > 0, \tag{6.2.16}$$

$$\liminf_{i\to\infty} \rho(T^i(x), T^i(\tilde{x})) = 0, \tag{6.2.17}$$

for each pair $x, \tilde{x} \in X', x \neq \tilde{x}$;

If $x, x' \in X', x \neq x'$, we call (x, x') a *Li–Yorke pair.*

One of the most effective ways to discover a chaos is to establish the topological conjugacy with the symbolic dynamics [8]. Consider the sequence space [14]

$$\Sigma_2 = \{s = (s_0 s_1 s_2 \ldots) : s_j = 0 \, or \, 1\}$$

with the metric

$$d[s, t] = \sum_{i=0}^{\infty} \frac{|s_i - \tilde{s}_i|}{2^i},$$

where $\tilde{s} = (\tilde{s}_0 \tilde{s}_1 \ldots) \in \Sigma_2$, and the shift map $\sigma : \Sigma_2 \to \Sigma_2$, such that $\sigma(s) = (s_1 s_2 \ldots)$. The pair (Σ_2, σ) is the symbolic dynamics. The map is continuous, $card\, Per_n(\sigma) = 2^n$, $Per(\sigma)$ is dense in Σ_2, and there exists a dense orbit in Σ_2.

Let $h : \Lambda \to \Lambda$, where Λ is a subset of the interval I, be a map topologically conjugate to σ. That is, there exists a homeomorphism $S : \Lambda \to \Sigma_2$ such that $S \circ h = \sigma \circ S$.

For every $t_0 \in \Lambda$ one can construct a sequence $\kappa(t_0)$ of real numbers $\kappa_i, i \geq 0$, such that $\kappa_{i+1} = h(\kappa_i)$ and $\kappa_0 = t_0$. Sequence $\zeta(t_0) = \{\zeta_i(t_0)\}$ in (6.2.12) is defined as $\zeta_i(t_0) = i + \kappa_i(t_0), i \geq 0$.

By applying the conjugacy of h and σ, we shall verify that map h has useful chaotic properties.

It is known [15] that σ has orbits periodic with every $p \in \mathbb{N}$. Now, consider an element $\omega_s = (s_0 01 s_0 s_1 0011 s_0 s_1 s_2 000111 \ldots) \in \Sigma_2, s = (s_0 s_1 \ldots) \in \Sigma_2$, and define a set $\Sigma_2' = \{\omega_s : s \in \Sigma_2\}$. One can easily check that Σ_2' is a scrambled set, and dynamics (Σ_2, σ) is Li–Yorke chaotic.

The proofs of the following lemmas are standard [14, 15].

Lemma 6.2 *If* (s, s') *is a Li–Yorke pair from* Σ_2', *then there exist sequences* $k_i, m_i \to \infty$, *as* $i \to \infty$, *such that* $s_{k_i+j} = s'_{k_i+j}$, $j = 0, 1, \ldots, m_i - 1$ *and* $s_{k_i+m_i} \neq s'_{k_i+m_i}$.

Lemma 6.3 *If* (s, s') *is a Li–Yorke pair from* Σ_2', *then there exists a sequence* $l_i \to \infty$, *as* $i \to \infty$, *such that* $d[\sigma^{l_i} s, \sigma^{l_i} s'] \geq 1$.

Now, the fact that S is a homeomorphism and compactness of Λ and Σ_2 imply easily that the following two assertions are valid.

Lemma 6.4 *If* $t, t' \in \Lambda'$, *then there exist sequences* $k_i, l_i \to \infty$, *as* $i \to \infty$, *such that* $\max\limits_{j=0,1,\ldots,l_i} |h^{k_i+j}(t) - h^{k_i+j}(t')| \to 0$ *as* $i \to \infty$.

Lemma 6.5 *For every pair* $t, t' \in \Lambda', t \neq t'$, *there exists a sequence* $m_i \to \infty$, *as* $i \to \infty$, *such that* $|h^{m_i}(t) - h^{m_i}(t')| \geq \delta$.

Lemmas 6.4 and 6.5 imply that h is a Li–Yorke chaotic, $\Lambda' = S^{-1}(\Sigma_2')$ is a scrambled subset of Λ. The dynamics (Λ', h) is Li–Yorke sensitive [20].

Let us fix $t_0 \in \Lambda$ and denote by $Z(t, s) = Z(t, s, t_0)$ the transition matrix [27] of the linear homogeneous system

$$\begin{aligned} z'(t) &= Az(t), \ t \neq \zeta_i \\ \Delta z|_{t=\zeta_i} &= Bz(\zeta_i), \end{aligned} \tag{6.2.18}$$

associated with (6.2.12).

Condition $(C5)$ in conjunction with Theorem 34, [27], implies that there exist two positive numbers N and ω such that for all $t_0 \in I$,

$$||Z(t, s, t_0)|| \leq N\mathrm{e}^{-\omega(t-s)}, \ t \geq s. \tag{6.2.19}$$

It is important that constants N and ω are common for all $t_0 \in \Lambda$.

Denote $\overline{m} = \max\limits_{|u|\leq 1} \|\mathrm{e}^{Au}\|$, $\underline{m} = \min\limits_{|u|\leq 1} \|\mathrm{e}^{Au}\|$. Fix a number $q \geq 3$, such that $\frac{1}{q} < \frac{2\underline{m}}{3\overline{m}}$. Condition $(C3)$ implies that $\eta = \min\limits_{\|x\|\leq M_1} (Bx + W(x)) > 0$. The following assumptions are also useful.

(C6) $NL\left[\frac{2}{\omega} + \frac{\mathrm{e}^{\omega}}{1-\mathrm{e}^{-\omega}}\right] < 1.$

(C7) $-\omega + NL + \ln(1 + NL) < 0.$

(C8) $L < \dfrac{\left[\frac{2\underline{m}}{3\overline{m}} - \frac{1}{q}\right]\underline{m}\eta}{2M_1(\overline{m} + \underline{m})}.$

Theorem 6.5 *If conditions* $(C1)$*–*$(C6)$ *are valid, and sequence* $\kappa_i(t_0)$, $t_0 \in \Lambda$, *is periodic with a period* $p \in \mathbb{N}$, *then:*

1. *Equation (6.2.12) admits a unique p-periodic solution* $\phi(t, t_0)$;
2. *all the periodic solutions are different.*

The verification of part 1. of this theorem replicates the proof of Theorem 58 [27]. It is easy to see that $(C3)$ implies that there exist infinitely many different periodic solutions, since the sequences of discontinuity moments of solutions with different periods do not intersect. Thus, part 2. is also proved.

The impulsive differential equations of type (6.2.12) play an increasingly important role in the investigation of the cardiovascular system, neural information processing in the brain, information communication, and population dynamics [33–40]. It is notable that the moments of time, where the impulses are performed, are chosen dependent on the initial data. This type of problems may occur if one considers an impulsively controlled process [27]. In [41], the author considered a system of impulsive differential equations with moments of impulses generated by a sensitive map, which depends on a parameter. Sensitivity was considered as a chaotic property. In the present section, we show that solutions of (6.2.12) have Li–Yorke chaos properties. They are similar to those formulated for maps [3, 14] with additional peculiarities caused by discontinuities occurring at different moments for different solutions.

6.2.2 Main Results

Let us fix an interval $J \subset [0, \infty)$, a positive number ε, and $t_0, t_1 \in I$. We introduce the distance $\|\zeta(t_0) - \zeta(t_1)\|_J = \sup_{\zeta_i(t_0), \zeta_i(t_1) \in J} |\zeta_i(t_0) - \zeta_i(t_1)|$, and we shall say that a function $\xi(t) \in \mathcal{PC}^1(t_0)$ is ε-equivalent to a function $\psi(t) \in \mathcal{PC}^1(t_1)$ on J and write $\xi(t)(\varepsilon, J)\psi(t)$, if ξ and ψ are defined on J, $\|\zeta(t_0) - \zeta(t_1)\|_J < \varepsilon$ and $\|\xi(t) - \psi(t)\| < \varepsilon$ for all t from J such that $t \notin \cup_{\zeta_i(t_0), \zeta_i(t_1) \in J} \widehat{[\zeta_i(t_0), \zeta_i(t_1)]}$. Here $\widehat{[a, b]}$, $a, b \in \mathbb{R}$, stands for an oriented interval, that is $\widehat{[a, b]} = [a, b]$ if $a \leq b$, and $\widehat{[a, b]} = [b, a]$, otherwise.

The ε-equivalence of two piecewise continuous functions with a small ε means that they have close discontinuity points, and the values of the functions are close at points that do not lie on intervals between the neighboring discontinuity points of these functions. The concept was developed in [27–29, 31, 42].

The following definitions were taken from [3, 14, 20] and adapted for (6.2.12).

Definition 6.10 A pair of solutions of (6.2.12) $z(t)=z(t, t_0, z_0)$, $z_1(t)=z(t, t_1, z_1)$, $t_0, t_1 \in \Lambda$, is proximal if for each $\varepsilon > 0$, $E > 0$ there exists an interval $J \subset [\max(t_0, t_1), \infty)$, with a length not less than E, such that $z_1(t)(\varepsilon, J)z(t)$.

Definition 6.11 The solutions of (6.2.12) $z(t) = z(t, t_0, z_0)$, $z_1(t) = z(t, t_1, z_1)$, t_0, $t_1 \in \Lambda$, are frequently $(\varepsilon_0, \varepsilon_1)$-separated if there exist positive numbers $\varepsilon_0, \varepsilon_1$ and infinitely many disjoint intervals of length not less than ε_1, such that $||z_1(t) - z(t)|| >$

ε_0 for each t from these intervals, and none of these intervals contain a discontinuity point of $z_1(t)$ or $z(t)$.

Definition 6.12 A couple $z(t) = z(t, t_0, z_0), z_1(t) = z(t, t_1, z_1), t_0, t_1 \in \Lambda$, of solutions of (6.2.12) is a Li–Yorke pair if they are proximal and $(\varepsilon_0, \varepsilon_1)$-separated for some positive $\varepsilon_0, \varepsilon_1$.

Definition 6.13 Problem (6.2.12) is Li–Yorke chaotic on Λ' if:

1. there exist solutions $\phi(t, t_0)$ with all periods $p \in \mathbb{N}$;
2. each couple of solutions $z(t) = z(t, t_0, z_0), z_1(t) = z(t, t_1, z_1)$, with $t_0, t_1 \in \Lambda', t_0 \neq t_1$, is Li–Yorke pair;

Definition 6.14 Problem (6.2.12) is Li–Yorke sensitive on Λ' if there exist positive numbers $\varepsilon_0, \varepsilon_1$, such that each couple of solutions $z(t) = z(t, t_0, z_0), z_1(t) = z(t, t_1, z_1)$, with $t_0, t_1 \in \Lambda', t_0 \neq t_1$, is frequently $(\varepsilon_0, \varepsilon_1)$-separated.

Lemma 6.6 *Assume that conditions $(C1)$–$(C7)$ are fulfilled. Then each couple of solutions of (6.2.12), $z(t) = z(t, t_0, z_0), z_1(t) = z(t, t_1, z_1)$, with $t_0, t_1 \in \Lambda', t_0 \neq t_1$, is proximal.*

Proof Fix numbers $t_0, t_1 \in \Lambda', t_0 \neq t_1$, solutions $z(t) = z(t, t_0, z_0)$, $z_1(t) = z(t, t_1, z_1)$, $z_0, z_1 \in \mathbb{R}^n$, of (6.2.12), and $E, \varepsilon > 0$.

Using the integral representation formula [27]

$$z(t) = Z(t, t_0)z_0 + \int_{t_0}^{t} Z(t, s)f(z(s))ds + \sum_{t_0 \le \zeta_i < t} Z(t, \zeta_i)W(z(\zeta_i)), \tag{6.2.20}$$

one can find that

$$||z(t)|| \le Ne^{-\omega(t-t_0)}||z_0|| + \int_{t_0}^{t} Ne^{-\omega(t-s)}M_0\, ds +$$

$$\sum_{\zeta_i < t} Ne^{-\omega(t-\zeta_i)}M_0 \le Ne^{-\omega(t-t_0)}||z_0|| + NM_0\left(\frac{1}{\omega} + \frac{e^{\omega}}{1-e^{-\omega}}\right).$$

Denote $\bar{T}(t_0, z_0) := t_0$, if $N\|z_0\| \le 1$, and $\bar{T}(t_0, z_0) := t_0 - \dfrac{1}{\omega}\ln\dfrac{1}{N\|z_0\|}$, if $N\|z_0\| > 1$.

From the last inequality, it follows that $\|z(t)\| \le M_1 = 1 + NM_0\left(\frac{1}{\omega} + \frac{e^{\omega}}{1-e^{-\omega}}\right)$ for $t \ge \bar{T}(t_0, z_0)$.

Similarly, one can find a number $T(t_1, z_1)$ such that $\|z_1(t)\| \le M_1$ follows from the inequality $t \ge \bar{T}(t_1, z_1)$. Thus, there exists a number $\bar{T}$ such that both solutions z, z_1 are in the tube with the radius M_1 if $t \ge \bar{T}$. By Lemma 6.4 there exist arbitrarily large numbers $\tilde{T} > \bar{T}, E > 0$, such that $\|\zeta(t_1) - \zeta(t_0)\|_Q < \delta$, where

$Q = (\tilde{T}, \tilde{T} + E)$. We shall find a sufficiently large E so that solutions $z(t), z_1(t)$ are ε-equivalent on $J = (\tilde{T} + \frac{1}{2}E, \tilde{T} + E)$.

Denote $Z_1(t, s) = Z(t, s, t_0)$ and $Z_2(t, s) = Z(t, s, t_1), t \geq s$. We have that

$$z(t) = Z_1(t, \tilde{T})z(\tilde{T}) + \int_{\tilde{T}}^{t} Z_1(t, s) f(z(s))ds + \sum_{\tilde{T} \leq \zeta_i < t} Z_1(t, \zeta_i(t_0))W(z(\zeta_i(t_0))),$$

$$z_1(t) = Z_2(t, \tilde{T})z_1(\tilde{T}) + \int_{\tilde{T}}^{t} Z_2(t, s) f(z_1(s))ds + \sum_{\tilde{T} \leq \zeta_i < t} Z_2(t, \zeta_i(t_1))W(z_1(\zeta_i(t_1))).$$

It is difficult to evaluate the difference between $z(t)$ and $z_1(t)$ using the last two expressions since the moments of discontinuity of $z(t)$ and $z_1(t)$ are distinct. For this reason, we assume that $\zeta_j(t_0) \leq \zeta_j(t_1)$ for a fixed integer j. The opposite case can be discussed similarly. We introduce the following map

$$W_j^1(z) = (\mathscr{I} + B)\Big[(\mathrm{e}^{A(\zeta_j(t_1)-\zeta_j(t_0))} - \mathscr{I})z + \int_{\zeta_j(t_0)}^{\zeta_j(t_1)} \mathrm{e}^{A(\zeta_j(t_1)-s)} f(z(s))ds\Big] +$$

$$W\left((\mathscr{I} + B)\left[\mathrm{e}^{A(\zeta_j(t_1)-\zeta_j(t_0))}z + \int_{\zeta_j(t_0)}^{\zeta_j(t_1)} \mathrm{e}^{A(\zeta_j(t_1)-s)} f(z(s))ds\right]\right) -$$

$$\int_{\zeta_j(t_0)}^{\zeta_j(t_1)} \mathrm{e}^{A(\zeta_i(t_1)-s)} f(\bar{z}(s))ds - W(z),$$

where $z(t), \bar{z}(t)$ are solutions of

$$z'(t) = Az(t), \tag{6.2.21}$$

such that $z(\zeta_j(t_0)) = z$ and $\bar{z}(\zeta_j(t_1)) = z(\zeta_j(t_1)+)$, where "+" indicates the right-side limit at moment $\zeta_j(t_1)$. Consider the following systems

$$\begin{aligned} v'(t) &= Av(t) + f(v), \\ \Delta v|_{t=\zeta_i(t_0)} &= Bv(\zeta_i(t_0)) + W(v(\zeta_i(t_0))) + W_i^1(v(\zeta_i(t_0))), \end{aligned} \tag{6.2.22}$$

and

$$\begin{aligned} z'(t) &= Az(t) + f(z), \\ \Delta|_{t=\zeta_i(t_1)} &= Bz(\zeta_i(t_1)) + W(z(\zeta_i(t_1))). \end{aligned} \tag{6.2.23}$$

One can easily see that $M_2 = \sup_{\|z\| \leq M_1, i \in \mathbb{Z}} \|W_i^1(z)\| < \infty$, and respective solutions of systems (6.2.22) and (6.2.23) with the same initial data coincide in the intersection of their domains only if $t \notin [\widehat{\zeta_i(t_0), \zeta_i(t_1)}], i \in \mathbb{Z}$. For details [28–30] can be referred. So, if $v(t), v(\tilde{T}) = z_1(\tilde{T})$, is the solution of (6.2.22), then $v(t) = z_1(t)$ for all $t \notin [\widehat{\zeta_i(t_0), \zeta_i(t_1)}], i \in \mathbb{Z}$.

We have that

$$v(t) = Z_1(t, \tilde{T})v(\tilde{T}) + \int_{\tilde{T}}^{t} Z_1(t, s)f(v(s))ds +$$

$$\sum_{\tilde{T} \leq \zeta_i < t} Z_1(t, \zeta_i(t_0))[W(v(\zeta_i(t_0)) + W_1(v(\zeta_i(t_0))].$$

Consequently,

$$\|z(t) - v(t)\| \leq \|z(\tilde{T}) - v(\tilde{T})\| \|Z_1(t, \tilde{T})\| + \int_{\tilde{T}}^{t} \|Z_1(t, s)\| L \|z(s) - v(s)\| ds +$$

$$\sum_{\tilde{T} \leq \zeta_j(t_0) < t} \|Z_1(t, \zeta_j(t_0))\| L \|z(\zeta_j(t_0)) - v(\zeta_j(t_0))\| +$$

$$\sum_{\tilde{T} \leq \zeta_j(t_0) < t} \|Z_1(t, \zeta_j(t_0))\| \|W_1(v(\zeta_j(t_0))\| \leq$$

$$2M_1 N + M_2 \frac{e^{\omega}}{1 - e^{-\omega}} + \int_{\tilde{T}}^{t} N e^{-\omega(t-s)} L \|z(s) - v(s)\| ds +$$

$$\sum_{\tilde{T} \leq \zeta_j < t} N e^{-\omega(t - \zeta_j(t_0))} L \|z(\zeta_j(t_0)) - v(\zeta_j(t_0))\|.$$

Now, applying the analogue of Gronwall–Bellman Lemma [27] for discontinuous functions, we find that

$$\|z(t) - v(t)\| \leq \left(2M_1 N + M_2 \frac{e^{\omega}}{1 - e^{-\omega}}\right) e^{(-\omega + NL)(t - \tilde{T})} \prod_{\tilde{T} \leq \zeta_j < t} (1 + NL) \leq$$

$$\left(2M_1 N + M_2 \frac{e^{\omega}}{1 - e^{-\omega}}\right) e^{(-\omega + NL + \ln(1+NL))(t - \tilde{T})}. \tag{6.2.24}$$

Last inequality implies that $\|z(t) - v(t)\| < \varepsilon$ if $t > \tilde{T} + \frac{1}{2}E$, $t \notin [\zeta_i(\widehat{t_0}), \widehat{\zeta_i(t_1)}]$, $i \in \mathbb{Z}$, where

$$E > \frac{2\ln\left(\frac{\varepsilon}{2M_1N + M_2 e^{\omega}(1 - e^{-\omega})^{-1}}\right)}{-\omega + NL + \ln(1 + NL)}$$

(we may assume that $\varepsilon < 2M_1N$, without loss of generality). That is why, if $J = (\tilde{T} + \frac{1}{2}E, \tilde{T} + E)$, then $z(t)(\varepsilon, J)\phi(t)$. The lemma is proved.

Lemma 6.7 *Assume that conditions* $(C1)$–$(C8)$ *are fulfilled. Then (6.2.12) is Li–Yorke sensitive on* Λ'.

Proof Consider a pair of solutions $z(t) = z(t, t_0, z_0)$, $z_1(t) = z(t, t_1, z_1)$, with $t_0, t_1 \in \Lambda'$, $t_0 \neq t_1$. Assume that $\|z_0\|, \|z_1\| > M_1$. The discussion of other cases is easier.

Denote $s^0 = S(t_0) = (s_0^0, s_1^0, \ldots)$ and $s^1 = S(t_1) = (s_0^1, s_1^1 \ldots)$. It is obvious that $s^0 \neq s^1$. By Lemmas 6.2 and 6.3, there exists a sequence of integers $m_i \to \infty$ as $i \to \infty$ such that $d[\sigma^{m_i - j}s^0, \sigma^{m_i - j}s^1] \geq \frac{1}{2^j}$, $0 \leq j \leq m_i/2$.

Similarly to the proof of the last lemma, one can choose $m_i > 2$ sufficiently large so that $\|z(t)\| < M_1$, $\|z_1(t)\| < M_1$, if $t > \dfrac{m_i + 1}{2}$. Let us fix this m_i.

Since S is a homeomorphism and set Σ_2 is compact, for a given $j, 0 \leq j \leq m_i$, the set

$$P_j = \left\{(\bar{s}, \tilde{s}) \in \Sigma_2 \times \Sigma_2 : d[\bar{s}, \tilde{s}] \geq \frac{1}{2^{m_i - j}}\right\}$$

is compact, and

$$\min_{(\bar{s},\tilde{s}) \in P_j} |S^{-1}(\bar{s}) - S^{-1}(\tilde{s})| = \mu_j > 0,$$

$P_{j+1} \subseteq P_j$, $\mu_{j+1} \geq \mu_j$, $0 \leq j < m_i - 1$. Fix $i_0 = m_i - 2$. Then $|\kappa_i(t_0) - \kappa_i(t_1)| \geq \mu_{i_0}$ if $i = i_0, i_0 + 1$.

Similarly, we also have that there exists a positive number $\mu_0 < 1$ such that $|\kappa_j(t_0) - \kappa_j(t_1)| \leq \mu_0$ if $0 \leq j < m_i$.

Next, we assume that $\kappa_j(t_0) < \kappa_j(t_1)$ for all j. It is easily seen that case $\kappa_j(t_0) > \kappa_j(t_1)$ can be analyzed similarly. Thus, there is a number k among $i_0, i_0 + 1$, such that $\kappa_k(t_1) - \kappa_k(t_0) > \mu_{i_0}$ and $\kappa_k(t_0) - \kappa_{k-1}(t_1) \geq \dfrac{1}{2}(1 - \mu_0)$.

Clearly, $(C8)$ implies that $\nu_1 = \dfrac{2}{3}\dfrac{\underline{m}\eta}{\overline{m}} - 2LM_1 > 0$ and $\nu_2 < \nu_1$, where $\nu_2 = \dfrac{2\overline{m}LM_1}{\underline{m}} + \dfrac{1}{q}\eta$.

We shall show that the constants $\varepsilon_0, \varepsilon_1$ of Definition 6.12 can be set as $\varepsilon_0 = \frac{1}{q}\underline{m}\eta$, $\varepsilon_1 = \min\{\mu_{i_0}, \frac{1}{2}(1 - \mu_0)\}$.

Assume that $\|z(\zeta_k(t_0)) - z_1(\zeta_k(t_0))\| < \nu_1$. Then, for $t \in [\zeta_k(t_0), \zeta_k(t_1)]$,

$$z(t) = e^{A(t-\zeta_k(t_0))}(\mathscr{I} + B)z((\zeta_k(t_0)) + \int_{\zeta_k(t_0)}^{t} e^{A(t-s)} f(z(s))ds + e^{A(t-\zeta_k(t_0))} W(z((\zeta_k(t_0))),$$

$$z_1(t) = e^{A(t-\zeta_k(t_0))} z_1((\zeta_k(t_0)) + \int_{\zeta_k(t_0)}^{t} e^{A(t-s)} f(z_1(s))ds.$$

We have that

$$\|z(t) - z_1(t)\| = \|e^{A(t-\zeta_k(t_0))}[Bz((\zeta_k(t_0)) + W(z((\zeta_k(t_0)))] + e^{A(t-\zeta_k(t_0))}[z((\zeta_k(t_0)) - z_1((\zeta_k(t_0))] + \int_{\zeta_k(t_0)}^{t} e^{A(t-s)}(f(z(s)) - f(z_1(s)))ds\| \geq \underline{m}\eta - \overline{m}(\nu_1 + 2LM_1) \geq \varepsilon_0.$$

If $\|z(\zeta_k(t_0)) - z_1(\zeta_k(t_0))\| > \nu_2$, then, for $t \in [\zeta_{k-1}(t_1), \zeta_k(t_0)]$,

$$z(t) = e^{A(t-\zeta_k(t_0))} z((\zeta_k(t_0)) + \int_{\zeta_k(t_0)}^{t} e^{A(t-s)} f(z(s))ds,$$

$$z_1(t) = e^{A(t-\zeta_k(t_0))} z_1((\zeta_k(t_0)) + \int_{\zeta_k(t_0)}^{t} e^{A(t-s)} f(z_1(s))ds.$$

and

$$\|z(t) - z_1(t)\| \geq \underline{m}\nu_2 - \overline{m}2LM_1 = \varepsilon_0.$$

The lemma is proved.

Lemmas 6.6 and 6.7 imply that (6.2.12) is Li–Yorke chaotic on Λ'.

In the main theorems of the original paper [3], properties of the chaotic behavior (i) and (ii) of Definition 6.9 are accompanied with the following one:

(iii) for each point $x \in X'$ and any periodic point $\bar{x} \in X$, one has that

$$\limsup_{i\to\infty} \rho(T^i(x), T^i(\bar{x})) > 0. \tag{6.2.25}$$

Then it was shown (see, for example [43], Lemma 28) that by omitting at most one point in the scrambled set one can have the chaos in the sense of Definition 6.9. The following assertion is about the analogue of (iii). Its proof is very similar to the proof of Lemma 6.7, if one uses property (iii) of map h.

Lemma 6.8 *Assume that conditions* $(C1)$–$(C8)$ *are fulfilled. Then every solution* $z(t) = z(t, t_1, z_0)$, $t_1 \in \Lambda'$, $z_0 \in \mathbb{R}^n$, *and every periodic solution* $\phi(t, t_0)$, $t_0 \in \Lambda$, *of (6.2.12) are* $(\varepsilon_0, \varepsilon_1)$*-separated,* ε_0 *and* ε_1 *being the same as in Lemma 6.7.*

Remark 6.2 It seems natural to consider the chaos only for uniformly bounded on $\mathbb{R}_+$ solutions, since the domain of chaos is always assumed to be a compact set. We consider the set of all solutions, where the chaos scenario starts at the moment when a solution reaches the region $\|z(t)\| \leq M_1$.

Example 6.2 Consider the following initial value problem

$$\begin{aligned} x_1' &= -1/3x_2 + f_1(x_1, x_2), \\ x_2' &= 1/3x_1 + f_2(x_1, x_2), t \neq \zeta_i(t_0), \\ \Delta x_1|_{t=\zeta_i(t_0)} &= W(x_1), \\ \Delta x_2|_{t=\zeta_i(t_0)} &= -\frac{1}{2}x_2, \end{aligned} \tag{6.2.26}$$

where $x_1, x_2 \in \mathbb{R}$, l is a positive constant, $f_1(s,u) = s\cos u$, $f_2(s,u) = s\sin u$, $W(s) = 1+s^2$, if $|s| \leq l$, and $f_1(s,u) = l\cos u$, $f_2(s,u) = l\sin u$, $W(s) = 1+l^2$, if $|s| > l$. One can easily see that all the functions are Lipschitzian with a constant proportional to l. The matrices of coefficients are

$$A = \begin{pmatrix} 0 - 1/3 \\ 1/3 \quad 0 \end{pmatrix}, \quad B = \begin{pmatrix} 0 & 0 \\ 0 - 1/2 \end{pmatrix}.$$

The matrices commute, and the eigenvalues of the matrix

$$A + \mathrm{Ln}\,(\mathcal{I} + B) = \begin{pmatrix} 0 & -1/3 \\ 1/3 & -\ln 2 \end{pmatrix}$$

are negative: $\lambda_{1,2} = -\ln 2/2 \pm \sqrt{\ln^2 2/4 - 1/9} < 0$.

Condition $(C3)$ is obvious, since function $W(s)$ is never equal to zero. All the other conditions required by the theorems could be easily checked with sufficiently small coefficient l. That is, the dynamics of (6.2.12) is Li–Yorke chaotic.

6.3 Li–Yorke Chaos in the System with Relay

In this section, we address a special initial value problem of a differential equation with relay function. The concept of Li–Yorke chaos [3] is considered.

6.3.1 Introduction and Preliminaries

In paper [44], Devaney's ingredients were indicated for a special initial value of a relay system with linear elements. In the present section, we attempt to shape the Li–Yorke chaos [3] for the multidimensional nonlinear relay system. The quasi-minimal set existence for this system has been proved in [45]. The fact, which makes Li–Yorke chaos attractive for applications, is that it can be developed for a multidimensional case [25]. So, in this section, an attempt to create other higher dimensional chaotic systems is made. The approach has been also used in [46] for impulsive differential equations.

Let us recall the definition of the chaos for maps. Consider a nonvoid compact metric space (X, ρ) with metric ρ and $T : X \to X$, a surjective continuous map. Li and Yorke call the map chaotic if:

1. the map has points with all periods $p \in \mathbb{N}$;
2. there exists an uncountable subset $X' \subseteq X$, the scrambled set, that does not contain periodic points and

$$\limsup_{i\to\infty} \rho(T^i(x), T^i(\tilde{x})) > 0, \tag{6.3.27}$$

$$\liminf_{i\to\infty} \rho(T^i(x), T^i(\tilde{x})) = 0, \tag{6.3.28}$$

   for each pair $x, \tilde{x} \in X', x \neq \tilde{x}$;
3. for each point $x \in X'$ and any periodic point $\bar{x} \in X$, one has that

$$\limsup_{i\to\infty} \rho(T^i(x), T^i(\bar{x})) > 0. \tag{6.3.29}$$

Consider the sequence space [14]

$$\Sigma_2 = \{s = (s_0 s_1 s_2 \ldots) : s_j = 0 \text{ or } 1\}$$

with the metric

$$d[s, t] = \sum_{i=0}^{\infty} \frac{|s_i - \tilde{s}_i|}{2^i},$$

where $\tilde{s} = (\tilde{s}_0\tilde{s}_1 \ldots) \in \Sigma_2$, and the shift map $\sigma : \Sigma_2 \to \Sigma_2$, such that $\sigma(s) = (s_1 s_2 \ldots)$. The pair (Σ_2, σ) is the symbolic dynamics. The map is continuous, *card* $Per_n(\sigma) = 2^n$, $Per(\sigma)$ is dense in Σ_2, and there exists a dense orbit in Σ_2.

In this part, we prove that every map which is topologically conjugate to the shift σ on Σ_2 is Li–Yorke chaotic. First we check if σ itself is chaotic, and then the general case will be considered.

Let us denote $h(t, \mu) \equiv \mu t(1 - t)$, the logistic map, and assume that $\mu > 4$. Then, [8, 14], there exists an invariant Cantor set $\bar{\Lambda} \subset I = [0, 1]$ for h. The map has a 3-period point, hence the map is Li–Yorke chaotic. There exists a homeomorphism $\bar{S}(t)$ between Λ and Σ_2, such that $\bar{S} \circ h = \sigma \circ \bar{S}$ [8]. That is, h and σ are topologically conjugate. Using the technique from [14] (see, for example, Theorem 5.1, there) one can show that $\Sigma_2' = S(\bar{\Lambda}')$ is a scrambled set. If $s, s' \in \Sigma_2', s \neq s'$, we call (s, s') a *Li–Yorke pair*.

The proofs of the following lemmas are standard [14].

Lemma 6.9 *If (s, s') is a Li–Yorke pair from Σ_2', then there exist sequences $k_i, m_i \to \infty$, as $i \to \infty$, such that $s_{k_i+j} = s'_{k_i+j}$, $j = 0, 1, \ldots, m_i - 1$ and $s_{k_i+m_i} \neq s'_{k_i+m_i}$. If $s \in \Sigma_2'$ and $s' \in \Sigma_2$ is a periodic point, then there exists a sequence $l_i \to \infty$, as $i \to \infty$, such that $s_{l_i} \neq s'_{l_i}$.*

Lemma 6.10 *If (s, s') is a Li–Yorke pair from Σ_2', then there exists a sequence $l_i \to \infty$, as $i \to \infty$, such that $d[\sigma^{l_i} s, \sigma^{l_i} s'] \geq 1$.*

That is, the dynamics (Σ_2, σ) is Li–Yorke sensitive [20].

Let $h : \Lambda \to \Lambda$, where Λ is a subset of the interval $[0, 1]$, be a map topologically conjugate to σ, and Λ' is an image of Σ_2' by the conjugacy.

For every $t_0 \in \Lambda$ one can construct a sequence $\kappa(t_0)$ of real numbers $\kappa_i, i \geq 0$, such that $\kappa_{i+1} = h(\kappa_i)$ and $\kappa_0 = t_0$. The sequence $\zeta(t_0) = \{\zeta_i(t_0)\}$ in (6.2.12) is defined as $\zeta_i(t_0) = i + \kappa_i(t_0), i \geq 0$.

By applying the conjugacy of h and σ, one can verify that map h has useful chaotic properties.

Lemma 6.11 *If $t, t' \in \Lambda'$, then there exist sequences $k_i, l_i \to \infty$, as $i \to \infty$, such that $\max_{j=0,1,\ldots,l_i} |h^{k_i+j}(t) - h^{k_i+j}(t')| \to 0$ as $i \to \infty$.*

Lemma 6.12 *For every pair $t, t' \in \Lambda', t \neq t'$, there exists a sequence $m_i \to \infty$, as $i \to \infty$, such that $|h^{m_i}(t) - h^{m_i}(t')| \geq \delta$.*

The last two lemmas imply that h is a Li–Yorke chaotic map, since the verification of the periodicity condition is simple.

6.3.2 The Li–Yorke Chaos

The main object of our investigation is the following special initial value problem

$$\begin{aligned} z'(t) &= Az(t) + f(z) + \nu(t, t_0), \\ z(t_0) &= z_0, \ (t_0, z_0) \in \Lambda \times \mathbb{R}^n, \end{aligned} \tag{6.3.30}$$

where $z \in \mathbb{R}^n$, $t \in \mathbb{R}_+ = [0, \infty)$, $i \geq 0$. Cantor set $\Lambda \subset I = [0, 1]$, the sequence $\zeta(t_0) = \{\zeta_i(t_0)\}$ of switching moments were described in the last subsection and

$$\nu(t, t_0) = \begin{cases} m_0 & \text{if} \ \ \zeta_{2i}(t_0) < t \leq \zeta_{2i+1}(t_0), i \in \mathbb{Z}, \\ m_1 & \text{if} \ \ \zeta_{2i-1}(t_0) < t \leq \zeta_{2i}(t_0), i \in \mathbb{Z}, \end{cases}$$

where $m_0,\ m_1 \in \mathbb{R}^n$ are different vectors. The function f satisfies the Lipshitz condition with a positive constant L, A is an $n \times n$ constant real-valued matrix with real parts of eigenvalues all negative. Denote the maximal of them $\alpha < 0$.

For a fixed $t_0 \in \Lambda$, system (6.3.30) is a differential equation with discontinuous right-hand side of a specific type when discontinuities happen on vertical planes in the (t, z)-space.

A function $z(t),\ z(t_0) = z_0$, is a solution of (6.3.30) on $[t_0, \infty)$ if: (i) $z(t)$ is continuous on $[t_0, \infty)$; (ii) the derivative $z'(t)$ exists at each point $t \in \mathbb{R}$ with the possible exception of the points $\zeta_i(t_0)$, where left-sided derivatives exist; (iii) Equation (6.3.30) is satisfied on each interval $(\zeta_i(t_0), \zeta_{i+1}(t_0)],\ i \geq 0$.

It can be easily verified that problem (6.3.30) has a unique solution $z(t, t_0, z_0)$ for each $t_0 \in \Lambda,\ z_0 \in \mathbb{R}^n$.

There exists a positive number N such that $\|e^{At}\| \leq N e^{\alpha t},\ t \geq 0$.

The solution $z(t) = z(t, t_0, z_0),\ t_0 \in \Lambda,\ z_0 \in \mathbb{R}^n$, of (6.3.30) satisfies the following integral equation

$$z(t) = e^{A(t-t_0)} z_0 + \int_{t_0}^{t} e^{A(t-s)}[f(z(s)) + \nu(s, t_0)]\, ds \tag{6.3.31}$$

In what follows we assume that $\sup_{\mathbb{R}^n} |f(z)| = M_0 < \infty,\ NL < \alpha$. Fix a sequence $\zeta(t_0),\ t_0 \in \Lambda$. Using the standard technique, one can verify that all solutions eventually, as t increases, enter the tube with the radius $M = M_0 \left[1 + \dfrac{N}{\alpha - NL}\right]$, $t \in \mathbb{R}$. Moreover, if the sequence $\kappa(t_0)$ is periodic with a period $p \in \mathbb{N}$, then there is a solution of (6.3.30) with the same period, and its integral curve is placed in the tube. One can easily see that all these solutions are different for different p. Let us, introduce the following distance. If $\phi,\ \psi$ are continuous on $\mathbb{R}$ functions, then denote $\|\phi(t) - \psi(t)\|_J = \sup_J \|\phi(t) - \psi(t)\|$, where J is an interval of $\mathbb{R}$.

We use the following definitions. They are taken from [3, 14, 20] and adapted for (6.3.30).

Definition 6.15 A pair of solutions of (6.3.30) $z(t)=z(t, t_0, z_0),\ z_1(t)=z(t, t_1, z_1)$, $t_0, t_1 \in \Lambda$, is proximal if for each $\varepsilon > 0,\ E > 0$ there exists an interval $J \subset [t_0, \infty)$ with length not less than E such that $\|z_1(t) - z(t)\|_E < \varepsilon$.

Definition 6.16 The solutions of (6.3.30) $z(t) = z(t, t_0, z_0),\ z_1(t) = z(t, t_1, z_1)$, $t_0, t_1 \in \Lambda$, are frequently $(\varepsilon_0, \varepsilon_1)$-separated if there exist positive numbers $\varepsilon_0, \varepsilon_1$ and infinitely many disjoint subintervals of $[t_0, \infty)$, of length not less than ε_1, such that $||z_1(t) - z(t)|| > \varepsilon_0$ for each t from these intervals.

Definition 6.17 A couple $z(t) = z(t, t_0, z_0),\ z_1(t) = z(t, t_1, z_1),\ t_0, t_1 \in \Lambda$, of solutions of (6.3.30) is a Li–Yorke pair if they are proximal and $(\varepsilon_0, \varepsilon_1)$-separated for some positive $\varepsilon_0, \varepsilon_1$.

Definition 6.18 Problem (6.3.30) is Li–Yorke chaotic on Λ' if:

1. there exist solutions $\phi(t, t_0)$ with all periods $p \in \mathbb{N}$;

2. each couple of solutions $z(t) = z(t, t_0, z_0), z_1(t) = z(t, t_1, z_1)$, with $t_0, t_1 \in \Lambda', t_0 \neq t_1$, is Li–Yorke pair;
3. every solution $z(t) = z(t, t_0, z_0), t_0 \in \Lambda'$, and every periodic solution $\phi(t) = \phi(t, t_0), t_0 \in \Lambda$, of (6.3.30) are frequently $(\varepsilon_0, \varepsilon_1)$-separated for some positive $\varepsilon_0, \varepsilon_1$.

Definition 6.19 Problem (6.3.30) is Li–Yorke sensitive on Λ' if there exist positive numbers $\varepsilon_0, \varepsilon_1$, such that each couple of solutions $z(t) = z(t, t_0, z_0), z_1(t) = z(t, t_1, z_1)$, with $t_0, t_1 \in \Lambda', t_0 \neq t_1$, is frequently $(\varepsilon_0, \varepsilon_1)$-separated.

By applying (6.3.31), Lemmas 6.11 and 6.12, one can prove that the following assertions are valid.

Lemma 6.13 *Problem (6.3.30) is Li–Yorke sensitive on Λ'.*

Theorem 6.6 *Problem (6.3.30) is Li–Yorke chaotic on Λ'.*

6.4 Dynamical Synthesis of Quasi-Minimal Sets

We address the quasilinear differential equation with a pulse function, whose moments of discontinuity depend on the initial moment. The existence of a Poisson stable trajectory dense in a quasi-minimal set is proved. An appropriate simulation of a chaotic attractor is presented.

6.4.1 Introduction

L. Shilnikov in [47] emphasizes that "... it seems quite reasonable the role of dynamical chaos orbits should be assigned to the Poisson stable trajectories," and "... we arrive at the following problem: how can one establish the existence of the Poisson stable trajectories in the phase space of a system?" One way of solving the problem is the method of *dynamical synthesis* [48, 49], which is a general technique of constructing dynamical systems with desired properties.

We use the map, which is topologically conjugate to symbolic dynamics, as the generator of moments of discontinuities in the multidimensional dissipative system to obtain the Poisson stable solutions. Since the main idea of this section is to obtain a quasi-minimal set by inserting the generator with a similar property into a dissipative system, one can say that the idea of dynamical synthesis is applied.

Another issue of relevance to the present section is the nonlinear dynamics of electric circuits, of mechanical models and of control systems [10, 50, 51] which convert discrete data into continuous output.

We believe that the approach can give a strong impact for applications, since one can investigate controllability of chaos [52, 53] based on similar properties of the generator function, which are already known or can be developed if needed. For example, it may give a tool to support the given degree of nonregularity, which is important for cardiac rhythm [54].

The main object of our investigation is the following special initial value problem

$$\begin{aligned} z'(t) &= Az(t) + f(z) + \nu(t, t_0), \\ z(t_0) &= z_0, \ (t_0, z_0) \in \Lambda \times \mathbb{R}^n, \end{aligned} \tag{6.4.32}$$

where $z \in \mathbb{R}^n, t \in \mathbb{R}, i \in \mathbb{Z}$, $\mathbb{R}$ and $\mathbb{Z}$ are sets of all real numbers and integers, respectively,

$$\nu(t, t_0) = \begin{cases} m_0 & \text{if } \ \zeta_{2i}(t_0) < t \leq \zeta_{2i+1}(t_0), i \in \mathbb{Z}, \\ 0 & \text{if } \ \zeta_{2i-1}(t_0) < t \leq \zeta_{2i}(t_0), i \in \mathbb{Z}. \end{cases}$$

where $m_0 \in \mathbb{R}^n$ is a nonzero vector. Cantor set $\Lambda \subset [0, 1]$, and the sequence $\zeta(t_0) = \{\zeta_i(t_0)\}, \ i \in \mathbb{Z}$, are described in Sect. 6.4.4. The function f satisfies the Lipshitz condition with a positive constant L, A is an $n \times n$ constant real-valued matrix with real parts of eigenvalues all negative. Denote the maximal of them $\alpha < 0$.

It is worth mentioning that we can consider other types of equations to obtain similar results, for instance, one may assume $\nu(t, t_0) \equiv 0$, and a function $f(t, z)$ with discontinuities of the first kind at the points of $\zeta(t_0)$.

For a fixed $t_0 \in \Lambda$, system (6.4.32) is a differential equation with discontinuous right-hand side of a specific type when discontinuities happen on vertical planes in the (t, z)-space.

In what follows, we use definition of solutions formulated in [17] (see, also, [16]). Most general results on existence and uniqueness of solutions for differential equation with discontinuous right-hand side can be found in [55].

A function $z(t), z(t_0) = z_0$, is a solution of (6.4.32) on $\mathbb{R}$ if: (i) $z(t)$ is continuous on $\mathbb{R}$; (ii) the derivative $z'(t)$ exists at each point $t \in \mathbb{R}$ with the possible exception of the points $\zeta_i(t_0)$, where left-sided derivatives exist; (iii) Equation (6.4.32) is satisfied on each interval $(\zeta_i(t_0), \zeta_{i+1}(t_0)], i \in \mathbb{Z}$.

It can be easily verified that problem (6.4.32) has a unique solution $z(t, t_0, z_0), t \in \mathbb{R}$, for each $t_0 \in \Lambda, z_0 \in \mathbb{R}^n$.

There exists a positive number N such that $\|e^{At}\| \leq Ne^{\alpha t}, t \geq 0$.

The solution $z(t) = z(t, t_0, z_0), t_0 \in \Lambda, z_0 \in \mathbb{R}^n$, of (6.4.32) satisfies the following integral equation

$$z(t) = e^{A(t-t_0)} z_0 + \int_{t_0}^{t} e^{A(t-s)} [f(z(s)) + \nu(s, t_0)] \, ds.$$

In the sequel, we assume that $\sup_{\mathbb{R}^n} |f(z)| = M_0 < \infty, NL < |\alpha|$. Fix a sequence $\zeta(t_0), t_0 \in \Lambda$. Using the standard technique one can verify that $z(t)$ is a bounded on $\mathbb{R}$ solution of (6.4.32) if and only if it satisfies the equation

$$z(t) = \int_{-\infty}^{t} e^{A(t-s)}[f(z(s)) + v(s, t_0)]\, ds,$$

and for each sequence $\zeta(t_0), t_0 \in \Lambda$, there exists a unique bounded on $\mathbb{R}$ solution $z(t, \zeta(t_0))$, and all these bounded solutions are placed in the tube with the radius $M = M_0[1 - \frac{N}{\alpha + NL}], t \in \mathbb{R}$. Moreover, if $z(t, t_0, z_0)$ is a solution of (6.4.32), then using Gronwall–Bellman Lemma one can obtain that

$$\|z(t, t_0, z_0) - z(t, \zeta(t_0))\| \leq N\|z_0 - z(t_0, \zeta(t_0))\|e^{(-\alpha+NL)(t-t_0)}.$$

That is, the bounded solution $z(t, \zeta(t_0))$ attracts all solutions of (6.4.32) with the same initial moment t_0, $t_0 \in \Lambda$.

Denote $\mathscr{CB} = \{z(t, \zeta(t_0)) : \zeta(t_0) \in \Pi\}$, where set Π of all sequences $\zeta(t_0)$, $t_0 \in \Lambda$, is described in Sect. 6.4.4. The set $\mathscr{CB}$ is placed in the tube with radius M, and it is an attractor for all solutions of (6.4.32). We shall show that the attractor is a quasi-minimal set in the next section.

6.4.2 Main Result

In the present section, we provide the definitions of the Poisson stable solution of non-autonomous differential equations, $\alpha-$ and $\omega-$ limit solutions. They can be compared with the definitions for non-autonomous equations in , Chap. 8 [56].

Let us, introduce the following distance. If ϕ, ψ are continuous on $\mathbb{R}$ functions, then denote $\|\phi(t) - \psi(t)\|_J = \sup_J \|\phi(t) - \psi(t)\|$, where J is an interval of $\mathbb{R}$.

We say that $z(t, \zeta(t^*)) \in \mathscr{CB}$, is positively Poisson stable (P_+ stable) if for each $\gamma \in \mathbb{R}$ there exist two sequences of real numbers β_n, E_n with $\beta_n, E_n \to \infty$, and $\|z(t + \beta_n, \zeta(t^*)) - z(t + \gamma, \zeta(t^*))\|_{(-E_n, E_n)} \to 0$, as $n \to \infty$.

We say that $z(t, \zeta(t^*)) \in \mathscr{CB}$, is negatively Poisson stable (P_- stable) if for each $\gamma \in \mathbb{R}$ there exist two sequences of real numbers β_n, E_n with $\beta_n, E_n \to -\infty$, and $\|z(t + \beta_n, \zeta(t^*)) - z(t + \gamma, \zeta(t^*))\|_{(E_n, -E_n)} \to 0$, as $n \to \infty$.

Solution $z(t, \zeta(t^*)) \in \mathscr{CB}$, is Poisson stable ($P$ stable) if it is P_- and P_+ stable.

We say that $z(t, \zeta(t)) \in \mathscr{CB}$, is ω-limit solution corresponding to $z(t, \zeta(t^*)) \in \mathscr{CB}$, if for each $\gamma \in \mathbb{R}$ there exist two sequences of real numbers β_n, E_n with $\beta_n, E_n \to \infty$, and $\|z(t + \beta_n, \zeta(t^*)) - z(t + \gamma, \zeta(t))\|_{(-E_n, E_n)} \to 0$, as $n \to \infty$.

We say that $z(t, \zeta(t)) \in \mathscr{CB}$, is α-limit solution corresponding to $z(t, \zeta(t^*)) \in \mathscr{CB}$, if for each $\gamma \in \mathbb{R}$ there exist two sequences of real numbers β_n, E_n with $\beta_n, E_n \to -\infty$, and $\|z(t + \beta_n, \zeta(t^*)) - z(t + \gamma, \zeta(t))\|_{(E_n, -E_n)} \to 0$, as $n \to \infty$.

Denote sets, which consist of all ω-limit solutions and α-limit solutions as Ω_{t^*} and A_{t^*}, respectively.

We say that $\mathscr{CB}$ is a quasi-minimal set if $\mathscr{CB} = \Omega_{t^*} = A_{t^*}$, where $z(t, \zeta(t^*)) \in \mathscr{CB}$ is a P stable solution.

Theorem 6.7 *$\mathscr{CB}$ is the quasi-minimal set.*

Proof Consider a sequence $\zeta(t^*)$, which has been defined in Theorem 6.8 (see Sect. 6.4.4), and fix the corresponding solution of (6.4.32), $z(t, \zeta(t^*)) \in \mathscr{C}\mathscr{B}$. We shall show that the solution is P_+ stable by considering $\gamma = 0$. For all other $\gamma \in \mathbb{R}$, the proof is very similar. Fix a positive ε. Moreover, fix a positive ε_1, whose dependence on ε will be described below. From Theorem 6.8 (1) we have that there exist sufficiently large natural numbers j and m such that $|\zeta_{i+m}(t^*) - \zeta_i(t^*)| < \varepsilon_1$ if $-j \le i \le j$. For the sake of simplicity below we shall write ζ_i instead of $\zeta_i(t^*)$.

We have that for $t \ge -j$,

$$z(t, \zeta(t^*)) = \mathrm{e}^{A(t-\zeta_j(t^*))} z(\zeta_j, \zeta(t^*)) + \int_{\zeta_j}^{t} \mathrm{e}^{A(t-s)}[f(z(s, \zeta(t^*))) + v(s, t_0)]ds,$$

and

$$z(t + \zeta_{-j+m} - \zeta_j) = \mathrm{e}^{A(t-\zeta_j)} z(\zeta_{-j+m}, \zeta(t^*)) +$$

$$\int_{\zeta_{-j+m}}^{t+\zeta_{-j+m}-\zeta_j} \mathrm{e}^{A(t+\zeta_{-j+m}-\zeta_j-s)}[f(z(s, \zeta(t^*))) + v(s, t_0)]ds = \mathrm{e}^{A(t-\zeta_j)} z((\zeta_{-j+m}) +$$

$$\int_{\zeta_j}^{t} \mathrm{e}^{A(t-s)}[f(z(s + \zeta_{-j+m} - \zeta_j)) + v(s + \zeta_{-j+m} - \zeta_j, t_0)]ds.$$

Subtract the last expression from the previous one to obtain that

$$\|z(t, \zeta(t^*)) - z(t + \zeta_{-j+m} - \zeta_j)\| \le 2MN\mathrm{e}^{-\zeta_j(t-\zeta_j)} +$$

$$\int_{\zeta_j}^{t} NL\mathrm{e}^{-\alpha(t-s)} \|z(s, \zeta(t^*)) - z(s + \zeta_{-j+m} - \zeta_j)\| ds + \int_{\zeta_j}^{t} N\mathrm{e}^{-\alpha(t-s)} \varepsilon_1 \|m_0\| ds.$$

Consider the following Lemma 2.2 from [57].

Lemma 6.14 *Let $u(t)$, $f(t)$ be nonnegative functions integrable over the interval $t_0 \le t \le t_0 + T$; let K be a positive constant. If the inequality*

$$u(t) \le f(t) + K \int_{t_0}^{t} u(s)ds, \; t_0 \le t \le t_0 + T,$$

is fulfilled then the following inequality holds

$$u(t) \le f(t) + K \int_{t_0}^{t} \mathrm{e}^{K(t-s)} f(s)ds.$$

Next, we denote $u(t) = \|z(t, \zeta(t^*)) - z(t + \zeta_{-j+m} - \zeta_j)\|\mathrm{e}^{\alpha t}$, and apply the last Lemma, to obtain that

$$\|z(t, \zeta(t^*)) - z(t + \zeta_{-j+m} - \zeta_j)\| \leq \frac{N\varepsilon_1 \|m_0\|}{\alpha}\left(1 + \frac{1}{\alpha - NL}\right) +$$

$$\frac{\alpha[2M(\alpha - NL) - \varepsilon_1\|m_0\|]}{\alpha L(\alpha - NL)} e^{(-\alpha+NL)(t-\zeta_j)} + \frac{\alpha[(NL-1)(2M\alpha - \varepsilon_1\|m_0\|)}{\alpha L} e^{-\alpha(t-\zeta_j)}.$$

On the basis of the last inequality one can easily see that $\|z(t, \zeta(t^*)) - z(t+\zeta_{-j+m} - \zeta_j)\| < \varepsilon$ if $t \in (-E, E)$, where $E = \frac{j}{2}$, j is sufficiently large, and ε_1 is a sufficiently small positive number. The number $\zeta_{-j+m} - \zeta_j$ is as large as m. Thus, we have proved that the solution is P_+ stable. Applying Theorem 6.8 (2) in a similar manner, one can show that it is P_- stable. Moreover, using Theorem 6.8 (3) and (4), we can show that $\mathscr{C}\mathscr{B} = \Omega_{t^*} = A_{t^*}$. The theorem is proved.

6.4.3 A Simulation Result

Consider the sequence $\zeta_i = i + \kappa_i$, $\kappa_i = 4\kappa_{i-1}(1 - \kappa_{i-1})$, $\kappa_0 = t_0$, $t_0 \in [0, 1]$, $i \geq 0$ and take into account the following system

$$\begin{aligned} x'' + 2x' + 1.5x &= \sin y, \\ y' &= -3y + v(t, t_0), \end{aligned} \tag{6.4.33}$$

where $v(t, t_0)$ is a scalar pulse function with $m_0 = 1$. The second equation is a drive equation and the first one, the pendulum equation. Using new variables $x_1 = x$, $x_2 = x'$, $x_3 = y$, one can reduce (6.4.33) to the system

$$\begin{aligned} x_1' &= x_2, \\ x_2' &= -1.5x_1 - 2x_2 + \sin x_3. \\ x_3' &= -3x_3 + v(t, t_0) \end{aligned} \tag{6.4.34}$$

One can easily verify that all eigenvalues of the matrix of coefficients have negative real parts. Fix $t_0 = 12/17$ and take a solution $(x_1(t), x_2(t), x_3(t))$ of the last system with the initial condition $x_1(t_0) = 0.02$, $x_2(t_0) = -0.025$, $x_3(t_0) = -0.02$. In Fig. 6.3 the chaotic attractor is shown using points $(x_1(n), x_2(n), x_3(n))$, $n = 1, 2, 3, \ldots, 75{,}000$, in x_1, x_2, x_3-space.

Remark 6.3 In [18] it was proved that the shadowing property, which is valid for uniformly hyperbolic maps [58, 59], can be extended if they are non-hyperbolic, particularly for the logistic map.Using the technique of the present section one can verify that system (6.4.32) inherits the shadowing property, if the generator function has it [60]. We suppose that the general problem "how numerical orbits of systems obtained by the dynamical synthesis can be shadowed by true orbits for long time" deserves to be considered in the future.

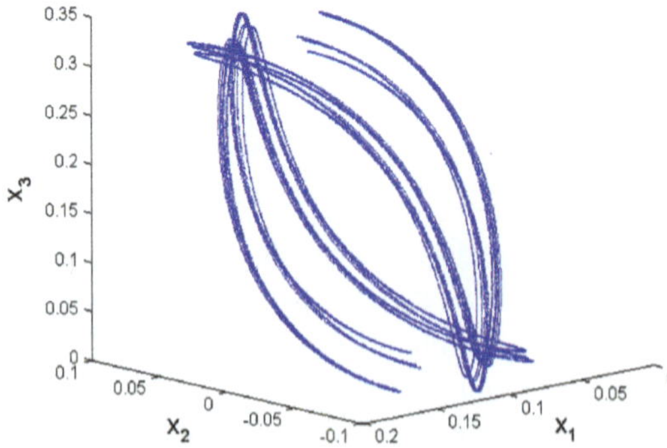

Fig. 6.3 The chaotic attractor by a stroboscopic sequence $(x_1(n), x_2(n), x_3(n))$, $1 \le n \le 75{,}000$, is observable

6.4.4 Appendix

Consider the sequence space $\Sigma_2 = \{s = (s_0 s_1 s_2 \ldots) : s_j = 0 \text{ or } 1\}$ with the metric $d[s, \bar{s}] = \sum\limits_{i=0}^{\infty} \frac{|s_i - \bar{s}_i|}{2^i}$, where $\bar{s} = (\bar{s}_0 \bar{s}_1 \ldots) \in \Sigma_2$, and the shift map $\sigma : \Sigma_2 \to \Sigma_2$, such that $\sigma(s) = (s_1 s_2 \ldots)$. The semidynamics (Σ_2, σ) is the symbolic dynamics [15].

Let us consider, also, the space of bi-sequences

$$\Sigma^2 = \{s = (\ldots s_{-n} \ldots s_{-1} s_0 s_1 s_2 \ldots s_n \ldots) : s_j = 0 \text{ or } 1\}.$$

At the first, an ordering should be introduced on Σ^2 [15]. If two finite sequences are given $s = \{s_1 \ldots s_k\}, \bar{s} = \{\bar{s}_1 \ldots \bar{s}_{k'}\}$, then it is said that $s < \bar{s}$ if $k < k'$, and if $k = k'$, then $s < \bar{s}$ if $s_i < \bar{s}_i$, where i is the first integer such that $s_i \neq \bar{s}_i$.

Thus, one denotes the sequences having length k as follows: $s_1^k < \cdots s_{2^k}^k$, where the superscript refers to the length of the sequence and the subscript refers to a particular sequence of length k which is uniquely specified by the above ordering scheme. Denote $s^* = (\ldots s_8^3 s_6^3 s_4^3 s_2^2 s_2^1 s_1^1 s_1^2 s_3^2 s_1^3 s_3^3 s_5^3 s_7^3 \ldots)$.

Introduce maps $B_i : \Sigma^2 \to \Sigma_2, i \in \mathbb{Z}$, such that $B_i(s) = (s_i, s_{i+1}, \ldots)$. From the method of construction of s^* it follows that the following assertion is valid.

Lemma 6.15 (1) *For a fixed $j \in \mathbb{Z}$ there exist two sequences of integers k_n, l_n with $k_n, l_n \to \infty$, and*

$$\lim_{n\to\infty} \max_{-l_n \le i \le l_n} d[B_{k_n+i}(s^*), B_{j+i}(s^*)] = 0.$$

(2) *For a fixed $j \in \mathbb{Z}$ there exist two sequences of integers k_n, l_n with $k_n, l_n \to -\infty$, and*

$$\lim_{n\to\infty} \max_{l_n \le i \le -l_n} d[B_{k_n+i}(s^*), B_{j+i}(s^*)] = 0.$$

(3) *For each* $s \in \Sigma^2$ *and* $j \in \mathbb{Z}$ *one can find sequences of integers* k_n, l_n *with* $k_n, l_n \to \infty$, *and*

$$\lim_{n\to\infty} \max_{-l_n \le i \le l_n} d[B_{k_n+i}(s^*), B_j(s)] = 0.$$

(4) *For each* $s \in \Sigma^2$ *and* $j \in \mathbb{Z}$ *one can find sequences of integers* k_n, l_n *with* $k_n, l_n \to -\infty$, *and*

$$\lim_{n\to\infty} \max_{l_n \le i \le -l_n} d[B_{k_n+i}(s^*), B_j(s)] = 0.$$

We assume that there exist a homeomorphism S between Σ_2 and a set $\Lambda \subset [0, 1]$, and a map $h : \Lambda \to \Lambda$, such that $S \circ h = \sigma \circ S$. That is h and σ are topologically conjugate. It is known that Σ_2 and Λ are Cantor sets: they are closed, perfect, and totally disconnected [15]. Obviously, they are compact. One of the most popular examples of the map h is the logistic map $\mu x(1-x)$, $\mu > 4$, considered on a subset of [0, 1] [14].

For every $t_0 \in \Lambda$, one can construct a sequence $\kappa(t_0)$ of real numbers κ_i, $i \in \mathbb{Z}$, in the following way. If $i \ge 0$, then $\kappa_{i+1} = h(\kappa_i)$ and $\kappa_0 = t_0$. Let us show, how the sequence is defined for negative i. Denote $s^0 = S(t_0)$, $s^0 = (s_0^0 s_1^0 \ldots)$. Consider elements $\underline{s} = (0 s_0^0 s_0^1 \ldots)$, $\overline{s} = (1 s_0^0 s_0^1 \ldots)$ of Σ_2, such that $\sigma(\underline{s}) = \sigma(\overline{s}) = s^0$ and $\underline{t} = S^{-1}(\underline{s})$, $\overline{t} = S^{-1}(\overline{s})$. The homeomorphism implies that $h(\overline{t}) = h(\underline{t}) = t_0$. Set $h^{-1}(t_0)$ may consist of not more than two elements $\overline{t}, \underline{t} \in \Lambda$. Each of these two values can be chosen as $\kappa_{-1}(t_0)$. Obviously, one can continue the process to $-\infty$, choosing always one element from the set h^{-1}. We have finalized the construction of the sequence, and, moreover, it is proved that $\kappa(t_0) \subset \Lambda$, $\kappa(t_0) = \{\kappa_i(t_0)\}$, $i \in \mathbb{Z}$. Thus, infinitely many sequences $\kappa(t_0)$ can be constructed for a given t_0. However, each of this type of sequence is unique for an increasing i. Fix one of the sequences and define a sequence $\zeta(t_0) = \{\zeta_i\}$, $\zeta_i = i + \kappa_i$, $i \in \mathbb{Z}$. If we denote by Π the set of all such sequences $\{\zeta_i\}$, $i \in \mathbb{Z}$, then a multivalued functional $w : I \to \Pi$ is defined. In this section, the sequence $\zeta(t_0)$ in (6.4.32) is considered to be a value of $w(t_0)$.

The above discussion shows that there exists an one-to-one correspondence between Σ^2 and Π. Denote $\zeta(t^*)$ the sequence, which corresponds to s^*. Then homeomorphism S, and Lemma 6.15 imply that the following assertion is correct.

Theorem 6.8 (1) *For a fixed* $j \in \mathbb{Z}$ *there exist two sequences of integers* k_n, l_n *with* $k_n, l_n \to \infty$, *and*

$$\lim_{n\to\infty} \max_{-l_n \le i \le l_n} |\zeta_{k_n+i}(t^*) - \zeta_{j+i}(t^*)| = 0.$$

(2) *For a fixed* $j \in \mathbb{Z}$ *there exist two sequences of integers* k_n, l_n *with* $k_n, l_n \to -\infty$, *and*

$$\lim_{n\to\infty} \max_{l_n \le i \le -l_n} |\zeta_{k_n+i}(t^*) - \zeta_{j+i}(t^*)| = 0.$$

(3) *For each* $\zeta(t) \in \Pi$ *and* $j \in \mathbb{Z}$ *one can find sequences of integers* k_n, l_n *with* $k_n, l_n \to \infty$, *and*

$$\lim_{n\to\infty} \max_{-l_n \le i \le l_n} |\zeta_{k_n+i}(t^*) - \zeta_{i+j}(t)| = 0.$$

(4) *For each* $\zeta(t) \in \Pi$ *and* $j \in \mathbb{Z}$ *one can find sequences of integers* k_n, l_n *with* $k_n, l_n \to -\infty$, *and*

$$\lim_{n\to\infty} \max_{l_n \le i \le -l_n} |\zeta_{k_n+i}(t^*) - \zeta_{i+j}(t)| = 0.$$

6.5 Hyperbolic Sets of Impact Systems

In this section, a hyperbolic set of bounded solutions is constructed on the basis of a quasilinear impulsive differential equation with a special initial condition.

The famous homoclinical structure of H. Poincaré initiated the fundamental investigations of the complex structure of dynamical systems by G. Birkhoff, M.L. Cartwright, J.E. Littlewood, N. Levinson, S. Smale [61, 62] and their followers. The results obtained proved to be universal and applicable as effective instruments of analysis. In this section, we investigate the structure of the bounded solutions set of a special initial value problem, which initial moments of time are from a Cantor set. The system was introduced in [28], where chaotic properties are discussed. Particularly, we prove that there exists a chaotic attractor with infinitely many periodic solutions. It is natural to expect that the attractor is a hyperbolic set. In the present section we develop the investigation. In [41], the interesting idea to generate sensitiveness of discontinuous motion using the parametric dependence of impulsive moments was considered.

Consider the sequence space [8] $\Sigma_2 = \{s = (s_0 s_1 s_2 \ldots) : s_j = 0 \text{ or } 1\}$. with the metric

$$d[s,t] = \sum_{i=0}^{\infty} \frac{|s_i - t_i|}{2^i},$$

where $t = (t_0 t_1 \ldots) \in \Sigma_2$, and the shift map $\sigma : \Sigma_2 \to \Sigma_2$, such that $\sigma(s) = (s_1 s_2 \ldots)$. The semidynamics (Σ_2, σ) is the symbolic dynamics [15].

The map is continuous, $card\, Per_n(\sigma) = 2^n$, $Per(\sigma)$ is dense in Σ_2.

We assume that there exist a homeomorphism S between Σ_2 and a set $\Lambda \subset I, I = [0, \bar{\omega}]$, where $\bar{\omega}$ is a fixed positive number, and a map $h : \Lambda \to \Lambda$, such that $S \circ h = \sigma \circ S$. That is h and σ are topologically conjugate. It is known that Σ_2 and Λ are Cantor sets [15]. Obviously, they are compact. One of the most popular examples of the map h is the logistic map $\mu x(1-x)$, $\mu > 4$, considered on a subset of [0, 1], [8].

Let us start with the description of the moments of impulses, as their generation is most important for the hyperbolic set construction.

For every $t_0 \in \Lambda$ one can construct a sequence $\kappa(t_0)$ of real numbers $\kappa_i, i \in \mathbb{Z}$, in the following way. If $i \ge 0$, then $\kappa_{i+1} = h(\kappa_i)$ and $\kappa_0 = t_0$. Let us show, how the sequence is defined for negative i. Denote $s^0 = S(t_0)$, $s^0 = (s_0^0 s_1^0 \ldots)$. Consider elements $\underline{s} = (0s_0^0 s_0^1 \ldots)$, $\overline{s} = (1s_0^0 s_0^1 \ldots)$ of Σ_2, such that $\sigma(\underline{s}) = \sigma(\overline{s}) = s^0$ and $\underline{t} = S^{-1}(\underline{s}), \overline{t} = S^{-1}(\overline{s})$. The homeomorphism implies that $h(\overline{t}) = h(\underline{t}) = t_0$. Set $h^{-1}(t_0)$ may consist of not more than two elements $\overline{t}, \underline{t} \in \Lambda$. Each of these two values can be chosen as $\kappa_{-1}(t_0)$. Obviously, one can continue the process to $-\infty$, choosing always one element from the set h^{-1}. We have finalized the construction of the sequence, and, moreover, it is proved that $\kappa(t_0) \subset \Lambda$. Thus, infinitely many sequences $\kappa(t_0)$ can be constructed for a given t_0. However, each of this type of sequence is unique for an increasing i. Fix one of the sequences and define a sequence $\zeta(t_0) = \{\zeta_i\}, \zeta_i = i\bar{\omega} + \kappa_i, i \in \mathbb{Z}$. The sequence has a *periodicity property* if there exists $p \in \mathbb{N}$ such that $\zeta_{i+p} = \zeta_i + p\bar{\omega}, \forall i \in \mathbb{Z}$. If we denote by Π the set of all such sequences $\zeta = \{\zeta_i\}, i \in \mathbb{Z}$, then a multivalued functional $w : \Lambda \to \Pi$ such that $\zeta(t_0) = w(t_0)$ is defined. In what follows, we assume, without loss of generality, that $\bar{\omega} = 1$.

The discussion indicates that homeomorphism S arranges one-to-one correspondence Φ between the space of bi-sequences
$\Sigma_2^2 = \{s = (\ldots s_{-2}s_{-1} \cdot s_0 s_1 s_2 \ldots) : s_j = 0 \ or \ 1\}$ and the elements of space Π such that if $\Phi(s) = \zeta$, then $\zeta_j = j\bar{\omega} + S((s_j s_{j+1} \ldots)), j \in \mathbb{Z}$.

Fix $s^* \in \Sigma_2^2$. We denote $W^s(s^*)$ a set of all elements $s \in \Sigma_2^2$, which entries agree with those of s^* to the right of some entry of s^*. Set $W^s(s^*)$ is the stable set of s^*. Similarly, we denote $W^u(s^*)$ a set of all elements $s \in \Sigma_2^2$, which entries agree with those of s^* to the left of some entry of s^*. Set $W^u(s^*)$ is the unstable set of s^*. A point $s \in \Sigma_2^2$, whose entries agree with those of s^* to the right and to the left of some entries of s^* is a homoclinic sequence to s^*. It is clear that the homoclinic sequence satisfies $s \in W^s(s^*) \cap W^u(s^*)$. If $s \in W^s(s^*) \cap W^u(s^{**})$, where $s, s^*, s^{**} \in \Sigma_2^2$, then s is a heteroclinic sequence to s^* and s^{**}. Since each point of Σ_2^2 is equipped with a stable and unstable set we call it the hyperbolic set. One can, now, formulate the following analogues for Π. We say that $\eta \in \Pi$ is homoclinic to $\zeta \in \Pi$ if $\Phi^{-1}(\eta)$ is homoclinic to $\Phi^{-1}(\zeta)$, and $\eta \in \Pi$ is heteroclinic to $\zeta, \phi \in \Pi$ if $\Phi^{-1}(\eta)$ is heteroclinic to $\Phi^{-1}(\zeta)$ and $\Phi^{-1}(\phi)$. Obviously, for a given $\zeta \in \Pi$, $W^s(\zeta) = \{\eta : \Phi^{-1}(\eta) \in W^s(\Phi^{-1}(\zeta))\}, W^u(\zeta) = \{\eta : \Phi^{-1}(\eta) \in W^u(\Phi^{-1}(\zeta))\}$. One can easily see that Π is a hyperbolic set. Moreover, using the compactness of sets Λ and Σ_2, the homeomorphism of S and the definition of Φ one can obtain that, if $\eta \in \Pi$ is homoclinic to $\zeta \in \Pi$, then $|\eta_i - \zeta_i| \to 0$, as $i \to \pm\infty$, and $|\eta_i - \zeta_i| \to 0$, as $i \to \infty$, $|\eta_i - \phi_i| \to 0$, as $i \to -\infty$, if η is heteroclinic to ζ and ϕ.

The following special initial value problem for the impulsive differential equation,

$$\begin{aligned} z'(t) &= Az(t) + f(z), t \neq \zeta_i(t_0), \\ \Delta|_{t=\zeta_i(t_0)} &= Bz(\zeta_i(t_0)) + W(z(\zeta_i(t_0))), \\ z(t_0) &= z_0, \ (t_0, z_0) \in \Lambda \times \mathbb{R}^n, \end{aligned} \tag{6.5.35}$$

where $z \in \mathbb{R}^n$, $t \in \mathbb{R}$, and $\zeta(t_0) = w(t_0)$, $t_0 \in \Lambda$, is the object, which will be mainly discussed in the present section.

We assume that:

(C1) A, B are $n \times n$ constant real-valued matrices, $\det(\mathscr{I} + B) \neq 0$, where $\mathscr{I}$ is the $n \times n$ identical matrix;
(C2) for all $x_1, x_2 \in \mathbb{R}^n$ the functions $f : \mathbb{R}^n \; to \; \mathbb{R}^n$, $W : \mathbb{R}^n \to \mathbb{R}^n$ satisfy

$$||f(x_1) - f(x_2)|| + \|W(x_1) - W(x_2)\| \leq L||x_1 - x_2||,$$

where $L > 0$ is a constant;
(C3) $Bx + W(x) \neq 0, \forall x \in \mathbb{R}^n$;
(C4) $\sup_{x \in \mathbb{R}^n} ||f(x)|| + \sup_{x \in \mathbb{R}^n} ||W(x)|| = M_0 < \infty.$

It implies from the last conditions that for each $z_0 \in \mathbb{R}^n$, $t_0 \in \Lambda$, there exists a unique solution of the problem, that is, a function which is continuous from the left with discontinuities of the first kind at the points $\zeta_i(t_0)$. Each solution is defined on $\mathbb{R}$.

Let us denote by $Z(t, s)$ the transition matrix of the associated with (6.5.35) linear homogeneous system

$$\begin{aligned} z'(t) &= Az(t), t \neq \zeta_i \\ \Delta z|_{t=\zeta_i} &= Bz(\zeta_i). \end{aligned} \tag{6.5.36}$$

We may assume that:

(C5) the matrices A and B commute, the real parts of all eigenvalues of the matrix $A + \ln(\mathscr{I} + B)$ are negative.

According to Theorem 34 [27], condition $(C5)$ implies that there exist positive numbers N and ω, which do not depend on θ, such that $||Z(t, s)|| \leq N\mathrm{e}^{-\omega(t-s)}, t \geq s$.

Assume additionally that

(C6) $NL\left[\frac{2}{\omega} + \frac{\mathrm{e}^{\omega}}{1 - \mathrm{e}^{-\omega}}\right] < 1.$

Theorems 37 and 89 from [27] imply that if conditions $(C1)$, $(C2)$, $(C4)$–$(C6)$ are valid, then for each $t_0 \in \Lambda$ there exists a unique bounded solution $z(t, t_0)$ of (6.5.35) and $\|z(t, t_0)\| < NM_0\left[\frac{1}{\omega} + \frac{\mathrm{e}^{\omega}}{1 - \mathrm{e}^{-\omega}}\right]$ for all $t \in \mathbb{R}$.

We denote the set of all bounded solutions $z(t, t_0)$, $t_0 \in \Lambda$, as $\mathscr{PCB}$.

Let us say that a solution $z(t, t_1) \in \mathscr{PCB}$ lies in the ε-neighborhood of a solution $z(t, t_0) \in \mathscr{PCB}$ on an interval $J \subset \mathbb{R}$ if $||z(t, t_1) - z(t, t_0)|| < \varepsilon$ for all t from J except, possibly, those from $(\zeta_i(t_0) - \varepsilon, \zeta_i(t_0) + \varepsilon)$, $i \in \mathbb{Z}$, and $|\zeta_i(t_0) - \zeta_i(t_1)| < \varepsilon$ for all i such that $\zeta_i(t_0) \in J$.

We shall say that a solution $z(t, t_1) \in \mathscr{PCB}$ belongs to the stable (unstable) set $W^s(z(t, t_0))(W^u(z(t, t_0)))$ of solution $z(t, t_0) \in \mathscr{PCB}$, $t_1 \neq t_0$, if for an arbitrary $\varepsilon > 0$ there exists a number $T(\varepsilon)$ such that $z(t, t_1)$ is in ε-neighborhood of $z(t, t_0)$ on $(T, \infty)((-\infty, T))$.

A solution $z(t, t_1) \in \mathscr{PCB}$ is homoclinic to solution $z(t, t_0) \in \mathscr{PCB}, t_1 \neq t_0$, if $z(t, t_1) \in W^s(z(t, t_0)) \cap W^u(z(t, t_0))$.

A solution $z(t, t_2) \in \mathscr{PCB}$ is heteroclinic to solutions $z(t, t_0), z(t, t_1) \in \mathscr{PCB}$, $t_2 \neq t_0, t_2 \neq t_1$, if $z(t, t_2) \in W^s(z(t, t_0)) \cap W^u(z(t, t_1))$.

The set $\mathscr{PCB}$ is hyperbolic if each solution of this set has stable and unstable sets.

We may assume that

(C7) $-\omega + NL + \ln(1 + NL) < 0.$

One can prove by using the technique of integral representation of bounded solutions of impulsive systems [27, 28] that the following theorem is valid.

Theorem 6.9 *Assume that conditions* $(C1)$, $(C2)$, $(C4)$–$(C7)$ *are fulfilled. Then,*

1. *Set* $\mathscr{PCB}$ *is hyperbolic;*
2. $W^s(z(t, t_0)) \cap W^u(z(t, t_0)) \neq \emptyset$, *for all* $t_0 \in \Lambda$;
3. *Each element of* $\mathscr{PCB}$ *is homoclinic to an element of* $\mathscr{PCB}$.

6.6 Chaos and Shadowing

6.6.1 Introduction and Preliminaries

The proof of the existence of chaotic attractors remains as an important and difficult problem, which is still not resolved fully, even for the Lorenz system [63–66]. In this section, a multidimensional chaos is generated by a special initial value problem for the non-autonomous impulsive differential equation. The existence of a chaotic attractor is shown, where density of periodic solutions, sensitivity of solutions, and existence of a trajectory, which is dense in the set of all orbits are observed. That is, we concentrate on the topological ingredients of the version proposed by Devaney [8]. An appropriate example is constructed, where a chaotic attractor is indicated, and the intermittency is observed.

The discontinuous system consists of an impulsive differential equation and of a discrete equation, which generates the moments of impacts.

We suppose that the generator is chaotic while the impulsive system is dissipative for all possible sequences of moments of discontinuities, and we prove that the system has a similar chaotic nature. Similarly, if the generator function has a shadowing property [14, 18, 58, 59], then the system admits an analogue of the property. The shadowing exists if the generator is uniformly hyperbolic on the invariant set of initial moments, or a non-hyperbolic map.

The results of this section illustrate that impulsive differential equations may play a special role in the investigation of the complex behavior of dynamical systems. The B-equivalence method is used to obtain the main results.

Let us consider a continuous map $H : I \to \mathbb{R}$, $I = [0, 1]$, with a positively invariant compact set $\Lambda \subseteq I$. Let $\kappa_{i+1} = H(\kappa_i)$, $\kappa_0 = t_0 \in \Lambda$, and the sequence $\zeta(t_0) = \{\zeta_i(t_0)\}$ be defined, where $\zeta_i(t_0) = i + \kappa_i(t_0)$, $i \geq 0$.

One may consider the logistic map $h(t, \mu) = \mu t(1-t)$, $\mu > 0$, as an example of H. The main object of discussion in this section is the following special initial value problem,

$$\begin{aligned} z'(t) &= Az(t) + f(z), \\ \Delta|_{t=\zeta_i(t_0)} &= Bz(\zeta_i(t_0)) + W(z(\zeta_i(t_0))), \\ z(t_0) &= z_0, \ (t_0, z_0) \in \Lambda \times \mathbb{R}^n, \end{aligned} \tag{6.6.37}$$

where $z \in \mathbb{R}^n$, $t_0 \in I$, $t \geq t_0$.

We shall need the following basic assumptions for the problem:

(C1) A, B are $n \times n$ constant real-valued matrices; $\det(\mathscr{I} + B) \neq 0$, where $\mathscr{I}$ is the identical matrix;

(C2) for all $x_1, x_2 \in \mathbb{R}^n$ the functions $f : \mathbb{R}^n \to \mathbb{R}^n$, $W : \mathbb{R}^n \to \mathbb{R}^n$ satisfy

$$||f(x_1) - f(x_2)|| + ||W(x_1) - W(x_2)|| \leq L||x_1 - x_2||,$$

where $L > 0$ is a constant;

(C3) $\sup\limits_{x \in \mathbb{R}^n} ||f(x)|| + \sup\limits_{x \in \mathbb{R}^n} ||W(x)|| = M_0 < \infty$;

(C4) the matrices A and B commute and the real parts of all eigenvalues of $A + \ln(\mathscr{I} + B)$ are negative.

Under these conditions a solution $z(t) = z(t, t_0, z_0)$, $z_0 \in \mathbb{R}^n$, of (6.6.37) exists and is unique on $[t_0, \infty)$.

Consider an unbounded and strictly increasing sequence θ with elements θ_i, $i - 1 < \theta_i < i + 2$, $i \in \mathbb{Z}$. Let us denote by $Z(t, s)$ the transition matrix of the linear homogeneous system

$$\begin{aligned} z'(t) &= Az(t), \\ \Delta z|_{t=\theta_i} &= Bz(\theta_i). \end{aligned} \tag{6.6.38}$$

Condition $(C4)$ implies that there exist positive numbers N and ω, which do not depend on θ, such that $||Z(t, s)|| \leq N\mathrm{e}^{-\omega(t-s)}$, $t \geq s$. In what follows, we shall denote by $Z(t, s, \xi)$ the transition matrix $Z(t, s)$ if $\theta = \zeta(\xi)$.

We shall need the following additional assumptions:

(C5) $NL\left[\dfrac{2}{\omega} + \dfrac{\mathrm{e}^{\omega}}{1 - \mathrm{e}^{-\omega}}\right] < 1$;

(C6) $-\omega + NL + \ln(1 + NL) < 0$.

The solution $z(t) = z(t, t_0, z_0)$ of (6.6.37) satisfies the following integral equation

$$z(t) = Z(t, t_0, t_0)z_0 + \int_{t_0}^{t} Z(t, s, t_0) f(z(s))ds + \sum_{t_0 \le \zeta_i < t} Z(t, \zeta_i(t_0), t_0) W(z(\zeta_i(t_0))).$$

Using the last formula and the technique of Chap. 7 [67], one can verify that all solutions eventually, as t increases, enter the tube with the radius $M = NM_0[\frac{1}{\omega} + \frac{e^{\omega}}{1-e^{-\omega}}]$, $t \in \mathbb{R}$. That is, the discussion of this section can be made assuming that all solutions are inside the tube. Moreover, if the sequence $\kappa(t_0)$ is periodic with a period $p \in \mathbb{N}$, then there is a solution of (6.6.37) with the same period, and its integral curve is placed in the tube.

We assume that:

(C7) $Bx + W(x) \neq 0$, if $\|x\| \le M$.

The last condition implies that periodic solutions are different for different p.

Denote by $\mathscr{PC}$ the set of all solutions $z(t) = z(t, t_0, z_0)$, $t_0 \in \Lambda$, $z_0 \in \mathbb{R}^n$, $t \ge t_0$ of (6.6.37), and denote $\mathscr{PCA} = \{z \in \mathscr{PC} : \|z(t_0)\| < M, t_0 \in \Lambda\}$. In the next subsection, we define conditions with which $\mathscr{PCA}$ is a chaotic attractor.

6.6.2 The Devaney's Chaos

Let us assume that the map H admits all Devaney's ingredients of chaos on the set Λ, that is:

1. there exists a positive δ_0 such that for each $t \in \Lambda$ and $\varepsilon > 0$ there is a point $\tilde{t} \in \Lambda$ with $|t - \tilde{t}| < \varepsilon$ and $|H^i(t) - H^i(\tilde{t})| \ge \delta_0$ for some positive integer i (sensitivity);
2. there exists an element $t^* \in \Lambda$ such that the set $H^i(t^*)$, $i \ge 0$, is dense in Λ (transitivity);
3. the set of period -p points, $p \ge 1$, is dense in Λ (density of periodic points).

Let us define the chaos for the discontinuous dynamics of (6.6.37).

Definition 6.20 We say that (6.6.37) is sensitive on Λ if there exist positive real numbers $\varepsilon_0, \varepsilon_1$ such that for each $t_0 \in \Lambda$, and $\delta > 0$ one can find a number $t_1 \in \Lambda$, $|t_0 - t_1| < \delta$, such that for each couple of solutions $z(t) = z(t, t_0, z_0)$, $z_1(t) = z(t, t_1, z_1)$, $z_0, z_1 \in \mathbb{R}^n$, there exists an interval $Q \subset [t_0, \infty)$ with the length not less than ε_1 such that $||z(t) - z_1(t)|| \ge \varepsilon_0$, $t \in Q$, and there are no points of discontinuity of $z(t)$, $z_1(t)$ in Q.

Definition 6.21 The set of all periodic solutions $\phi(t) = \phi(t, t_0)$, $t_0 \in \Lambda$, of (6.6.37) is called dense in $\mathscr{PC}$ if for every solution $z(t) \in \mathscr{PC}$ and each $\varepsilon > 0$, $E > 0$, there exist a periodic solution $\phi(t, t^*)$, $t^* \in \Lambda$, and an interval $J \subset [t_0, \infty)$ with the length E such that $\phi(t)(\varepsilon, J)z(t)$.

Definition 6.22 A solution $z_*(t) \in \mathscr{PC}$ of (6.6.37) is called dense in the set of all orbits of $\mathscr{PC}$ if for every solution $z(t) \in \mathscr{PC}$ of (6.6.37), and each $\varepsilon > 0$, $E > 0$, there exist an interval $J \subset [0, \infty)$ with the length E and a real number ξ such that $z_*(t+\xi)(\varepsilon, J)z(t)$.

Definition 6.23 The problem (6.6.37) is chaotic if: (i) it is sensitive; (ii) the set of all periodic solutions $\phi(t, t_0)$, $t_0 \in \Lambda$, is dense in $\mathscr{PC}$; (iii) there exists a solution $z_*(t)$, which is dense in $\mathscr{PC}$.

Remark 6.4 Definitions of the chaotic ingredients have been worked out in detail issuing from the two reasons: the considered system is non-autonomous and consequently we analyze integral curves, but not trajectories; the system is impulsive and different solutions have different points of discontinuity that necessitates the B-topology.

Theorem 6.10 *Assume that conditions* $(C1)$–$(C6)$ *are fulfilled. Then the set of all periodic solutions* $\phi(t, t_0)$, $t_0 \in \Lambda$, *of (6.6.37) is dense in* $\mathscr{PC}$.

Proof Fix $t_1 \in \Lambda$ and $E, \varepsilon > 0$. The density of periodic points of H and uniform continuity of this map imply that for an arbitrary large number $\tilde{T}$ there exists a sequence $\zeta(t_0)$, defined by a periodic sequence $\kappa(t_0)$, such that $\|\zeta(t_1)-\zeta(t_0)\|_Q < \varepsilon$, where $Q = (t_1, t_1 + \tilde{T} + E)$. We shall find the number $\tilde{T}$ so large that solution $z(t) = z(t, t_1, z_1)$, $\|z_1\| < M$, is ε-equivalent to $\phi(t, t_0)$ on $J = (t_1+\tilde{T}, t_1+\tilde{T}+E)$.

Denote by $Z_1(t, s) = Z(t, s, t_1)$ and $Z_2(t, s) = Z(t, s, t_0)$, $t \geq s$, the transition matrices. We have that

$$z(t) = Z_1(t, 1)z(1) + \int\limits_{c1}^{t} Z_1(t, s)f(z(s))ds + \sum_{1 \leq \zeta_i < t} Z_1(t, \zeta_i(t_1))W(z(\zeta_i(t_1))),$$

$$\phi(t) = Z_2(t, 1)\phi(1) + \int\limits_{c1}^{t} Z_2(t, s)f(\phi(s))ds + \sum_{1 \leq \zeta_i < t} Z_2(t, \zeta_i(t_0))W(\phi(\zeta_i(t_0))).$$

The difference between $z(t)$ and $\phi(t)$ cannot be evaluated by using the last two expressions since the moments of discontinuities do not coincide. The method of B-equivalence is helpful here. Introduce the following B-maps

$$W_i^1(z) = (\mathscr{I} + B)\Big[(\mathrm{e}^{A(\zeta_i(t_1)-\zeta_j(t_0))} - \mathscr{I})z + \int_{\zeta_j(t_0)}^{\zeta_i(t_1)^{*}} \mathrm{e}^{A(\zeta_i(t_1)-s)} f(z(s))ds\Big] +$$

$$W((\mathscr{I} + B)[\mathrm{e}^{A(\zeta_i(t_1)-\zeta_j(t_0))}z + \int_{\zeta_j(t_0)}^{\zeta_i(t_1)} \mathrm{e}^{A(\zeta_i(t_1)-s)} f(z(s))ds]) -$$

$$\int_{\zeta_j(t_0)}^{\zeta_i(t_1)} \mathrm{e}^{A(\zeta_i(t_1)-s)} f(z_1(s))ds - W(z),$$

where $z(t), z_1(t), z(\zeta_i(t_0)) = z, z_1(\zeta_i(t_1)) = z(\zeta_i(t_1)+)$, are solutions of the equation $z' = Az$. One can easily verify that $M_1 = \sup_{\|z\|\le M, i\in\mathbb{Z}} \|W_i^1(z)\| < \infty$. Consider the following system

$$\begin{aligned} v'(t) &= Av(t) + f(v), t \neq \zeta_i(t_0), \\ \Delta v|_{t=\zeta_i(t_0)} &= Bv(\zeta_i(t_0)) + W(v(\zeta_i(t_0))) + W_i^1(v(\zeta_i(t_0))), \end{aligned} \tag{6.6.39}$$

together with the system

$$\begin{aligned} z'(t) &= Az(t) + f(z), t \neq \zeta_i(t_1), \\ \Delta|_{t=\zeta_i(t_1)} &= Bz(\zeta_i(t_1)) + W(\zeta_i(t_1)), \end{aligned} \tag{6.6.40}$$

where t_0, t_1 are the numbers under discussion.

Systems (6.6.39) and (6.6.40) are B-equivalent. That is, their solutions with the same initial condition coincide on the common domain if $t \notin (\widehat{\zeta_i(t_0), \zeta_i(t_1)}], i \in \mathbb{Z}$. So, if $v(t), v(1) = z(1)$, is the solution of (6.6.39), then $v(t) = z(t)$ for all $t \notin (\widehat{\zeta_i(t_0), \zeta_i(t_1)}], i \in \mathbb{Z}$. For $v(t)$ we have that

$$v(t) = Z_2(t,1)v(1) + \int_{c1}^{t} Z_2(t,s) f(v(s))ds +$$
$$\sum_{1\le\zeta_i<t} Z_2(t,\zeta_i(t_0))[W(v(\zeta_i(t_0)) + W_1(v(\zeta_i(t_0)))].$$

Thus,

$$\|\phi(t) - v(t)\| \le \|\phi(1) - v(1)\| \|Z_2(t,1)\| + \int_{c1}^{t} \|Z_2(t,s)\| L \|\phi(s) - v(s)\| ds +$$
$$\sum_{1\le\zeta_j(t_0)<t} \|Z_2(t,\zeta_j(t_0))\| L \|\phi(\zeta_j(t_0)) - v(\zeta_j(t_0))\| +$$
$$\sum_{1\le\zeta_j(t_0)<t} \|Z_2(t,\zeta_j(t_0))\| \|W_1(v(\zeta_i(t_0))\| \le$$
$$2MN + M_1 \frac{e^{\omega}}{1 - e^{-\omega}} + \int_{c1}^{t} N e^{-\omega(t-s)} L \|z(s) - v(s)\| ds +$$
$$\sum_{1\le\zeta_j<t} N e^{-\omega(t-\zeta_j(t_0))} L \|v(\zeta_j(t_0)) - v(\zeta_j(t_0))\|.$$

Now, applying Lemma 2.5.1 [67], we can find that

$$\|z(t) - v(t)\| \le \left(2MN + M_1 \frac{e^{\omega}}{1 - e^{-\omega}}\right) e^{(-\omega + NL + \ln(1+NL))(t-1)}.$$

The last inequality implies that $\|z(t)-v(t)\| < \varepsilon$ if $t > \tilde{T}, t \notin [\widehat{\zeta_i(t_0), \zeta_i(t_1)}], i \geq 0$, where $\tilde{T} = 1 + \ln\left(\dfrac{\varepsilon}{2MN + M_1 e^{\omega}(1-e^{-\omega})^{-1}}\right)(-\omega + NL + \ln(1+NL))^{-1}$, (we may assume that $\varepsilon < 2M$). That is why, $z(t)(\varepsilon, J)\phi(t)$ if $J = (t_1 + \tilde{T}, t_1 + \tilde{T} + E)$. The theorem is proved.

Theorem 6.11 *Assume that conditions* $(C1)$*–*$(C6)$ *are fulfilled. Then there exists a solution of (6.6.37), which is dense in* $\mathscr{PC}$.

Proof Fix positive E, ε, and $t^* \in \Lambda$ such that the orbit of t^* is dense in Λ. Set $z_*(t) = z(t, t^*, z^*)$, $\|z^*\| < M$. Let us prove that $z_*(t)$ is the dense solution.

Consider an arbitrary solution $z(t) = z(t, t_0, z_0) \in \mathscr{PC}$. Consider an interval $J_1 = (0, E_1)$, where E_1 is an arbitrarily large positive number. By density of the orbit of t^* and uniform continuity of H there exists a natural m such that

$$\|\zeta(t_1) - \zeta(t^*, m)\|_{J_1} < \varepsilon, \tag{6.6.41}$$

where $\zeta(t^*, m) = \{\zeta_{i+m}(t^*)\}$.

We have

$$z_*(t+m) = Z_*(t+m, 1+m)z_*(1+m) + \int_{1+m}^{t+m} Z_*(t+m, u)f(z_*(u))du +$$

$$\sum_{1+m \leq \zeta_i(t_0) < t+m} Z_*(t+m, \zeta_i(t_0))W(z_*(\zeta_i(t_0))) = Z_*(t+m, 1+m)z_*(1+m) +$$

$$\int_1^t Z_*(t, u)f(z_*(u+m))du + \sum_{1+m \leq \zeta_i(t_0) < t+m} Z_*(t+m, \zeta_i(t_0))W(z_*(\zeta_i(t_0))),$$

and

$$z_1(t) = Z_1(t, 1)z_1(1) + \int_1^t Z_1(t, u)f(z_1(u))du + \sum_{1 \leq \zeta_i(t_1) < t} Z_1(t, \zeta_i(t_1))W(z_1(\zeta_i(t_1))),$$

where Z_* and Z_1 are fundamental matrices corresponding to points t_* and t_1, respectively. Now, by means of the last two formulas, similarly to proof of Theorem 6.10, using (6.6.41) and the B-equivalence technique, we can find a sufficiently large number $E_1 > 2E$, and a natural number m such that $z_*(t+m)$ and $z_1(t)$ are ε-equivalent on $J = (E_1/2, E_1)$. The theorem is proved.

Let $\overline{m} = \max_{|u| \leq 1} \|e^{Au}\|$, $\underline{m} = \min_{|u| \leq 1} \|e^{Au}\|$. Condition $(C7)$ implies that

$$\eta = \min_{\|x\| \leq M} (Bx + W(x)) > 0.$$

From now on we make the assumption:

(C8) $L < \frac{\underline{m}\eta}{2\overline{m}M} \min\left\{1, \frac{\underline{m}\overline{m}}{\overline{m}+\underline{m}}\right\}.$

Theorem 6.12 *Assume that conditions $(C1)$–$(C8)$ are fulfilled. Then (6.6.37) is sensitive on $\mathscr{PC}$.*

Proof Fix a solution $z(t) = z(t, t_0, z_0), t_0 \in \Lambda, z_0 \in \mathbb{R}^n$, and a positive δ. By sensitivity of H there exist $t_1 \in \Lambda, k > 0$, such that $|t_0 - t_1| < \delta$, $|\zeta_k(t_0) - \zeta_k(t_1)| \geq \delta_0$. Consequently, by uniform continuity of H, there exist numbers δ_1, δ_2, which do not depend on k and $t_0, t_1 \in \Lambda$, such that $|\zeta_{k-1}(t_0) - \zeta_{k-1}(t_1)| \geq \delta_1, |\zeta_{k-2}(t_0) - \zeta_{k-2}(t_1)| \geq \delta_2$. Obviously, one can assume that $k > 3$. Moreover, uniform continuity of H implies that k can be an arbitrarily large number. Take arbitrary $z_1 \in \mathbb{R}^n$ and solution $z_1(t) = z(t, t_1, z_1)$.

Now, let us prove the sensitiveness through the solution $z_1(t)$.

Condition $(C8)$ implies that there exists a positive number ν such that

$$\frac{2\overline{m}M}{\underline{m}\eta} < \nu < \frac{\underline{m}\eta - 2\overline{m}ML}{\overline{m}}.$$

We shall show that the constants ε_0, ε_1 for Definition 6.20 can be taken equal to $\varepsilon_0 = \min(\underline{m}\eta - \overline{m}(\nu + 2LM), \underline{m}\nu - \overline{m}2LM)$, $\varepsilon_1 = \min(\underline{\delta}, \frac{1}{2}(1 - \overline{\delta}))$, where $\overline{\delta} = \max(\delta_0, \delta_1, \delta_2)$, $\underline{\delta} = \min(\delta_0, \delta_1, \delta_2)$. One can easily see that among numbers k and $k-1$ there exists one , let us say k itself, such that $|\zeta_k(t_0) - \zeta_k(t_1)| \geq \varepsilon_1$ and interval $[\zeta_k(t_0) - \varepsilon_1, \zeta_k(t_0))$ does not have points of discontinuity from $\zeta(t_0)$ and $\zeta(t_1)$. Assume that $\|z(\zeta_k(t_0)) - z_1(\zeta_k(t_0))\| < \nu$. Then, for $t \in [\zeta_k(t_0), \zeta_k(t_1)]$,

$$\begin{aligned} z(t) &= e^{A(t-\zeta_k(t_0))}(\mathscr{I} + B)z((\zeta_k(t_0)) \\ &+ \int_{\zeta_k(t_0)}^{t} e^{A(t-s)} f(z(s))ds + e^{A(t-\zeta_k(t_0))} W(z((\zeta_k(t_0))), \\ z_1(t) &= e^{A(t-\zeta_k(t_0))} z_1((\zeta_k(t_0)) + \int_{\zeta_k(t_0)}^{t} e^{A(t-s)} f(z_1(s))ds. \end{aligned}$$

We have that

$$\|z(t) - z_1(t)\| = \|e^{A(t-\zeta_k(t_0))}[Bz(\zeta_k(t_0)) + W(z(\zeta_k(t_0)))] + e^{A(t-\zeta_k(t_0))}[z((\zeta_k(t_0)) -$$

$$z_1((\zeta_k(t_0))] + \int_{\zeta_k(t_0)}^{t} e^{A(t-s)}(f(z(s)) - f(z_1(s)))ds\| \geq \underline{m}\eta - \overline{m}(\nu + 2LM) \geq \varepsilon_0.$$

If $\|z(\zeta_k(t_0)) - z_1(\zeta_k(t_0))\| > \nu$, then, for $t \in [\zeta_k(t_0) - \varepsilon_1, \zeta_k(t_0))$,

$$z(t) = e^{A(t-\zeta_k(t_0))} z((\zeta_k(t_0)) + \int_{\zeta_k(t_0)}^{t} e^{A(t-s)} f(z(s))ds,$$

$$z_1(t) = e^{A(t-\zeta_k(t_0))} z_1((\zeta_k(t_0)) + \int_{\zeta_k(t_0)}^{t} e^{A(t-s)} f(z_1(s))ds.$$

and $\|z(t) - z_1(t)\| \geq \underline{m}\nu - \overline{m}2LM \geq \varepsilon_0$. The theorem is proved.

On the basis of Theorems 6.10–6.12, we can conclude that (6.6.37) admits the Devaney's chaos.

It seems natural to consider the chaos only for uniformly bounded solutions on $[0, \infty)$, since the domain of chaos is always assumed to be a compact set, but we consider chaotic properties of all solutions, since the chaotic scenario for these unbounded solutions starts at the moment they reach the region where solutions from $\mathscr{PCA}$ are placed. This set is a chaotic attractor as it is easily seen that $\mathscr{PCA}$ admits defined above all ingredients of Devaney's chaos.

6.6.3 Shadowing Property

In this subsection, we give definitions of shadowing property for the flow of system (6.6.37), and prove it for this system if the generator map has the property. A corollary of the result for a map H with the hyperbolic set Λ is obtained.

Assume that the generator map, $H(t)$, is defined in a neighborhood of the unit interval I.

The following definitions are from [14, 59, 68, 69] and are adapted for our system.

A sequence $\{\kappa_i\}_0^N$, $N \leq \infty$, is said to be a *true* trajectory of H, if $\kappa_0 \in \Lambda$ and $\kappa_{i+1} = H(\kappa_i), 0 \leq i < N$.

A sequence $\{\pi_i\}_0^N$, $N \leq \infty$, is said to be a *κ-pseudo-orbit*, $\kappa > 0$, of H, if $|\pi_{i+1} - H(\pi_i)| < \kappa$, and $|p_i - \lambda| < \kappa$ for all $0 \leq i < N$, and $\lambda \in \Lambda$.

The true orbit $\{\kappa_i\}_0^N$ δ-shadows the pseudo-orbit $\{\pi_i\}_0^N$ if $|\kappa_i - \pi_i| < \delta$ for all i.

A sequence $\{z_i\}_0^N$ is said to be a *true discrete orbit* of (6.6.37) if $z_{i+1} = z(\zeta_{i+1}, \zeta_i, z_i)$, where $\zeta_i = i + \kappa_i$ for all $0 \leq i < N$. Let δ be a positive number, and k a positive integer. A sequence y_{ik} such that $0 \leq ik \leq N$ if $N < \infty$, and $i \geq 0$, if $N = \infty$, is said to be a *discrete δ-pseudo-orbit* for the problem (6.6.37) with associated sequence $\{p_i\}_0^N$ if $\|y_{(i+1)k} - w(p_{(i+1)k})\| < \delta$ for all admissible i, and the solution $w(t)$ of the initial value problem

$$\begin{aligned} w'(t) &= Aw(t) + f(w), \\ \Delta|_{t=p_i} w &= Bw(p_i) + W(w(p_i)), \\ w(p_{ik}) &= y_{ik}. \end{aligned} \tag{6.6.42}$$

A discrete δ-pseudo-orbit y_{ik} of problem (6.6.37) is said to be ε-shadowed by a true orbit $\{z_i\}_0^N$ of (6.6.37) if $\|z_{ik} - y_{ik}\| < \varepsilon$, and $|\zeta_{ik} - p_{ik}| < \varepsilon$ for all i such that $0 \leq ik \leq N$ if $N < \infty$, and $i \geq 0$, if $N = \infty$. Consider the logistic function $h(x, \mu) \equiv \mu x(1 - x)$ with coefficient $\mu = 3.8$. It is proved in [18] that for $\varepsilon = 10^{-8}$, $N = 10^7$, $p_0 = 0.4$, the pseudo-orbit p_i, $i = 0$ to N, is ε-shadowed by a true orbit, if $\delta = 3 \times 10^{-14}$. Several values of μ were claimed to be proper for the shadowing.

Taking into account this result as well as results from [14, 58, 59, 70, 71] the following assertion is very useful.

Theorem 6.13 *Assume that conditions $(C1)$–$(C6)$ are fulfilled. Then, given $\varepsilon > 0$, there exists $0 < \delta < \varepsilon$ and a positive integer k such that a δ-pseudo-orbit y_{ik} of problem (6.6.37) is ε-shadowed by a true orbit $\{z_i\}_0^N$ of (6.6.37) if $p_i = i + \pi_i$, and π_i is δ-shadowed by $\{\kappa_i\}_0^N$.*

Proof Fix positive ε and nonnegative integer i. We assume that $\|z_{ik} - y_{ik}\| < \varepsilon$, and we will find δ and k, such that $\|z_{(i+1)k} - y_{(i+1)k}\| < \varepsilon$. Assume, without loss of generality, that $\zeta_{ik} < p_{ik}$, and let $z(t) = z(t, \zeta_{ik}, z_{ik})$. We have that

$$\begin{aligned}
&\|z(p_{ik}) - y_{ik}\| \leq \|z(p_{ik}) - z_{ik}\| + \|z_{ik} - y_{ik}\| = \\
&\left\| e^{A(p_{ik}-\zeta_{ik})} z_{ik} + \int_{\zeta_{ik}}^{p_{ik}} e^{A(t-s)} f(z(s))ds \right\| + \|z_{ik} - y_{ik}\| \leq \\
&\|[\mathscr{I} - e^{A(p_{ik}-\zeta_{ik})}]\| \|z_{ik}\| + \delta N M_0 + \varepsilon = \delta\phi(\delta) + \varepsilon,
\end{aligned}$$

where $\phi(s)$ is a bounded function.

Similarly to the proof of Theorem 6.10 we find that (6.6.37) is B-equivalent to the following system

$$\begin{aligned}
v'(t) &= Av(t) + f(v), \\
\Delta v|_{t=p_i} &= Bv(p_i) + W(v(p_i)) + \tilde{W}_i^1(v(p_i)), \\
v(t_0) &= z_0, \ (t_0, z_0) \in \Lambda \times \mathbb{R}^n,
\end{aligned} \tag{6.6.43}$$

with $M_2 = \sup\limits_{\|z\| \leq M, i \in \mathbb{Z}} \|\tilde{W}_i^1(z)\| < \infty$.

Then we can obtain that

$$\|z(t) - w(t)\| \leq \left[N(\delta\phi(\delta) + \varepsilon) + M_2 \frac{e^{\omega}}{1 - e^{-\omega}} \right] e^{(-\omega + NL + \ln(1+NL))(t-1)},$$

if $t \notin \widehat{[p_i, \zeta_i)}$. Now, choose k sufficiently large, and δ small for the right side of the last inequality to be less than $\varepsilon/3$ at $t = (i+1)k - 1$, and $\delta \max(1, \phi(\delta)) < \varepsilon/3$. Then $\|z_{(i+1)k} - y_{(i+1)k}\| < \|z_{(i+1)k} - z(p_{(i+1)k})\| + \|z(p_{(i+1)k}) - w(p_{(i+1)k})\| + \|y_{(i+1)k} - w(p_{(i+1)k})\| < \varepsilon$. The theorem is proved.

Now, by using the Shadowing Theorem [59, 68, 69] one can easily prove that the following assertion is true.

Theorem 6.14 *Assume that conditions $(C1)$–$(C6)$ are fulfilled and H has a compact positively invariant hyperbolic set $\Lambda \subset I$. Then, given $\varepsilon > 0$, there exist $0 < \delta < \varepsilon$,* *and a positive integer k such that a* δ*-pseudo-orbit* $\{y_{ik}\}_0^\infty$, *of problem (6.6.37) is* ε*-shadowed by a true orbit* $\{z_i\}_0^\infty$ *of (6.6.37) if* $\pi_i = p_i - i, i \geq 0$, *is a* δ*-pseudo-orbit of* H.

6.6.4 Simulations

Consider the following initial value problem

$$\begin{aligned}
x_1' &= 2/5x_2 + l\sin^2 x_2,\\
x_2' &= 2/5x_1 + l\sin^2 x_1, t \neq \zeta_i(t_0),\\
\Delta x_1|_{t=\zeta_i(t_0)} &= -\frac{4}{3}x_1,\\
\Delta x_2|_{t=\zeta_i(t_0)} &= -\frac{4}{3}x_2 + W(x_2),
\end{aligned} \tag{6.6.44}$$

where $W(s) = 1 + s^2$, if $|s| \leq l$, l is a positive constant, and $W(s) = 1 + l^2$, if $|s| > l$. One can easily see that all the functions are Lipschitzian with a constant proportional to l. The matrices of coefficients

$$A = \begin{pmatrix} 0 & 2/5 \\ 2/5 & 0 \end{pmatrix}, \quad B = \begin{pmatrix} -4/3 & 0 \\ 0 & -4/3 \end{pmatrix}$$

commute, and the eigenvalues of the matrix

$$A + \mathrm{Ln}\,(\mathscr{I} + B) = \begin{pmatrix} -\ln 3 & 2/5 \\ 2/5 & -\ln 3 \end{pmatrix}$$

are negative: $\lambda_{1,2} = -\ln 3 \pm 2/5 < 0$.

The results of the last section make possible the following appropriate simulations.

Choose $\mu = 3.8$ and $l = 10^{-2}$ in (6.6.44) and consider the solution $x(t) = (x_1, x_2)$ with initial moments $t_0 = 7/9$ and the initial value $x(t_0) = (0.005, 0.002)$.

If one considers the sequence $(x_1(n), x_2(n))$, $n = 1, 2, 3, \ldots, 75{,}000$, in $x_1 - x_2$ plane, then the attractor seen in Fig. 6.4 is obtained. To approve that the attractor is chaotic, we verify the conditions of the chaotic theorems in the following way. If $|s| \leq l$, then $-\frac{4}{3}s + W(s) = s^2 - \frac{4}{3}s + 1$, and it is never equal to zero. If $|s| > l$, then $-\frac{4}{3}s + W(s) = l^2 - \frac{4}{3}s + 1$. For the last expression to be zero, we need, $s = \frac{3}{4}(1 + l^2)$. It is seen in the figure that the second coordinate takes values between 0.32 and 0.42. This is the region where $-\frac{4}{3}s + W(s)$ does not have zeros. All the other conditions required by theorems of this section could be easily checked with sufficiently small coefficient l.

Now, consider (6.6.44) with $\mu = 3.8282$. Then the phenomenon of intermittency, i.e., irregular switching between periodic and chaotic behavior, for the solution $x(t)$

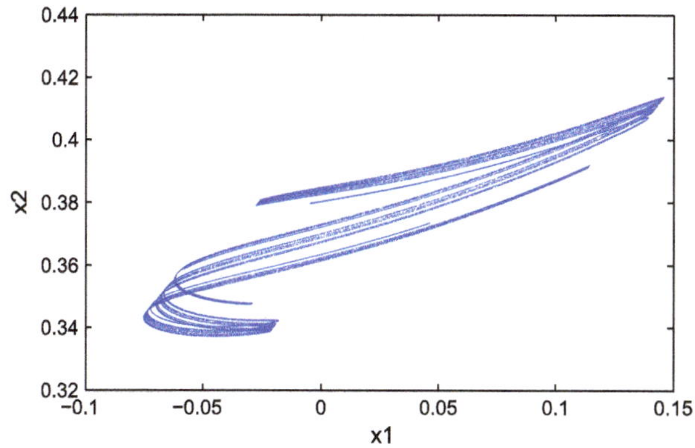

Fig. 6.4 The chaotic attractor by a stroboscopic sequence $(x_1(n), x_2(n))$, $1 \le n \le 75{,}000$

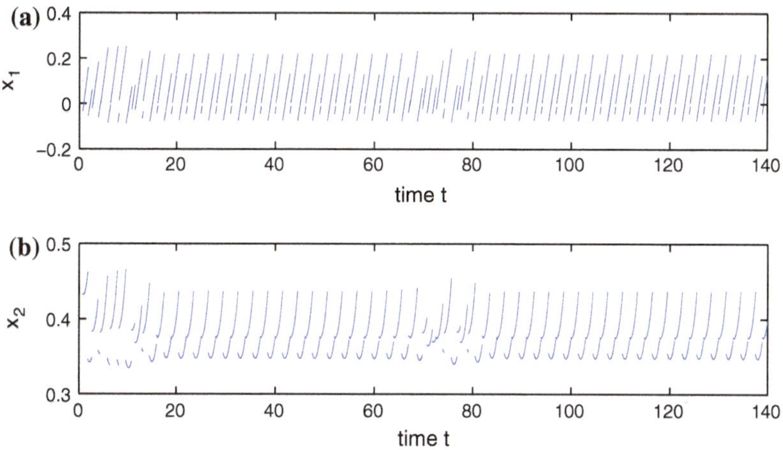

Fig. 6.5 The intermittency of the both coordinates $x_1(t)$, $x_2(t)$ is observable

can be observed in Fig. 6.5. The coefficient's value is such that the logistic map admits intermittency [8].

6.7 Chaos in the Forced Duffing Equation

6.7.1 Introduction and Preliminaries

The Duffing equation is a second-order differential equation of the type

$$x'' + c_1 x' + c_2 x + c_3 x^3 = B\cos(\omega_0 t), \tag{6.7.45}$$

where x is a function of t and B, c_1, c_2, c_3, ω_0 are fixed real numbers [72].

Ueda examined chaos in an electrical circuit with a nonlinear inductor [73, 74] using the Duffing equation, and gave description for the parameters of these type of equations [75]. Moreover, in [76–80], the Duffing equations have been used to model physical systems. Further, Thompson and Stewart [81] provided many more details on the equation (6.7.45).

In the last decades, the effect of non-smoothness and discontinuity for the chaos phenomena was widely investigated and realized [12, 44–46, 67, 82–87]. Non-smooth nonlinear characteristics are often encountered within the system components while considering real-world problems and commonly used in control systems, such as mechanical, hydraulic, magnetic, biomedical, and physical systems [88, 89]. Moreover, these nonlinearities limit the system performance and it is known that they vary with time [89]. For systems with non-smooth characteristics, the control problem is very complicated and becomes even more difficult to handle in the case of unknown time-varying parameters [88, 89]. There have been developed control techniques to diminish the effects of unknown non-smooth nonlinearities [90, 91].

One of the important applications of nonlinear oscillators subjected to non-smooth perturbations is the vibro-impact oscillators which has a wide spectrum of studies among scientists and engineers. In the presence or absence of friction, the motion of vibro-impact systems is usually described by non-smooth nonlinear differential equations [27, 92–104]. Such systems have a complex dynamic structure that comprises chaotic motions, subharmonic oscillations, and coexistence of different attractors for the same excitation and system parameters under different initial data [12, 88–91, 96, 102]. In general, these systems involve multiple impact interactions in the form of jumps in the state space. On the other hand, vibro-impact dynamics has applications on lumped systems such as bouncing ball on a vibrating platform, mass-spring-dashpot systems, and on continuous systems such as strings and beams, which differ from lumped systems [96]. In papers [105, 106], feedback-based control of impact oscillators under asymmetric double-sided barriers is proposed and it was shown that chaotic impact oscillators can be controlled and kept in a desired position using a synchronization scheme. The OGY control method is applied to impact oscillators and stabilized their chaotic attractor on period-1 and period-2 orbits using small time-dependent perturbations of the driving frequency [107]. Moreover, some results pertaining to chaotic motions in a periodically forced impacting system, which is analogous to the version of Duffing equations with negative linear stiffness have been presented in [108].

Our investigation demonstrates that processes comprising discontinuity phenomena is convenient to generate rigorously approved chaotic motions from the theoretical point of view. This is not surprising since the same we have already for discrete equations such as the logistic map and the Hénon map [109, 110]. But in our case, we have proved assertively the presence of the chaos for continuous dynamics. We want to emphasize that despite the most popular and well-known examples of chaos are the Lorenz systems and the Van der Pol equations, there are not definitely proved results of the chaos for them. Most advanced result of the Lorenz systems is given by J. Guckenheimer [64], where he considers not the system itself but the geometric approach. Similarly, the proof for the Van der Pol equations has been made by

Levi [111] for the simplified version of the equation. On the other hand, for the Duffing equations, the occurrence of chaotic period-doubling is discussed by making simulations of bifurcation diagrams, but not proved mathematically [87, 112, 113]. Consequently, the problem of discovering chaos rigorously with precise indication which kind of chaos is admitted continuous to be very actual for the nonlinear science. For modified systems, in our papers [44–46, 67, 82–84], we provide the method which allows to analyze the problems rigorously. Of course we do not pretend that our results even are begin of the solution for the already discussed equation. But we hope this constructive approach may give a light on solutions of the problems in future.

Formation of chaos in systems with arbitrary large dimension is one of the significant consequences of the present section. More precisely, our results show that the chaos of one-dimensional maps can be extended to multidimensional systems. In addition to this, extension of chaos control techniques for low-dimensional maps to multidimensional systems is another result. Therefore, the present section leads for the applications of theoretical results for one-dimensional maps to high dimensional systems. In this sense, it is a continuation of our investigations which we start in [44–46, 67, 82–84, 114, 115].

In the paper [116], besides the familiar period-doubling scenario to chaos, intermittent and quasiperiodic routes to chaos as well as period-adding sequences and Farey sequences are introduced in a nonlinear non-autonomous circuit, and verified experimentally and through simulations. On the other hand, a control method without feedback is developed for controlling a Duffing equation which admits chaos through the period-doubling cascade [117, 118]. Two different modifications of the OGY control method [119], which can lead to a better performance of the control and the method presented by Pyragas [120] are applied to the classical Duffing oscillator [72, 119], but in these cases the nature of chaos is not precise. Oppositely, in our results, we prove the type of the existing chaos theoretically and use the OGY method not for the classical Duffing equation but for the one which involves a pulse function, such that we emphasis it to be considered as a primary object of analysis.

Switching systems have important applications in high dimensional systems and hybrid systems [82, 121, 122], and the system taken into account in this section can be considered as one example. Moreover, the systems with impacts are convenient for simulations. The method and solutions that we present can be applied to hybrid systems in the future, for instance to impulsive systems [27, 67]. In this section, we construct chaos with prescribed properties such that chaos developed by using the logistic map with slightly deviated characteristics. Consequently, it can be effectively used for the security of communications and information using our chaos to mask and unmask [39, 117, 123, 124]. Since we have the chaos with known properties, it can also be used in master-slave systems and correspondingly to control these type of systems [125, 126]. Moreover, the research in the artificial neural networks emphasize that the deterministic chaos is a powerful mechanism for the storage and retrieval of information in the dynamics of artificial neural networks [127–130]. Therefore, our results are also applicable to neuroscience.

The main object of the present section is the following modified Duffing equation

$$x'' + d_1 x' + d_2 x + d_3 x^3 = D\cos(k\pi t) + \nu(t, t_0, \mu), \tag{6.7.46}$$

where d_1, d_2, d_3, D are real numbers and k is a natural number, the scalar pulse function $\nu(t, t_0, \mu)$ is defined below.

Using the new variables $x_1 = x$ and $x_2 = x'$, one can reduce the differential equation (6.7.46) to the system

$$\begin{aligned} x_1' &= x_2 \\ x_2' &= -d_1 x_2 - d_2 x_1 - d_3 x_1^3 + D\cos(k\pi t) + \nu(t, t_0, \mu). \end{aligned} \tag{6.7.47}$$

Let $\mathbb{R}$ and $\mathbb{N}$ denote the sets of real numbers and natural numbers, respectively, and I the unit interval $[0, 1]$.

In this section, we will investigate also the system

$$\begin{aligned} z'(t) &= Az(t) + f(t, z) + \nu(t, t_0, \mu) \\ z(t_0) &= z_0, \ \ (t_0, z_0) \in I \times \mathbb{R}^n, \end{aligned} \tag{6.7.48}$$

which is the general form of the system (6.7.47).

In system (6.7.48), $z \in \mathbb{R}^n, t \in \mathbb{R}_+ = [0, \infty)$, the $n \times n$ constant real-valued matrix A has real parts of eigenvalues all negative. The function $f(t, z)$ satisfies the periodicity condition $f(t+2, z) = f(t, z), t \in \mathbb{R}_+$, and is Lipschitzian with respect to z with the Lipschitz constant L.

Let us now, introduce the function $\nu(t, t_0, \mu)$ as follows

$$\nu(t, t_0, \mu) = \begin{cases} m_0, & \text{if } \zeta_{2i}(t_0, \mu) < t \le \zeta_{2i+1}(t_0, \mu) \\ m_1, & \text{if } \zeta_{2i+1}(t_0, \mu) < t \le \zeta_{2i+2}(t_0, \mu), \end{cases} \tag{6.7.49}$$

where i is a nonnegative integer and $m_0, m_1 \in \mathbb{R}^n$, such that $m_0 \neq m_1$. The sequence $\zeta(t_0, \mu) = \{\zeta_i(t_0, \mu)\}, \ i \ge 0$, is defined through the equation $\zeta_i(t_0, \mu) = i + \kappa_i(t_0, \mu)$, with $\kappa_{i+1}(t_0, \mu) = h(\kappa_i(t_0, \mu), \mu), \kappa_0(t_0, \mu) = t_0$, and $h(s, \mu) = \mu s(1 - s)$ is the logistic map, the central auxiliary instrument in the present section.

We shall need those values of the parameter μ, which are between 3.57 and 4, such that the period-doubling cascade accumulates there to provide the chaotic structure [8, 14] for the logistic map, $h(s, \mu)$. In paper [65], it was proved that the measure of such μ is positive. In the sequel, we fix one of them, and notate it as μ_∞. Moreover, we will not indicate the dependence on the parameter μ, if there is no need to specify it. Thus, for every $t_0 \in I$, the sequence $\kappa(t_0)$ of real numbers $\kappa_i, i \ge 0, \kappa(t_0) \subset I$ is defined. The sequence $\zeta(t_0)$ has the periodicity property if there exists a natural number p such that $\zeta_{i+p} = \zeta_i + p$, for all $i \ge 0$. In other words, if $\kappa_{i+p} = \kappa_i, i \ge 0$. The main object of the present section is to stabilize the periodic solutions of the chaotic structure generated by the differential equation (6.7.46).

We should point out that the adjoint linear equation of the nonperturbed Duffing equation

$$x'' + d_1 x' + d_2 x + d_3 x^3 = D\cos(k\pi t) \tag{6.7.50}$$

has eigenvalues both with negative real parts. The logistic map, which has the positive Lyapunov exponent [131, 132], gives rise to the emergence of chaos in the main equation (6.7.46) and generates the switching moments. That is, the chaotic scenario in our model is developing "along" the time axis.

We suppose that the main reason of dealing an equation of the type of Eq. (6.7.46) is that the generated chaos can give the way of analysis of systems with discontinuous perturbations, which is unfortunately far of to be complete [112].

The section is organized as follows. In Sect. 6.7.2, the existence of the chaotic attractor is proved, through the period-doubling cascade. Section 6.7.3 contains the results of the OGY control of the chaos.

6.7.2 The Chaos Emergence

6.7.2.1 The Cascade: The Analysis Results

Let us start with the analysis of system (6.7.48). In what follows we assume that

$$\sup_{z\in\mathbb{R}^n, t\in\mathbb{R}_+} \|f(t,z)\| = M_0 < \infty$$

and we denote the maximum of the real parts of the eigenvalues of matrix A by σ. Note that σ is negative.

There exist a positive number N and a negative number $\alpha \geq \sigma$ such that $\left\|e^{At}\right\| \leq Ne^{\alpha t}$, for $t \geq 0$. Therefore, we can find a natural number p_0 such that $\left\|e^{Ap_0}\right\| \leq Ne^{\alpha p_0} < 1$. For $p \geq p_0$, we have

$$\left\|(I - e^{Ap})^{-1}\right\| \leq \frac{1}{1 - Ne^{\alpha p}} \leq \frac{1}{1 - Ne^{\alpha p_0}}.$$

Let us denote

$$K = \max\left\{\max_{1\leq i\leq p_0-1}\left\|(I - e^{Ai})^{-1}\right\|, \frac{1}{1 - Ne^{\alpha p_0}}\right\}, \tag{6.7.51}$$

and in the sequel we assume also that

$$\frac{-KNL}{\alpha} < 1. \tag{6.7.52}$$

A function $z(t)$, $z(t_0) = z_0$ is a solution of (6.7.48) on $[t_0, \infty)$, $t_0 \in I$ if: (i) $z(t)$ is continuous on $[t_0, \infty)$, (ii) the derivative $z'(t)$ exists at each point $t \in [t_0, \infty)$ with the possible exception of the points $\zeta_i(t_0)$, $i \geq 0$, where left-sided derivatives exist, (iii) Equation (6.7.48) is satisfied on each interval $(\zeta_i(t_0), \zeta_{i+1}(t_0)]$, $i \geq 0$ [44].

In [44–46, 67, 83, 84], we develop the approach, when a system of differential equations inserted with a chaotic element, the generator of switching moments, produces a chaotic attractor. It is proved that the attractor presents Li–Yorke [46, 83] and Devaney [44] chaos, as well as a quasi-minimal set [45]. In the same time, it is known that both Li–Yorke and Devaney scenario of chaos emergence are difficult in the simulation with the logistic map. Moreover, speaking generally, period-doubling cascade route to the chaos is most celebrated in simulations. That is why, in the present section we consider the route to identify a chaotic structure for the equation. One must say, also that, it is a difficult task to observe chaos in multidimensional systems, exceptionally with clear theoretically supported properties. The next result is suitable for systems with arbitrary finite dimension.

Consider the sequence of period-doubling bifurcation values $\{\mu_m\}$, $\mu_m \to \mu_\infty$ as $m \to \infty$ for the logistic map $h(s, \mu) = \mu s(1 - s)$ [133].

We shall say that the system (6.7.48) has a chaos through the period-doubling cascade at $\mu = \mu_\infty$, if for each p-periodic sequence $\{\kappa_i(t_0, \mu)\}$, $p \in \mathbb{N}$, where $t_0 \in I$, and μ is equal either to μ_m, $m \in \mathbb{N}$ or μ_∞, there exists a unique periodic solution, $z_p(t)$, of the system (6.7.48) with the same μ. Moreover, all trajectories of these solutions lie in a bounded domain. This definition is natural since periodic solutions, which correspond to different sequences κ, do not coincide, and consequently, the Eq. (6.7.48) with $\mu = \mu_\infty$ has infinitely many periodic solutions.

The principal result of this section is the following theorem.

Theorem 6.15 *System (6.7.48) admits the chaos through period-doubling cascade at* μ_∞.

Proof Fix μ and $t_0 \in I$ such that the sequence $\{\kappa_i(t_0, \mu)\}$ is p-periodic, $p \in \mathbb{N}$. It is easily seen that to verify the theorem, one needs to prove that the system (6.7.48) with the same μ admits a periodic solution, $z_p(t)$, and the norms of all these periodic solutions with all the possible μ, are bounded with one and the same positive number.

Set $\rho_0 = \max\{\|m_0\|, \|m_1\|\}$, and pick a number $H = \dfrac{-KN}{\alpha}(M_0 + \rho_0)$, where the number K is defined by the formula (6.7.51). One can see that H does not depend on p.

We shall consider the cases in which p is even and odd. Let us start with p is even. Using the standard technique [134], one can verify that the solution $z_p(t)$, if exists, satisfies the integral equation

$$z_p(t) = \int_0^p \left(I - e^{Ap}\right)^{-1} e^{A(p-s)} \left[f(t+s, z_p(t+s)) + \nu(t+s, t_0, \mu)\right] ds.$$

Introduce the set $\mathscr{B}_1$ of continuous functions $\varphi : [t_0, \infty) \to \mathbb{R}^n$ such that $\varphi(t + p) = \varphi(t)$, $t \geq t_0$ and $\|\varphi\|_1 \leq H$, where $\|\varphi\|_1 = \sup_{t \geq t_0} \|\varphi(t)\|$.

Define an operator S on the set $\mathscr{B}_1$ through the equation

$$S(\varphi)(t) = \int_0^p \left(I - e^{Ap}\right)^{-1} e^{A(p-s)} \left[f(t+s, \varphi(t+s)) + \nu(t+s, t_0, \mu)\right] ds.$$

First of all, we shall check that $S(\mathscr{B}_1) \subseteq \mathscr{B}_1$.

Since p is even, we have for $\varphi \in \mathscr{B}_1$ that $f(t+p+s, \varphi(t+p+s)) = f(t+s, \varphi(t+s))$ and $\nu(t+p+s, t_0, \mu) = \nu(t+s, t_0, \mu)$ for each $t \geq t_0$ and $s \in [0, p]$. Therefore, $S(\varphi)(t+p) = S(\varphi)(t)$ for all $t \geq t_0$.

Let us define $\overline{M} = \max_{s\in[0,p]} \left\| \left(I - e^{Ap}\right)^{-1} e^{A(p-s)} \right\|$. Take $\varphi \in \mathscr{B}_1$, and fix $\bar{t} \in [t_0, \infty)$ and an arbitrary $\varepsilon > 0$. Because the functions $f(t, z)$ and $\varphi(t)$ are continuous in all their arguments, the function $f(t, \varphi(t))$ is also continuous. Therefore, there exists a number $\delta_1 > 0$ such that for any $s \in [0, p]$ the inequality

$$\left\| f(t+s, \varphi(t+s)) - f(\bar{t}+s, \varphi(\bar{t}+s)) \right\| < \frac{\varepsilon}{2p\overline{M}}$$

holds, provided that $\left|t - \bar{t}\right| < \delta_1$.

Set $\delta = \min \left\{ \delta_1, \dfrac{\varepsilon}{2p\overline{M} \left\|m_0 - m_1\right\|} \right\}$. In the case that $\left|t - \bar{t}\right| < \delta$, one can verify that

$$\int_0^p \left\| \nu(t+s, t_0, \mu) - \nu(\bar{t}+s, t_0, \mu) \right\| ds < p\delta \left\|m_0 - m_1\right\|,$$

since there are at most p subintervals of $[0, p]$, each with a length less than δ, such that in each of these subintervals the functions $\nu(t+s, t_0, \mu)$ and $\nu(\bar{t}+s, t_0, \mu)$, $s \in [0, p]$, are different from each other.

Thus, if $\left|t - \bar{t}\right| < \delta$, then we obtain that

$$\begin{aligned}
\left\| S(\varphi)(t) - S(\varphi)(\bar{t}) \right\| &= \left\| \int_0^p \left(I - e^{Ap}\right)^{-1} e^{A(p-s)} \Big[f(t+s, \varphi(t+s)) \right. \\
&\left. + \nu(t+s, t_0, \mu) - f(\bar{t}+s, \varphi(\bar{t}+s)) - \nu(\bar{t}+s, t_0, \mu) \Big] ds \right\| \\
&\leq \overline{M} \int_0^p \left\| f(t+s, \varphi(t+s)) - f(\bar{t}+s, \varphi(\bar{t}+s)) \right\| ds \\
&+ \overline{M} \int_0^p \left\| \nu(t+s, t_0, \mu) - \nu(\bar{t}+s, t_0, \mu) \right\| ds \\
&< \frac{\varepsilon}{2} + p\delta\overline{M} \left\|m_0 - m_1\right\| \\
&\leq \varepsilon.
\end{aligned}$$

Hence, $S(\varphi)(t)$ is continuous on the interval $[t_0, \infty)$. On the other hand, for $\varphi \in \mathscr{B}_1$, one can attain for all $t \geq t_0$ that

$$\begin{aligned}
\|S(\varphi)(t)\| &\le \int_0^p \left\| \left(I - e^{Ap}\right)^{-1} \right\| \left\| e^{A(p-s)} \right\| \|f(t+s, \varphi(t+s)) + \nu(t+s, t_0, \mu)\| \, ds \\
&\le KN(M_0 + \rho_0) \int_0^p e^{\alpha(p-s)} ds \\
&= \frac{-KN}{\alpha}(M_0 + \rho_0)(1 - e^{\alpha p}) \\
&\le H.
\end{aligned}$$

The last inequality implies that $\|S(\varphi)\|_1 \le H$. Consequently, $S(\mathscr{B}_1) \subseteq \mathscr{B}_1$.

Next, we shall show that the operator S is a contraction. For $\varphi_1, \varphi_2 \in \mathscr{B}_1$, we have that

$$\begin{aligned}
&S(\varphi_1)(t) - S(\varphi_2)(t) \\
&= \int_0^p \left(I - e^{Ap}\right)^{-1} e^{A(p-s)} \left[f(t+s, \varphi_1(t+s)) - f(t+s, \varphi_2(t+s))\right] ds.
\end{aligned}$$

Therefore,

$$\begin{aligned}
&\|S(\varphi_1)(t) - S(\varphi_2)(t)\| \\
&\le \int_0^p \left\| \left(I - e^{Ap}\right)^{-1} \right\| \left\| e^{A(p-s)} \right\| \|f(t+s, \varphi_1(t+s)) - f(t+s, \varphi_2(t+s))\| \, ds \\
&\le K \int_0^p NLe^{\alpha(p-s)} \|\varphi_1(t+s) - \varphi_2(t+s)\| \, ds \\
&\le \frac{-KNL}{\alpha}(1 - e^{\alpha p}) \|\varphi_1 - \varphi_2\|_1 \\
&\le \frac{-KNL}{\alpha} \|\varphi_1 - \varphi_2\|_1,
\end{aligned}$$

and hence $\|S(\varphi_1) - S(\varphi_2)\|_1 \le \dfrac{-KNL}{\alpha} \|\varphi_1 - \varphi_2\|_1$.

Since $\dfrac{-KNL}{\alpha} < 1$, the operator S is a contraction. Thus, there exists a unique fixed point of S, and for each p-periodic $\{\kappa_i(t_0, \mu)\}$, there exists a unique solution of the system (6.7.48) with the same period, provided that p is even.

In the case that p is an odd natural number, due to its definition, the relay function $\nu(t, t_0, \mu)$ is $2p$-periodic. Therefore, if $z_p(t)$ exists, it satisfies the integral equation

$$z_p(t) = \int_0^{2p} \left(I - e^{2Ap}\right)^{-1} e^{A(2p-s)} \left[f(t+s, z_p(t+s)) + \nu(t+s, t_0)\right] ds.$$

Introduce the set $\mathscr{B}_2$ of continuous functions $\varphi : [t_0, \infty) \to \mathbb{R}^n$ such that $\varphi(t + 2p) = \varphi(t)$, $t \ge t_0$ and $\|\varphi\|_1 \le H$, and define an operator $\overline{S} : \mathscr{B}_2 \to \mathscr{B}_2$ by means of the equation

Table 6.1 Correlation between p and the period of $z_p(t)$

Range of μ	p	Period of $z_p(t)$
$1 < \mu < 3$	1	2
$3 < \mu < 3.4494$	2	2
$3.4494 < \mu < 3.5440$	4	4
$3.5440 < \mu < 3.5644$	8	8
$3.5644 < \mu < 3.5687$	16	16
$3.5687 < \mu < 3.5696$	32	32
...	...	...
$3.6265 < \mu < 3.6304$	6	6
...	...	...
$3.7382 < \mu < 3.7411$	5	10
...	...	...
$3.8284 < \mu < 3.8415$	3	6
...	...	...

$$\overline{S}(\varphi)(t) = \int_0^{2p} \left(I - e^{2Ap}\right)^{-1} e^{A(2p-s)} \left[f(t+s, \varphi(t+s)) + \nu(t+s, t_0, \mu)\right] ds.$$

Similar to the case of even p, it can be proved that $\overline{S}$ is a contraction. Therefore, for each p-periodic sequence $\{\kappa_i(t_0, \mu)\}$, where p is odd, there exists a unique $2p$-periodic solution $z_p(t)$ of the system (6.7.48) such that $\|z_p(t)\| \leq H$ for all $t \geq t_0$. Consequently, system (6.7.48) admits the chaos through period-doubling cascade at μ_∞. □

As a result of the proof of Theorem 6.15 and making use of various parameter values of period-doubling bifurcations for the logistic map $h(s, \mu) = \mu s(1 - s)$ [5, 135], Table 6.1 is constructed. The table indicates the periodicity dependence between a p-periodic $\{\kappa_i(t_0, \mu)\}$ and the unique periodic solution $z_p(t)$ of system (6.7.48) with the same μ. In the table, we also specify the values of the parameter μ for which the p-periodic $\{\kappa_i(t_0, \mu)\}$ is stable, likewise the periodic solution $z_p(t)$ of system (6.7.48).

If system (6.7.48) is compared with the system

$$\begin{aligned} z'(t) &= Az(t) + \nu(t, t_0, \mu) \\ z(t_0) &= z_0, \, (t_0, z_0) \in I \times \mathbb{R}^n, \end{aligned} \tag{6.7.53}$$

one can see that the difference is the presence of the function $f(t, z)$, and the old theorems from [44] can be repeated almost identically for system (6.7.48) by taking into account the Lipschitz condition on the function $f(t, z)$.

6.7.2.2 The Duffing Equation's Chaotic Behavior

In this part, we consider both the Duffing equation (6.7.46) and the corresponding system (6.7.47) with the coefficients $d_1 = 0.18, d_2 = 2, d_3 = 0.00004, D = 0.02, k = 2$, and $m_0 = 2, m_1 = 1$.

The bifurcation diagram of equation (6.7.46) with the specified coefficients is shown in Fig. 6.6. In the range of μ values greater than 3.57, correlatively to the behavior of the logistic map [93, 135], successive intervals of chaos and intervals of stable periodic solutions, called the periodic windows, appear in the diagram.

At $\mu = 3$, for which the period-doubling bifurcation for the logistic map occurs for the first time [5, 133], splitting occurs in the bifurcation diagram of equation (6.7.46) with the appointed coefficients, but period-doubling does not occur at this parameter value. That is, up to the second bifurcation value $\mu = 3.4494$, all periodic solutions of the Duffing equation have period 2. This is a prospective behavior, since the periodicity of the periodic solution of the Duffing equation corresponding to a p-periodic sequence $\{\kappa_i(t_0, \mu)\}$ is $2p$ in the case of p is an odd integer.

If we denote by $\{r_m\}$ the sequence of the values of the parameter μ at which the period-doubling bifurcations for the Duffing equation (6.7.46) with the given coefficients occur, it is numerically observed that this sequence coincides with the sequence $\{\mu_m\}$, which has been defined above for the cascade of the logistic map, except the

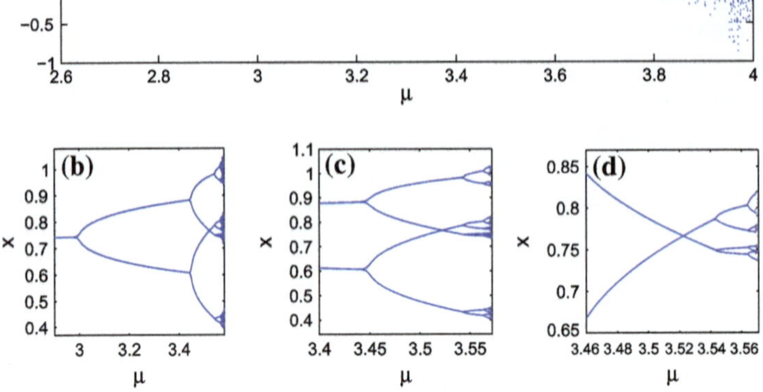

Fig. 6.6 Bifurcation diagrams of the Duffing equation perturbed with a pulse function $x''+0.18x'+2x+0.00004x^3-0.02\cos(2\pi t) = \nu(t, t_0, \mu)$, where $m_0 = 2$ and $m_1 = 1$. **a** The bifurcation diagram where the parameter μ varies between 2.6 and 4.0. **b** Magnification of (**a**) where μ is between 2.90 and 3.58. **c** Magnification of (**b**) where μ is between 3.400 and 3.572. **d** Magnification of (**c**) where μ changes from 3.460 to 3.571

first term. That is, $r_m = \mu_{m+1}$, $m \geq 1$. Consequently, when $\lim\limits_{m\to\infty} \dfrac{r_m - r_{m+1}}{r_{m+1} - r_{m+2}}$ is evaluated, the universal constant known as the Feigenbaum number 4.6692016... is achieved [81, 132, 136].

In the regions where stable periodic solutions exist, for a fixed value of the parameter μ, the bifurcation diagram represents the values of the stable periodic solutions of Eq. (6.7.46) at time $t = \zeta_0 \in I$, where ζ_0 is the initial term of the periodic sequence $\{\zeta_i\}$ corresponding to the same value of μ. We note that, for $\mu_m < \mu < \mu_{m+1}$, there are 2^m different choices for the periodic sequence $\{\zeta_i\}$ with periodicity 2^m, and this is the reason for the observation of 2^m different stable periodic solutions for these values of the parameter.

A stable periodic solution in turn becomes unstable and is replaced by a new couple of stable solutions as the parameter μ increases through the bifurcation values. A stable solution is replaced by a couple of stable periodic solutions of twice its period, except at the parameter values corresponding to a p-periodic $\{\kappa_i\}$ with p odd and the process continues in this way. For such values of μ, the periodicity does not change, by the same reasoning explained as above. In the intervals of chaos, all existing periodic solutions are unstable.

In Fig. 6.7, one can see the larger image of the periodic window which starts at $\mu = 3.8284$, and its magnification for the parameter values between 3.8350 and 3.8600. It is observed that a similar copy of the whole bifurcation diagram reappears in this region.

Now, let us check that the conditions of the last theorem are true for the system (6.7.47). The matrix of coefficients of the system (6.7.47) with the assumed coefficients is $A = \begin{pmatrix} 0 & 1 \\ -2 & -0.18 \end{pmatrix}$.

The eigenvalues of the matrix A are $a \mp ib$, where $a = -0.09$ and $b = \sqrt{2 - 0.09^2}$. The real Jordan form of the matrix A is given by $J = \begin{pmatrix} a & -b \\ b & a \end{pmatrix}$ and the identity $P^{-1}AP = J$ is satisfied where $P = \begin{pmatrix} 0 & 1 \\ b & a \end{pmatrix}$ and $P^{-1} = \dfrac{1}{b}\begin{pmatrix} -a & 1 \\ b & 0 \end{pmatrix}$.

Evaluating the exponential matrix e^{At} we have

$$e^{At} = e^{at} P \begin{pmatrix} cos(bt) & -sin(bt) \\ \sin(bt) & cos(bt) \end{pmatrix} P^{-1}. \tag{6.7.54}$$

Denote by $\|.\|$ the matrix norm which is induced by the usual Euclidean norm in $\mathbb{R}^n$. That is,

$$\|\Gamma\| = \max\left\{\sqrt{\lambda} : \lambda \; is \; an \; eigenvalue \; of \, \Gamma^T \Gamma\right\}$$

for any $n \times n$ matrix Γ with real entries, and Γ^T denotes the transpose of the matrix Γ [137].

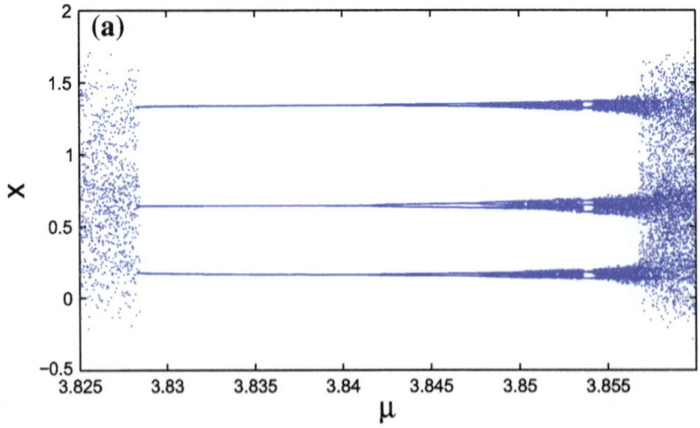

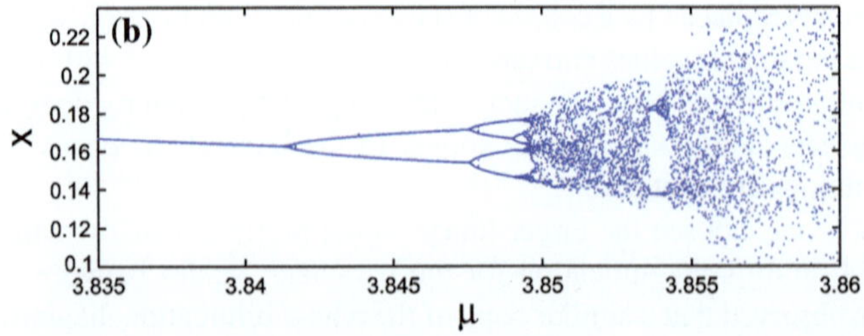

Fig. 6.7 The periodic window which starts at $\mu = 3.8284$ in the bifurcation diagram of the Duffing equation perturbed with a pulse function $x''+0.18x'+2x+0.00004x^3-0.02\cos(2\pi t) = \nu(t, t_0, \mu)$, where $m_0 = 2$ and $m_1 = 1$. **a** The bifurcation diagram where μ is between 3.8250 and 3.8600. **b** Magnification of (**a**) where μ changes from 3.8350 to 3.8600

One can see that

$$\|P\| = \left(\frac{3}{2} + \frac{\sqrt{1+0.18^2}}{2}\right)^{1/2},$$

and

$$\left\|P^{-1}\right\| = \frac{1}{\sqrt{2-0.09^2}}\left(\frac{3}{2} + \frac{\sqrt{1+0.18^2}}{2}\right)^{1/2}.$$

Therefore, using (6.7.54), we obtain $\left\|e^{At}\right\| \leq Ne^{\alpha t}$ where $N = \dfrac{3+\sqrt{1+0.18^2}}{\sqrt{8-0.18^2}}$ and $\alpha = -0.09$.

In what follows, we use approximation with accuracy of 7 digits in the decimal part.

For $p_0 = 4$, $Ne^{\alpha p_0} = \left(\dfrac{3+\sqrt{1+0.18^2}}{\sqrt{8-0.18^2}}\right)e^{-0.36} \cong 0.9926395 < 1$. One can easily evaluate that

$$\max_{1\le i\le 3}\left\{\left\|\left(I-e^{Ji}\right)^{-1}\right\|\right\} = \left\|\left(I-e^{J1}\right)^{-1}\right\| \cong 0.8045044.$$

Then, using the matrix identity $\left(I-e^{At}\right)^{-1} = P\left(I-e^{Jt}\right)^{-1}P^{-1}$, the inequality

$$\max_{1\le i\le 3}\left\{\left\|\left(I-e^{Ai}\right)^{-1}\right\|\right\} \le \|P\|\left\|P^{-1}\right\| \max_{1\le i\le 3}\left\{\left\|\left(I-e^{Ji}\right)^{-1}\right\|\right\} \cong 1.1446324$$

is obtained. On the basis of above evaluations, one can find that

$$K = \max\left\{\max_{1\le i\le 3}\left\|(I-e^{Ai})^{-1}\right\|, \frac{1}{1-Ne^{4\alpha}}\right\} \cong 135.8619956.$$

System (6.7.47) with the prescribed coefficients has the nonlinear term

$$f(t,x_1,x_2) = \begin{bmatrix}0 & -0.00004x_1^3\end{bmatrix}^T.$$

The Lipschitz constant L for this function can be taken as 0.0003468 since the x_1 values of the chaotic attractor satisfies the condition $|x_1| \le 1.7$. Thus $\dfrac{-KNL}{\alpha} \cong 0.7448557$ and the condition (6.7.52) is also satisfied.

We end up this part, by simulating a solution (x_1, x_2) of system (6.7.47) with initial data $x_1(0.5) = 0.01$, $x_2(0.5) = 0.025$ and $\mu_\infty = 3.8$. In Fig. 6.8, the chaotic behavior of the solution is revealed.

6.7.2.3 Lyapunov Exponents

The Lyapunov exponent is a measure of divergence of state trajectories, and is one of the most important features of deterministic chaos [136]. There are well-developed results for Lyapunov exponents of maps, and it is technically difficult for continuous dynamics [72, 132, 133, 138]. Evaluation procedures of Lyapunov exponents for continuous dynamics are, in general, provided for low-dimensional systems [139, 140]. Our system, despite there is discontinuity property, evolves along continuous-time. Therefore, to work with Lyapunov exponents, we should consider mainly the results for continuous dynamics. More exactly, our systems involve continuous and discrete dynamics such that the space variables change continuously while the switching moments of time satisfy discrete equations, that is, they belong to the class of hybrid systems [82, 121, 122]. Consequently, we have to evaluate the divergence of solutions by continuous as well as discrete Lyapunov exponents. Moreover, our

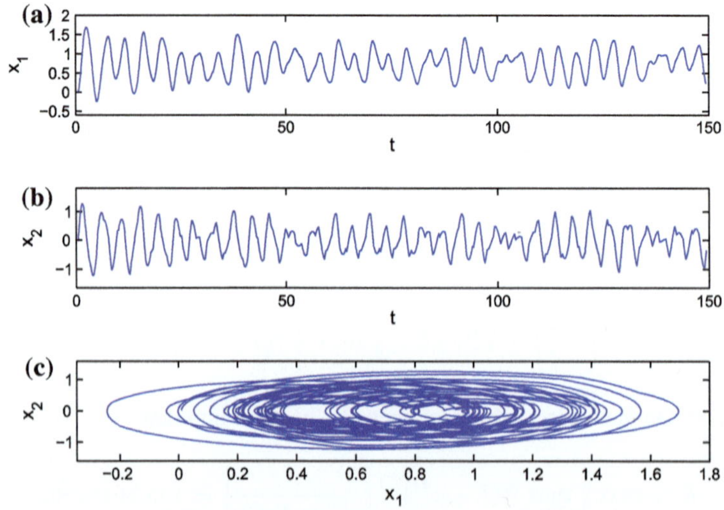

Fig. 6.8 Simulation results of the Duffing equation perturbed with a pulse function $x'' + 0.18x' + 2x + 0.00004x^3 - 0.02cos(2\pi t) = \nu(t, t_0, \mu_\infty)$, where $m_0 = 2, m_1 = 1$ and $\mu_\infty = 3.8$. The pictures in **a** and **b** show the graphs of the x_1 and x_2 coordinates, respectively, while the picture in **c** represents the trajectory of the solution $(x_1(t), x_2(t))$

systems are essentially non-autonomous. That is why one has to consider the method of Lyapunov exponents for non-autonomous systems [140]. For chaos development, the positive Lyapunov exponent is appropriate. So, one can conclude that the positiveness of one of the Lyapunov exponents is an indicator of chaos if the system is considered in a bounded region. That is why for the general case of our analysis in this section, it is sufficient to find that the Lyapunov exponent is positive for the logistic map, the generator of the switching moments.

To illustrate the general discussions, let us consider the following example.

Example Let the equation

$$x' = -2x + \nu(t, t_0, \mu_\infty) \tag{6.7.55}$$

be given with $\mu_\infty = 3.8$.

If we consider two solutions of Eq. (6.7.55) with the same t_0, they are both bounded and approach to each other with exponent -2, that is, -2 is an eigenvalue. Since the equation is non-autonomous then it needs a special treatment [140]. When the time variable is considered as a spatial one, one can transform Eq. (6.7.55) to a system as

$$\begin{aligned} \frac{dx}{dt} &= -2x + \nu(\tau, t_0, \mu_\infty) \\ \frac{d\tau}{dt} &= 1 \\ \zeta_{i+1} &= i + 1 + h(\zeta_i - i, \mu_\infty). \end{aligned} \tag{6.7.56}$$

The second equation in (6.7.56) provides us the zero Lyapunov exponent [140]. Since our system involves the discrete equation, the logistic map with $\mu_\infty = 3.8$, it admits the third Lyapunov exponent which is approximately 0.432 [141]. This Lyapunov exponent describes the divergence of solutions with different initial moments along the time axis. Finally, we have obtained that the divergence of solutions of Eq. (6.7.55) is described through three Lyapunov exponents $\lambda_1 = 0.432$, $\lambda_2 = 0$, $\lambda_3 = -2$.

6.7.3 Controlling Results

6.7.3.1 The Logistic Map

We stabilize the periodic solutions by control of the switching moments of the pulse function, which are defined through the logistic map. Therefore, one will need the description of the OGY method for the map [142].

Suppose that the parameter μ, in the map, can be finely tuned in a small range around the value $\mu_\infty = 3.8$, that is, μ is allowed to vary in the range $[\mu_\infty - \delta, \mu_\infty + \delta]$, where δ is small. Denote the target period-p orbit to be controlled as $\kappa^{(i)}(t_0, \mu_\infty)$, $i = 1, 2, \ldots, p$ where t_0 belongs to the unit interval $I = [0, 1]$, $\kappa^{(i+1)}(t_0, \mu_\infty) = h(\kappa^{(i)}(t_0, \mu_\infty), \mu_\infty)$ and $\kappa^{(p+1)}(t_0, \mu_\infty) = \kappa^{(1)}(t_0, \mu_\infty)$. The logistic map, $h(s, \mu) = \mu s(1 - s)$, in the neighborhood of a periodic orbit can be approximated by a linear equation expanded around the periodic orbit. If we denote $\bar{\mu}_j - \mu_\infty = \Delta\bar{\mu}_j$, and $\kappa_{j+1}(t_1, \bar{\mu}_j) = h(\kappa_j(t_1, \bar{\mu}_j), \bar{\mu}_j)$, $t_1 \in I$, we get

$$\begin{aligned} \kappa_{j+1} - \kappa^{(i+1)} &= \frac{\partial h}{\partial s}[\kappa_j - \kappa^{(i)}] + \frac{\partial h}{\partial \mu}\Delta\bar{\mu}_j \\ &= \mu_\infty[1 - 2\kappa^{(i)}][\kappa_j - \kappa^{(i)}] + \kappa^{(i)}[1 - \kappa^{(i)}]\Delta\bar{\mu}_j, \end{aligned} \tag{6.7.57}$$

where partial derivatives are evaluated at $s = \kappa^{(i)}(t_0, \mu_\infty)$ and $\mu = \mu_\infty$. We require $\kappa_{j+1}(t_1, \bar{\mu}_j)$ to stay in the neighborhood of $\kappa^{(i+1)}(t_0, \mu_\infty)$. Therefore, if we set

$$\kappa_{j+1}(t_1, \bar{\mu}_j) - \kappa^{(i+1)}(t_0, \mu_\infty) = 0,$$

then we obtain that

$$\Delta\bar{\mu}_j = \mu_\infty \frac{[2\kappa^{(i)} - 1][\kappa_j - \kappa^{(i)}]}{\kappa^{(i)}[1 - \kappa^{(i)}]} \tag{6.7.58}$$

or equivalently

$$\bar{\mu}_j = \mu_\infty \left(1 + \frac{[2\kappa^{(i)} - 1][\kappa_j - \kappa^{(i)}]}{\kappa^{(i)}[1 - \kappa^{(i)}]}\right). \tag{6.7.59}$$

This equation holds only when the trajectory κ_j enters a small neighborhood of the period -p orbit, hence the required parameter perturbation $\Delta\bar{\mu}_j$ is small. When the trajectory is outside the neighborhood of the target periodic orbit, we do not apply any parameter perturbation, so the system evolves at its nominal parameter value μ_∞. Hence, we set $\bar{\mu}_j = \mu_\infty$, when $\left|\Delta\bar{\mu}_j\right| > \delta$.

Suppose that $t_0 \in I$ is fixed such that the sequence $\{\kappa_i(t_0, \mu_\infty)\}$ is p-periodic. Thus, for given $\delta > 0$, there exist $\varepsilon > 0$ and $i_0, j_0 \in \mathbb{N}$ such that for all $i, i_0 \leq i \leq i_0 + j_0$, we have $|\Delta\bar{\mu}_i| \leq \delta$ and $|\kappa_i(t_1, \bar{\mu}_i) - \kappa_i(t_0, \mu_\infty)| < \varepsilon$ [53, 72, 142, 143], where $\kappa_{i+1}(t_1, \bar{\mu}_i) = h(\kappa_i(t_1, \bar{\mu}_i), \bar{\mu}_i)$. The number j_0 is, in general, finite, since the nonlinearity is not included in (6.7.57). We will use the numbers ε, i_0 and j_0 for Theorem 6.16.

We note that the control of chaos is not achieved immediately after switching on the control mechanism, rather, there is a transient time before the logistic map is controlled. The transient time increases if the δ decreases [72, 143].

Now, we consider a simulation for the stabilization of the logistic map. Namely, of the sequence $\{\kappa_i\}$, where $\kappa_{i+1} = 3.8\kappa_i(1 - \kappa_i)$, $i \geq 0$ and $\kappa_0 = t_1 = 0.5$. If the OGY control method is applied around the fixed point 2.8/3.8, that is the period-1 orbit of the logistic equation $h(s, 3.8) = 3.8s(1 - s)$, we obtain the result that is shown in Fig. 6.9. We used the value $\delta = 0.19$. The control starts at the iteration number $i = 25$ and ends at $i = 60$. Despite the control was switched off at 60th iteration, the stabilization prolongs till the 110th iteration.

6.7.3.2 The General System Control

From the description made above, it is seen that the control by OGY method means construction of a sequence of the parameter's value μ near a chaotic value of the

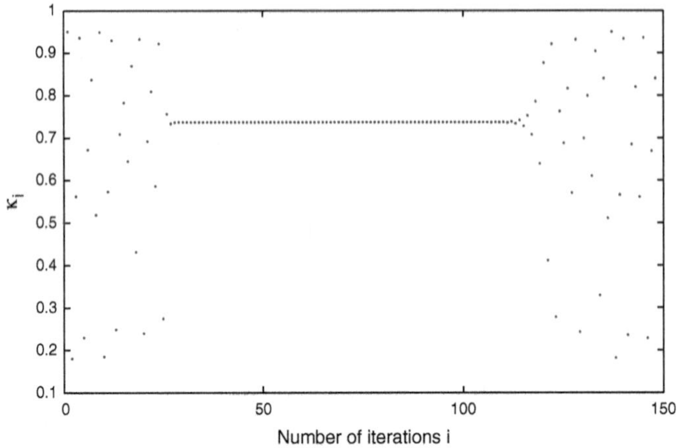

Fig. 6.9 The OGY control method applied to the sequence $\{\kappa_i\}$, where $\kappa_{i+1} = 3.8\kappa_i(1 - \kappa_i)$, $\kappa_0 = 0.5$, around the fixed point 2.8/3.8 of the logistic map with $\delta = 0.19$. The control is switched on at the iteration number $i = 25$ and switched off at $i = 60$

parameter, μ_∞, to generate a solution, which is close to the chosen periodic one. It is obvious that similar control problem can be formulated for the system (6.7.47), and consequently, for Eq. (6.7.46).

To control system (6.7.47), we replace the parameter μ by the control sequence $\{\mu^i\}$ and define

$$\nu(t, t_1, \mu^i) = \begin{cases} m_0, & \text{if } \zeta_{2i}(t_1, \mu^i) < t \le \zeta_{2i+1}(t_1, \mu^i) \\ m_1, & \text{if } \zeta_{2i+1}(t_1, \mu^i) < t \le \zeta_{2i+2}(t_1, \mu^i), \end{cases} \tag{6.7.60}$$

where $i \ge 0$ is an integer, $m_0, m_1 \in \mathbb{R}^n$, the same as for the function $\nu(t, t_0, \mu)$ in (6.7.49). The sequence $\zeta(t_1, \mu^i) = \{\zeta_i(t_1, \mu^i)\}$, $i \ge 0$, is defined through the equation $\zeta_i(t_1, \mu^i) = i + \kappa_i(t_1, \mu^i)$, with $\kappa_{i+1}(t_1, \mu^i) = h(\kappa_i(t_1, \mu^i), \mu^i)$, $\kappa_0(t_1, \mu^i) = t_1$.

Consider, now, the system,

$$\begin{aligned} z'(t) &= Az(t) + f(t, z) + \nu(t, t_1, \mu^i) \\ z(t_1) &= z_1, \ (t_1, z_1) \in I \times \mathbb{R}^n, \end{aligned} \tag{6.7.61}$$

which is the control system conjugate to the system (6.7.48).

Our aim is to determine the sequence $\{\mu^i\}$ which stabilizes the periodic solutions of (6.7.48) and in the next theorem a convenient choice for this sequence is indicated.

By $\phi(t, \bar{t}, \bar{z})$, $\bar{t} \in I$, $\bar{z} \in \mathbb{R}^n$, we denote a solution of (6.7.61) with $t_1 = \bar{t}$ and $z_1 = \bar{z}$.

In the following theorem we shall use the numbers ε, i_0, and j_0, which were mentioned above for the stabilization of the logistic map.

Suppose that $z_p(t)$, $p \in \mathbb{N}$, denotes the periodic solution of (6.7.48) with $z_p(t_0) = z_0$ and $\mu = \mu_\infty$. Take $z_1 \in \mathbb{R}^n$ and consider the solution $z(t) = \phi(t, t_1, z_1)$ of system (6.7.61). If $z(\zeta_{i_0}(t_0, \mu_\infty))$ is not equal to $z_p(\zeta_{i_0}(t_0, \mu_\infty))$, then suppose that the number $T(\varepsilon, z_1)$ is the maximum of the numbers $\zeta_{i_0}(t_0, \mu_\infty)$ and

$$\zeta_{i_0}(t_0, \mu_\infty) + \frac{1}{NL + \alpha} \ln\left(\frac{1 - e^{\alpha\varepsilon}}{N \left\| z(\zeta_{i_0}(t_0, \mu_\infty)) - z_p(\zeta_{i_0}(t_0, \mu_\infty)) \right\|} \right).$$

Set $T(\varepsilon, z_1) = \zeta_{i_0}(t_0, \mu_\infty)$ in the case that $z(\zeta_{i_0}(t_0, \mu_\infty))$ and $z_p(\zeta_{i_0}(t_0, \mu_\infty))$ are equal to each other. The number $T(\varepsilon, z_1)$ will be needed in the following theorem, which is one of the main results of this section.

In the proof of the following theorem, we assume without loss of generality that $i_0 = 0$. In this case, $\zeta_{i_0}(t_0, \mu_\infty) = t_0$ and $\zeta_{i_0}(t_1, \bar{\mu}_i) = t_1$. It is worth saying that since $\frac{-NL}{\alpha} < \frac{-KNL}{\alpha} < 1$, we have $NL + \alpha < 0$.

Theorem 6.16 *Assume that $T(\varepsilon, z_1) < i_0 + j_0$. Then the sequence $\{\bar{\mu}_i\}$ stabilizes the periodic solution $z_p(t)$ such that*

$$\|\phi(t,t_1,z_1)-z_p(t)\| < \left(1-\frac{Ne^{-\alpha}\|m_0-m_1\|}{(NL+\alpha)(1-e^{\alpha})}\right)\left(1-e^{\alpha\varepsilon}\right),$$

if $t\in[T(\varepsilon,z_1),i_0+j_0]$.

Proof Without loss of generality, assume that $t_1 \le t_0$. The solution $z(t) = \phi(t,t_1,z_1)$, $t_1 \in I$, $z_1 \in \mathbb{R}^n$, of (6.7.61) can be continued up to $t = t_0$. Let us denote $z(t_0) = \eta_1$ and $z_p(t_0) = z_0$. In this case, the integral equations

$$z(t) = e^{A(t-t_0)}\eta_1 + \int_{t_0}^{t} e^{A(t-s)}[f(s,z(s)) + \nu(s,t_1,\bar{\mu}_i)]ds$$

and

$$z_p(t) = e^{A(t-t_0)}z_0 + \int_{t_0}^{t} e^{A(t-s)}[f(s,z_p(s)) + \nu(s,t_0,\mu_\infty)]ds$$

are satisfied. Therefore, for $t\ge t_0$ we have

$$\begin{aligned} z(t)-z_p(t) &= e^{A(t-t_0)}(\eta_1-z_0) + \int_{t_0}^{t} e^{A(t-s)}[f(s,z(s)) - f(s,z_p(s))]ds \\ &+ \int_{t_0}^{t} e^{A(t-s)}[\nu(s,t_1,\bar{\mu}_i) - \nu(s,t_0,\mu_\infty)]ds. \end{aligned} \tag{6.7.62}$$

Since for each i, $0\le i\le j_0$, the inequality

$$\left|\zeta_i(t_0,\mu_\infty)-\zeta_i(t_1,\overline{\mu}_i)\right| = |\kappa_i(t_0,\mu_\infty)-\kappa_i(t_1,\bar{\mu}_i)| < \varepsilon$$

holds, one can verify that

$$\left|\int_{\zeta_i(t_0,\mu_\infty)}^{\zeta_i(t_1,\overline{\mu}_i)} e^{\alpha(t-s)}ds\right| < \left(\frac{-1}{\alpha}\right)\left(1-e^{\alpha\varepsilon}\right)e^{\alpha(\lfloor t\rfloor-1-i)}, \tag{6.7.63}$$

where $\lfloor t\rfloor$ denotes the greatest integer which is not larger than t. On the other hand, by means of the inequality (6.7.63) we have that

$$\begin{aligned} &\left\|\int_{t_0}^{t} e^{A(t-s)}\left[\nu(s,t_1,\overline{\mu}_i)-\nu(s,t_0,\mu_\infty)\right]ds\right\| \\ &\le \int_{t_0}^{t} Ne^{\alpha(t-s)}\left\|\nu(s,t_1,\overline{\mu}_i)-\nu(s,t_0,\mu_\infty)\right\|ds \\ &\le \sum_{i=1}^{\lfloor t\rfloor}\left|\int_{\zeta_i(t_0,\mu_\infty)}^{\zeta_i(t_1,\overline{\mu}_i)} Ne^{\alpha(t-s)}\|m_0-m_1\|ds\right| \end{aligned}$$

$$< \left(\frac{-N}{\alpha}\right)\left(1-e^{\alpha\varepsilon}\right)\|m_0-m_1\|\sum_{i=1}^{\lfloor t\rfloor} e^{\alpha(\lfloor t\rfloor-1-i)}$$

$$< \frac{-Ne^{-\alpha}\|m_0-m_1\|}{\alpha\left(1-e^{\alpha}\right)}\left(1-e^{\alpha\varepsilon}\right).$$

Using Eq. (6.7.62) together with the last inequality one can obtain that

$$\left\|z(t)-z_p(t)\right\| \le Ne^{\alpha(t-t_0)}\|\eta_1-z_0\| + \frac{-Ne^{-\alpha}\|m_0-m_1\|}{\alpha\left(1-e^{\alpha}\right)}\left(1-e^{\alpha\varepsilon}\right)$$
$$+\int_{t_0}^{t} NLe^{\alpha(t-s)}\left\|z(s)-z_p(s)\right\| ds.$$

Now, let $u(t)=\left\|z(t)-z_p(t)\right\|e^{-\alpha t}$. Under the circumstances we have

$$u(t)\le Ne^{-\alpha t_0}\|\eta_1-z_0\| + \frac{-Ne^{-\alpha(t+1)}\|m_0-m_1\|}{\alpha\left(1-e^{\alpha}\right)}\left(1-e^{\alpha\varepsilon}\right) + NL\int_{t_0}^{t} u(s)ds.$$

Applying Lemma 2.2 [57] we attain that

$$u(t)\le Ne^{-\alpha t_0}\|\eta_1-z_0\| + \frac{-Ne^{-\alpha(t+1)}\|m_0-m_1\|}{\alpha\left(1-e^{\alpha}\right)}\left(1-e^{\alpha\varepsilon}\right)$$
$$+N^2L\|\eta_1-z_0\|e^{-\alpha t_0}\int_{t_0}^{t} e^{NL(t-s)}ds$$
$$+\frac{-N^2Le^{-\alpha}\|m_0-m_1\|}{\alpha\left(1-e^{\alpha}\right)}\left(1-e^{\alpha\varepsilon}\right)\int_{t_0}^{t} e^{NL(t-s)}e^{-\alpha s}ds$$

Making use of the equations

$$\int_{t_0}^{t} e^{NL(t-s)}ds = \frac{1}{NL}\left(e^{NL(t-t_0)}-1\right),$$

and

$$\int_{t_0}^{t} e^{NL(t-s)}e^{-\alpha s}ds = \left(\frac{-1}{NL+\alpha}\right)e^{-\alpha t}\left(1-e^{(NL+\alpha)(t-t_0)}\right)$$

it can be verified that

$$u(t)\le N\|\eta_1-z_0\|e^{-\alpha t_0}e^{NL(t-t_0)} - \frac{Ne^{-\alpha(t+1)}\|m_0-m_1\|}{\alpha\left(1-e^{\alpha}\right)}\left(1-e^{\alpha\varepsilon}\right)$$
$$+\frac{N^2Le^{-\alpha(t+1)}\|m_0-m_1\|}{(NL+\alpha)\alpha\left(1-e^{\alpha}\right)}\left(1-e^{\alpha\varepsilon}\right)\left(1-e^{(NL+\alpha)(t-t_0)}\right)$$

$$
\begin{aligned}
&< N \left\|\eta_1 - z_0\right\| e^{-\alpha t_0} e^{NL(t-t_0)} - \frac{Ne^{-\alpha(t+1)} \left\|m_0 - m_1\right\|}{\alpha\left(1 - e^{\alpha}\right)} \left(1 - e^{\alpha\varepsilon}\right) \\
&+ \frac{N^2 L e^{-\alpha(t+1)} \left\|m_0 - m_1\right\|}{(NL+\alpha)\alpha\left(1 - e^{\alpha}\right)} \left(1 - e^{\alpha\varepsilon}\right) \\
&= N \left\|\eta_1 - z_0\right\| e^{-\alpha t_0} e^{NL(t-t_0)} - \frac{Ne^{-\alpha(t+1)} \left\|m_0 - m_1\right\|}{(NL+\alpha)\left(1 - e^{\alpha}\right)} \left(1 - e^{\alpha\varepsilon}\right).
\end{aligned}
$$

Multiplication of both sides of the last inequality by $e^{\alpha t}$ implies that

$$
\left\|z(t) - z_p(t)\right\| < N \left\|\eta_1 - z_0\right\| e^{(NL+\alpha)(t-t_0)} - \frac{Ne^{-\alpha} \left\|m_0 - m_1\right\|}{(NL+\alpha)\left(1 - e^{\alpha}\right)} \left(1 - e^{\alpha\varepsilon}\right).
$$

It is clear that if $\eta_1 = z_0$, then the conclusion of the theorem is true. Suppose that $\eta_1 \neq z_0$. If $t \in [T(\varepsilon, z_1), j_0]$, then one can easily verify that

$$
e^{(NL+\alpha)(t-t_0)} \leq \frac{1 - e^{\alpha\varepsilon}}{N \left\|\eta_1 - z_0\right\|}.
$$

Consequently, the inequality

$$
\left\|z(t) - z_p(t)\right\| < \left(1 - \frac{Ne^{-\alpha} \left\|m_0 - m_1\right\|}{(NL+\alpha)\left(1 - e^{\alpha}\right)}\right) \left(1 - e^{\alpha\varepsilon}\right)
$$

holds, for $t \in [T(\varepsilon, z_1), j_0]$.

The theorem is proved.□

Implementation of Theorem 6.16 to system (6.7.47) is mentioned in the next part.

6.7.3.3 The Duffing Equation Control

Let us consider the main system (6.7.47) with $\mu_\infty = 3.8$ and $d_1 = 0.18$, $d_2 = 2$, $d_3 = 0.00004$, $D = 0.02$, $k = 2$, $m_0 = 2$, $m_1 = 1$, again. The system satisfies the conditions for existence of chaos and admits the chaos at $\mu_\infty = 3.8$. Theorem 6.16 is applicable to (6.7.47). The control system (6.7.61) has, in this case, the form

$$
\begin{aligned}
x_1' &= x_2 \\
x_2' &= -2x_1 - 0.18x_2 - 0.00004x_1^3 + 0.02\cos(2\pi t) + v(t, t_1, \bar{\mu}_i).
\end{aligned} \tag{6.7.64}
$$

To simulate the result, let us take $t_1 = 0.5$, $t_0 = 2.8/3.8$ and the solution (x_1, x_2) of system (6.7.64) with the initial condition $x_1(t_1) = 0.01$, $x_2(t_1) = 0.025$. Its graph is seen in Fig. 6.10 and it approximates the 2-periodic solution $z_1(t)$. The value $\delta = 0.19$ is used, and the control starts at time $t = \zeta_{25}$ and ends at $t = \zeta_{60}$. Here, we note that since the OGY control method is applied to the logistic map, the iteration

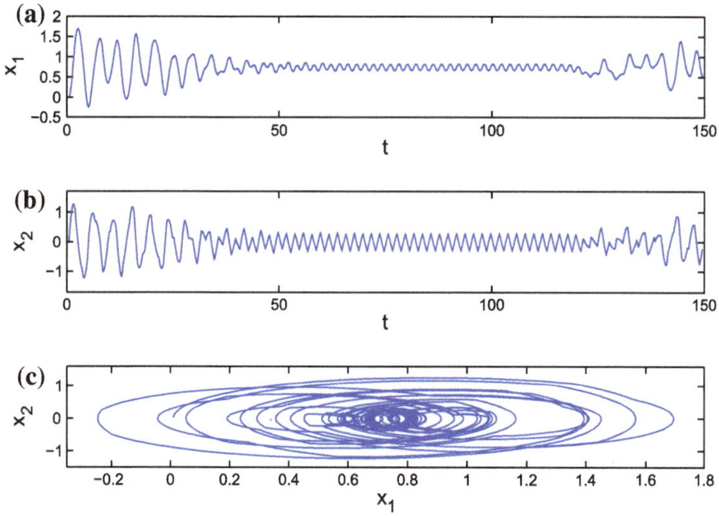

Fig. 6.10 The OGY control method applied to the Duffing equation perturbed with a pulse function $x'' + 0.18x' + 2x + 0.00004x^3 - 0.02\cos(2\pi t) = \nu(t, t_0, \mu_\infty)$, where $m_0 = 2, m_1 = 1$ and $\mu_\infty = 3.8$. The control starts at time $t = \zeta_{25}$ and ends at $t = \zeta_{60}$. **a** The graph of the x_1-coordinate. **b** The graph of the x_2-coordinate. **c** The trajectory of the solution $(x_1(t), x_2(t))$

moment $\bar{i}$ when the control is switched on corresponds to the time moment $t = \zeta_{\bar{i}}$, and a similar argument is valid for the moment when the control ends.

We use the same interval of stabilization for the logistic map and the Duffing equation. But the interval of periodicity for the map is larger in the former, approximately 80 and 60, respectively. The reason is that the chaos of the equation is secondary with respect to the chaos of the logistic map. Likewise the control of the logistic map, the chaos transient time increases if the δ decreases.

To discuss our main assumptions, let us arrange the following simulations. Consider the following Duffing equation in the standard form [81]

$$x'' + 0.05x' + x^3 = 7.5\cos t. \tag{6.7.65}$$

To convert this equation to a suitable form for which our theorem can be applied, we use the change of variables $u = t/\pi$ and $y(u) = x(t)$. Using these new variables and relabeling u as t, one can reduce (6.7.65) to the differential equation

$$y'' + 0.05\pi y' + \pi^2 y^3 = 7.5\pi^2\cos(\pi t). \tag{6.7.66}$$

Defining new variables $x_1 = y$ and $x_2 = y'$ we can reduce (6.7.66) to the system

$$\begin{aligned} x_1' &= x_2 \\ x_2' &= -0.05\pi x_2 - \pi^2 x_1^3 + 7.5\pi^2\cos(\pi t). \end{aligned} \tag{6.7.67}$$

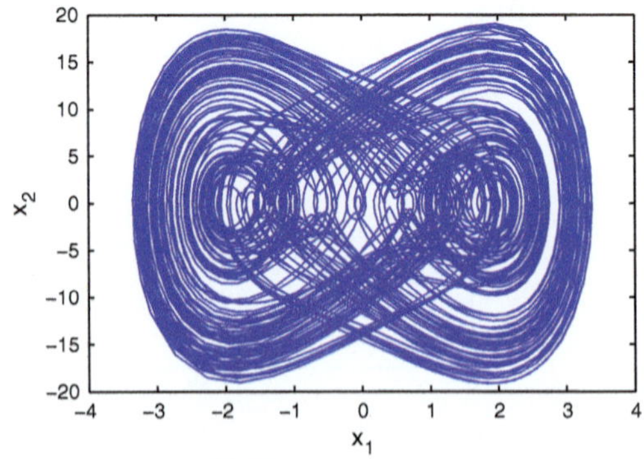

Fig. 6.11 The trajectory of the solution $(x_1(t), x_2(t))$ for system (6.7.67)

The eigenvalues for this system are 0 and -0.05π. Since one of the eigenvalues is zero, one can expect that our results are not applicable to system (6.7.67). That is, the system is not controllable with our method. Take a solution of system (6.7.67) with $x_1(0.5) = 1, x_2(0.5) = 2$. The chaotic behavior is seen in Fig. 6.11.

Now, we apply the method developed in the previous part to the equation

$$y'' = -0.05\pi y' - \pi^2 y^3 + 7.5\pi^2 \cos(\pi t) + \nu(t, t_0, \mu_\infty). \tag{6.7.68}$$

The corresponding control system is

$$\begin{aligned} x_1' &= x_2 \\ x_2' &= -0.05\pi x_2 - \pi^2 x_1^3 + 7.5\pi^2 \cos(\pi t) + \nu(t, t_1, \bar{\mu}_i). \end{aligned} \tag{6.7.69}$$

Let $t_1 = 0.5$, $t_0 = 2.8/3.8$ and $\delta = 0.19$. We take the solution of the last system with $x_1(t_1) = 1$ and $x_2(t_1) = 2$. The control is switched on at $t = \zeta_{25}$ and switched off at $t = \zeta_{60}$. The simulation result is seen in Fig. 6.12.

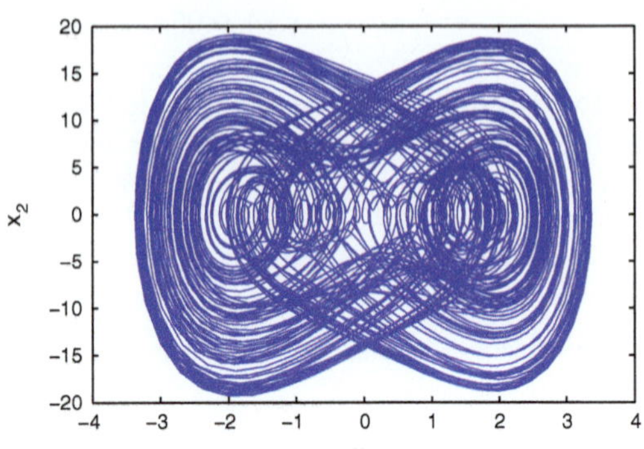

Fig. 6.12 The trajectory of the solution $(x_1(t), x_2(t))$ for the control system (6.7.69), where $m_0 = 2$ and $m_1 = 1$

One can see that our way of application of the OGY method does not work for the system (6.7.69). The reason is that the corresponding nonperturbed Duffing equation to this system has the zero eigenvalue.

6.7.4 Morphogenesis and the Logistic Map

In Sect. 6.7.2.2, we demonstrated that the Duffing equation perturbed with a pulse function

$$x'' + 0.18x' + 2x + 0.00004x^3 = 0.02\cos(2\pi t) + \nu(t, t_0, \mu_\infty), \quad (6.7.70)$$

with the coefficients $m_0 = 2$, $m_1 = 1$ and $\mu_\infty = 3.8$, admits the chaos through period-doubling cascade on the time interval $[0, \infty)$ and obeys the Feigenbaum universal behavior [144].

By favor of the new variables $x_1 = x$ and $x_2 = x'$, Eq. (6.7.70) can be reduced to the system

$$\begin{aligned} x_1' &= x_2 \\ x_2' &= -0.18x_2 - 2x_1 - 0.00004x_1^3 + 0.02\cos(2\pi t) + \nu(t, t_0, \mu_\infty). \end{aligned} \quad (6.7.71)$$

For the illustration of chaos extension, we will make use of the relay system (6.7.71) as the generator, in the role of a core as displayed in Fig. 2.5, and attach three replicator systems with coordinates $x_3 - x_4$, $x_5 - x_6$ and $x_7 - x_8$ to obtain the 8-dimensional result-relay-system

$$\begin{aligned} x_1' &= x_2 \\ x_2' &= -0.18x_2 - 2x_1 - 0.00004x_1^3 + 0.02\cos(2\pi t) + \nu(t, t_0, \mu_\infty) \\ x_3' &= x_4 - 0.1x_1 \\ x_4' &= -10x_3 - 6x_4 - 0.03x_3^3 + 4x_2 \\ x_5' &= x_6 + 2x_1 \\ x_6' &= -2x_5 - 2x_6 + 0.007x_5^3 + 0.6x_2 \\ x_7' &= x_8 - 0.5x_2 \\ x_8' &= -5x_7 - 4x_8 - 0.05x_7^3 + 2.5x_1, \end{aligned} \quad (6.7.72)$$

where again $m_0 = 2$, $m_1 = 1$ and $\mu_\infty = 3.8$.

The theoretical results mentioned in Chap. 2 reveal that system (6.7.72), as well as the replicators, admit the chaos through period-doubling cascade and obey the universal behavior of Feigenbaum. Figure 6.13 shows the two-dimensional projections on the $x_1 - x_2$, $x_3 - x_4$, $x_5 - x_6$ and $x_7 - x_8$ planes of the trajectory of the result-relay-system (6.7.72) with initial data $x_1(0) = 1.37$, $x_2(0) = -0.05$, $x_3(0) = 0.05$, $x_4(0) = -0.1$, $x_5(0) = 1.09$, $x_6(0) = -0.81$, $x_7(0) = 0.08$ and $x_8(0) = 0.21$. The picture seen in Fig. 6.13, (a) is the attractor of the generator (6.7.71) and accordingly Fig. 6.13, (b)–(d) represent the attractors of the first, second and the third replicator

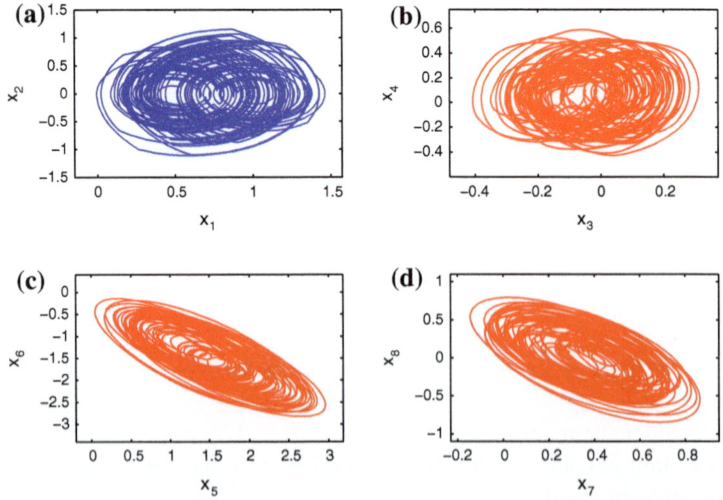

Fig. 6.13 2-dimensional projections of the chaotic attractor of the result-system (6.7.72). The pictures in **a**, **b**, **c** and **d** represent the projections on the $x_1 - x_2$, $x_3 - x_4$, $x_5 - x_6$ and $x_7 - x_8$ planes, respectively. The picture in (**a**) shows the attractor of the prior chaos produced by the generator (6.7.71), which is a relay system, and in (**b**)–(**d**) the chaotic attractors of the replicator systems are observable. The illustrations in (**b**)–(**d**) repeated the structure of the attractor shown in (**a**), and the mimicry between these pictures is an indicator of the replication of chaos

systems, respectively. It can be easily verified that all replicators used inside the system (6.7.72) satisfy condition $(A7)$ of Chap. 2. The resemblance of the chaotic attractors of the generator and the replicators is a consequence of morphogenesis of chaos.

Now, let us continue with the control of morphogenesis of chaos by means of the OGY control method. In order to stabilize the unstable periodic solutions of system (6.7.71), we consider the system

$$\begin{aligned}
x_1' &= x_2 \\
x_2' &= -0.18x_2 - 2x_1 - 0.00004x_1^3 + 0.02\cos(2\pi t) + \nu(t, t_1, \bar{\mu}_i) \\
x_3' &= x_4 - 0.1x_1 \\
x_4' &= -10x_3 - 6x_4 - 0.03x_3^3 + 4x_2 \\
x_5' &= x_6 + 2x_1 \\
x_6' &= -2x_5 - 2x_6 + 0.007x_5^3 + 0.6x_2 \\
x_7' &= x_8 - 0.5x_2 \\
x_8' &= -5x_7 - 4x_8 - 0.05x_7^3 + 2.5x_1,
\end{aligned} \tag{6.7.73}$$

which is the control system conjugate to the result-relay-system (6.7.72), where $m_0 = 2$ and $m_1 = 1$.

To simulate the control results, we make use of the values $\delta = 0.19$, $t_1 = 0.5$, $t_0 = 2.8/3.8$ and the trajectory of system (6.7.73) with the initial data $x_1(0) = 1.37$,

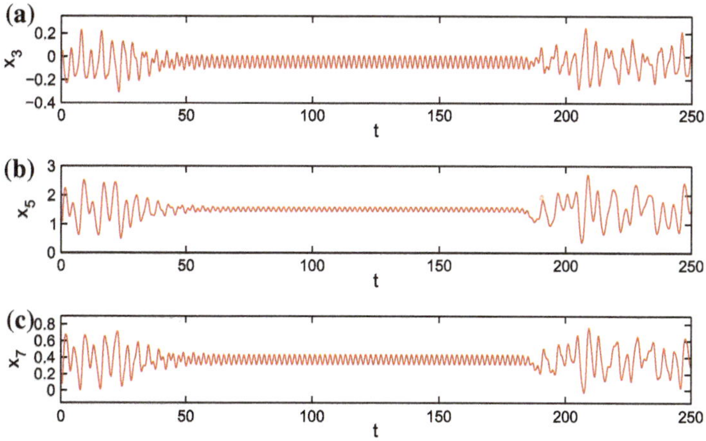

Fig. 6.14 OGY control method applied to the result-relay-system (6.7.72). **a** The graph of the x_3-coordinate, **b** The graph of the x_5-coordinate, **c** The graph of the x_7-coordinate

$x_2(0) = -0.05$, $x_3(0) = 0.05$, $x_4(0) = -0.1$, $x_5(0) = 1.09$, $x_6(0) = -0.81$, $x_7(0) = 0.08$, $x_8(0) = 0.21$. Taking the value $t_0 = 2.8/3.8$ means that the control mechanism is applied around the fixed point of the logistic map, and consequently stabilizes the 2-periodic solutions of the generator and the existing replicators. We switch on the control mechanism at the iteration number $i = 25$ for the logistic map, such that for the continuous-time system this moment corresponds to $t = \zeta_{25}$, and switch off at $i = 125$ which corresponds to the time moment $t = \zeta_{125}$. The graphs of the coordinates x_3, x_5 and x_7 are pictured in Fig. 6.14, and it is possible to obtain similar illustrations for the remaining ones, which are not just simulated here. It is observable that the 2-periodic solutions of the replicators and hence of the result-relay-system (6.7.72) are stabilized. In other words, the extended chaos is controlled, and the result of Theorem 2.6 is validated one more time. One can see in Fig. 6.14 that after approximately 60 iterations when the control is switched off, the chaos becomes dominant again and irregular motion reappears.

6.7.5 Miscellany

6.7.5.1 Pyragas Control

A chaotic attractor contains an infinite number of unstable periodic orbits. The control of chaos is the stabilization of one of these orbits, by means of small perturbations applied to the system. One of the important applications of nonlinear oscillators subjected to non-smooth perturbations is the vibro-impact systems, and such systems can exhibit chaotic motions [12, 88–91, 96, 102]. The pioneering paper [53] provides the famous OGY method of the control, and there have been proposed many other ideas to control chaos [120, 145–151]. The parameters of the Duffing equation can be chosen such that it alternatively admits only regular motions or a chaotic attractor [72–75, 81, 133, 143, 152]. In the present section, the Duffing equation is modified with a pulse function such that it admits the period-doubling cascade of chaos. This idea of insertion of chaotic non-smooth elements in such systems to obtain chaos has been realized in [44–46, 67, 83, 84].

One can find that to control chaos of the system (6.7.48), unstable periodic orbits of the logistic equation must be necessarily controlled. There are several other methods to control chaos of the logistic map such as the method proposed by Pyragas [120] and the extended time delayed auto synchronization method [136]. The main idea of the Pyragas method applied to logistic map is the usage of a perturbation in the form of a delay, that is, a perturbation of the form $\gamma(\kappa_{i-j} - \kappa_i)$. Here, the parameter γ represents the strength of the perturbation and the positive integer j is the order of the desired unstable periodic orbit [120, 136].

To show the results of Pyragas method applied to the system (6.7.47) with the coefficients $d_1 = 0.18$, $d_2 = 2$, $d_3 = 0.00004$, $D = 0.02$, $k = 2$, and $m_0 = 2$, $m_1 = 1$, $\mu_\infty = 3.8$, we use the method around the period-1 orbit, that is the fixed point, of the logistic map $h(s, \mu) = \mu s(1 - s)$ and construct the following control system

$$\begin{aligned}
&x_1' = x_2\\
&x_2' = -2x_1 - 0.18x_2 - 0.00004x_1^3 + 0.02cos(2\pi t) + \nu(t, t_1, \mu_\infty)\\
&\zeta_{i+1}(t_1, \mu_\infty) = i + 1 + h(\zeta_i(t_1, \mu_\infty) - i, \mu_\infty) + \gamma(\zeta_{i-1}(t_1, \mu_\infty)\\
&-\zeta_i(t_1, \mu_\infty) + 1).
\end{aligned} \tag{6.7.74}$$

If we simulate a solution of the last system with $t_1 = 0.5$ and $x_1(t_1) = 0.01$, $x_2(t_1) = 0.025$, the result seen in Fig. 6.15 is obtained. It approximates the 2-periodic solution $z_1(t)$ of system (6.7.47). We use the value $\gamma = -0.5$ and the control starts at time $t = \zeta_{30}$ and ends at $t = \zeta_{100}$.

6.7.5.2 High-Frequency Oscillations and Chaos

Now, let us analyze through simulation an interesting question if large own frequency of unperturbed Duffing equation may suppress the chaos appearance in the perturbed system. With this aim, consider the system

$$\begin{aligned}
&x_1' = x_2\\
&x_2' = -50x_1 - 0.18x_2 - 0.00004x_1^3 + 0.02\cos(2\pi t) + \nu(t, t_0, \mu_\infty),
\end{aligned} \tag{6.7.75}$$

which is in the form of system (6.7.47), where $m_0 = 2$, $m_1 = 1$ and $\mu_\infty = 3.8$. The eigenvalues for this system are $-0.09 \mp i\sqrt{50 - 0.09^2}$. Take a solution (x_1, x_2) of the system with initial data $x_1(0.5) = 0.01$ and $x_2(0.5) = 0.025$. One can see

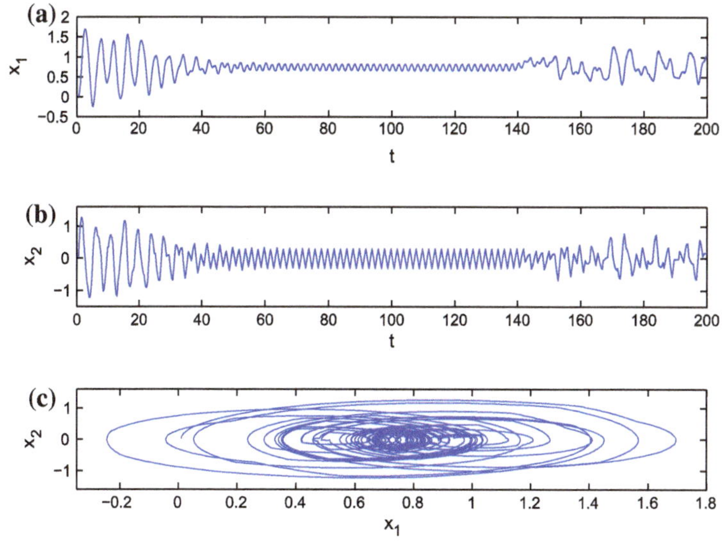

Fig. 6.15 The Pyragas control method applied to the Duffing equation perturbed with a pulse function $x'' + 0.18x' + 2x + 0.00004x^3 - 0.02\cos(2\pi t) = \nu(t, t_0, \mu_\infty)$, where $m_0 = 2, m_1 = 1$ and $\mu_\infty = 3.8$. The control starts at time $t = \zeta_{30}$ and ends at $t = \zeta_{100}$. **a** The graph of the x_1 coordinate. **b** The graph of the x_2 coordinate. **c** The trajectory of the solution $(x_1(t), x_2(t))$

that the frequency is high, but the simulation seen in Fig. 6.16 shows that the chaos appearance is persistent since conditions of our theorems are fulfilled for the system.

We have one more confirmation of our theoretical results. The control environment is sustained for the system (6.7.75) as seen in Fig. 6.17. In this simulation, we take $t_1 = 0.5$, $t_0 = 2.8/3.8$, $\delta = 0.19$, and consider the solution (x_1, x_2) of the control system

$$\begin{aligned} x_1' &= x_2 \\ x_2' &= -50x_1 - 0.18x_2 - 0.00004x_1^3 + 0.02\cos(2\pi t) + \nu(t, t_1, \bar{\mu}_i), \end{aligned} \tag{6.7.76}$$

with the initial condition $x_1(t_1) = 0.01$, $x_2(t_1) = 0.025$. The control starts at time $t = \zeta_{25}$ and ends at $t = \zeta_{60}$.

Figure 6.17 supports our results such that the depicted solution approximates the 2-periodic solution $z_1(t)$ of system (6.7.75). Therefore, one can say that the chaos control results are valid even if the frequency is high. The Pyragas control method can also be used in the case of high frequency.

As the simulation results show, our proposals of generation of chaos and consequently control of it can be extended by the rich diversity of results for discrete maps. Exceptional interest is expected for development of security of communication systems [52, 72, 143, 152]. We suppose also that direct extension of the results can be done on the basis of works, which consider control of chaos generated by the logistic map [153, 154] and uses the map as an instrument of ciphering and

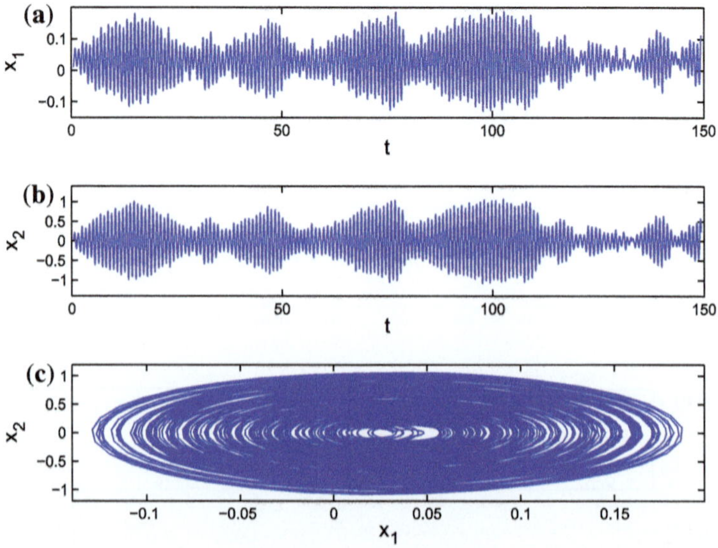

Fig. 6.16 Simulation results of the perturbed Duffing equation $x'' + 0.18x' + 50x + 0.00004x^3 - 0.02\cos(2\pi t) = \nu(t, t_0, \mu_\infty)$, where $m_0 = 2, m_1 = 1$ and $\mu_\infty = 3.8$. **a** The graph of the x_1 coordinate. **b** The graph of the x_2 coordinate. **c** The trajectory of the solution $(x_1(t), x_2(t))$

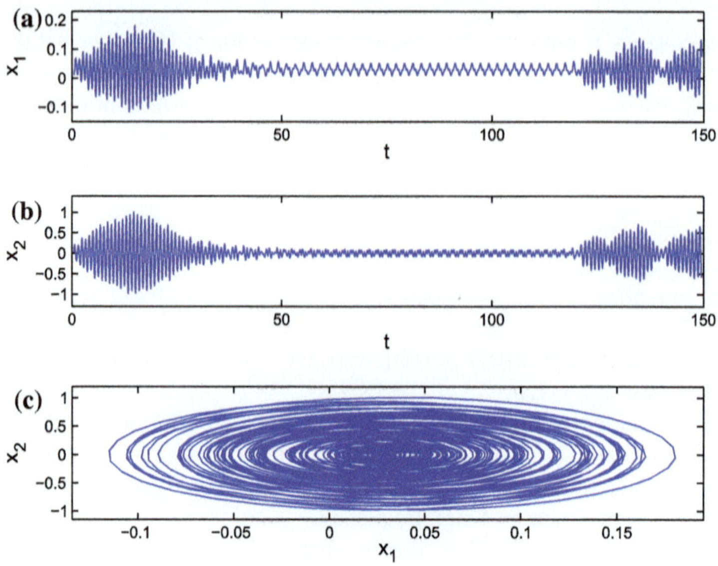

Fig. 6.17 The OGY control method applied to the Duffing equation perturbed with a pulse function $x'' + 0.18x' + 50x + 0.00004x^3 - 0.02\cos(2\pi t) = \nu(t, t_0, \mu_\infty)$, where $m_0 = 2, m_1 = 1$ and $\mu_\infty = 3.8$. The control starts at time $t = \zeta_{25}$ and ends at $t = \zeta_{60}$. **a** The graph of the x_1 coordinate. **b** The graph of the x_2 coordinate. **c** The trajectory of the solution $(x_1(t), x_2(t))$

deciphering [155]. Next extension of investigation can be done by the discussion of low-dimensional discrete dynamics [156, 157].

6.8 Notes

In Sect. 6.1 we developed an approach to form Devaney chaos in a non-autonomous system. The method can be applied to differential equations with a piecewise constant argument of generalized type [16], and to quasilinear systems with a pulse function. One can say that a new method to form the multidimensional chaos is proposed. The existence of a chaotic attractor of the initial value problem is proved. The chaos is observed not only on the attractor, but also in the set of all solutions.The results of Sect. 6.1 were published in the paper [44].

In Sect. 6.2 we defined the features of the analogue of Li–Yorke chaos for the multidimensional discontinuous dynamics. Apparently, the question of "whether theoretical results obtained for systems with low-dimensionality are still applicable for high or infinite dimensional systems" [158] has been partially answered. Taking into account Theorem 6.5 the period-doubling route to chaos can be obtained. The main modeling novelty of Sect. 6.2, which is useful for applications, is that the moments of discontinuity are prescribed by the choice of the initial moment. The present results were published in [46] and they could be effectively used in mechanics, electronics, control theory and economics, and they could be developed further, by using more delicate properties of one-dimensional maps [159].

In Sect. 6.3, a special initial value problem of a differential equation with relay function is addressed. In paper [44], Devaney's ingredients were indicated for a special initial value problem of a relay system with linear elements. In Sect. 6.3, we attempt to shape the Li–Yorke chaos for the multidimensional nonlinear relay system. Li–Yorke chaos is attractive for applications, as it can be developed for a multidimensional case [25]. In Sect. 6.3, another attempt to create higher dimensional Li–Yorke chaotic systems is made.

The existence of a Poisson stable trajectory dense in a quasi-minimal set is proved in Sect. 6.4 for the quasilinear differential equation with a pulse function, whose moments of discontinuity depend on the initial moment. An appropriate simulation of a chaotic attractor is presented. The results of Sects. 6.3 and 6.4 were published in the papers [45, 83]

A hyperbolic set of bounded solutions is constructed in Sect. 6.5, on the basis of a quasilinear impulsive differential equation with a special initial condition. We investigate the structure of the bounded solutions set of a special initial value problem, in which initial moments of time are from a Cantor set. The system was introduced in [28], where chaotic properties were discussed. Particularly, we prove that there exists a chaotic attractor with infinitely many periodic solutions.

The investigation of Sect. 6.6 is inspired by the discontinuous dynamics of the neural information processing in the brain, information communication, and population dynamics [13, 34, 37–39, 41, 160]. While there are many interesting papers concerned with the complex behavior generated by impulses, the rigorous theory of chaotic impulsive systems remains far from being complete. Our goal is to develop further the theoretical foundations of this area of research. The complex dynamics is obtained using Devaney's definition for guidance. The main results of Sect. 6.6 were published in [84], where simulations for a pendulum were provided. Applications of the present approach to the analysis of the cardiovascular system were considered in [33, 161]. More of our results on chaos excitability can be found in [44–46].

We have proved in Sect. 6.7 that the OGY control of the logistic map stabilizes the unstable periodic solutions embedded in the attractor. The exceptional result is that an arbitrary solution of the system (6.7.61) approaches to the controlled periodic solution eventually, if the initial moment is chosen properly. Thus, the way is found, which extends control and chaos of low-dimensional maps to continuous systems with arbitrary large dimension. This method can be useful for the construction and stabilization of mechanical systems and electric circuits with chaotic features. The results of Sect. 6.7 were published in the paper [115].

Concerning the Lyapunov exponents, we must say that the dynamics of the system (6.7.48) consist of the continuous dynamics of the differential equation itself and of the discrete dynamics of the switching moments. That is, one can say that our system is a hybrid system [82, 121, 122]. The important fact is that the Lyapunov exponent of the discrete part of the system is a positive one, since it is the Lyapunov exponent of the logistic map [132].

References

1. L.O. Chua, M. Komuro, T. Matsumoto, The double scroll family, parts I and II. IEEE Trans. Circuit Syst. **CAS-33**, 1072–1118 (1986)
2. M. Hénon, A two-dimensional mapping with a strange attractor. Commun. Math. Phys. **50**(1), 69–77 (1976)
3. T.Y. Li, J.A. Yorke, Period three implies chaos. Am. Math. Mon. **82**, 985–992 (1975)
4. E.N. Lorenz, Deterministic nonperiodic flow. J. Atmos. Sci. **20**, 130–141 (1963)
5. R.M. May, Simple mathematical models with very complicated dynamics. Nature **261**, 459–467 (1976)
6. D. Ruelle, Sensitive dependence on initial condition and turbulent behavior of dynamical systems, in *Bifurcation Theory and Applications in Scientific Disciplines*, ed. by O. Gurel, O.E. Rössler (New York Academy of Sciences, New York, 1979), pp. 408–446
7. S. Smale, Differentiable dynamical systems. Bull. Am. Math. Soc. **73**, 747–817 (1967)
8. R.L. Devaney, *An Introduction to Chaotic Dynamical Systems* (Addison-Wesley, Menlo Park, 1989)
9. N. Minorsky, *Theory of Nonlinear Control Systems* (McGraw-Hill Book Company, New York, 1969)
10. Ya.Z. Tsypkin, *Sampling Systems Theory and Its Application*, vols. 1,2 (The Macmillan Company, New York, 1964)
11. A.A. Andronov, S.E. Chaikin, *Theory of Oscillations* (Princeton University Press, Princeton, 1949)
12. J. Awrejcewicz, C.H. Lamarque, *Bifurcation and Chaos in Nonsmooth Mechanical Systems* (World Scientific Publishing, Singapore, 2003)

13. A.C.J. Luo, *Global Transversality, Resonance and Chaotic Dynamics* (World Scientific, Hackensack, 2008)
14. C. Robinson, *Dynamical Systems: Stability, Symbolic Dynamics, and Chaos* (CRC Press, Boca Raton, 1995)
15. S. Wiggins, *Global Bifurcations and Chaos: Analytical Methods* (Springer, New York, 1988)
16. M.U. Akhmet, On the reduction principle for differential equations with piecewise constant argument of generalized type. J. Math. Anal. Appl. **336**, 646–663 (2007)
17. J. Wiener, *Generalized Solutions of Functional Differential Equations* (World Scientific, Singapore, 1993)
18. S.M. Hammel, J.A. Yorke, C. Grebogi, Do numerical orbits of chaotic dynamical processes represent true orbits? J. Complex. **3**, 136–145 (1987)
19. S.H. Strogatz, *Nonlinear Dynamics and Chaos With Applications to Physics, Biology, Chemistry, and Engineering* (Perseus Books, New York, 1994)
20. E. Akin, S. Kolyada, Li-Yorke sensitivity. Nonlinearity **16**, 1421–1433 (2003)
21. A.N. Sharkovskii, Coexistence of cycles of a continuous map of a line into itself (Russian). Ukr. Mat. Zh. **16**, 61–71 (1964)
22. W. Huang, X. Ye, Devaney's chaos or 2-scattering implies Li-Yorkes chaos. Topol. Appl. **117**, 259–272 (2002)
23. P. Kloeden, Z. Li, Li-Yorke chaos in higher dimensions: a review. J. Differ. Equ. Appl. **12**, 247–269 (2006)
24. P. Li, Z. Li, W.A. Halang, G. Chen, Li-Yorke chaos in a spatiotemporal chaotic system. Chaos Solitons Fractals **33**(2), 335–341 (2007)
25. F.R. Marotto, Snap-back repellers imply chaos in $\mathbb{R}^n$. J. Math. Anal. Appl. **63**, 199–223 (1978)
26. Y. Shi, P. Yu, Chaos induced by regular snap-back repellers. J. Math. Anal. Appl. **337**(2), 1480–1494 (2008)
27. A.M. Samoilenko, N.A. Perestyuk, *Impulsive Differential Equations* (World Scientific, Singapore, 1995)
28. M.U. Akhmet, Perturbations and Hopf bifurcation of the planar discontinuous dynamical system. Nonlinear Anal. **60**, 163–178 (2005)
29. M.U. Akhmet, On the general problem of stability for impulsive differential equations. J. Math. Anal. Appl. **288**, 182–196 (2003)
30. M.U. Akhmetov, N.A. Perestyuk, The comparison method for differential equations with impulse action. Differ. Equ. **26**, 1079–1086 (1990)
31. A. Halanay, D. Wexler, *Qualitative Theory of Impulsive Systems (Romanian)* (Edit. Acad. RPR, Bucuresti, 1968)
32. V. Lakshmikantham, D.D. Bainov, P.S. Simeonov, *Theory of Impulsive Differential Equations* (World Scientific, Singapore, 1989)
33. M.U. Akhmet, G.A. Bekmukhambetova, A prototype compartmental model of the blood pressure distribution. Nonlinear Anal.: RWA **11**, 1249–1257 (2010)
34. L. Glass, M.C. Mackey, A simple model for phase locking of biological oscillators. J. Math. Biol. **7**, 339–352 (1979)
35. F.C. Hoppensteadt, C.S. Peskin, *Mathematics in Medicine and in the Life Sciences* (Springer, New York, 1992)
36. J.P. Keener, F.C. Hoppensteadt, J. Rinzel, Integrate-and-fire models of nerve membrane response to oscillatory input. SIAM J. Appl. Math. **41**, 503–517 (1981)
37. W. Lin, J. Ruan, Chaotic dynamics of an integrate-and-fire circuit with periodic pulse-train input. IEEE Trans. Circuits Syst.-I: Fundam. Theory Appl. **50**, 686–693 (2003)
38. R.E. Mirollo, S.H. Strogatz, Synchronization of pulse-coupled biological oscillators. SIAM J. Appl. Math. **50**, 1645–1662 (1990)
39. T. Yang, L.O. Chua, Impulsive control and synchronization of nonlinear dynamical systems and application to secure communication. Int. J. Bifurc. Chaos **7**, 645–664 (1997)
40. J. Zhen, Z. Ma, M. Han, The existence of periodic solutions of n-species Lotka-Volterra competition systems with impulses. Chaos Solitons Fractals **22**, 181–188 (2004)

41. W. Lin, Description of complex dynamics in a class if impulsive differential equations. Chaos Solutions Fractals **25**, 1007–1017 (2005)
42. A.N. Kolmogorov, On the Skorokhod convergence (Russian). Teor. Veroyatnost. i Primenen. **1**, 239–247 (1956)
43. L.S. Block, W.A. Coppel, *Dynamics in One Dimension* (Springer, Berlin, 1991)
44. M.U. Akhmet, Devaney's chaos of a relay system. Commun. Nonlinear Sci. Numer. Simul. **14**, 1486–1493 (2009)
45. M.U. Akhmet, Dynamical synthesis of quasi-minimal sets. Int. J. Bifurc. Chaos **19**, 2423–2427 (2009)
46. M.U. Akhmet, Li-Yorke chaos in the system with impacts. J. Math. Anal. Appl. **351**, 804–810 (2009)
47. L. Shilnikov, Bifurcations and strange attractors, in *Proceedings of the International Congress of Mathematicians*, vol. III (Higher Education Press, Beijing, 2002), pp. 349–372
48. R. Brown, L. Chua, Dynamical synthesis of Poincaré maps. Int. J. Bifurc. Chaos **3**, 1235–1267 (1993)
49. R. Brown, L. Chua, Chaos or turbulence? Int. J. Bifurc. Chaos **2**, 1005–1009 (1992)
50. P. Atherton, *Nonlinear Control Engineering* (Van Nostrand Reinhold Company, New York, 1982)
51. A.S. Elwakil, Nonautonomous pulse-driven chaotic oscillator based on Chua's circuit. Microelectron. J. **33**, 479–486 (2002)
52. A.L. Fradkov, *Cybernetical Physics: From Control of Chaos to Quantum Control* (Springer, Berlin, 2007)
53. E. Ott, C. Grebogi, J.A. Yorke, Controlling chaos. Phys. Rev. Lett. **64**, 1196–1199 (1990)
54. A. Garfinkel, M.L. Spano, W.L. Ditto, J.N. Weiss, Controlling cardiac chaos. Science **257**, 1230–1233 (1992)
55. A.F. Filippov, *Differential Equations with Discontinuous Right Hand Sides* (Kluwer, Dordrecht, 1988)
56. G.R. Sell, *Topological Dynamics and Ordinary Differential Equations* (Van Nostrand Reinhold Company, New York, 1971)
57. E.A. Barbashin, *Introduction to the Theory of Stability* (Wolters-Noordhoff Publishing, Groningen, 1970)
58. D.V. Anosov, Geodesic flows and closed Riemannian manifolds with negative curvature. Proc. Steklov Inst. Math. **90**, 209 (1967)
59. R. Bowen, ω—limit sets for Axiom A diffeomorfisms. J. Differ. Equ. **18**, 333–339 (1975)
60. M.U. Akhmet, Shadowing property of hybrid systems (in progress)
61. G.D. Birkhoff, *Dynamical Systems* (American Mathematical Society, Providence, 1966)
62. S. Smale, Diffeomorphisms with many periodic points, *Differential and Combinatorial Topology: A Symposium in Honor of Marston Morse* (Princeton University Press, Princeton, 1965), pp. 63–70
63. M. Benedicks, L. Carleson, The dynamics of the Hénon map. Ann. Math. **133**, 73–169 (1991)
64. J. Guckenheimer, R.F. Williams, Structural stability of Lorenz attractors. Publ. Math. **50**, 307–320 (1979)
65. M.V. Jakobson, Absolutely continuous invariant measures for one-parameter families of one-dimensional maps. Commun. Math. Phys. **81**, 39–88 (1981)
66. I. Stewart, The Lorenz attractor exists. Nature **406**, 948–949 (2000)
67. M. Akhmet, *Principles of Discontinuous Dynamical Systems* (Springer, New York, 2010)
68. K. Palmer, *Shadowing in Dynamical Systems: Theory and Applications* (Kluwer Academic Publishers, Dordrecht, 2000)
69. S.Yu. Pilugin, *Shadowing in Dynamical Systems* (Springer, Berlin, 1999)
70. E.M. Coven, I. Kan, J.A. Yorke, Pseudo-orbit shadowing in the family of tent maps. Trans. Am. Math. Soc. **308**, 227–241 (1988)
71. H.E. Nusse, J.A. Yorke, Is every approximate trajectory of some process near an exact trajectory of a nearby process? Commun. Math. Phys. **114**, 363–379 (1988)

72. J.M. Gonzáles-Miranda, *Synchronization and Control of Chaos* (Imperial College Press, London, 2004)
73. Y. Ueda, Randomly transitional phenomena in the system governed by Duffing's equation. J. Stat. Phys. **20**, 181–196 (1979)
74. Y. Ueda, Steady motions exhibited by Duffing's equation: a picture book of regular and chaotic motions, in *New Approaches to Nonlinear Problems in Dynamics*, ed. by P.J. Holmes (SIAM, Philadelphia, 1980)
75. Y. Ueda, Explosion of strange attractors exhibited by Duffing's equation. Ann. N.Y. Acad. Sci. **357**, 422–434 (1980)
76. K.R. Asfar, K.K. Masoud, On the period-doubling bifurcations in the Duffing's oscillator with negative linear stiffness. Trans. ASME J. Vib. Acoust. **114**, 489–494 (1992)
77. V. Brunsden, J. Cortell, P.J. Holmes, Power spectra of chaotic vibrations of a buckled beam. J. Sound Vib. **130**, 1–25 (1989)
78. J.A. Gottwald, L.N. Virgin, E. Dowell, Experimental mimicry of Duffing's equation. J. Sound Vib. **158**, 447–467 (1992)
79. F.C. Moon, Experiments on chaotic motions of a forced nonlinear oscillator: strange attractors. Trans. ASME J. Appl. Mech. **47**, 638–644 (1980)
80. F.C. Moon, P.J. Holmes, A magnetoelastic strange attractor. J. Sound Vib. **65**, 275–296 (1979)
81. J.M.T. Thompson, H.B. Stewart, *Nonlinear Dynamics and Chaos* (Wiley, Chichester, 2002)
82. M. Akhmet, *Nonlinear Hybrid Continuous/Discrete-Time Models* (Atlantis Press, Paris, 2011)
83. M.U. Akhmet, Creating a chaos in a system with relay. Int. J. Qual. Theory Differ. Equ. Appl. **3**, 3–7 (2009)
84. M.U. Akhmet, Shadowing and dynamical synthesis. Int. J. Bifurc. Chaos **19**, 3339–3346 (2009)
85. J. Awrejcewicz, M.M. Holicke, *Smooth and Nonsmooth High Dimensional Chaos and the Melnikov Type Methods* (World Scientific Publishing, Singapore, 2007)
86. T. Kousaka, T. Ueta, H. Kawakami, Controlling chaos in a state-dependent nonlinear system. Int. J. Bifurc. Chaos **12**, 1111–1119 (2002)
87. A. Venkatesan, S. Parthasarathy, M. Lakshmanan, Occurrence of multiple period-doubling bifurcation route to chaos in periodically pulsed chaotic dynamical systems. Chaos Solitons Fractals **18**, 891–898 (2003)
88. G. Tao, F.L. Lewis (eds.), *Adaptive Control of Nonsmooth Dynamic Systems* (Springer, London, 2001)
89. J. Zhou, C. Wen, *Adaptive Backstepping Control of Uncertain Systems, Nonsmooth Nonlinearities, Interactions or Time-Variations* (Springer, Berlin, 2008)
90. H. Cho, E.W. Bai, Convergence results for an adaptive dead zone inverse. Int. J. Adapt. Control Signal Process. **12**, 451–466 (1998)
91. A.J. Kurdila, G. Webb, Compensation for distributed hysteresis operators in active structural systems. J. Guid. Control Dyn. **20**, 1133–1140 (1997)
92. V.I. Babitsky, *Theory of Vibro-Impact Systems and Applications* (Springer, Berlin, 1998)
93. M. di Bernardo, C.J. Budd, A.R. Champneys, P. Kowalczyk, *Piecewise-Smooth Dynamical Systems* (Springer, London, 2008)
94. B. Brogliato, *Impacts in Mechanical Systems-Analysis and Modeling* (Springer, New York, 2000)
95. J. Guckenheimer, P.J. Holmes, *Nonlinear Oscillations, Dynamical Systems and Bifurcations of Vector Fields* (Springer, New York, 1997)
96. R.A. Ibrahim, *Vibro-Impact Dynamics* (Springer, Berlin, 2009)
97. A.E. Kobrinskii, A.A. Kobrinskii, *Vibro-Shock Systems (Russian)* (Nauka, Moscow, 1971)
98. R.F. Nagaev, *Mechanical Processes with Repeated and Decaying Impacts (Russian)* (Nauka, Moscow, 1985)
99. A.B. Nordmark, Existence of periodic orbits in grazing bifurcations of impacting mechanical oscillators. Nonlinearity **14**, 1517–1542 (2001)
100. H.E. Nusse, E. Ott, J.A. Yorke, Border-collision bifurcations: an explanation for observed bifurcation phenomena. Phys. Rev. E **49**, 1073–1076 (1994)

101. F. Peterka, Part I: Theoretical analysis of n-multiple $(1/n)$-impact solutions. CSAV Acta Technica **26**, 462–473 (1974)
102. A.F. Vakakis, L.I. Manevitch, Y.V. Mikhlin, V.N. Plipchuk, A.A. Zevin, *Normal Modes and Localization in Nonlinear Systems* (Wiley, New York, 1996)
103. L.A. Wood, K.P. Byrne, Analysis of a random repeated impact process. J. Sound Vib. **82**, 329–345 (1981)
104. V.F. Zhuravlev, A method for analyzing vibration-impact systems by means of special functions. Mech. Solids **11**, 23–27 (1976)
105. J.Y. Lee, J.J. Yan, Control of impact oscillator. Chaos Solitons Fractals **28**(1), 136–142 (2006)
106. J.Y. Lee, J.J. Yan, Position control of double-side impact oscillator. Mech. Syst. Signal Process. **21**(2), 1076–1083 (2007)
107. J.R. Kalagnanam, Controlling chaos, the example of an impact oscillator. ASME J. Dyn. Syst. Meas. Control **116**, 557–564 (1994)
108. S.W. Shaw, A.G. Haddow, S.R. Hsieh, Properties of cross-well chaos in an impacting systems. Philos. Trans. R. Soc. Lond. A **347**, 391–410 (1994)
109. M.R. Joglekar, E. Sander, J.A. Yorke, Fixed points indices and period-doubling cascades. J. Fixed Point Theory Appl. **8**, 151–176 (2010)
110. E. Sander, J.A. Yorke, Period-doubling cascades for large perturbations of Hénon families. J. Fixed Point Theory Appl. **6**, 153–163 (2009)
111. M. Levi, *Qualitative Analysis of the Periodically Forced Relaxation Oscillations* (Memoirs of the American Mathematical Society, Providence, 1981)
112. C. Grebogi, J.A. Yorke, *The Impact of Chaos on Science and Society* (United Nations University Press, Tokyo, 1997)
113. E. Sander, J.A. Yorke, Connecting period-doubling cascades to chaos. Int. J. Bifurc. Chaos **22**, 1–16 (2012)
114. M.U. Akhmet, Homoclinical structure of the chaotic attractor. Commun. Nonlinear Sci. Numer. Simul. **15**, 819–822 (2010)
115. M.U. Akhmet, M.O. Fen, Chaotic period-doubling and OGY control for the forced Duffing equation. Commun. Nonlinear Sci. Numer. Simul. **17**, 1929–1946 (2012)
116. K. Thamilmaran, M. Lakshmanan, Rich variety of bifurcations and chaos in a variant of Murali-Lakshmanan-Chua circuit. Int. J. Bifurc. Chaos **10**, 1781–1785 (2000)
117. T. Kapitaniak, *Controlling Chaos: Theoretical and Practical Methods in Non-linear Dynamics* (Butler and Tanner Ltd., Frome, 1996)
118. T. Kapitaniak, Controlling chaotic oscillators without feedback. Chaos Solitons Fractals **2**, 519–527 (1992)
119. U. Dressler, G. Nitsche, Controlling chaos using time delay coordinates. Phys. Rev. Lett. **68**, 1–4 (1992)
120. K. Pyragas, Continuous control of chaos by self-controlling feedback. Phys. Rev. A **170**, 421–428 (1992)
121. A.V. Savkin, R.J. Evans, *Hybrid Dynamical Systems: Controller and Sensor Switching Problems* (Birkhäuser, Boston, 2002)
122. A.J. van der Schaft, J.M. Schumacher, *An Introduction to Hybrid Dynamical Systems* (Springer, London, 2000)
123. A. Khadra, X. Liu, X. Shen, Application of impulsive synchronization to communication security. IEEE Trans. Circuits Syst.-I, Fundam. Theory Appl. **50**, 341–351 (2003)
124. G. Kolumban, P. Kennedy, L.O. Chua, The role of synchronization in digital communications using chaos-part II: chaotic modulation and chaotic synchronization. IEEE Trans. Circuit Syst. **45**, 1129–1140 (1998)
125. L.M. Pecora, T.L. Carroll, Synchronization in chaotic systems. Phys. Rev. Lett. **64**, 821–825 (1990)
126. C. Tresser, P.A. Worfolk, H. Bass, Master-slave synchronization from the point of view of global dynamics. Chaos **5**, 693–699 (1995)
127. N.T. Crook, C.H. Dobbyn, T. olde Scheper, Chaos as a desirable stable state of artificial neural networks, *Advances in Soft Computing: Soft Computing Techniques and Applications* (Physica-Verlag, Heidelberg, 2000), pp. 52–60

128. N. Crook, T. olde Scheper, A novel chaotic neural network architecture, in *ESANN'2001 Proceedings—European Symposium on Artificial Neural Networks Bruges (Belgium), D-Facto Public* (2001), pp. 295–300
129. C. Lourenco, A. Babloyantz, Control of spatiotemporal chaos in neural networks. Int. J. Neural Syst. **7**, 507–517 (1996)
130. I. Tsuda, A new type of self-organization associated with chaotic dynamics in neural networks. Int. J. Neural Syst. **7**, 451–459 (1996)
131. D. Gulick, *Encounters with Chaos* (University of Maryland, College Park, 1992)
132. H.G. Schuster, W. Just, *Deterministic Chaos, An Introduction* (Wiley-VCH, Federal Republic of Germany, 2005)
133. E. Ott, *Chaos in Dynamical Systems* (Cambridge University Press, New York, 1993)
134. J.K. Hale, *Ordinary Differential Equations* (Krieger Publishing Company, Malabar, 1980)
135. K.T. Alligood, T.D. Sauer, J.A. Yorke, *Chaos: An Introduction to Dynamical Systems* (Springer, New York, 1996)
136. I. Zelinka, S. Celikovsky, H. Richter, G. Chen (eds.), *Evolutionary Algorithms and Chaotic Systems* (Springer, Berlin, 2010)
137. R.A. Horn, C.R. Johnson, *Matrix Analysis* (Cambridge University Press, Cambridge, 1992)
138. E. Ott, T. Sauer, J.A. Yorke, *Coping with Chaos: Analysis of Chaotic Data and the Exploitation of Chaotic Systems* (Wiley, New York, 1994)
139. K. Ramasubramanian, M.S. Sriram, A comparative study of computation of Lyapunov spectra with different algorithms. Phys. D **139**, 72–86 (2000)
140. J.C. Sprott, *Chaos and Time-Series Analysis* (Oxford University Press, New York, 2003)
141. Y.V. Andreyev, A.S. Dmitriev, E.V. Efremova, Dynamic separation of chaotic signals in the presence of noise. Phys. Rev. E **65**, 046220 (2002)
142. H.G. Schuster, *Handbook of Chaos Control* (Wiley-VCH, Weinheim, 1999)
143. E. Schöll, H.G. Schuster, *Handbook of Chaos Control* (Wiley-VCH, Weinheim, 2008)
144. M.J. Feigenbaum, Universal behavior in nonlinear systems, Los Alamos Sci./Summer 4–27 (1980)
145. F.H. Abed, H.O. Wang, R.C. Chen, Stabilization of period doubling bifurcations and implications for control of chaos. Phys. D **70**, 154–164 (1994)
146. Y. Braiman, I. Goldhirsch, Taming chaotic dynamics with weak periodic perturbations. Phys. Rev. Lett. **66**, 2545–2548 (1991)
147. G. Chen, X. Yu, On time delayed feedback control of chaos. IEEE Trans. Circuits Syst.-I **46**, 767–772 (1999)
148. R. Lima, M. Pettini, Suppression of chaos by resonant parametric perturbations. Phys. Rev. A **41**, 726–733 (1990)
149. J.E.S. Socolar, D.W. Sukow, D.J. Gauthier, Stabilizing unstable periodic orbits in fast dynamical systems. Phys. Rev. E **50**, 3245–3248 (1994)
150. M.S. Vieira, A.J. Lichtenberg, Controlling chaos using nonlinear feedback with delay. Phys. Rev. E **54**, 1200–1207 (1996)
151. L. Yang, Z. Liu, J. Mao, Controlling hyperchaos. Phys. Rev. Lett. **84**, 67–70 (2000)
152. A. Pikovsky, M. Rosenblum, J. Kurths, *Synchronization: A Universal Concept in Nonlinear Sciences* (Cambridge University Press, New York, 2001)
153. J. McGuire, M.T. Batchelor, B. Davies, Linear and optimal non-linear control of one-dimensional maps. Phys. Lett. A **233**, 361–364 (1997)
154. P. Melby, J. Kaidel, N. Weber, A. Hübler, Adaptation to the edge of chaos in the self-adjusting logistic map. Phys. Rev. Lett. **84**, 5991–5993 (2000)
155. M.S. Baptista, Cryptography with chaos. Phys. Lett. A **240**, 50–54 (1998)
156. N.J. Corron, An exactly solvable chaotic differential equation. Dyn. Contin. Discret. Impuls. Syst. Ser. A: Math. Anal. **16**, 777–788 (2009)
157. W. Melo, S. Strien, *One-Dimensional Dynamics* (Springer, Berlin, 1993)
158. M. Ding, C. Grebogi, J.A. Yorke, *Chaotic Dynamics, The Impact of Chaos on Science and Society, 1991* (United Nations University Press, Tokyo, 1997), pp. 1–17

159. P. Collet, J.P. Eckmann, *Iterated Maps on the Interval as Dynamical Systems* (Birkhäuser, Basel, 1980)
160. A. Khadra, X. Liu, X. Shen, Impulsive control and synchronization of spatiotemporal chaos. Chaos Solitons Fractals **26**, 615–636 (2005)
161. M.U. Akhmet, The complex dynamics of the cardiovascular system. Nonlinear Anal.: Theory Methods Appl. **71**, e1922–e1931 (2009)

Chapter 7
Economic Models with Exogenous Continuous/Discrete Shocks

7.1 Chaos in Economic Models with Equilibria

In this section, we investigate the generation of chaos in economic models with equilibria through exogenous shocks. The perturbation is formulated as a pulse function where either values or instants of discontinuity are chaotically behaved. We provide a rigorous proof of the existence of chaos in the perturbed model. The analytical results are applied to Kaldor–Kalecki-type models of the aggregate economy. Simulations are used to demonstrate the emergence and the control of chaos. Our results shed light on a novel source of chaos in economic models and have important implications for policy-making.

7.1.1 Introduction

Irregularity is an inherent feature of economic reality. Regularity, as reflected in a constant solution of a model or a periodic and even almost periodic motion in mathematical sense, is a good assumption in engineering and natural science applications, but less so in economic models. This was pointed out in early scientific work and has been widely discussed in recent years [1–7]. One way of introducing irregularity in economics is by allowing for stochastic processes. A different approach is generating chaos in deterministic differential equations.[1] The main property of chaos is *sensitivity*, which can be interpreted as unpredictability in real-world problems. This is also known as the *butterfly effect* [9]. Devaney [10] proposed that *sensitivity* in conjunction with other properties, namely *transitivity* and *density of periodic solutions*,

[1]There exists a third approach, which is somewhere in between the two, where Iterated Function Systems generated by the optimal policy functions for a class of stochastic growth models converge to invariant distributions with support over fractal sets [8].

M. Akhmet and M.O. Fen, *Replication of Chaos in Neural Networks, Economics and Physics*, Nonlinear Physical Science,
DOI 10.1007/978-3-662-47500-3_7

be considered as ingredients of chaos. Another popular way to prove theoretically the presence of chaos is by observing the *period-doubling cascade* [11].

The main theoretical contribution of the present section lies in demonstrating that *exogenous chaotic perturbations* can produce irregular motions in economic models. In mathematical terms, we augment the right-hand side of otherwise regular differential equations with chaotic terms and verify the intuitive idea that the resulting models admit chaotic solutions. Previous works have considered the "endogenous" appearance of chaos in economic models, where the presence of chaos hinges on some crucial parameters (e.g., [12, 13], and papers cited there). The principal novelty of our investigation is that we create an exogenous chaotic perturbation, plug it in a regular dynamical system, and find that similar chaos is inherited by the solutions of the new system. Such an approach has been widely used for differential equations before, but for regular disturbance functions. That is, it has been shown that an (almost) periodic perturbation function implies the existence of an (almost) periodic solution of the system. While the literature on chaos synchronization has also produced methods of generating chaos in a system by plugging in special terms that are chaotic, it relies on the asymptotic convergence between the chaotic exogenous terms and the solutions of the system [14, 15]. Instead, we provide a direct verification of the ingredients of chaos for the perturbed system. Currently, we study cases where the shocks enter the system additively, but future investigations may involve more complicated forms, where the disturbance enters as an argument of the main functions.

One can think of two types of shocks exogenous to a given economic system, say a macroeconomic model of a country. Shocks of the first type are generated by global forces that are either completely outside of human control (for example, weather phenomena) or are shaped in some worldwide marketplace (for example, commodity prices which are determined in the world markets). Zhou et al. [16] demonstrate that the flood series in the Huaihe river basin in China over the last 500 years exhibits chaotic dynamics. Decoster et al. [17] found evidence of chaotic motion in daily silver, copper, sugar, and coffee futures prices. Wei and Leuthold [18] show that futures prices of corn, soybeans, wheat, hogs, and coffee are chaotic processes. Panas and Ninni [19] provide strong support to the presence of chaos in daily oil product prices in the Rotterdam and Mediterranean petroleum markets. These works employ tests developed by Brock [20] and Brock et al. [21], among others, that aim to distinguish between random and chaotic deterministic series. While it is in general very difficult to do so, especially for high-dimensional systems and for short economic time series [22, 23], this only implies that just as there is as yet no definite proof of the chaotic nature of economic variables, there is no definite proof of their random nature, either. Moreover, it is plausible that a hybrid of the two types of processes generates some economic data.

The second type of exogenous shocks that could affect a given economic system is shocks generated outside the system but endogenous to some other system that is linked with the former through financial, trade, and information flows. In this case we can talk of the transmission of chaos from one economy to another. Multiple papers investigating the emergence of endogenous chaos in economic models have been produced. Many of them study Kaldor–Kaleckian or Keynesian models of the macroeconomy, as in [12, 13, 24], where real output is determined along with other economic variables, such as capital stock or money supply. Suppose real output in a foreign economy affects the level of demand by this economy for the exports of the home country, and exports to the foreign economy influence the economic activity at home. Then exports to the foreign country may be viewed as an exogenous shock to the home economic system. The present section points out that if real output abroad is chaotic, then the variables at home will be chaotic, as well, that is, chaos is transmitted through chaotic export shocks from the foreign to the home economy. Lorenz [25] produces chaos in a system consisting of three similar economies linked through international trade, a six-dimensional system altogether. However, his goal is to show that multidimensional systems of the kind that generate chaos are plausible in economics, rather than to study the transmission of chaos internationally.

There is also a literature that studies the emergence of endogenous chaos in economic models with microfoundations, such as standard models of overlapping generations and models with infinitely lived representative agents [2, 26–30]. Benhabib and Day [27] give several examples of utility functions that generate chaotic consumption trajectories in a standard, deterministic, overlapping generations model. Among others, they derive a logistic map as the optimal consumption function. Boldrin and Montrucchio [2] and Deneckere and Pelikan [29] show that in dynamic optimisation problems satisfying the standard continuity and convexity assumptions, the optimal policy function can be chaotic. In these investigations, the discount factor plays an important role. Nishimura et al. [31] and Nishimura and Yano [32] show that chaotic optimal solutions can be obtained in these models even for a discount factor arbitrarily close to 1. Mitra and Sorger [33] prove that the logistic map can be the optimal policy function of a regular dynamic optimisation problem if and only if the discount factor does not exceed $1/16$. We rely on the results of Benhabib and Day [27] and Mitra and Sorger [33], among others, to motivate our use of the logistic map in what follows.

An implication of our results is that detecting the source of chaos in an economic system is crucial for effective control of said chaos. The complex nature of economic systems implied by the presence of chaos may suggest that the evolution of economic variables is not only unpredictable, but also uncontrollable. To borrow a citation from Mendes and Mendes [34], the common view until early 1990 s was that "A chaotic motion is generally neither predictable, nor controllable. It is unpredictable because a small disturbance will produce exponentially growing perturbation of the motion. Is it uncontrollable because small disturbances lead only to other chaotic motions and not to any stable and predictable alternative" [35]. As a corollary, it may seem that "any improvement in the functioning of these economies would require a radical change to their basic structures, because the crises and booms associated with the dynamics of capitalist structures, by being chaotic manifestations, can be neither controllable nor predictable" [34].

However, developments in the study of chaos since early 1990s have provided theoretical tools to effectively control chaos [36–42]. These methods rely on the sensitivity of chaotic systems to small changes, by fine-tuning the parameters of the system to nudge the dynamics toward a desired trajectory. "In the case of chaotic systems, as these are sensitive to very small changes in the parameters, a small butterfly effect in one of them is (in most cases) all that is required to control their outcome, without changing the very nature of the controlled system in any relevant way," while "conventional classical control techniques control the dynamics of nonlinear processes through the use of brute force, having in fact frequently to change the nature of the very system that is subject to control" [34]. As a result, the cost of these control instruments is likely to be small, as well.

While the application of chaos control methods to real-world economic policy-making remains an open question, numerous papers have demonstrated the potential implementation of these techniques in various economic settings. Holyst et al. [43], Holyst and Urbanowicz [44], Ahmed and Hassan [45], Salarieh and Alasty [46], and Chen and Chen [47] control chaos in microeconomic models of firm competition, such as Cournot duopoly/oligopoly and Behrens–Feichtinger model of two competing firms [48, 49]. Kaas [50] and Bala et al. [51] implement chaos control in macroeconomic disequilibrium models, and Kopel [52] does so in a disequilibrium model of firms with bounded rationality. Haag et al. [53] stabilize a chaotic urban system, Mendes and Mendes [34] control chaos in an overlapping generations model (OLG), and Wieland and Westerhoff [54] demonstrate the possible control of chaotic exchange rate dynamics by a central bank. In all these applications control is carried out by varying the values of parameters that have a clear economic interpretation and that can be plausibly set at will by either the government or private actors, such as firms. For example, in [50] the government varies income tax rates or government expenditures to stabilize an unstable Walrasian equilibrium, in [54] the central bank intervenes in the foreign exchange market by varying the value of the foreign exchange buy orders, and in [46] chaos can be controlled either through government production tax/subsidy imposed on firms or through firms' adjustment of their production quantities.

The literature on the control of chaos originated with Ott et al. [39]. Their method (commonly known as the OGY method) relies on the observation that a chaotic set contains an infinite number of unstable periodic orbits. One can select the most desirable unstable periodic orbit, wait until the system approaches it sufficiently and apply a slight nudge to an appropriate parameter to keep the system on that orbit. Notice that the controller can choose which orbit out of infinitely many orbits to target. Particularly, the policy-makers may pick a trajectory that delivers the highest welfare, based on the preferences for levels and volatility of the variables of interest. The implementation of the method requires an observation of a slice of the chaotic attractor (called the Poincaré section). This can be done for most economic variables, data on which is collected by governments and other agencies. Finally, Ott et al. [39] show that their approach is effective if a random noise is introduced into the system, as long as the noise variable assumes extreme values very infrequently, i.e., is sufficiently bounded. This is very convenient for the hybrid case of both deterministic chaos and random shocks present in a model.

We will focus on the OGY method due to the advantages mentioned above. We argue that correctly identifying the source of chaos in an economic system can have a significant impact on the implementation of the OGY method. The construction of the

Poincaré map, an essential step, demands the knowledge of the solutions in analytical form and this is an unsolvable problem in many cases, since chaotic dynamics are nonlinear. Therefore, it is extremely convenient to be able to isolate and apply the OGY control directly to an exogenous shock that is driving the chaotic dynamics in a system and whose Poincaré map can be constructed (either as an analytical solution to a differential equation or through empirical analysis). This would be less resource-consuming, since it would involve modifying fewer parameters, and in some cases could be the only feasible solution.

Moreover, our findings emphasize the cost-effectiveness and importance of international cooperation in economic policy. In the case of the first type of exogenous shocks mentioned above, such as commodity prices (oil, gold, silver, etc.), whose values are determined in a global marketplace, cooperation between the major players in the market could allow to control the chaotic dynamics of these variables and would translate into control of chaos in all economies affected. For the second type of shocks, it is plausible that controlling chaos in one economy can be done most effectively with the cooperation of another country that is the source of exogenous chaotic shocks to the home economy. In the extreme case, controlling chaos in one economy can help control chaos in another economy, which in turn helps control chaos in a third country, and so on and so forth. Properly identifying the source of global chaos in one economy and controlling chaos there would then be the most effective way to control chaos worldwide and turns out to be very cheap relative to the scale of the ultimate effect.

The rest of the section is organized as follows. In Sect. 7.1.2 we discuss the particular class of exogenous disturbance that we consider, i.e., exogenous perturbations that take the form of a pulse function. The mathematical investigation of the perturbed system is presented in Sect. 7.1.3. Simulations of a Kaldor–Kalecki model with pulsative disturbances of two types—with chaotically behaved values and chaotically behaved discontinuity instants—are shown in Sect. 7.1.4. We also demonstrate the application of the OGY control method to these models.

7.1.2 Modeling the Exogenous Shock

Theoretically, it is clear that we need a chaotic function to model the disturbance, but in practice there is not a ready supply of such functions. For this reason, we have to either use solutions of differential equations that are known for their chaotic properties, or create functions with chaotic elements. In this section, we employ the latter approach. In future work, we will consider how one can use actual economic time series, such as those of commodity prices, that have been tested for deterministic chaos [16–20, 55, 56], as exogenous shocks.

We investigate exogenous perturbations that take the form of a pulse function. Consider a strictly increasing sequence of real numbers $\{\theta_i\}$ such that $|\theta_i| \to \infty$ as $|i| \to \infty$. We say that a function $p(t) : \mathbb{R} \to \mathbb{R}^n$ is a *pulse function* if for each integer i there is $p_i \in \mathbb{R}^n$ such that $p(t) = p_i$ either on the interval $(\theta_i, \theta_{i+1}]$ or on the interval $[\theta_i, \theta_{i+1})$.

Consider a general form of economic models,

$$\dot{v} = H(v), \tag{7.1.1}$$

where $v : \mathbb{R} \to \mathbb{R}^n$ is a function of time, $v(t)$, and $H : \mathbb{R}^n \to \mathbb{R}^n$ is continuously differentiable in its arguments.

Perturb the model chaotically (this will be explained later) with a pulse function $\widetilde{d}_{[\frac{t}{h}]}$, where h is a fixed positive real number, $[s]$ denotes the largest integer that is not greater than s, so that $\widetilde{d}_{[\frac{t}{h}]} = \widetilde{d}_i \in \mathbb{R}^n$, if $ih \leq t < (i+1)h$, i is an integer. We obtain the following model:

$$\dot{v} = H(v) + \widetilde{d}_{[\frac{t}{h}]}. \tag{7.1.2}$$

Assume that the pulse function has only one nonzero coordinate, that is only one equation in system (7.1.1) is chaotically perturbed. This assumption will be justified in Sect. 7.1.3. This is a specific case only, and the more general case can be investigated in a similar manner. Suppose that

$$\widetilde{d}_i = (g(d_i), 0, 0, \ldots, 0) \tag{7.1.3}$$

for all integers i, where $g : \mathbb{R} \to \mathbb{R}$ is a continuous function.

We construct the values of the disturbance using a solution of a discrete equation:

$$d_i = F\left(d_{i-1}\right), \tag{7.1.4}$$

where the function $F : \mathbb{R} \to \mathbb{R}$ is continuous, and (7.1.4) generates chaotic exogenous shocks. Two definitions of chaos are used: Devaney chaos and chaos through period-doubling cascade.

While it is an intuitive conclusion, one has to verify rigorously whether system (7.1.2) admits chaos. This is the objective of the present section. One of the most convenient ways of analysis in dynamics is to consider a problem near an equilibrium. So, assume that (7.1.1) admits a steady state at $v = v^*$. Transform the state variables $x = v - v^*$ in (7.1.2). Then, near the equilibrium point the linearized model takes the form

$$\begin{aligned} \dot{x} &= Ax + f(x) + \widetilde{d}_{[\frac{t}{h}]}, \\ d_{[\frac{t}{h}]} &= F\left(d_{[\frac{t}{h}]-1}\right), \end{aligned} \tag{7.1.5}$$

where $t \in \mathbb{R}$, $x \in \mathbb{R}^n$, $f(0) = 0$. Assume that A is a matrix, all of whose eigenvalues have negative real parts, and $f : \mathbb{R}^n \to \mathbb{R}^n$ is a function continuously differentiable in its arguments.

We call the type of disturbances just described as pulsative perturbations with chaotically behaved values and prove that the perturbed system exhibits chaos. This is the first time that such disturbances are introduced in the chaos literature. There are many applications for shocks with chaotically behaved values in economics. Consider economic time series such as commodity prices, productivity indices, and international trade indicators, all of which are examples of exogenous shocks in some economic models. These are usually gauged by economists at regular discrete intervals, no matter how disaggregated (year, month, day, minute, second), and it is their value that is unpredictable and irregular. Another interpretation is that some variables truly change values only at fixed times, for example, the government budget that is determined once a year, earnings of a farm that sells its produce in accordance with the seasons, a firm's capital equipment that changes with periodical investment. All of these quantities vary at regular instants of time, but their values may be irregular. Thus, pulsative perturbations with chaotically behaved values are a good approximation of reality. In this section, we provide rigorous verification of the presence of chaos in system (7.1.5).

To model F, one can consider the logistic map of P.-F. Verhulst [57],

$$F(d_{i-1}) = \mu d_{i-1}(1 - d_{i-1}). \tag{7.1.6}$$

It is known that if $0 < \mu \leq 4$, then the unit interval $[0, 1]$ is invariant under the iterations of the map, and there are values of the parameter μ such that the map is chaotic. The logistic map plays a very important role in many fields of science, and economics is not an exception. Good examples of the applications of the logistic map and its historical background are provided in [58].

Bala et al. [51] show that for particular forms of the utility functions in a simple discrete-time model of an exchange economy with two goods under Walrasian tatonnement, the evolution of the price of the non-numeraire good is described with a logistic map. This result can be used to model commodity prices, such as prices of oil, gold, silver, etc., using a logistic map. Benhabib and Day [27] obtain a logistic map as the law of motion of consumption in a simple overlapping generations model with quadratic utility function, and Mitra and Sorger [33] verify that the logistic map can be the optimal policy function of a regular dynamic optimization problem, i.e., one satisfying some regularity assumptions, if and only if the discount factor does not exceed $1/16$. We use these results to motivate our use of the logistic map to model the export shock in the Kaldor–Kalecki model of the aggregate economy in Sect. 7.1.4, where exports to a foreign country are a function of consumption levels there. Since consumption in the foreign country can be thought of as the solution to a regular dynamic optimization problem of foreign consumers, we describe it with a logistic map. Of course, the logistic map is only an illustrative example of a wide range of chaotic dynamics that exogenous shocks can follow. Modeling the shocks in any other way in our simulations would not alter the main message of the present section.

An alternative way to generate a pulsative chaotic perturbation is considering a pulse function with chaotically behaved discontinuity instants, the θ_i in the definition of the pulse function above (as opposed to chaotically behaved values):

$$\nu(t,d) = \begin{cases} m_0, & \text{if } \theta_{2i}(d) < t \le \theta_{2i+1}(d) \\ m_1, & \text{if } \theta_{2i-1}(d) < t \le \theta_{2i}(d), \end{cases} \tag{7.1.7}$$

where i is an integer and m_0, m_1 are real numbers such that $m_0 \neq m_1$. The sequence $\{\theta_i(d)\}$, which defines the discontinuity instants of the function $\nu(t,d)$, is introduced through the equation $\theta_i(d) = i + d_i$, where $d = \{d_i\}$ is a solution of Eq. (7.1.4). Examples of this type of shocks are natural disasters and extreme events in general, such as market crashes. They take a finite number of values (an earthquake either happens or not), but their timing is irregular.

The original system (in its linearized form) then becomes

$$\dot{x} = Ax + f(x) + \nu(t,d). \tag{7.1.8}$$

The theory of the systems of the form (7.1.8) is described in [59, 60]. In this work, we will present simulations of chaos in a Kaldor–Kalecki model subjected to such shocks. We will also demonstrate the application of the OGY control method for both types of shocks described.

7.1.3 Mathematical Investigation of System (7.1.5)

In this subsection, we study differential equations perturbed by a pulse function with chaotically behaved values. We first give a complete description of the perturbation, and then consider the space of all bounded solutions of the system.

We shall make use of the uniform norm $\|\Gamma\| = \sup\limits_{\|v\|=1} \|\Gamma v\|$ for any matrix Γ.

Since all eigenvalues of the constant $n \times n$ real-valued matrix A have negative real parts, one can verify the existence of positive real numbers N and ω such that the inequality $\left\|e^{At}\right\| \le Ne^{-\omega t}$ is valid for all $t \ge 0$.

The following four assumptions are needed throughout this section:

(C1) There exist positive real numbers M_f and M_g such that $\sup\limits_{x\in\mathbb{R}^n} \|f(x)\| = M_f$, $\sup\limits_{s\in\mathbb{R}} |g(s)| = M_g$;

(C2) There exists a positive real number L_f such that the inequality

$$\|f(x_1) - f(x_2)\| \le L_f \|x_1 - x_2\|$$

holds for all $x_1, x_2 \in \mathbb{R}^n$;

(C3) There exist positive real numbers L_1 and L_2 such that the inequality $L_1 |s_1 - s_2| \le |g(s_1) - g(s_2)| \le L_2 |s_1 - s_2|$ holds for all $s_1, s_2 \in \mathbb{R}$;

(C4) $NL_f - \omega < 0$.

Condition $(C4)$ ensures that system (7.1.2) is weakly nonlinear. We assume that Eq. (7.1.4) admits a set of bounded solutions, defined for all integers. More precisely, assume that there exists a bounded set Λ of real numbers such that the values of bounded solutions are in this set. Notice that in the case of the logistic map (7.1.6), with $0 < \mu \leq 4$, the set Λ can be taken as the unit interval $[0, 1]$. We shall denote by $\mathscr{D}$ the set of all bounded solutions.

To solve system (7.1.5), one has to solve the discrete equation (7.1.4), given initial value d_0, obtain a sequence $\{d_i\}$ as a solution, build a function $d_{[\frac{t}{h}]} = d_i$, if $t \in [ih, (i+1)h)$, and substitute this function in (7.1.5). The resulting system is

$$\dot{x} = Ax + f(x) + \widetilde{d}_{[\frac{t}{h}]}, \tag{7.1.9}$$

where $\widetilde{d}_{[\frac{t}{h}]} = \left(g(d_{[\frac{t}{h}]}), 0, 0, \ldots, 0\right) \in \mathbb{R}^n$. If the Lipschitz constant L_f is sufficiently small so that condition (C4) is satisfied, then for a given $d \in \mathscr{D}$ this system admits a unique, bounded on the entire real axis, solution, denoted by $\phi_d(t)$ [61]. Let us denote by X the set of such solutions for all possible $d \in \mathscr{D}$. One can show that $\phi_d(t)$ satisfies the relation [62]

$$\phi_d(t) = \int_{-\infty}^{t} e^{A(t-s)} \left(f(\phi_d(s)) + \widetilde{d}_{[\frac{s}{h}]}\right) ds. \tag{7.1.10}$$

Let us denote $M = M_f + M_g$, where the numbers M_f and M_g are discussed in condition $(C1)$. For any $x(t) \in X$, we have $\sup_{t \in \mathbb{R}} \|x(t)\| \leq H_0$, where $H_0 = \frac{NM}{\omega}$. That is, all bounded solutions of system (7.1.9) lie in a tube with radius H_0.

In what follows, for fixed $d \in \mathscr{D}$, the function $x_d(t, x_0)$, $x_0 \in \mathbb{R}^n$, will stand for the unique solution of system (7.1.9) with the initial condition $x_d(0, x_0) = x_0$. Notice that this solution is not necessarily bounded.

We say that a sequence $\{d_i\} \in \mathscr{D}$ is p-periodic if there exists a natural number p such that $d_{i+p} = d_i$ for each integer i. Suppose that system (7.1.4) admits infinitely many periodic solutions, and let us denote the set of all such solutions by $\mathscr{P}$, which is a subset of $\mathscr{D}$.

By applying the standard technique [61], common for quasilinear ordinary differential equations, one can prove the following two assertions. We omit their verification.

Lemma 7.1 *For every $d \in \mathscr{D}$ and $x_0 \in \mathbb{R}^n$, the inequality $\|x_d(t, x_0) - \phi_d(t)\| \leq N \|x_0 - \phi_d(0)\| e^{(NL_f - \omega)t}$ holds for all $t \geq 0$.*

Using the last lemma together with condition $(C3)$, one can show that for every $d \in \mathscr{D}$ and any $x_0 \in \mathbb{R}^n$, $\|x_d(t, x_0) - \phi_d(t)\| \to 0$ as $t \to \infty$, and consequently $x_d(t, x_0)$ eventually enters the tube with radius H_0.

The proof of the next lemma uses representation (7.1.10).

Lemma 7.2 *Suppose that p is a natural number. If $d \in \mathscr{P}$ is a p-periodic sequence, then the solution $\phi_d(t)$ of system (7.1.9) is ph-periodic, and vice versa.*

Now, we demonstrate the chaotic properties of Eq. (7.1.4). We use two exact mathematical descriptions of chaos: Devaney chaos and chaos through period-doubling cascade. The former is the most theoretical known type of chaos, and the latter is convenient for simulations.

The following are the ingredients of chaos [10], adapted for our needs. They hold for any map which is topologically conjugate to symbolic dynamics [62].

(i) The set $\mathscr{D}$ is called sensitive if there exists a positive real number $\overline{\varepsilon}$ such that, for each sequence $\{d_i\} \in \mathscr{D}$ and an arbitrary positive real number δ, there exist a sequence $\{c_i\} \in \mathscr{D}$ and a natural number j such that $|c_i - d_i| < \delta$, for all $i \leq 0$ and $\left|c_j - d_j\right| > \overline{\varepsilon}$.

(ii) The set $\mathscr{D}$ is called transitive if there exists a sequence $\left\{d_i^*\right\} \in \mathscr{D}$ such that for each $\{d_i\} \in \mathscr{D}$, an arbitrarily small positive number ε and an arbitrarily large natural number E, there exist a natural number m and an integer n such that $\left|d_i - d_{i+m}^*\right| < \varepsilon$ for each integer i between n and $n + E$.

(iii) The set of all periodic solutions $\mathscr{P}$ of Eq. (7.1.4) is called dense in $\mathscr{D}$ if for each sequence $\{d_i\} \in \mathscr{D}$, an arbitrarily small positive number ε and an arbitrarily large natural number E, there exist a periodic sequence $\{c_i\} \in \mathscr{P}$ and an integer n such that $|c_i - d_i| < \varepsilon$, for each integer i between n and $n + E$.

In our discussions of chaos, we will suppose that the set $\mathscr{D}$ is sensitive, transitive, and admits a dense set of periodic solutions.

We will make use of the number $\tau = \min\left\{\frac{h}{2}, \frac{L_1\overline{\varepsilon}h}{4(H_0\,\|A\| + M)[2 + h(L_f + \|A\|)]}\right\}$ in the next lemma, where $\overline{\varepsilon}$ is that from the definition (i) of sensitivity of the set $\mathscr{D}$.

Lemma 7.3 *Suppose that the set $\mathscr{D}$ is sensitive. In this case, there exists a positive number ε_0 such that for each sequence $d \in \mathscr{D}$ and an arbitrary positive real number δ, there exist $c \in \mathscr{D}$ and an interval $J \subset [0, \infty)$ of length τ such that $\|\phi_c(0) - \phi_d(0)\| < \delta$ and $\|\phi_c(t) - \phi_d(t)\| > \varepsilon_0$, for all $t \in J$. That is, X is sensitive.*

Proof Fix an arbitrary sequence $d \in \mathscr{D}$ and an arbitrary positive number δ. Let us take a sufficiently small positive real number δ_0 which satisfies the inequality $\left(1 + \frac{NL_2}{\omega - NL_f}\right)\delta_0 < \delta$ and a negative real number R such that $\frac{2MN}{\omega}e^{(\omega - NL_f)R} < \delta_0$.

Since the set $\mathscr{D}$ is sensitive, there exists a positive number $\overline{\varepsilon}$ such that both of the inequalities $|c_i - d_i| < \delta_0$, $i \leq 0$, and $\left|c_j - d_j\right| > \overline{\varepsilon}$ hold for some sequence $c \in \mathscr{D}$ and a natural number j.

First of all, we will show that $\|\phi_c(0) - \phi_d(0)\| < \delta$. According to the relation (7.1.10), the functions $\phi_c(t)$ and $\phi_d(t)$ satisfy the following couple of integral equations

$$\phi_c(t) = \int_{-\infty}^{t} e^{A(t-s)}\left(f\left(\phi_c(s)\right)) + \widetilde{c}_{[\frac{s}{h}]}\right) ds,$$

$$\phi_d(t) = \int_{-\infty}^{t} e^{A(t-s)} \left(f(\phi_d(s))) + \widetilde{d}_{[\frac{s}{h}]} \right) ds,$$

where $\widetilde{c}_{[\frac{s}{h}]} = \left(g(c_{[\frac{s}{h}]}), 0, 0, \ldots, 0 \right) \in \mathbb{R}^n$ and $\widetilde{d}_{[\frac{s}{h}]} = \left(g(d_{[\frac{s}{h}]}), 0, 0, \ldots, 0 \right) \in \mathbb{R}^n$.

Using these equations one can obtain for $R \le t \le 0$ that

$$e^{\omega t} \left\| \phi_c(t) - \phi_d(t) \right\| \le \frac{2MN}{\omega} e^{\omega R} + \frac{NL_2\delta_0}{\omega} \left(1 - e^{-\omega(t-R)} \right) e^{\omega t}$$
$$+ NL_f \int_R^t e^{\omega s} \left\| \phi_c(s) - \phi_d(s) \right\| ds.$$

Applying Gronwall's Lemma [61] to the last inequality, one can find that

$$\left\| \phi_c(0) - \phi_d(0) \right\| \le \frac{NL_2\delta_0}{\omega - NL_f} + \frac{2MN}{\omega} e^{(\omega - NL_f)R} < \delta.$$

In the remaining part of the proof, we shall determine an interval $J \subset [0, \infty)$ of length τ such that the inequality $\left\| \phi_d(t) - \phi_c(t) \right\| > \varepsilon_0$ is valid for all $t \in J$.

For $t \in [jh, (j+1)h]$, the functions $\phi_c(t)$ and $\phi_d(t)$ satisfy the equation

$$\phi_c(t) - \phi_d(t) = (\phi_c(jh) - \phi_d(jh)) + \int_{jh}^{t} A(\phi_c(s) - \phi_d(s)) ds$$
$$+ \int_{jh}^{t} [f(\phi_c(s)) - f(\phi_d(s))] ds + \int_{jh}^{t} \left(\widetilde{c}_{[\frac{s}{h}]} - \widetilde{d}_{[\frac{s}{h}]} \right) ds,$$

and evaluating at $t = (j+1)h$, one can produce the inequality

$$\left\| \phi_c((j+1)h) - \phi_d((j+1)h) \right\| \ge \left| g(c_j) - g(d_j) \right| h - \left\| \phi_c(jh) - \phi_d(jh) \right\|$$
$$- \int_{jh}^{(j+1)h} \left(L_f + \|A\| \right) \left\| \phi_c(s) - \phi_d(s) \right\| ds.$$

By means of the last inequality, we have

$$\max_{t \in [jh, (j+1)h]} \left\| \phi_c(t) - \phi_d(t) \right\| \ge \left\| \phi_c((j+1)h) - \phi_d((j+1)h) \right\|$$
$$> L_1 \overline{\varepsilon} h - [1 + h(L_f + \|A\|)] \max_{t \in [jh, (j+1)h]} \left\| \phi_c(t) - \phi_d(t) \right\|.$$

Therefore, $\max_{t \in [jh, (j+1)h]} \left\| \phi_c(t) - \phi_d(t) \right\| > \dfrac{L_1 \overline{\varepsilon} h}{[2 + h(L_f + \|A\|)]}$.

Suppose that on the interval $[jh, (j+1)h]$, the real-valued function $\left\| \phi_c(t) - \phi_d(t) \right\|$ takes its maximum value at the point η. Let us define the number

$$\xi = \begin{cases} \eta, & \text{if } \eta \le jh + \frac{h}{2} \\ \eta - \tau, & \text{if } \eta > jh + \frac{h}{2} \end{cases},$$

and let $J = [\xi, \xi + \tau]$, which is an interval of length τ. We note that the interval J is a subset of the interval $[jh, (j+1)h]$ and depends on the sequences c and d, but its length remains the same for different sequences.

By virtue of the inequality

$$\|\phi_c(t) - \phi_d(t)\| \ge \|\phi_c(\eta) - \phi_d(\eta)\| - \left| \int_\eta^t \|A\| \, \|\phi_c(s) - \phi_d(s)\| \, ds \right|$$
$$- \left| \int_\eta^t \left\| f(\phi_c(s)) - f(\phi_d(s)) + \widetilde{c}_{[\frac{s}{h}]} - \widetilde{d}_{[\frac{s}{h}]} \right\| ds \right|,$$

for $t \in J$, one has $\|\phi_c(t) - \phi_d(t)\| > \varepsilon_0$, where $\varepsilon_0 = \dfrac{L_1 \overline{\varepsilon} h}{2[2 + h(L_f + \|A\|)]}$.

The proof is finalized. □

We shall proceed to the next ingredient of Devaney chaos. In the case when Eq. (7.1.4) possesses a dense sequence $d^* \in \mathscr{D}$, the following assertion is valid.

Lemma 7.4 *Suppose the set $\mathscr{D}$ is transitive. Then there exists a solution $\phi_{d^*}(t) \in X$, $d^* \in \mathscr{D}$ such that for each solution $\phi_d(t) \in X$, $d \in \mathscr{D}$, an arbitrarily small positive real number ε and an arbitrarily large natural number E, there exist a positive real number ζ and an interval $J \subset \mathbb{R}$ of length Eh, such that $\|\phi_d(t) - \phi_{d^*}(t + \zeta)\| < \varepsilon$, for all $t \in J$.*

Proof Fix an arbitrarily small positive number ε and an arbitrarily large natural number E. Let $d \in \mathscr{D}$ be a given solution of Eq. (7.1.4) and suppose that $\gamma = \dfrac{\omega(\omega - NL_f)}{2MN(\omega - NL_f) + NL_2\omega}$. Since $\mathscr{D}$ is transitive, there exist a natural number m and an integer n such that $\left\| d_i - d^*_{i+m} \right\| < \gamma\varepsilon$, for all integers i satisfying $n \le i \le n+2E$.

Let $\zeta = mh$. Equation (7.1.10) implies that

$$\phi_{d^*}(t + \zeta) = \int_{-\infty}^t e^{A(t-s)} \left(f(\phi_{d^*}(s + \zeta)) + \widetilde{d^*}_{[\frac{s}{h}]+m} \right) ds.$$

Therefore, for $t \in [nh, (n+2E)h]$, the equation

$$\phi_d(t) - \phi_{d^*}(t + \zeta)$$
$$= \int_{-\infty}^{nh} e^{A(t-s)} \left(f(\phi_d(s)) - f(\phi_{d^*}(s + \zeta)) + \widetilde{d}_{[\frac{s}{h}]} - \widetilde{d^*}_{[\frac{s}{h}]+m} \right) ds$$
$$+ \int_{nh}^t e^{A(t-s)} \left(f(\phi_d(s)) - f(\phi_{d^*}(s + \zeta)) \right) ds + \int_{nh}^t e^{A(t-s)} \left(\widetilde{d}_{[\frac{s}{h}]} - \widetilde{d^*}_{[\frac{s}{h}]+m} \right) ds$$

holds, where

$$\widetilde{d}_{[\frac{s}{h}]} = \left(g(d_{[\frac{s}{h}]}), 0, 0, \ldots, 0\right)$$

and

$$\widetilde{d^*}_{[\frac{s}{h}]} = \left(g(d^*_{[\frac{s}{h}]}), 0, 0, \ldots, 0\right)$$

are n-dimensional vectors in $\mathbb{R}^n$.

Making use of the last equation we obtain that

$$e^{\omega t} \|\phi_d(t) - \phi_{d^*}(t+\zeta)\| \leq \frac{2MN}{\omega} e^{\omega nh} + \frac{NL_2\gamma\varepsilon}{\omega}\left(e^{\omega t} - e^{\omega nh}\right)$$
$$+ \int_{nh}^{t} NL_f e^{\omega s} \|\phi_d(s) - \phi_{d^*}(s+\zeta)\| \, ds.$$

Applying Gronwall's Lemma [61], one can arrive at the following inequality:

$$\|\phi_d(t) - \phi_{d^*}(t+\zeta)\| \leq \frac{2MN}{\omega} e^{(NL_f-\omega)(t-nh)} + \frac{NL_2\gamma\varepsilon}{\omega - NL_f}\left(1 - e^{(NL_f-\omega)(t-nh)}\right).$$

Suppose that the natural number E is large enough so that

$$Eh > \frac{1}{\omega - NL_f} \ln\left(\frac{1}{\gamma\varepsilon}\right)$$

and let the interval J be defined as $J = [(n+E)\,h, (n+2E)\,h]$. We note that the length of the interval J is Eh. For $t \geq (n+E)\,h$, it is the case that

$$e^{(NL_f-\omega)(t-nh)} \leq e^{(NL_f-\omega)Eh} < \gamma\varepsilon.$$

Consequently, for $t \in J$ we have

$$\|\phi_d(t) - \phi_{d^*}(t+\zeta)\| < \left(\frac{2MN}{\omega} + \frac{NL_2}{\omega - NL_f}\right)\gamma\varepsilon = \varepsilon.$$

The proof of the lemma is completed. □

For the case when the set of all periodic solutions $\mathscr{P}$ of Eq. (7.1.4) is dense in $\mathscr{D}$, we shall formulate an important ingredient of chaos for the set X, which states that an arbitrary function chosen from this set can be approximated by periodic functions from the set X_P of all periodic solutions of system (7.1.9) on intervals of arbitrary lengths. In other words, the density property of the set $\mathscr{P}$ is inherited by the set X_P. We call this property the density of the set X_P in X. The next assertion can be proved similarly to the previous lemma.

Lemma 7.5 *Suppose that the set of all periodic solutions $\mathscr{P}$ of Eq. (7.1.4) is dense in $\mathscr{D}$. Then for every solution $\phi_d(t) \in X$, $d \in \mathscr{D}$, an arbitrarily small positive ε and an arbitrarily large positive E, one can find a periodic solution $\phi_c(t) \in X_P$, $c \in \mathscr{P}$, and an interval $J \subset \mathbb{R}$ of length Eh, such that $\|\phi_c(t) - \phi_d(t)\| < \varepsilon$ for $t \in J$.*

We call the combined properties expressed in Lemmas 7.3–7.5 as chaos in the sense of Devaney for the set X, and we can formulate the following theorem:

Theorem 7.1 *The set X is chaotic in the sense of Devaney provided that the set $\mathscr{D}$ is sensitive, transitive, and possesses a dense set of periodic sequences.*

Next, let us describe chaos for Eq. (7.1.4) as obtained through period-doubling cascade.

Let us consider the equation

$$d_i = G(d_{i-1}, \mu), \tag{7.1.11}$$

where i is an integer and the function $G : \mathbb{R} \times \mathbb{R} \to \mathbb{R}$ satisfies for all $x \in \mathbb{R}$ the property that $F(x) = G(x, \mu_\infty)$, for some finite value μ_∞ of the parameter μ, which will be explained below.

Suppose that there exist a natural number k_0 and a sequence of period-doubling bifurcation values $\{\mu_m\}$ of the parameter μ, such that for each natural number m, as the parameter μ increases or decreases through μ_m, system (7.1.11) undergoes a period-doubling bifurcation and the previously existing stable $k_0 2^{m-1}$-periodic sequence becomes unstable and is replaced by a stable periodic sequence of period $k_0 2^m$. Moreover, the sequence $\{\mu_m\}$ of parameter values converges to a finite value μ_∞ as $m \to \infty$ and as a result, at $\mu = \mu_\infty$, there exist infinitely many unstable periodic solutions of Eq. (7.1.11), and consequently of Eq. (7.1.4), all lying in a bounded region. In this case, we say that Eq. (7.1.4) admits chaos through period-doubling cascade [63].

Since chaos through period-doubling cascade is based on the existence of infinitely many periodic solutions, if Eq. (7.1.4) admits chaos through period-doubling cascade, then by Lemma 7.2 the same is true for system (7.1.9), as stated in the following theorem. The instability of these periodic solutions can be proved using the same technique as in the proof of Lemma 7.3.

Theorem 7.2 *If Eq. (7.1.4) is chaotic through period-doubling cascade, then the same is true for (7.1.9).*

From the above discussion, Eq. (7.1.9), like Eq. (7.1.11), undergoes period-doubling bifurcations as the parameter μ increases or decreases through the values μ_m, $m \in \mathbb{N}$. In other words, the sequence $\{\mu_m\}$ of bifurcation parameters is exactly the same for both equations. It is worth pointing out that if Eq. (7.1.11) obeys the universality of Feigenbaum [63], one can conclude that the same holds for Eq. (7.1.9). That is, when $\lim\limits_{m\to\infty} \dfrac{\mu_m - \mu_{m+1}}{\mu_{m+1} - \mu_{m+2}}$ is evaluated, the universal constant known as the Feigenbaum number 4.6692016... is achieved, and this universal number is the same for both equations, and consequently for Eq. (7.1.5).

7.1.4 Chaos in a Kaldor–Kalecki Model

Consider the model of the aggregate economy of a given country:

$$\begin{aligned}\dot{Y} &= \alpha[I(Y,K) - S(Y,K)],\\ \dot{K} &= I(Y,K) - \delta K,\end{aligned} \tag{7.1.12}$$

where Y is income, K is capital stock, I is gross investment, and S is savings. Income changes proportionally to the excess demand in the goods market, and the second equation is a standard capital accumulation equation. The constant depreciation rate δ and the adjustment coefficient α are positive. This model was studied in detail by Lorenz [12] and Zhang [13]. It admits a stable equilibrium under certain conditions on the functions involved. We will show how perturbing it with a chaotic disturbance affects the resulting dynamics. For this purpose, let us consider the following specification of system (7.1.12) with $I(Y,K) = Y - aY^3 + bK$, $S(Y,K) = sY$,

$$\begin{aligned}\dot{Y} &= \alpha[(1-s)Y - aY^3 + bK],\\ \dot{K} &= Y - aY^3 + bK - \delta K,\end{aligned} \tag{7.1.13}$$

where the constant parameters satisfy $\alpha > 0$, $a > 0$, $b < 0$, $0 < s < 1$ and $0 < \delta < 1$. We present the following modified systems:

$$\begin{aligned}\dot{Y} &= \alpha[(1-s)Y - aY^3 + bK] + g(d_{[t]}),\\ \dot{K} &= Y - aY^3 + bK - \delta K,\\ d_{[t]} &= \mu d_{[t]-1}(1 - d_{[t]-1}).\end{aligned} \tag{7.1.14}$$

and

$$\begin{aligned}\dot{Y} &= \alpha[(1-s)Y - aY^3 + bK] + \nu(t,d),\\ \dot{K} &= Y - aY^3 + bK - \delta K,\end{aligned} \tag{7.1.15}$$

where the function $\nu(t,d)$ is defined in (7.1.7), and d is a solution of (7.1.4) and (7.1.6).

We introduce the perturbation only in the equation for income Y, since the equation for capital stock K can be viewed as a mechanical relation between investment and capital stock, where there is little room for exogenous influences. Income of a given country, on the other hand, is subject to many possible exogenous disturbances, such as productivity shocks and global economic fluctuations. This explains why we investigated the case of a perturbation with only one nonzero coordinate in the theoretical part. Of course, as we emphasized above, a more general case can be considered in a similar manner.

We model these perturbations as pulse functions, with chaotically behaved values in (7.1.14) and with chaotically behaved discontinuity instants in (7.1.15). Both types of disturbances are plausible: the first one is relevant if income shocks "pulsate" at regular time intervals (that is, they change values monthly, daily, etc.). Many economic time series that are good examples of exogenous disturbances to output, such as productivity indices, international trade indicators and commodity prices, can be modeled in this way. The second case is applicable if the disturbances admit a finite number of values, but the timing of these is chaotic. For example, output shocks due to natural disasters and weather fluctuations could be described with a finite set of values (e.g., the "Atlas of the Flood/Dryness in China for the last 500-year period" distinguishes between flood, wetness, normal level, dryness, and aridity [16]), but their timing is irregular. Of course, there may be a third case, where both the values of the shocks and the instants of discontinuity evolve irregularly. We can provide simulations for this scenario, as well, and the resulting chaotic behavior would be similar to the other cases.

One can see that a steady state of (7.1.13) with positive coordinates

$$Y^* = \sqrt{\frac{\delta(1-s)+bs}{a\delta}}, \quad K^* = \frac{s}{\delta}\sqrt{\frac{\delta(1-s)+bs}{a\delta}},$$

exists only if

$$\delta s < \delta + bs. \tag{7.1.16}$$

In the remaining part of the section, set $\alpha = 1$, $s = \delta = 1/2$, $b = -7/16$ to obtain $Y^* = \dfrac{1}{4\sqrt{a}}$, $K^* = \dfrac{1}{4\sqrt{a}}$. Now, the transformations $Y = y + Y^*$, $K = k + K^*$, applied to (7.1.13), give us the system

$$\begin{aligned} \dot{y} &= (5/16)y - (7/16)k - ay^3 - (3/4)\sqrt{a}y^2, \\ \dot{k} &= (13/16)y - (15/16)k - ay^3 - (3/4)\sqrt{a}y^2. \end{aligned} \tag{7.1.17}$$

The eigenvalues for the associated linear system are $\lambda_1 = -1/2$, $\lambda_2 = -1/8$. In this case, the equilibrium is asymptotically stable, if the number a is chosen to be sufficiently small.

7.1.4.1 Perturbation with Chaotically Behaved Values

Suppose that the home economy exports goods to a foreign country. The export flows are a function of consumption levels in the foreign country (normalized so that they lie in the interval $[0, 1]$ and denoted by d), which evolve according to a logistic map: $d_{[t]} = \mu d_{[t]-1}(1 - d_{[t]-1})$. That is, foreign consumption is a pulse function, where the unit of time can be chosen as fine as desired (year, month, day, minute, etc.). We motivate our choice of the logistic map with the results of Benhabib and Day [27] and Mitra and Sorger [33], who show that in some standard optimization problems the optimal policy function is a logistic map, under certain conditions. In general, any other way of modeling chaotic shocks could be implemented. Assume that the export flows to the foreign country are determined by a cubic function, $ex_{[t]} \equiv 0.0005(d_{[t]} + d_{[t]}^3)$, and $\mu = 3.8$. We multiply the export flows by the same multiplier as excess demand in the domestic goods market in the output equation, α, and since in our case $\alpha = 1$, we obtain the following system of type (7.1.14) introduced above:

$$\begin{aligned} \dot{Y} &= (1-s)Y - aY^3 + bK + 0.0005(d_{[t]} + d_{[t]}^3), \\ \dot{K} &= Y - aY^3 + bK - \delta K, \\ d_{[t]} &= 3.8 d_{[t]-1}(1 - d_{[t]-1}), \end{aligned} \tag{7.1.18}$$

with $g(d_{[t]}) \equiv 0.0005(d_{[t]} + d_{[t]}^3)$.

Let us take $a = 0.02$ in system (7.1.18). Transforming the variables as $Y = y + Y^*$, $K = k + K^*$, where $Y^* = \frac{5}{2\sqrt{2}}$, $K^* = \frac{5}{2\sqrt{2}}$, one can obtain

$$\begin{aligned} \dot{y} &= (5/16)y - (7/16)k - ay^3 - (3/4)\sqrt{a}y^2 + 0.0005(d_{[t]} + d_{[t]}^3), \\ \dot{k} &= (13/16)y - (15/16)k - ay^3 - (3/4)\sqrt{a}y^2, \\ d_{[t]} &= 3.8 d_{[t]-1}(1 - d_{[t]-1}). \end{aligned} \tag{7.1.19}$$

The sequence $\{d_i\}$ with $d_0 = 0.219$ is chaotic [41], and according to Theorem 7.2, the solution with $y(0) = 0.0078$, $k(0) = 0.0067$, is chaotic. In Figs. 7.1 and 7.2 the chaotic behavior of the solution is observable. Notice that both coordinates are positive.

We proceed by briefly explaining the OGY control method. Suppose that the parameter μ in the logistic map (7.1.6) is allowed to vary in the range $[3.8-\varepsilon, 3.8+\varepsilon]$, where ε is a given small number. That is, it is not possible (say, it is prohibitively costly or practically infeasible) to simply shift the value of μ to a level that generates

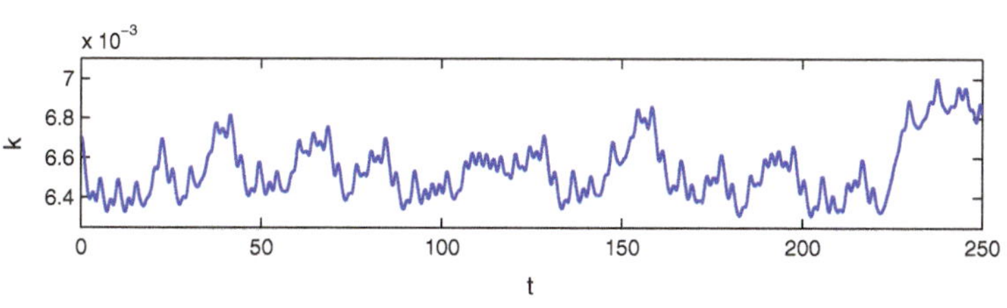

Fig. 7.1 The graphs of the y and k coordinates of the chaotic solution of system (7.1.19)

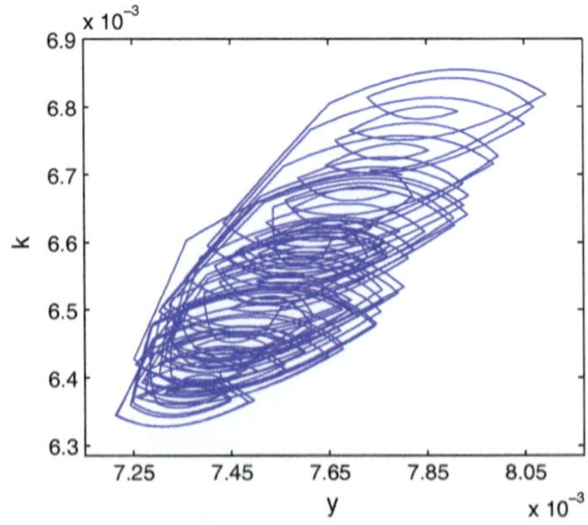

Fig. 7.2 The chaotic trajectory of system (7.1.19)

nonchaotic dynamics. Let us consider an arbitrary solution $\{d_i\}$, $d_0 \in [0, 1]$, of the map and denote by $d^{(j)}$, $j = 1, 2, \ldots, p$, the target unstable p-periodic orbit to be stabilized. In the OGY control method [41], at each iteration step i after the control mechanism is switched on, we consider the logistic map with the parameter value $\mu = \bar{\mu}_i$, where

$$\bar{\mu}_i = 3.8\left(1 + \frac{[2d^{(j)} - 1][d_i - d^{(j)}]}{d^{(j)}[1 - d^{(j)}]}\right), \tag{7.1.20}$$

provided that the number on the right-hand side of the formula (7.1.20) belongs to the interval $[3.8 - \varepsilon, 3.8 + \varepsilon]$. In other words, we apply a perturbation in the amount of $\dfrac{3.8[2d^{(j)} - 1][d_i - d^{(j)}]}{d^{(j)}[1 - d^{(j)}]}$ to the parameter $\mu = 3.8$ of the logistic map, if the trajectory $\{d_i\}$ is sufficiently close to the target periodic orbit. This perturbation makes the map behave regularly so that at each iteration step the orbit d_i is forced to be located in a small neighborhood of a previously chosen periodic orbit $d^{(j)}$. Unless the parameter perturbation is applied, the orbit d_i moves away from $d^{(j)}$ due to the instability. If $\left|\dfrac{3.8[2d^{(j)} - 1][d_i - d^{(j)}]}{d^{(j)}[1 - d^{(j)}]}\right| > \varepsilon$, we set $\bar{\mu}_i = 3.8$, so that the system evolves at its original parameter value, and wait until the trajectory $\{d_i\}$ enters a sufficiently small neighborhood of the periodic orbit $d^{(j)}$, $j = 1, 2, \ldots, p$, such that the inequality $-\varepsilon \le \dfrac{3.8[2d^{(j)} - 1][d_i - d^{(j)}]}{d^{(j)}[1 - d^{(j)}]} \le \varepsilon$ holds. If this is the case, the control of chaos is not achieved immediately after switching on the control mechanism. Instead, there is a transition time before the desired periodic orbit is stabilized. The transition time increases if the number ε decreases [14].

An unstable p-periodic solution of system (7.1.19) can be stabilized by controlling the corresponding p-periodic solution of the third equation, that is, the p-periodic solution of the logistic map. In the next example, a 2-periodic solution of system

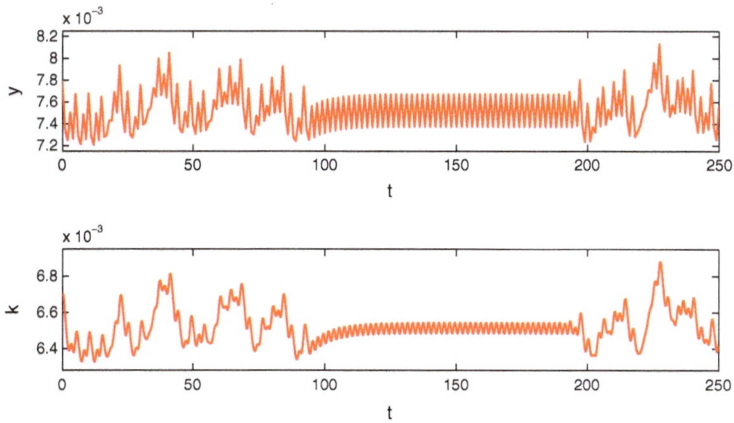

Fig. 7.3 OGY control method applied to system (7.1.19). It is seen in both panels that the 2-periodic solution of system (7.1.19) is stabilized

(7.1.19) is stabilized by applying the OGY control around the 2-periodic orbit $d^{(1)} \approx 0.3737$, $d^{(2)} \approx 0.8894$ of the logistic map. Notice that there exist two different 2-periodic solutions of system (7.1.19). One of them corresponds to the 2-periodic solution $\{c_i\}$ of the logistic map with $c_0 = d^{(1)}$, and the other corresponds to the 2-periodic solution $\{\bar{c}_i\}$ with $\bar{c}_0 = d^{(2)}$.

We consider the solution of system (7.1.19) with the initial data $y(0) = 0.0078$, $k(0) = 0.0067$, and $d_0 = 0.219$ again and apply the OGY control method around the 2-periodic solution $\{\bar{c}_i\}$ of the logistic map, with $c_0 = d^{(2)} \approx 0.8894$. Figure 7.3 shows the simulation results for $\varepsilon = 0.04$. The control mechanism is switched on at $t = 50$ and switched off at $t = 130$. The control becomes dominant approximately at $t = 100$ and its effect lasts approximately until $t = 190$, after which the instability becomes dominant and irregular behavior develops again.

The OGY control has to be applied to the logistic map, i.e., to the consumption function of the foreign economy. This highlights the importance of international economic cooperation between countries. Since foreign consumption is out of direct control of the home policy-makers, its adjustment can be only done by the foreign country in response to international negotiations. Additionally, the application of the OGY control to differential equations is feasible if a Poincaré map can be constructed, but this requires the knowledge of analytical solutions, which is a difficult task in general. Properly recognizing the unidimensional export shock as the source of the chaotic motion in the home economy will lead to locating the most effective and least costly way of stabilizing this dynamics, and in more general (and realistic) cases of higher dimensional models of home economy may prove to be the only way of controlling chaos.

7.1.4.2 Perturbation with Chaotically Behaved Discontinuity Instants

Now suppose that the home economy modeled in (7.1.13) is perturbed with an exogenous weather shock $\nu(t, d)$, such as rainfall, that affects agricultural output and therefore the entire economy. We model rainfall as taking one of two values, where the higher value is normal rainfall, and the lower value is drought, which leads to lower agricultural production and slower output growth:

$$\nu(t, d) = \begin{cases} 0.024, & \text{if } \theta_{2i}(d) < t \leq \theta_{2i+1}(d) \\ 0.007, & \text{if } \theta_{2i-1}(d) < t \leq \theta_{2i}(d), \end{cases} \tag{7.1.21}$$

where d is a solution of (7.1.6):

$$d_i = F(d_{i-1}) = \mu d_{i-1}(1 - d_{i-1}).$$

Consider system (7.1.15) with $a = 10^{-5}$. Transforming the variables as described for system (7.1.19), one can reduce system (7.1.15) to the following:

$$\begin{aligned} \dot{y} &= (5/16)y - (7/16)k - ay^3 - (3/4)\sqrt{a}y^2 + \nu(t, d), \\ \dot{k} &= (13/16)y - (15/16)k - ay^3 - (3/4)\sqrt{a}y^2. \end{aligned} \tag{7.1.22}$$

The bifurcation diagram of system (7.1.22), for $2.6 \leq \mu \leq 4$, is depicted in Fig. 7.4, where successive intervals of chaos and stable periodic solutions can be observed. In the regions of stability, for fixed μ, the bifurcation diagram represents the values of the stable periodic solutions of (7.1.22) at $t = \theta_0(d)$, where d is a periodic sequence. Therefore, in such regions, the number of intersection points of the graph with a vertical line through a given value of μ gives the number of stable periodic solutions for that μ. For example, for $\mu = 3.2$, a vertical line intersects the diagram at two points, which means that for this value of the parameter there exist two stable periodic solutions. Even though the diagram indicates regions where

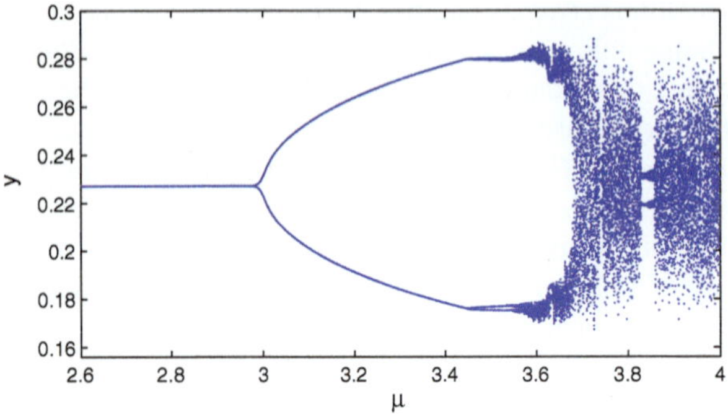

Fig. 7.4 Bifurcation diagram of system (7.1.22) with $a = 10^{-5}$

stable solutions exist, it does not inform us about the periods of these solutions. A theoretical discussion for the periods can be found in [60]. System (7.1.22) undergoes period-doubling bifurcations at the same parameter values as the logistic map and obeys the Feigenbaum universality [63]. One can observe from the diagram that system (7.1.22) is chaotic for $\mu = 3.8$.

Let us take into account system (7.1.22) with the parameter value $\mu = 3.8$. The time series of the y and k coordinates of system (7.1.22) corresponding to the initial data $y(0.39) = 0.21$, $k(0.39) = 0.18$ and $d_0 = 0.39$, are graphed in Fig. 7.5. The trajectory of the same solution is shown in Fig. 7.6. In both figures, the chaotic behavior of the solution can be observed. Notice that both coordinates of the solution, y and k, are positive.

Similarly to the previous case, the unstable periodic solutions of system (7.1.22) can be stabilized by controlling the chaos of the logistic map, particularly, using the OGY control method [60]. Of course, the policy-makers cannot control rainfall directly. To implement this method, they would need to regulate the *timing of the impact* of the rainfall on the economy, by providing assistance to farmers

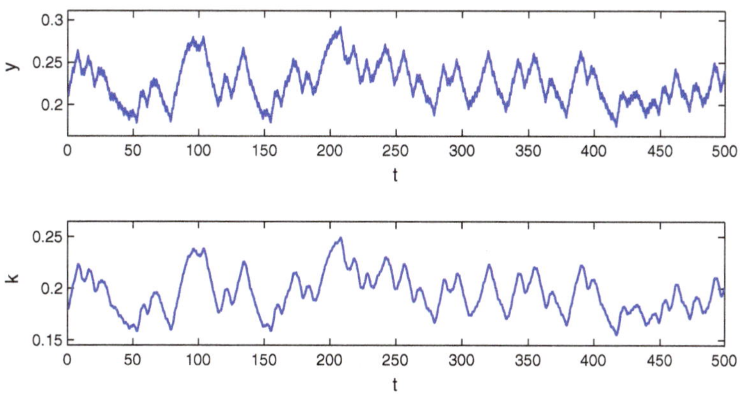

Fig. 7.5 The graphs of the y and k coordinates of system (7.1.22)

Fig. 7.6 The chaotic trajectory of system (7.1.22)

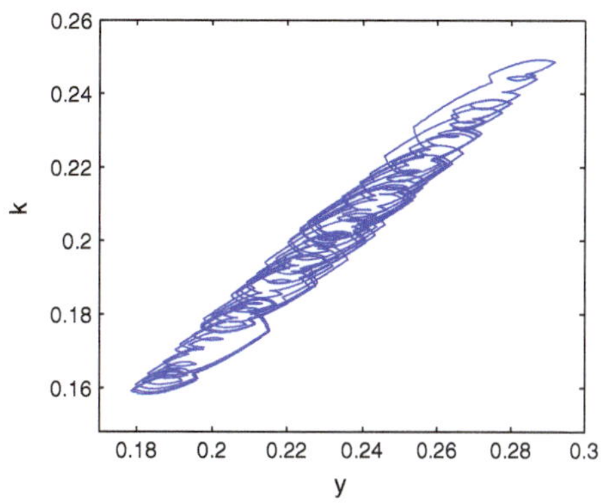

in order to stimulate their demand for goods and employ workers that are unemployed/underemployed due to the shortfall in agricultural production in the times of drought, and conversely tax farmers and/or agricultural workers in other times. Again, applying OGY control directly to the exogenous shock would be less costly than applying it to the entire system, and in high-dimensional models of the economy would be the only feasible approach.

7.2 Chaotic Business Cycles

In this section, we propose a novel approach to generate chaotic business cycles in a deterministic setting. Rather than producing chaos endogenously, we consider aggregate economic models with limit cycles and/or equilibriums, subject them to chaotic exogenous shocks and obtain chaotic cyclical motions. Thus, we emphasize that chaotic cycles, which are inevitable in economics, are not only interior properties of economic models, but also can be considered as a result of interaction of several economical systems. This provides a comprehension of chaos (unpredictability, lack of forecasting) and control of chaos as a global economic phenomenon from the deterministic point of view.

We suppose that the results of this section contribute to the mixed exogenous–endogenous theories of business cycles in classification by P.A. Samuelson [64]. Moreover, they demonstrate that the irregularity of the extended chaos can be structured, and this distinguishes them from the generalized synchronization. The advantage of the knowledge of the structure is that by applying instruments, which already have been developed for deterministic chaos, one can control the chaos, emphasizing a parameter or a type of motion. For the globalization of cyclic chaos phenomenon we utilize new mechanisms such that entrainment by chaos, attraction of chaotic cycles by equilibriums and bifurcation of chaotic cycles developed in our earlier papers.

7.2.1 Introduction

Scientists are interested in chaos theory due to the fact that the theory could offer new controlling strategies which have some particularly interesting insights for economic policies. There was a opinion among economists that dynamics of chaos is neither predictable nor controllable because of sensitivity. Results of Ott et al. [39] showed that control of chaos can be made by very small corrections of parameters [14, 44]. This achievement has been widely used in economics by Kaas, Kopel, Holyst, Urbanovwics [43, 50, 52, 65], and many others. In [1] it is written that the role of chaos is "... of revealing sources of uncertainty, and enriching the list of recognized *possible* developments." The control of chaos makes the possible developments *realistic*.

In the classic book [64] it is observed that while forced oscillator systems naturally emerge in theoretical investigations of several technical and physical devices, economic examples for this special family of functions have only rarely been provided. The main reason for this deficiency may lie in the fact that the necessary periodicity of the dynamic forcing may not be obvious in most economic applications. Our proposals are to apply *deterministic and chaotic* exogenous shocks to economic models and make them more realistic.

One may consider that chaos (the lack of forecasting) is undesirable in economics, but unavoidable. Thus, one can say that an economic model is realistic if there are chaotic motions of the system. We suggest to consider the presence of chaos in a model not only as indication of adequacy to economic real motions, but also as a measure of a power for an economic model. Indeed, chaos means that the model generates infinitely many motions with different periods and aperiodic motions, which are unstable, and consequently easy to be altered by control such that they can be sustained in a desirable mode. In other words, deterministic chaos is essential for economical models to function at many levels of activity. These is true, exceptionally, for the modern economies, with their intensive connections and globalization.

The principal novelty of our investigation is that we create a chaotic perturbation, plug it in a regular dynamic system, and find that similar chaos is inherited by the solutions of the new system. We call this as *the input–output analysis of chaos generation*. Such approach has been widely used for differential equations before, but with regular inputs. In papers [59, 60, 66–68], the mechanisms are provided for generating chaos in systems with asymptotically stable equilibriums. Previously, unpredictability in solutions of differential equations has been considered as a result of random perturbations with small probability [69–72].

P.A. Samuelson [64] accepts purely endogenous theory as "self-generating" cycle. Following this opinion we understand chaos as endogenous if it self-generated by an economic model. One can find detailed analysis of the endogenous chaos in books [7, 12, 13] and paper [1], which are very seminal sources on the subject. The dynamics arise in duopoly models [73], in simple ad hoc macroeconomic models [74, 75]. By applying the Li–Yorke theorem it is shown in [26, 27] that an overlapping generations model of the Gale type could generate endogenous chaotic cycles. Discrete equations have been applied to investigate the presence of the chaos in papers [3, 76], where models representing a capital stock with a maximum capital–labor ratio and a Malthusian agrarian economy are investigated. In [3, 77, 78] endogenous chaotic cycles are demonstrated in growth cycle models. The multiplier–accelerator model of Samuelson has been modified for generation of chaotic endogenous cycles and investigated in [79–81]. Investigations in Kaldor's type models, which are originated from [79, 82] and finalized in [83] showed that they could generate endogenous chaos.

Economists of the first half of the last century already filled a strong need of the developed theory of irregularities. In the classical book [64] one can find that "... in a physical system there are grand conservation laws of nature, which guarantee that the system must fall on the thin line between stability and instability. But there is nothing in the economic world corresponding to these laws ...". That is why, considering endogenous models P.A. Samuelson [64] suggests that "It is to be stressed that the exogenous impulses which keep the cycle alive need not themselves be even quasi-oscillatory in character." Thus, it was stressed at first that irregularities are significantly proper for economic dynamics, and at second that they are mostly because of *irregular exogenous* shocks.

In the present section, we realize observations of the famous economist. He suggested to consider mixed exogenous–endogenous theories, recognizing that "most economists are eclectic and prefer a combination of endogenous and exogenous theories." Accordingly, we use economical models, which admit *endogenous* business cycles to be perturbed by *exogenous* chaotic inputs. Examples of models possessing limit cycles are Kaldor–Kalecki model, Lienard type equations with relaxation oscillations which are also popular in economics, etc. Moreover, the systems are subject to *exogenous* chaotic disturbances, sensitive, and with infinitely many unstable periodic solutions.

We suggest two scenarios to obtain exogenous chaotic cycles as solutions of differential equations. For the first one an economical model with limit cycle is perturbed chaotically to produce a chaotic business cycle. For the second one, we consider a system with an equilibrium, perturb it by cyclic chaos and observe that the chaotic business cycle emerges in the case too. The first scenario was theoretically approved in our paper [84]. The second method of cyclic chaos generation is a new one, and in this section we demonstrate it by simulations. Currently, we study cases where the shocks enter the system additively, but future investigations may involve more complex forms, where the disturbance enters the main functions (e.g., the investment function, savings function, etc.). Both ways of chaotic cycles generation are applied in discussion of chaotic business cycles as global phenomena. Indeed, in the basis of any process of extension an input–output mechanism has to be considered.

Goodwin [4] argues that the apparent unpredictability of economic systems is due to deterministic chaos as much as to exogenous shocks. In this sense, our results can be interpreted as the *transmission of unpredictability* from one economic system to another, and even models that do not admit irregularity in isolation can eventually be contaminated with chaos. Thus, we provide support to the idea that unpredictability is a *global phenomenon* in economics, and demonstrate one of the mechanisms for this contagion. Considering the current extensive globalization process, this is a good depiction of reality.

Our results demonstrate that the control may become not local (applied to an isolated model) but global phenomenon with strong effectiveness such that control applied to a model, which is realizable easily (for example, the logistic map or Feichtinger's generic model), can be sufficient to rule the process in all models joined with the one controlled. Another benefit of our studies is that in the literature controls are applied to those systems, which are simple and low dimensional. It is worth to mention that control of chaos (unstable periodic motions) becomes difficult if dimensions of systems increase and the construction of Poincaré sections is complexified. For this reason the idea to control the generated chaos by controlling the exogenous shocks became useful for applications. In this section it is demonstrated through the application of the OGY control to the logistic map. A chaos control cannot be realized if we do not know the period of unstable motion to be controlled. In our case the control is applicable to models with arbitrary dimension if only the basic period of the generator is known. It is obvious that our methods provide us a scheme of investigation, which can be accompanied with detailed investigation in the future. Control of chaos nowadays is a synonym to the suppression of chaos. Thus our results give another way of suppression of chaos. If we find the controllable link (member) in a chain (collection) of connected chaotic systems, then we can suppress chaos in the whole chain. This is an effective consequence of our studies.

The section is organized in the following way. In Sect. 7.2.2 we describe the input–output mechanism, which is in the basis of the chaos extension, and formulate two theorems, which are theoretical approvement for the following economical and simulation discussions. In Sect. 7.2.3 economical models with regular motions—stable equilibrium and orbitally stable cycle are introduced. The models will be perturbed chaotically in the next subsection to obtain the main economical dynamics. So, Sect. 7.2.4 considers a stellar consisting of five economical models connected unilaterally. Extension of chaos near an equilibrium attractor, entrainment of a business cycle by chaos, bifurcation of a chaotic economical cycle, and attraction of a chaotic cycle are considered in economical models. Effects of the chaos generation for global economies which utilize methods of the OGY control [39] is also demonstrated in Sect. 7.2.4. Section 7.2.5 contains simulation evidence of irregular business cycle generation. It is demonstrated that entrainment by chaos is possible also for limit cycles of economical models of retarded type. Detailed comparison of our method of chaos generation with that on the basis of synchronization of chaos [15, 85, 86] is made in Sect. 7.2.6. In particular, we provide an argument that the dynamics of the obtained chaotic business cycles cannot be explained by the synchronization theory of chaotic systems. In Sect. 7.3 we discuss our results from the point of view of self-organization and synergetics of Haken [87].

7.2.2 The Input–Output Mechanism and Applications

To explain the input–output mechanism of chaos generation, let us introduce systems, which we call *the base system, the replicator* and *the generator*. They are intensively throughout the section. Consider the following system of differential equations,

$$\frac{dz}{dt} = B(z), \tag{7.2.23}$$

where $B : \mathbb{R}^n \to \mathbb{R}^n$ is a continuously differentiable function. The system (7.2.23) is called *the base system.*

Next, we subdue the base system to a perturbation, $I(t)$, which will be called an *input* and obtain the following system,

$$\frac{dy}{dt} = B(y) + I(t), \tag{7.2.24}$$

which will be called as *the replicator.*

Suppose that the input I admits a certain property, let us say, it is a bounded function. We assume then that there exists a unique solution, $y(t)$, of the last equation, the replicator, with the boundedness property. This solution is considered as an *output*. The process for obtaining the solution $y(t)$ of the replicator system by applying the perturbation $I(t)$ to the base system (7.2.23) is called the *input–output mechanism*. It is known that for certain base systems, if the input is periodic, almost periodic, bounded function, then there exists an output, which is also periodic, almost periodic, bounded function. In this section, we consider inputs of the new nature: chaotic functions and set of cyclic chaotic functions. The motions which are in the chaotic attractor of the Lorenz system [9] considered altogether provide us an example of a chaotic set of functions. Each element of this set is considered as a chaotic function. Both of these types of inputs will be used effectively. We have to say the input can be a set of functions as well as a single function. The same is true for the output.

In this section, we apply base systems of two types, (1) with asymptotically stable equilibriums, (2) with limit cycles. In the former case we will say about attraction of chaos by equilibriums, in particular, attraction of cyclic chaos by equilibrium. If the base system admits a limit cycle, then we say about entrainment of limit cycles by chaos or just about entrainment by chaos. If the limit cycle in a base system is a result of the Hopf bifurcation, we will say also about bifurcation of the cyclic chaos.

The main source of chaos in theory are difference and differential equations. For this reason we consider, inputs which are solutions of some systems of differential or discrete equations. These systems will be called *generators* in this section.

Thus, we can consider the following system of differential equations,

$$\frac{dx}{dt} = G(x), \tag{7.2.25}$$

and it is assumed that this system possesses chaos. We shall call this system a *generator*. If $x(t)$ is a solution of the system from the chaotic attractor, that is, it is a chaotic solution, then we notate $I(t) = \varepsilon\phi(x(t))$ and use the function $I(t)$ in the Eq. (7.2.24). Here, ε is a nonzero real number and the functions $G : \mathbb{R}^m \to \mathbb{R}^m$, $\phi : \mathbb{R}^m \to \mathbb{R}^n$ are continuous.

In this section, we will utilize also the logistic map as a generator.

Next, we will give the formulation of two assertions, the main mathematical recourse of the section.

The following conditions are required:

(A1) System (7.2.23) admits a nonconstant and *orbitally stable* periodic solution,
(A2) System (7.2.25) possesses sensitivity and is chaotic through period-doubling cascade,
(A3) Functions B, ϕ are bounded,
(A4) There exists a positive number L_B such that

$$\|B(y_1) - B(y_2)\| \leq L_B \|y_1 - y_2\|,$$

for all $y_1, y_2 \in \mathbb{R}^n$,

(A5) There exists a positive number L_ϕ such that

$$\|\phi(x_1) - \phi(x_2)\| \geq L_\phi \|x_1 - x_2\|,$$

for all $x_1, x_2 \in \mathbb{R}^m$.

The following assertion is based on the results in [84].

Theorem 7.3 *If conditions (A1)–(A5) holds and $|\varepsilon|$ is sufficiently small, there exists a neighborhood $\mathcal{U}$ of the orbitally stable limit cycle of (7.2.23) such that solutions of (7.2.24) which start inside $\mathcal{U}$ behave chaotically around the limit cycle. That is, the solutions are sensitive and there are infinitely many unstable periodic solutions.*

If one assumes that the limit cycle is with zero amplitude, that is an equilibrium, then immediately the theorem for attraction of chaos by equilibrium [68] can be obtained. This mathematical result has been utilized in our paper [88] for the economic dynamics analysis.

Theorem 7.4 [88] *If Eq. (7.2.25) is chaotic through period-doubling cascade, then the same is true for (7.2.24).*

7.2.3 Economic Models: The Base Systems

In what follows, to arrange the procedure of chaotic business cycles generation we shall need regular systems to be perturbed chaotically, i.e., models with asymptotically stable equilibriums or limit cycles. In this subsection, we suggest three economical models as the base systems.

7.2.3.1 Kaldor–Kalecki Model with a Steady Equilibrium

Consider the model of the aggregate economy of a given country:

$$\begin{aligned} Y' &= \alpha[I(Y,K) - S(Y,K)], \\ K' &= I(Y,K) - \delta K, \end{aligned} \tag{7.2.26}$$

where Y is income, K is capital stock, I is gross investment, and S is savings. Income changes proportionally to the excess demand in the goods market, and the second equation is a standard capital accumulation equation. The constant depreciation rate δ and the adjustment coefficient α are positive. This model was studied in detail in [12, 13]. It admits a stable equilibrium under certain conditions on the functions involved.

Let us consider the following specification of system (7.2.26) with $I(Y, K) = Y - aY^3 + bK$, $S(Y, K) = sY$,

$$\begin{aligned} Y' &= \alpha[(1-s)Y - aY^3 + bK], \\ K' &= Y - aY^3 + bK - \delta K, \end{aligned} \tag{7.2.27}$$

where the constant parameters satisfy $a > 0$, $b < 0$, $0 < s < 1$, and $0 < \delta < 1$.

One can see that a steady state of (7.2.27) with positive coordinates

$$Y^* = \sqrt{\frac{\delta(1-s)+bs}{a\delta}}, \quad K^* = \frac{s}{\delta}\sqrt{\frac{\delta(1-s)+bs}{a\delta}},$$

exists only if $\delta s < \delta + bs$.

The transformations $Y = y + Y^*$, $K = k + K^*$, applied to (7.2.27), give us the system

$$\begin{aligned} y' &= \alpha\left[\left(2(s-1) - \frac{3bs}{\delta}\right)y + bk - ay^3 - 3\sqrt{\frac{a\delta(1-s)+abs}{\delta}}y^2\right], \\ k' &= \left(3s - 2 - \frac{3bs}{\delta}\right)y + (b-\delta)k - ay^3 - 3\sqrt{\frac{a\delta(1-s)+abs}{\delta}}y^2. \end{aligned} \tag{7.2.28}$$

7.2.3.2 The Model with Business Cycle

We also investigate the idealized macroeconomic model with foreign capital investment,

$$\begin{aligned} S' &= \alpha Y + pS(k - Y^2), \\ Y' &= v(S + F), \\ F' &= mS - rY, \end{aligned} \tag{7.2.29}$$

where $S(t)$ are savings of households, $Y(t)$ is Gross Domestic Product (GDP), $F(t)$ is foreign capital inflow, k is potential GDP, and t is time. If k is set to 1, then Y, S, F are measured as multiples of potential output. The parameters represent corresponding ratios: α is the variation of the marginal propensity to save, p is the ratio of capitalized profit, $\frac{1}{v}$ is the capital–output ratio, m is the capital inflow–savings ratio, and r is the debt refund–output ratio.

Consider system (7.2.29) with specified coefficients,

$$\begin{aligned} S' &= \alpha Y + 0.1S(1 - Y^2), \\ Y' &= 0.5(S + F), \\ F' &= 0.19S - 0.25Y. \end{aligned} \tag{7.2.30}$$

According to [89], the system (7.2.30) admits Hopf bifurcation at $\alpha = \alpha_0 \equiv 0.25$ and an orbitally stable cycle appears as α decreases.

7.2.3.3 Kaldor–Kalecki Model with Time Delay

Let us take into account the system,

$$\begin{aligned} Y' &= 1.5[\tanh(Y) - 0.25K - (4/3)Y], \\ K' &= \tanh(Y(t-\tau)) - 0.5K. \end{aligned} \tag{7.2.31}$$

System (7.2.31) is the Kaldor–Kalecki model with time delay. According to [90], the model admits an orbitally stable limit cycle for $\tau > 5.4$. More precisely, the periodic solution appearance follows the Hopf bifurcation, such that the origin is asymptotically stable if $\tau < 5.4$, and the origin loses its stability and the cycle bifurcates from the origin for $\tau > 5.4$. One can find additional information on the models with delay in papers [91, 92].

7.2.4 Chaos in a Stellar of Economical Models

To provide a comprehensible discussion we will consider a stellar consisting of five unilaterally connected economical models A_k, $k = 1, \ldots, 5$. The topology of the connection is seen in Fig. 7.7, and equations of the connected models are given in system (7.2.32) below. It can be easily seen through the dependent variables migration from one system to another that the systems are connected unilaterally. We will show that the chaos appeared in A_1 spreads among all of models, such that A_2 serves as a replicator of chaos in A_1 and also as a generator of chaos for A_3 and A_4. The model A_4 is a replicator of chaos in A_2 and, in the same time, it is a generator for A_5.

The following is a system of the five unidirectionally coupled models $A_1, \ldots, A_5$:

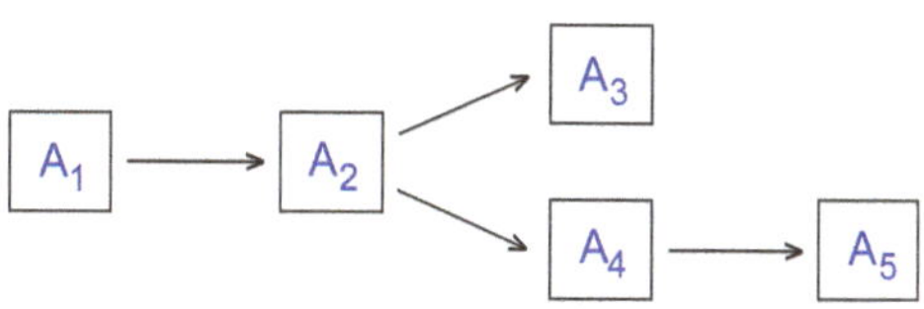

Fig. 7.7 The connection topology of the systems A_1–A_5

$$
\begin{aligned}
&\left.\kappa_{j+1} = \mu\kappa_j(1-\kappa_j),\right\} A_1\\
&\left.\begin{aligned}
y_1' &= (1/8)y_1 - (5/16)k_1 - a_1y_1^3 - \frac{3\sqrt{a_1}}{2}y_1^2 + \nu_1(t,\theta),\\
k_1' &= (1/4)y_1 - (3/8)k_1 - a_1y_1^3 - \frac{3\sqrt{a_1}}{2}y_1^2,
\end{aligned}\right\} A_2\\
&\left.\begin{aligned}
y_2' &= (1/3)y_2 - k_2 - a_2y_2^3 - \frac{\sqrt{6a_2}}{2}y_2^2 + 0.6y_1(t) + \nu_2(t,\zeta),\\
k_2' &= (1/2)y_2 - (5/4)k_2 - a_2y_2^3 - \frac{\sqrt{6a_2}}{2}y_2^2,
\end{aligned}\right\} A_3\\
&\left.\begin{aligned}
S' &= 0.23Y + 0.1S(1-Y^2),\\
Y' &= 0.5(S+F) + 2(y_1(t)+0.5),\\
F' &= 0.19S - 0.25Y,
\end{aligned}\right\} A_4\\
&\left.\begin{aligned}
y_3' &= (3/5)y_3 - (4/5)k_3 - a_3y_3^3 - \frac{3\sqrt{a_3}}{\sqrt{10}}y_3^2 + 0.01Y(t),\\
k_3' &= (7/10)y_3 - (9/10)k_3 - a_3y_3^3 - \frac{3\sqrt{a_3}}{\sqrt{10}}y_3^2,
\end{aligned}\right\} A_5
\end{aligned}
\tag{7.2.32}
$$

where the piecewise constant functions $\nu_1(t,\theta)$ and $\nu_2(t,\zeta)$ are defined as follows:

$$
\nu_1(t,\theta) = \begin{cases} 0.019, & \text{if } \theta_{2j} < t \le \theta_{2j+1},\\ 0.002, & \text{if } \theta_{2j-1} < t \le \theta_{2j}, \end{cases} \tag{7.2.33}
$$

and

$$
\nu_2(t,\zeta) = \begin{cases} 0.0006, & \text{if } \zeta_{2j} < t \le \zeta_{2j+1},\\ 0.0017, & \text{if } \zeta_{2j-1} < t \le \zeta_{2j}, \end{cases}. \tag{7.2.34}
$$

The sequences $\theta = \{\theta_j\}$ and $\zeta = \{\zeta_j\}$ are described immediately in the next subsection.

Examples of shocks of the form (7.2.33) and (7.2.34) are natural disasters and extreme events in general, such as market crashes. They take a finite number of values (an earthquake either happens or not), but their timing is irregular or regular.

7.2.4.1 Description of the Models A_1 to A_5

A_1: Equation A_1 is the logistic map, which will be used as the main source of chaos in system (7.2.32). The interval [0, 1] is invariant under the iterations of the map for the parameter values $\mu \in (0, 4]$, and for $\mu = 3.8$ it is chaotic through period-doubling cascade [93]. The equation A_1 is the generator of chaos for the global system (7.2.32) and as we mentioned above, a generator can be not only with continuous dynamics, but the with discrete dynamics, and even hybrid, which combines both continuous and dicrete cases. In fact the whole model (7.2.32) is an example of a hybrid system. Next, immediately, we will show how chaotic solutions of the logistic map can be transformed to the chaotic continuous perturbations of the differential equation.

A_2: System A_2 is a perturbed Kaldor–Kalecki model (7.2.28) such that in the absence of the perturbation function $\nu_1(t,\theta)$ the system possesses an asymptotically stable equilibrium point provided that the number a_1 is sufficiently small. The nonperturbed system is obtained by taking $\alpha = 1, s = 1/8, \delta = 1/16$ and $b = -5/16$ in the Kaldor–Kalecki model (7.2.28). One can evaluate that the associated linear system admits the complex conjugate eigenvalues $(-1 \pm i)/8$. The sequence $\theta = \{\theta_j\}$, $j \in \mathbb{Z}$, of the discontinuity instants of the function (7.2.33) satisfies the relation $\theta_j = j + \kappa_j$, where the sequence $\{\kappa_j\}$ is a solution of the logistic map A_1 with $\kappa_0 \in [0, 1]$. According to Theorem 7.4, the system A_2 is chaotic through period-doubling cascade. That is, it admits infinitely many unstable periodic solutions and sensitivity is present. The chaotic change in the discontinuity instants of the function $\nu_1(t, \theta)$ gives rise to the appearance of chaos in the system. For each natural number p, the system possesses an unstable periodic solutions with period $2p$. Next, in its own turn system A_2 is the generator for the systems A_3 and A_4.

A_3: System A_3 is obtained by using the coefficients $\alpha = 1$, $s = 1/6$, $\delta = 1/4$, $b = -1$ in the Kaldor–Kalecki model (7.2.28) and by perturbing it with the solutions of A_2 as well as with the periodic function (7.2.34). The associated linear system has the eigenvalues $(-11 \pm \sqrt{73})/24$. If the number a_2 is sufficiently small, then the system admits an asymptotically stable equilibrium point in the case that the perturbation terms $0.6y_1(t)$ and $\nu_2(t, \zeta)$ are absent. The sequence $\zeta = \{\zeta_j\}$, $j \in \mathbb{Z}$, of the discontinuity instants of (7.2.34) satisfies the relation $\zeta_j = 2\sqrt{2}j$ for each j. Since the function (7.2.34) and the perturbations from the system A_2 have incommensurate periods, one can confirm by using the results of the paper [68] that the system A_3 is chaotic with infinitely many quasiperiodic solutions in basis.

A_4: System A_4 is obtained by influencing system (7.2.30) with the solutions of A_2. It is a replicator with respect to system A_2. The term $2(y_1(t) + 0.5)$ is the used perturbation. It is worth noting that, in the absence of the perturbation, A_4 possesses an orbitally stable limit cycle [89]. Theorem 7.3 implies that the system A_4 admits chaotic business cycles. Since the orbitally stable cycle of system (7.2.30) occurs through the bifurcation one can say that there is the phenomenon of *chaotic cycles bifurcation*.

A_5: System A_5 is constructed by perturbing the Kaldor–Kalecki model (7.2.28) with the solutions of A_4. That is, A_5 is a replicator with respect to A_4. The eigenvalues of the associated linear system are $-1/5$ and $-1/10$. In the absence of the perturbation term $0.01Y(t)$, the system possesses an asymptotically stable equilibrium point if the number a_3 is chosen to be sufficiently small. We will make use of system A_5 to demonstrate the *attraction of chaotic business cycles*.

7.2.4.2 Simulations

In this part, we will demonstrate numerically the chaotic behavior in system (7.2.32). In what follows, we will use $a_1 = 3 \times 10^{-6}$, $a_2 = 10^{-6}$, $a_3 = 5 \times 10^{-6}$, $\mu = 3.8$ and $\kappa_0 = 0.63$.

Let us start with system A_2. Making use of the initial data $y_1(t_0) = 0.12$, $k_1(t_0) = 0.08$, where $t_0 = 0.63$, we represent in Fig. 7.8 the graphs of the y_1 and k_1 coordinates of system A_2. It is seen that both of the coordinates behave chaotically.

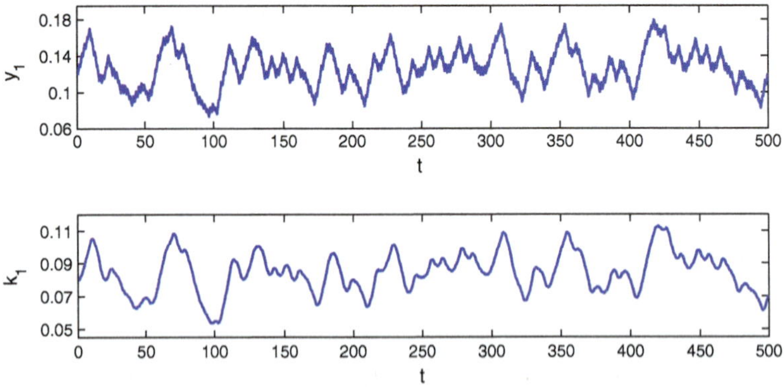

Fig. 7.8 The graphs of the y_1 and k_1 coordinates of system A_2

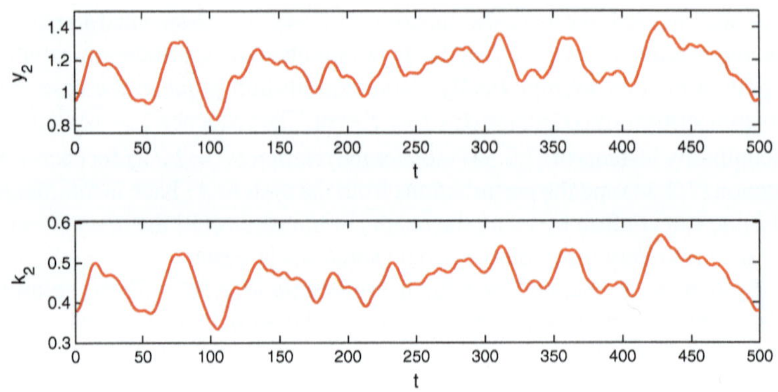

Fig. 7.9 Extension of chaos by system A_3

Now, to show the extension of chaos by system A_3, we make use of the solution shown in Fig. 7.8 as perturbation in system A_3 and represent in Fig. 7.9 the graphs of the y_2 and k_2 coordinates of A_3. The initial data $y_2(t_0) = 0.95$, $k_2(t_0) = 0.38$, where $t_0 = 0.63$, is used in the simulation. Figure 7.9 reveals that the chaos of system A_2 is extended such that the system A_3 also possesses chaos. In order to confirm the extension of chaos one more time, we depict in Fig. 7.10 the projection of the trajectory of the coupled Kaldor–Kalecki system A_2–A_3 corresponding to the same initial data on the $y_1 - k_1 - y_2$ space.

Next, we continue with system A_4. We take into account system A_4 with the solution of A_2 that is shown in Fig. 7.8, and represent the trajectory of A_4 with $S(t_0) = 1.67$, $Y(t_0) = 0.94$, $F(t_0) = -5.15$, where $t_0 = 0.63$, in Fig. 7.11. One can observe in Fig. 7.11 that the system A_4 admits a chaotic business cycle.

In order to observe the attraction of the cyclic chaos of system A_4, we use the solution represented in Fig. 7.11 and depict in Fig. 7.12 the trajectory of system A_5 with $y_3(t_0) = 0.72$, $k_3(t_0) = 0.56$. It is seen in Fig. 7.12 that the chaotic business cycle of A_4 is attracted by A_5 and the cyclic irregular behavior is extended.

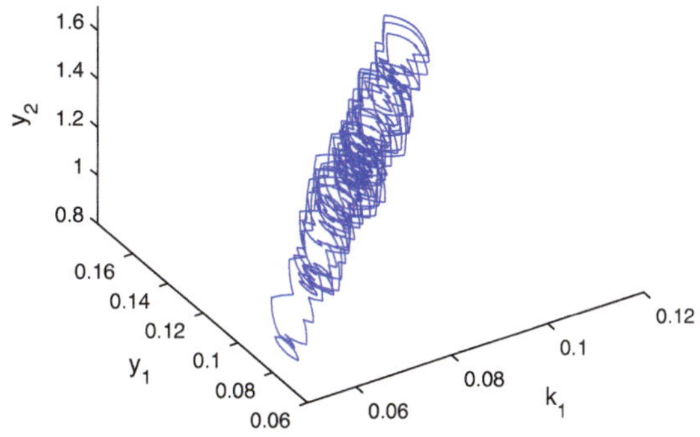

Fig. 7.10 The projection of the chaotic trajectory of the coupled Kaldor–Kalecki system $A_2 - A_3$ on the $y_1 - k_1 - y_2$ space

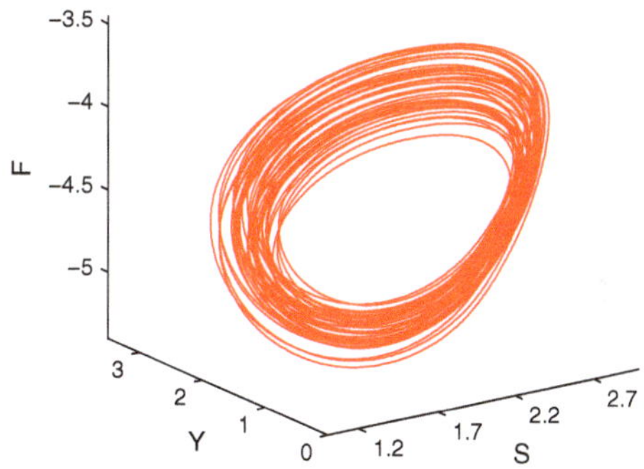

Fig. 7.11 Chaotic business cycle of system A_4

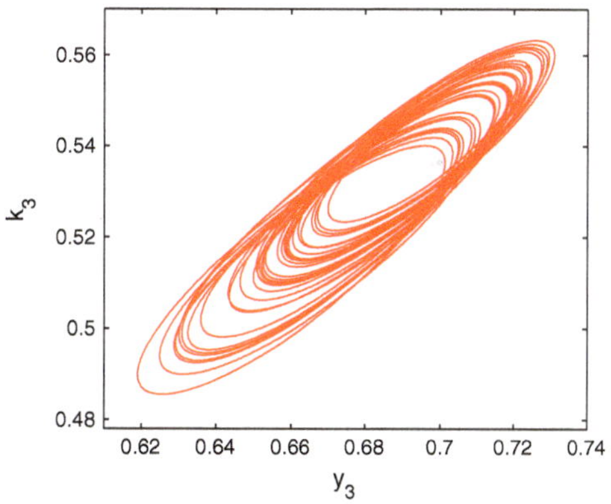

Fig. 7.12 Attraction of cyclic chaos by system A_5

7.2.4.3 Control of Extended Chaos

In system (7.2.32), the source of the chaotic motions is the logistic map A_1. Therefore, to control the chaos of the whole system, one has to stabilize an unstable periodic solution of the logistic map. The OGY control method [39] is one of the possible ways to do this. We proceed by briefly explaining the method.

Suppose that the parameter μ in the logistic map A_1 is allowed to vary in the range $[3.8-\varepsilon, 3.8+\varepsilon]$, where ε is a given small number. That is, it is not possible (say, it is prohibitively costly or practically infeasible) to simply shift the value of μ to a level that generates nonchaotic dynamics. Let us consider an arbitrary solution $\{\kappa_j\}$, $\kappa_0 \in [0,1]$, of the map and denote by $\kappa^{(q)}$, $q = 1, 2, \ldots, p$, the target unstable p-periodic orbit to be stabilized. In the OGY control method [93], at each iteration step j after the control mechanism is switched on, we consider the logistic map with the parameter value $\mu = \bar{\mu}_j$, where

$$\bar{\mu}_j = 3.8\left[1 + \frac{(2\kappa^{(q)}-1)(\kappa_j - \kappa^{(q)})}{\kappa^{(q)}(1-\kappa^{(q)})}\right], \tag{7.2.35}$$

provided that the number on the right-hand side of the formula (7.2.35) belongs to the interval $[3.8-\varepsilon, 3.8+\varepsilon]$. In other words, we apply a perturbation in the amount of $\dfrac{3.8(2\kappa^{(q)}-1)(\kappa_j - \kappa^{(q)})}{\kappa^{(q)}(1-\kappa^{(q)})}$ to the parameter $\mu = 3.8$ of the logistic map, if the trajectory $\{\kappa_j\}$ is sufficiently close to the target periodic orbit. This perturbation makes the map behave regularly so that at each iteration step the orbit κ_j is forced to be located in a small neighborhood of a previously chosen periodic orbit $\kappa^{(q)}$. Unless the parameter perturbation is applied, the orbit κ_j moves away from $\kappa^{(q)}$ due to the instability. If $\left|\dfrac{3.8(2\kappa^{(jq)}-1)(\kappa_j - \kappa^{(q)})}{\kappa^{(q)}(1-\kappa^{(q)})}\right| > \varepsilon$, we set $\bar{\mu}_j = 3.8$, so that the system evolves at its original parameter value, and wait until the trajectory $\{\kappa_j\}$ enters a sufficiently small neighborhood of the periodic orbit $\kappa^{(q)}$, $q = 1, 2, \ldots, p$, such that the inequality $-\varepsilon \le \dfrac{3.8(2\kappa^{(q)}-1)(\kappa_j - \kappa^{(q)})}{\kappa^{(q)}(1-\kappa^{(q)})} \le \varepsilon$ holds. If this is the case, the control of chaos is not achieved immediately after switching on the control mechanism. Instead, there is a transition time before the desired periodic orbit is stabilized. The transition time increases if the number ε decreases [14].

The chaos of system A_2 can be stabilized by controlling an unstable periodic orbit of the logistic map A_1, since the map gives rise to the presence of chaos in the system. By applying the OGY control method around the fixed point 2.8/3.8 of the logistic map, we stabilize the corresponding unstable 2-periodic solution of system A_2. The simulation result is seen in Fig. 7.13. We used the same initial data as in Fig. 7.8. It is seen in Fig. 7.13 that the OGY control method successfully controls the chaos of system A_2. The control is switched on at $t = \theta_{50}$ and switched off at $t = \theta_{280}$. The values $\kappa_0 = 0.63$ and $\varepsilon = 0.08$ are utilized in the simulation. The control becomes dominant approximately at $t = 150$ and its effect lasts approximately until $t = 340$, after which the instability becomes dominant and irregular behavior develops again.

Next, we will demonstrate the stabilization of an unstable quasiperiodic solution of system A_3. We suppose that an unstable quasiperiodic solution of A_3 can be stabilized by controlling the chaos of system A_2. We use the solution shown in

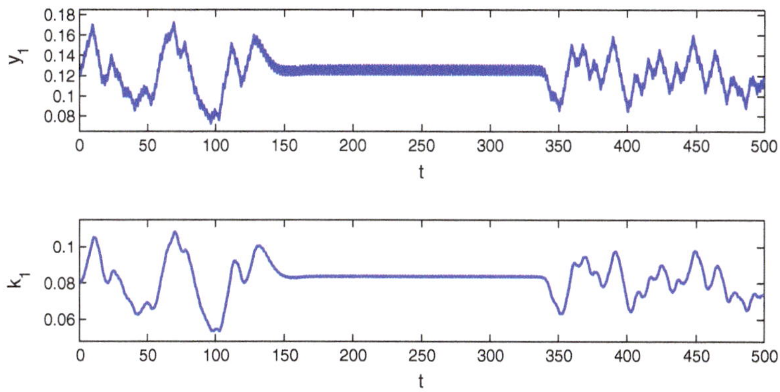

Fig. 7.13 The chaos control of system A_2. The OGY control method is applied around the fixed point 2.8/3.8 of the logistic map. The value $\varepsilon = 0.08$ is used

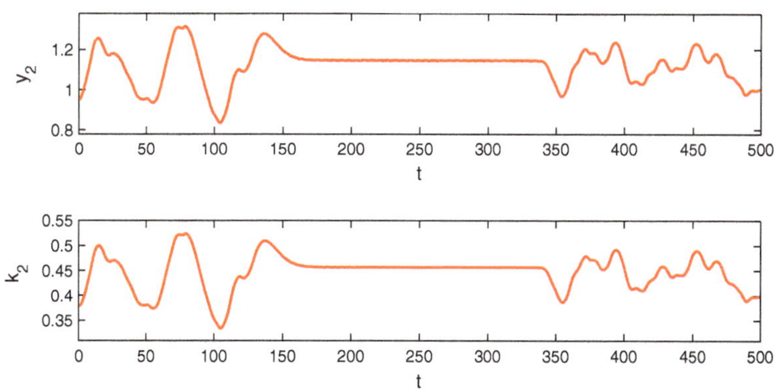

Fig. 7.14 The chaos control of system A_3. It is observable in the figure that controlling the chaos of system A_3 makes the chaos of system A_2 to be also controlled

Fig. 7.13 as the perturbation in system A_3, and represent in Fig. 7.14 the solution of A_3 with $y_2(t_0) = 0.95$, $k_2(t_0) = 0.38$, where $t_0 = 0.63$. Similarly to system A_2, it seen in the figure that the chaos of A_3 is controlled approximately for $150 \leq t \leq 340$.

To reveal that the stabilized solution is indeed quasiperiodic, we depict in Fig. 7.15 the graph of the same solution for $200 \leq t \leq 300$. Figure 7.15 manifests that the application of the OGY control method to system A_2 makes an unstable quasiperiodic solution of A_3 to be stabilized. On the other hand, the stabilized torus of system A_3 is shown in Fig. 7.16).

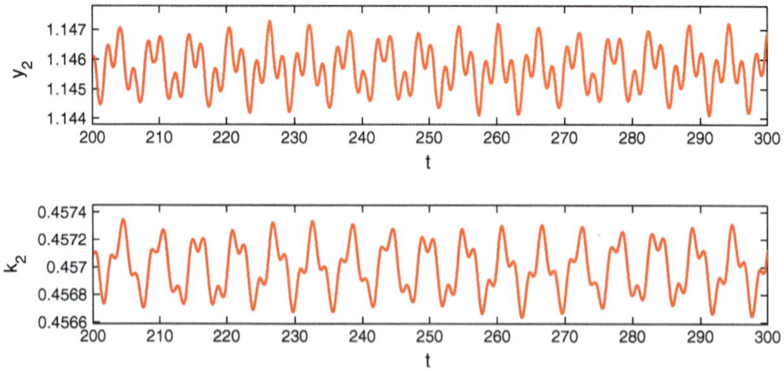

Fig. 7.15 The stabilized quasiperiodic solution of system A_3

7.2.5 Kaldor–Kalecki Model with Time Delay

This subsection considers the phenomenon of chaos extension by utilizing an economical model with time lag (7.2.36). We decided to separate the model from the previous discussion, since the result is not theoretically approved. The chaos extension can be only observed numerically in the example, but one can prove in the future the entrainment of the limit cycle by chaos for functional differential equations by considering our results. In this subsection, we will demonstrate numerically the formation of chaotic business cycles in the Kaldor–Kalecki model with time delay.

Let us take into account the system,

Fig. 7.16 The stabilized torus of system A_3

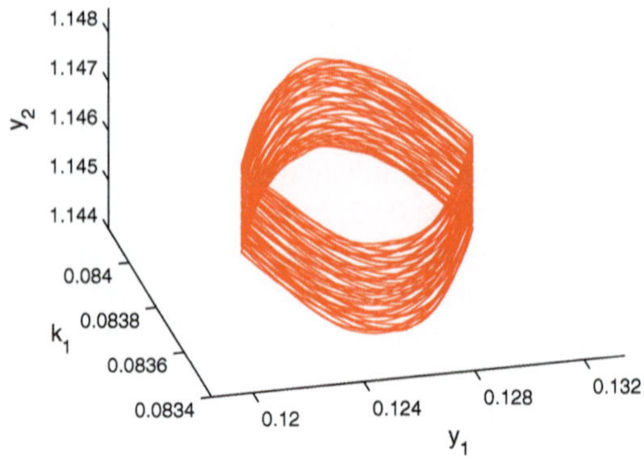

$$
\begin{aligned}
&x'' + 5(x^2 - 1)x' + x = 5\cos(2.467t), \} \; B_1 \\
&\left.\begin{aligned} Y' &= 1.5\,[\tanh(Y) - 0.25K - (4/3)Y] + 0.0045x(t), \\ K' &= \tanh(Y(t-\tau)) - 0.5K. \end{aligned}\right\} B_2
\end{aligned}
\tag{7.2.36}
$$

The description of the subsystems B_1 and B_2 are as follows.

B_1: Equation B_1 is the chaotic Van der Pol oscillator, which is used as the generator system in (7.2.36). Van der Pol type equations have played a role in economic modeling [4, 12, 94]. It is shown in [95] that equation B_1 is chaotic through period-doubling cascade. The process of period-doubling is described in [96]. This implies that there are infinitely many *unstable* periodic solutions of B_1, all with different periods. Due to the absence of stability, any solution that starts near the periodic motions behaves *irregularly*.

B_2: System B_2 is the Kaldor–Kalecki model with time delay and it is a result of the perturbation of the model (7.2.31). Here, the term $0.0045x(t)$ is the perturbation provided by the solutions the generator system B_1. Thus, one can observe numerically entrainment of the limit cycle of system (7.2.31) by chaos. In other words, the appearance of a chaotic business cycle will be seen in the next simulations.

The motion generated by the Van der Pol equation B_1 corresponding to the initial data $x(0) = 1.1008$, $x'(0) = -1.5546$ is depicted in Fig. 7.17, which confirms that the equation possesses chaos.

Next, we will demonstrate the presence of business cycles in system B_2 numerically. Let us take $\tau = 5.5$ in B_2 so that the system possesses an orbitally stable limit cycle in the absence of perturbation. We make use of the solution $x(t)$ of B_1 shown in Fig. 7.17 as the perturbation in system B_2, and represent in Fig. 7.18 the solution of B_2 with the initial condition $Y(t) = u(t)$ and $K(t) = v(t)$ for $t \in [-\tau, 0]$, where $u(t) = -0.057$ and $v(t) = 0.063$ are constant functions. Figure 7.18 reveals that chaotic business cycles takes place in the dynamics of B_2. This result shows that our theory of chaotic business cycles can be extended to systems with time delay.

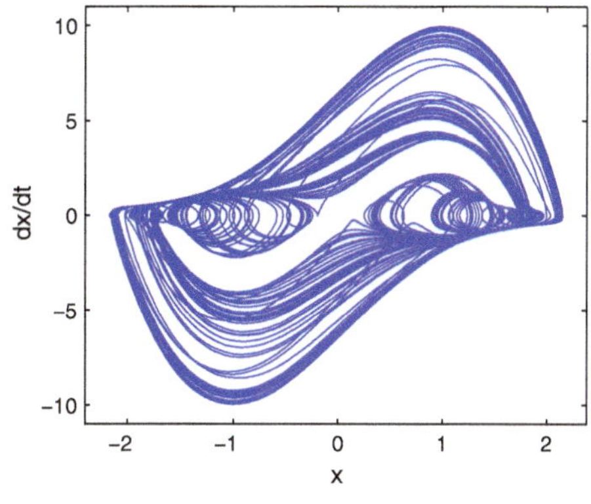

Fig. 7.17 Chaotic dynamics of the Van der Pol oscillator B_1

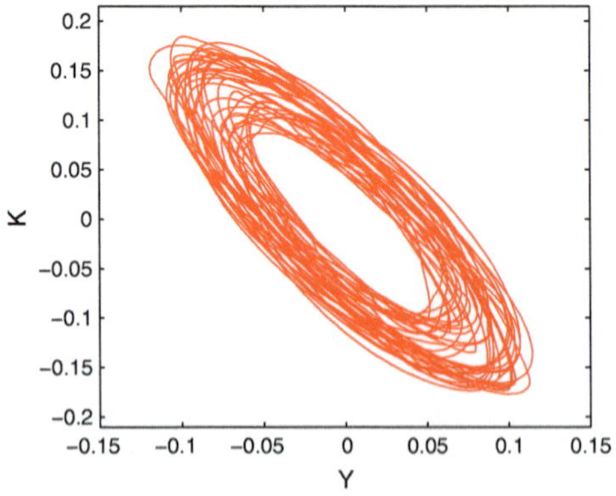

Fig. 7.18 The appearance of chaotic business cycle in the Kaldor–Kalecki model B_2

7.2.6 Chaos Extension Versus Synchronization

Generalized synchronization characterizes the dynamics of a response system that is driven by the output of a chaotic driving system [14, 85, 86, 97, 98]. Suppose that the dynamics of the drive and response are governed by the following systems with a skew product structure

$$x' = D(x) \tag{7.2.37}$$

and

$$y' = R(y, K(x)), \tag{7.2.38}$$

respectively, where $x \in \mathbb{R}^p$, $y \in \mathbb{R}^q$. Synchronization [86] is said to occur if there exist sets I_x, I_y of initial conditions and a transformation ϕ, defined on the chaotic attractor of (7.2.37), such that for all $x(0) \in I_x$, $y(0) \in I_y$ the relation

$$\lim_{t\to\infty} \|y(t) - \phi(x(t))\| = 0$$

holds. In this case, a motion that starts on $I_x \times I_y$ collapses onto a manifold $M \subset I_x \times I_y$ of synchronized motions. The transformation ϕ is not required to exist for the transient trajectories. When ϕ is the identity, the identical synchronization takes place [14, 15].

It is formulated by [85] that generalized synchronization occurs if and only if for all $x_0 \in I_x$, $y_{10}, y_{20} \in I_y$, the following asymptotic stability criterion holds:

$$\lim_{t\to\infty} \|y(t, x_0, y_{10}) - y(t, x_0, y_{20})\| = 0,$$

where $y(t, x_0, y_{10})$, $y(t, x_0, y_{20})$ denote the solutions of (7.2.38) with the initial data $y(0, x_0, y_{10}) = y_{10}$, $y(0, x_0, y_{20}) = y_{20}$ and the same $x(t)$, $x(0) = x_0$.

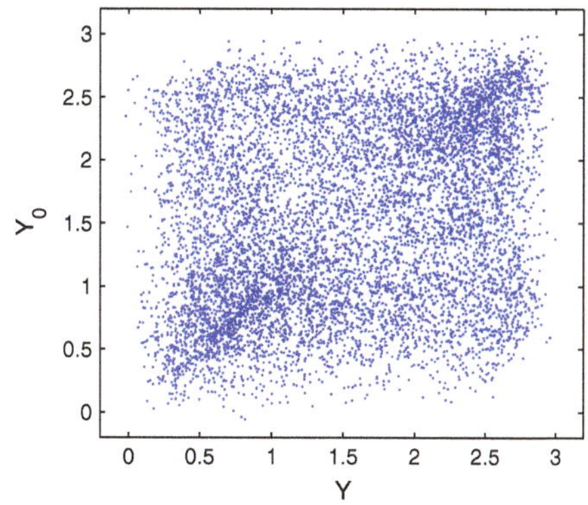

Fig. 7.19 The auxiliary system approach shows that the systems A_2 and A_4 are not synchronized in the generalized sense

A numerical method that can be used to investigate coupled systems for generalized synchronization is the auxiliary system approach [14, 97]. Let us investigate the coupled economic model $A_2 - A_4$ for generalized synchronization by means of the auxiliary system approach.

Consider the auxiliary system

$$\begin{aligned} S_0' &= 0.23Y_0 + 0.1S_0(1 - Y_0^2), \\ Y_0' &= 0.5(S_0 + F_0) + 2y_3(t), \\ F_0' &= 0.19S_0 - 0.25Y_0. \end{aligned} \tag{7.2.39}$$

System (7.2.39) is an identical copy of system A_4.

By marking the trajectory of system A_2–A_4-(7.2.39) with the initial data $y_1(t_0) = 0.12$, $k_1(t_0) = 0.08$, $S(t_0) = 1.67$, $Y(t_0) = 0.94$, $F(t_0) = -5.15$, $S_0(t_0) = 2.63$, $Y_0(t_0) = 0.84$, $F_0(t_0) = -2.89$ at times $t = \theta_j$ and omitting the first 500 iterations, we obtain the stroboscopic plot whose projection on the Y–Y_0 plane is shown in Fig. 7.19. Since the plot is not placed on the line $Y_0 = Y$, we conclude that generalized synchronization does not occur in the couple A_2–A_4.

It is worth noting that generalized synchronization does not take place also in the dynamics of the unidirectionally coupled subsystems B_1 and B_2, which are mentioned in Sect. 7.2.5, and this can be verified by means of the auxiliary system approach [14, 97] as well.

7.3 The Global Unpredictability, Self-organization and Synergetics

We have to say that the idea of the transition of chaos from one system to another as well as the arrangement of chaos in an ordered way can be considered as another level of self-organization [87, 99]. Durrenmatt [100] indicated that "... a system is self-organizing if it acquires a spatial, temporal or functional structure without specific interference from the outside. By "specific" we mean that the structure of functioning is not impressed on the system, but the system is acted upon from the outside in a nonspecific fashion. "There are three approaches to self-organization, namely thermodynamic (dissipative structures), synergetic and the autowaves. For the theory of dynamical systems (e.g., differential equations) the phenomenon means that an autonomous system of equations admits a regular and stable motion (periodic, quasiperiodic, almost periodic). This is what in the literature is called autowaves processes [101] or self-excited oscillations [102]. We are inclined to add to the list one more phenomenon—the chaos extension. For example, consider the collection of systems, $A_1, A_2, \ldots, A_5$, once again, where A_1 is the original generator of chaos. Because of the connections and the conditions discovered in our analysis, all the other subsystems, $S_i, i = 2, 3, \ldots, 9$, are also chaotic. We suppose that this is a self-organization phenomenon. That is a coherent behavior of a large number of systems [87]. One can interpret the chaos extension as global unpredictability in economics.

The German theoretical physicist Haken [87] introduced a new interdisciplinary field of science, synergetics, which deals with the origins and the evolution of spatiotemporal structures. The profound part of synergetics is based on the dynamical systems theory. One of the main features of systems in synergetics is self-organization, which have been discussed above. According to Haken [87], the central question in synergetics is whether there are general principles which govern the self-organized formation of structures and/or functions. The main principles by the founder of the theory are instability, order parameters, and slaving [87]. Instability is understood as the formation or collapse of structures (patterns) [99]. This is very common in fluid dynamics, lasers, chemistry and biology [87, 99, 103, 104]. A number of examples of instability can be found in the literature about morphogenesis [105] and the pattern formation examples can be found in fluid dynamics. The phenomenon is called as instability because the former state of fluid transforms to a new one, loses its ability to persist, and becomes unstable. One can see the instability in our results as formation of chaos in systems A_2, A_3, A_4, A_5. Despite the fact that processes in finite dimensional spaces are considered, we have to say that chaotic attractors are assumed to be not single trajectories, but collections of infinitely many trajectories with complex topologies. This allows to say that they are somehow between objects of ordinary differential equations and partial differential equations. This provides an opportunity to say also about dissipative structures [99], and we hope that this makes our investigations more attractive, due to the "density" of the chaotic trajectories in the space. For differential equations theory, order parameters mean those phase variables, whose behavior formate the main properties of a macroscopic structure, which dominate over all other variables in the formation such that they can even depend on the order parameters functionally. The dependence that is proved (discovered) mathematically is what we call as slaving. It is not difficult to see that the variables of the system A_1 are order parameters, and they determine chaotic behavior of the joined systems' variables.

7.4 Notes

This chapter highlights a novel source of chaos in economic models. Unlike previous literature that studies endogenous chaos emergence, we allow chaotic exogenous shocks perturbing a system with a stable equilibrium to generate chaos there. We focus on exogenous disturbances that take the form of a pulse function. The pulsative shocks may have chaotically behaved values or chaotically behaved instants of discontinuity. Both types of shocks are plausible in economics, as is the hybrid of the two. We rigorously verify that the system perturbed with a pulsative disturbance with chaotically behaved values admits chaos. The results are applied to a model of the aggregate economy of *a country subject to export shocks*, which are determined by the chaotic consumption levels of a foreign economy. We show simulations of the chaotic motion, as well as the stabilized periodic solutions, obtained by implementing the OGY control method [39]. We also demonstrate chaos in a model of the aggregate economy *perturbed by rainfall shocks* that behave as a pulse function with chaotically behaved instants of discontinuity. The theory of this type of perturbations was developed in [59, 60].

According to Baumol and Benhabib [1], "apparently random behavior may not be random at all", but a product of deterministic chaos. We argue that what we used to interpret as endogenous chaotic behavior may not be endogenous at all, but a product of exogenous chaotic shocks. For the purposes of economic policy-making, the control of chaos needs to be carried out in a very different way once its source in an exogenous chaotic perturbation is recognized. The OGY control can be applied directly to the exogenous disturbance, rather than the entire system. This will significantly reduce the costs of the policy, and in most instances will be the only feasible approach.

Our results also illustrate the *transmission of unpredictability* from one economic system to another, so that even economies that do not admit irregularity in isolation can eventually be contaminated with chaos. Thus, we provide support to the idea that unpredictability is a *global phenomenon* in economics and demonstrate a mechanism for this contagion. Considering the current extensive globalization process, this is a good depiction of reality.

We provide an example of a model of aggregate economy, based on a system of differential and discrete equations, where the main variables exhibit cycle-like motion with chaotic perturbations. Thus, we obtain an irregular business cycle in a deterministic setting. This provides a modeling alternative to the business cycle literature relying on stochastic variation in the economy. Additionally, our investigation highlights the variety of ways of generating chaos in an economic model. Previous work has focused on generating chaos and, in particular, chaotic business cycles *endogenously* (see [7, 12, 13, 106, 107]). Our method of creating chaos has its own relevance for economics, since we show the role of *exogenous* shocks in the appearance of chaos in models that otherwise do not exhibit irregular behavior. It can also be said that our work fills a missing link in the research on the origins of irregularities in economic time series. While the literature on endogenous chaos was a response to the view that exogenous stochastic shocks are the source of fluctuations in the economy (see [1]), Sect. 7.2 is a response to the former, in that it provides a role for exogenous chaotic disturbances in producing these fluctuations, and thus completes the circle.

In paper [1], the essence of chaos for economics was remarked as "Chaos theory has at least equal power in providing caveats for both the economic analysis and the policy designer. For example, it warns us that apparently random behavior may not be random at all. It demonstrates dramatically the dangers of extrapolation and the difficulties that can beset economic forecasting generally. It provides the basis for the construction of simple models of the behavior of rational agents, showing how even these can yield extremely complex developments. It has served as the basis for models of learning behavior and has been shown to arise naturally in a number of standard equilibrium models. It offers additional insights about the economic source of oscillations in a number of economic models."

Applications of chaos theory have illustrated the possibility of producing complex dynamics in deterministic settings [4, 12, 13, 89, 106–108], with some papers specifically focusing on building "chaotic business cycles" [109]. Chaos is generated endogenously, and its appearance hinges on the values of some crucial parameters of the model. The main novelty of Sect. 7.2 is that we start with a model that is not endogenously complex. We assume that there exists a limit cycle, where the limit cycle is understood to be a closed orbit that is also an attractor [110]. We then subject the model to chaotic exogenous shocks and obtain a perturbed system that admits chaotic motions. This approach is based on rigorous mathematical theory [59, 60], and we provide numerical simulations. The chaos emerging around the original limit cycle is cycle-like, and therefore can be called a chaotic business cycle.

Our goal is to show that it is possible to produce a chaotic business cycle in a very natural way—take a system of differential equations with a limit cycle as a point of departure, and introduce a chaotic exogenous disturbance. An example of an exogenous disturbance is a technology shock to the economy which affects output, holding all other variables constant. We describe it using solutions of chaos generator models. We use them to demonstrate the proposed approach, and other formulations can be studied in future work. For example, one can use actual economic time series, such as commodity prices, that have been tested for deterministic chaos [19, 55, 56, 111]. Moreover, shocks other than technology shocks can be considered, in view of the ongoing debate between two literatures supporting and rejecting the importance of technology shocks for generating business cycles [69, 112, 113]. The results of Sect. 7.1 are published in the paper [88], while the results of Sect. 7.2 and Sect. 7.3 are partially published in [114].

Our results give more theoretical lights on the processes, as we suggest a mathematical apparatus, which describe rigorously the *extension of chaos*, increases its complexity, and provides new structures for the *effective control* of whole clusters of economic models.

References

1. W.J. Baumol, J. Benhabib, Chaos: significance, mechanism, and economic applications. J. Econ. Perspect. **3**, 77–105 (1989)
2. M. Boldrin, L. Montrucchio, On the indeterminacy of capital accumulation paths. J. Econ. Theory **40**, 26–39 (1986)
3. R.H. Day, The emergence of chaos from classical economic growth. Q. J. Econ. **98**, 201–213 (1983)
4. R. Goodwin, *Chaotic Economic Dynamics* (Oxford University Press, Oxford, 1990)
5. T.R. Malthus, *An Essay on the Principle of Population, As It Affects the Future Improvement of Society, with Remarks on the Speculations of Mr. Godwin, M. Condorcet, and Other Writers* (J. Johnson, London, 1798)
6. A. Marshall, *Principles of Economics* (Macmillan, London, 1920)
7. J.B. Rosser Jr, *From Catastrophe to Chaos: A General Theory of Economic Discontinuities*, 2nd edn. (Kluwer Academic Publishers, Norwell, 2000)
8. T. Mitra, F. Privileggi, On Lipschitz continuity of the Iterated Function System in a stochastic optimal growth model. J. Math. Econ. **45**, 185–198 (2009)
9. E.N. Lorenz, Deterministic nonperiodic flow. J. Atmos. Sci. **20**, 130–141 (1963)
10. R.L. Devaney, *An Introduction to Chaotic Dynamical Systems* (Addison-Wesley, Menlo Park, 1989)
11. J. Gleick, *Chaos: The Making of a New Science* (Viking, New York, 1987)
12. H.W. Lorenz, *Nonlinear Dynamical Economics and Chaotic Motion* (Springer, New York, 1993)
13. W.B. Zhang, *Differential Equations, Bifurcations, and Chaos in Economics* (World scientific, Singapore, 2005)
14. J.M. Gonzáles-Miranda, *Synchronization and Control of Chaos* (Imperial College Press, London, 2004)
15. L.M. Pecora, T.L. Carroll, Synchronization in chaotic systems. Phys. Rev. Lett. **64**, 821–825 (1990)
16. Y. Zhou, M. Zhiyuan, L. Wang, Chaotic dynamics of the flood series in the Huaihe river basin for the last 500 years. J. Hydrol. **258**, 100–110 (2002)
17. G.P. Decoster, W.C. Labys, D.W. Mitchell, Evidence of chaos in commodity futures prices. J. Futures Markets **12**(3), 291–305 (1992)
18. A. Wei, R.M. Leuthold, *Long agricultural futures prices: ARCH, long memory, or chaos processes? OFOR Paper 98–03*, University of Illinois at Urbana-Champaign, Urbana, 1998
19. E. Panas, V. Ninni, Are oil markets chaotic? A non-linear dynamic analysis. Energy Econ. **22**, 549–568 (2000)
20. W.A. Brock, Distinguishing random and deterministic systems: abridged version. J. Econ. Theory **40**, 168–195 (1986)
21. W.A. Brock, W. Dechert, J.A. Scheinkman, B. LeBaron, A test for independence based on the correlation dimension. Econ. Rev. **15**, 197–235 (1996)
22. J. Benhabib, Chaotic dynamics in economics. Forthcoming in *The New Palgrave Dictionary of Economics*, ed. by S.N. Durlauf, L.E. Blume (Macmillan, Palgrave, 2005)
23. H. Sakai, H. Tokumaru, Autocorrelations of a certain chaos. IEEE Trans. Acoust. Speech Signal Process **28**, 588–590 (1980)
24. A. Medio, G. Gallo, Chaotic dynamics, *Theory and Applications to Economics* (Cambridge University Press, Cambridge, 1992)
25. H.W. Lorenz, International trade and the possible occurrence of chaos. Econ. Lett. **23**, 135–138 (1987)
26. J. Benhabib, R.H. Day, Erratic accumulation. Econ. Lett. **6**, 113–117 (1980)
27. J. Benhabib, R.H. Day, A characterization of erratic dynamics in the overlapping generations model. J. Econ. Dyn. Control **4**, 37–55 (1982)
28. J. Benhabib, K. Nishimura, The Hopf bifurcation and the existence and stability of closed orbits in multisector models of optimal economic growth. J. Econ. Theory **21**, 421–444 (1979)

29. R. Deneckere, S. Pelikan, Competitive chaos. J. Econ. Theory **40**, 13–25 (1986)
30. J.M. Grandmont, On endogenous competitive business cycles. Econometrica **53**, 995–1045 (1985)
31. K. Nishimura, G. Sorger, M. Yano, Ergodic chaos in optimal growth models with low discount rates. Econ. Theory **4**, 705–717 (1994)
32. K. Nishimura, M. Yano, Non-linear dynamics and chaos in optimal growth: an example. Econometrica **63**, 981–1001 (1995)
33. T. Mitra, G. Sorger, On the existence of chaotic policy functions in dynamic optimization. Jpn. Econ. Rev. **50**(4), 470–484 (1999)
34. D.A. Mendes, V. Mendes, Control of chaotic dynamics in an OLG economic model. J. Phys.: Conf. Ser. **23**, 158–181 (2005)
35. F. Dyson, *Infinite in All Directions* (Harper & Row, New York, 1988)
36. G. Chen, B. Raton (eds.), *Controlling Chaos and Bifurcation in Engineering Systems* (CRS Press, West Palm Beach, 2000)
37. G. Chen, X. Yu (eds.), Chaos control, *Theory and Applications* (Springer, Berlin, 2003)
38. T. Kapitaniak, *Controlling Chaos: Theoretical and Practical Methods in Non-linear Dynamics* (Butler and Tanner Ltd., Frome, 1996)
39. E. Ott, C. Grebogi, J.A. Yorke, Controlling chaos. Phys. Rev. Lett. **64**, 1196–1199 (1990)
40. K. Pyragas, Continuous control of chaos by self-controlling feedback. Phys. Rev. A **170**, 421–428 (1992)
41. E. Schöll, H.G. Schuster, *Handbook of Chaos Control* (Wiley, Weinheim, 2008)
42. T. Shinbrot, E. Ott, C. Grebogi, J.A. Yorke, Using chaos to direct trajectories to targets. Phys. Rev. Lett. **65**, 3215–3218 (1990)
43. J.A. Holyst, T. Hagel, G. Haag, W. Weidlich, How to control a chaotic economy? J. Evolut. Econ. **6**(1), 31–42 (1996)
44. J.A. Holyst, K. Urbanowicz, Chaos control in economical model by time delayed feedback method. Phys. A: Stat. Mech. Appl. **287**(3–4), 587–598 (2000)
45. E. Ahmed, S.Z. Hassan, On controlling chaos in Cournot games with two and three competitors. Nonlinear Dyn. Psychol. Life Sci. **4**, 189–194 (2000)
46. H. Salarieh, A. Alasty, Chaos control in an economic model via minimum entropy strategy. Chaos Solut. Fractals **40**, 839–847 (2009)
47. L. Chen, G. Chen, Controlling chaos in an economic model. Phys. A: Stat. Mech. Appl. **374**(1), 349–358 (2007)
48. D. Behrens, *Two and three-dimensional models of the army races*, Diplomarbeit, Institut fur Okonometrie, Operations Research and Systemtheorie, Technische Universitat Wien (1992)
49. G. Feichtinger, Nonlinear threshold dynamics: further examples for chaos in social sciences, in *Economic Evolution and Demographic Change*, ed. by G. Haag, U. Mueller, K.G. Troitzsh (Springer, Berlin, 1992)
50. L. Kaas, Stabilizing chaos in a dynamic macroeconomic model. J. Econ. Behav. Organ. **33**, 313–332 (1998)
51. V. Bala, M. Majumdar, T. Mitra, A note on controlling a chaotic tatonnement. J. Econ. Behav. Organ. **33**, 411–420 (1998)
52. M. Kopel, Improving the performance of an economic system: controlling chaos. J. Evol. Econ. **7**, 269–289 (1997)
53. G. Haag, T. Hagel, T. Sigg, Active stabilization of a chaotic urban system. Discrete Dyn. Nat. Soc. **1**, 127–134 (1997)
54. C. Wieland, F.H. Westerhoff, Exchange rate dynamics, central bank interventions and chaos control methods. J. Econ. Behav. Organ. **4**(2), 189–194 (2005)
55. W.A. Barnett, P. Chen, The aggregation-theoretic monetary aggregates are chaotic and have strange attractors: An econometric application of mathematical chaos, 199–246, in *Proceedings of the Third International Symposium in Economic Theory and Econometrics*, ed. by W.A. Barnett, E. Berndt, H. White (Cambridge University Press, Cambridge, 1988)
56. M. Frank, T. Stengos, Measuring the strangeness of gold and silver rates of return. Rev. Econ. Stud. **56**, 553–567 (1989)

57. J. Guckenheimer, P.J. Holmes, Nonlinear oscillations, *Dynamical Systems and Bifurcations of Vector Fields* (Springer, New York, 1997)
58. M. Ausloos, M. Dirickx (eds.), *The Logistic Map and the Route to Chaos: From the Beginnings to Modern Applications (Understanding Complex Systems)* (Springer, Berlin, 2010)
59. M.U. Akhmet, Devaney's chaos of a relay system. Commun. Nonlinear Sci. Numer. Simulat. **14**, 1486–1493 (2009)
60. M.U. Akhmet, M.O. Fen, Chaotic period-doubling and OGY control for the forced Duffing equation. Commun. Nonlinear. Sci. Numer. Simulat. **17**, 1929–1946 (2012)
61. J.K. Hale, *Ordinary Differential Equations* (Krieger Publishing Company, Malabar, 1980)
62. M. Akhmet, *Nonlinear Hybrid Continuous/Discrete-Time Models* (Atlantis Press, Paris, 2011)
63. M.J. Feigenbaum, Universal behavior in nonlinear systems. Los Alamos Sci./Summer **1**, 4–27 (1980)
64. P.A. Samuelson, *Foundations of Economic Analysis* (Harvard University Press, Cambridge, 1947)
65. M. Allais, The economic science of today and global disequilibrium, in *Global Disequilibrium in the World Economy*, ed. by M. Baldassarry, J. McCallum, R.A. Mundell (Macmillan, Basingstoke, 1992)
66. M. Akhmet, *Principles of Discontinuous Dynamical Systems* (Springer, New York, 2010)
67. M.U. Akhmet, Li-Yorke chaos in the system with impacts. J. Math. Anal. Appl. **351**, 804–810 (2009)
68. M.U. Akhmet, M.O. Fen, Replication of chaos. Commun. Nonlinear. Sci. Numer. Simulat. **18**, 2626–2666 (2013)
69. F.E. Kydland, E.C. Prescott, Time to build and aggregate fluctuations. Econometrica **50**, 1345–1370 (1982)
70. B. Liu, *Uncertainty Theory: A Branch of Mathematics for Modeling, Human Uncertainty* (Springer, Berlin, 2010)
71. J.B. Long Jr, C.I. Plosser, Real business cycles. J. Polit. Econ. **91**, 39–69 (1983)
72. G. Mircea, M. Neamt, D. Opris, The KaldorKalecki stochastic model of business cycle. Nonlinear Anal.: Model. Control **16**, 191–205 (2011)
73. D. Rand, Exotic phenomena in games and duopoly models. J. Math. Econ. **5**, 173–184 (1978)
74. R.H. Day, W.J. Shafer, Keynesian chaos. J. Macroecon. **7**, 277–295 (1985)
75. M. Stutzer, Chaotic dynamics and bifurcations in a macro model. J. Econ. Dyn. Control **2**, 353–376 (1980)
76. R.H. Day, Irregular growth cycles. Am. Econ. Rev. **72**, 406–414 (1982)
77. R.A. Dana, P. Malgrange, The dynamics of a discrete version of a growth cycle model, in *Analyzing the Structure of Econometric Models*, ed. by J.P. Ancot (M. Nijhoff, Amsterdam, 1984)
78. M.T. Pohjola, Stable, cyclic and chaotic growth: a dynamics of a discrete time version of Goodwin's growth cycle model. Zeitsschrift fur Nationalekonomie **41**, 27–38 (1981)
79. J.M. Blatt, *Dynamic economic systems: a post-Keynesian approach* (M.E. Sharpe, Armonk, 1983)
80. G. Gabish, Nonlinear models of business cycle theory, in *Selected Topics in Operations Research and Mathematical Economics*, ed. by G. Hammer, D. Pallaschke (Springer, Berlin, 1984), pp. 205–222
81. H. Nusse, Asymptotically periodic behavior in the dynamics of chaotic mappings. SIAM J. Appl. Math. **47**, 498–515 (1987)
82. J.R. Hicks, *A Contribution to the Theory of the Trade Cycle* (Oxford University Press, Oxford, 1950)
83. W.A. Brock, Hicksian nonlinearity, SSRI Paper No. 8815, University of Wisconsin-Madison (1988)
84. M.U. Akhmet, M.O. Fen, Entrainment by chaos. J. Nonlinear Sci. **24**, 411–439 (2014)
85. L. Kocarev, U. Parlitz, Generalized synchronization, predictability, and equivalence of unidirectionally coupled dynamical systems. Phys. Rev. Lett. **76**(11), 1816–1819 (1996)

86. N.F. Rulkov, M.M. Sushchik, L.S. Tsimring, H.D.I. Abarbanel, Generalized synchronization of chaos in directionally coupled chaotic systems. Phys. Rev. E **51**(2), 980–994 (1995)
87. H. Haken, *Advanced Synergetics: Instability Hierarchies of Self-Organizing Systems and Devices* (Springer, Berlin, 1983)
88. M. Akhmet, Z. Akhmetova, M.O. Fen, Chaos in economic models with exogenous shocks. J. Econ. Behav. Organ. **106**, 95–108 (2014)
89. L. Pribylova, Bifurcation routes to chaos in an extended Van der Pol's equation applied to economic models. Electron. J. Diff. Equ. **2009**, 1–21 (2009)
90. C. Zhang, J. Wei, Stability and bifurcation analysis in a kind of business cycle model with delay. Chaos Solitons Fractals **22**, 883–896 (2004)
91. M. Szydlowski, A. Krawiec, J. Tobola, Nonlinear oscillations in business cycle model with time lags. Chaos Solitons Fractals **12**, 505–517 (2001)
92. L. Wang, X.P. Wu, Bifurcation analysis of a Kaldor-Kalecki model of business cycle with time delay. Electron. J. Qual. Theory Diff. Equ. (27), 1–20 (2009)
93. H.G. Schuster, *Handbook of Chaos Control* (Wiley, Weinheim, 1999)
94. R.M. Goodwin, The nonlinear accelerator and the persistence of business cycles. Econometrica **19**, 1–17 (1951)
95. U. Parlitz, W. Lauterborn, Period-doubling cascades and devil's staircases of the driven Van der Pol oscillator. Phys. Rev. A. **36**, 1428–1434 (1987)
96. J.M.T. Thompson, H.B. Stewart, *Nonlinear Dynamics and Chaos* (Wiley, Chichester, 2002)
97. H.D.I. Abarbanel, N.F. Rulkov, M.M. Sushchik, Generalized synchronization of chaos: the auxiliary system approach. Phys. Rev. E **53**, 4528–4535 (1996)
98. B.R. Hunt, E. Ott, J.A. Yorke, Differentiable generalized synchronization of chaos. Phys. Rev. E **55**(4), 4029–4034 (1997)
99. G. Nicolis, I. Prigogine, *Exploring Complexity: An Introduction* (W.H. Freeman, New York, 1989)
100. F. Durrenmatt, *The Physicists* (Grove, New York, 1964)
101. A.A. Andronov, A.A. Vitt, C.E. Khaikin, *Theory of Oscillations* (Pergamon Press, Oxford, 1966)
102. F.C. Moon, *Chaotic Vibrations: An Introduction For Applied Scientists and Engineers* (Wiley, Hoboken, 2004)
103. J.D. Murray, *Mathematical biology II: spatial models and biomedical applications* (Springer, New York, 2003)
104. M.A. Vorontsov, W.B. Miller, *Self-organization in Optical Systems and Applications in Information Technology* (Springer, Berlin, 1998)
105. A.M. Turing, The chemical basis of morphogenesis. Philos. Trans. R. Soc. Lond., Ser. B, Biol. Sci. **237**, 37–72 (1952)
106. S. Bouali, Feedback loop in extended Van der Pol's equation applied to an economic model of cycles. Int. J. Bifurc. Chaos **9**, 745–756 (1999)
107. S. Bouali, A. Buscarino, L. Fortuna, M. Frasca, L.V. Gambuzza, Emulating complex business cycles by using an electronic analogue. Nonlinear Anal.: Real World Appl. **13**, 2459–2465 (2012)
108. G. Gabisch, H.-W. Lorenz, *Business Cycle Theory* (Springer, New York, 1987)
109. L. Fanti, P. Manfredi, Chaotic business cycles and fiscal policy: an IS-LM model with distributed tax collection lags. Chaos Solitons Fractals **32**, 736–744 (2007)
110. M.W. Hirsch, S. Smale, *Differential Equations, Dynamical Systems and Linear Algebra* (Academic Press, New York, 1974)
111. W.A. Brock, Distinguishing random and deterministic system: abridged version. J. Econ. Theory **40**, 168–195 (1986)
112. V. Atella, M. Centoni, G. Cubadda, Technology shocks, structural breaks and the effects on the business cycle. Econ. Lett. **100**, 392–395 (2008)
113. J. Gali, Technology, employment and the business cycle: do technology shocks explain aggregate fluctuations? Am. Econ. Rev. **89**, 249–271 (1999)
114. M. Akhmet, Z. Akhmetova, M.O. Fen, Exogenous versus endogenous for chaotic business cycles. Interdiscip. J. Discontin. Nonlinearity Complex. (in press)

Chapter 8
Chaos by Neural Networks

Interests of researchers to neural networks originate, first of all, from the fact that principles of functioning of neural networks are based on well-known biological processes about the methods of information processing by the brain. In the basis of brain functioning, there are collectives of huge number of, respectively, simple elements called neurons. The joint activities of neuron collectives are guaranteed by the giant number of connections between them.

Nowadays, neural networks have wide applications to solve problems connected with image processing, artificial intelligence, associate memories, signal processing, different types of forecast predictions, optimization problems, modeling of chemical reactions, information processing, and other areas.

Each neuron in a neural network is capable of receiving input signals, processing them, and sending an output signal. Neural signals consist of short electrical pulses called action potentials or spikes. A chain of action potentials emitted by a single neuron is called a spike train. Action potentials in a spike train are usually well separated, and it is impossible to excite a second spike during or immediately after a first one [1]. On the other hand, according to the switching phenomenon, frequency changes or other sudden noises, the states of the electronic networks are often subject to instantaneous perturbations and experience abrupt changes at certain instants [2–4]. That is why the discontinuity phenomenon is a natural property of neural networks, and models with discontinuities are more accurate to describe the evolutionary processes of neural networks.

The main subject of the present chapter is the investigation of chaos extension in collectives of neural networks. The extension is realized for Hopfield neural networks [5] and shunting inhibitory cellular neural networks [6], but the presented techniques are applicable to other kinds of neural networks as well. Our results reveal that chaos appears not only as an intrinsic property of neural networks, but also through transmissions between them.

M. Akhmet and M.O. Fen, *Replication of Chaos in Neural Networks, Economics and Physics*, Nonlinear Physical Science,
DOI 10.1007/978-3-662-47500-3_8

8.1 SICNNs with Chaotic External Inputs

Taking advantage of external inputs, in the present section, it is shown that shunting inhibitory cellular neural networks (SICNNs) behave chaotically. The analysis is based on the Li–Yorke definition of chaos. Appropriate illustrations which support the theoretical results are depicted.

Cellular neural networks have been paid much attention in the past two decades. Exceptional role in psychophysics, speech, perception, robotics, adaptive pattern recognition, vision, and image processing has been played by SICNNs. Chaotic dynamics is an object of great interest in neural networks theory. This is natural since chaotic outputs have been obtained for several types of neural networks. According to the design of neural networks, solutions of some of them can be used as an input for another ones. In this section, we realize this idea by considering SICNNs to obtain chaos through chaotic external inputs. This is the first time that a theoretically approved chaos is obtained in SICNNs.

8.1.1 Introduction

A class of cellular neural networks, introduced by Bouzerdoum and Pinter [6], is the shunting inhibitory cellular neural networks, which have been extensively applied in psychophysics, speech, perception, robotics, adaptive pattern recognition, vision, and image processing [7–13].

The model in its most original formulation [6] is as follows. Consider a two-dimensional grid of processing cells, and let C_{ij}, $i = 1, 2, \ldots, m$, $j = 1, 2, \ldots, n$, denote the cell at the (i, j) position of the lattice. Denote by $N_r(i, j)$ the r-neighborhood of C_{ij}, such that

$$N_r(i, j) = \{C_{kl} : \max\{|k - i|, |l - j|\} \le r,\ 1 \le k \le m,\ 1 \le l \le n\}.$$

In SICNNs, neighboring cells exert mutual inhibitory interactions of the shunting type. The dynamics of the cell C_{ij} is described by the nonlinear ordinary differential equation

$$\frac{dx_{ij}}{dt} = -a_{ij}x_{ij} - \sum_{C_{kl} \in N_r(i,j)} C_{ij}^{kl} f(x_{kl}(t))x_{ij} + L_{ij}(t), \tag{8.1.1}$$

where x_{ij} is the activity of the cell C_{ij}; $L_{ij}(t)$ is the external input to C_{ij}; the constant a_{ij} represents the passive decay rate of the cell activity; $C_{ij}^{kl} \ge 0$ is the connection or coupling strength of postsynaptic activity of the cell C_{kl} transmitted to the cell C_{ij}; and the activation function $f(x_{kl})$ is a positive continuous function representing the output or firing rate of the cell C_{kl}. For our theoretical discussions, we will consider continuous external inputs.

The existence and the stability of periodic, almost periodic and antiperiodic solutions of SICNNs have been published in papers [14–23]. The main novelty of the present section is the verification of the chaotic behavior in SICNNs. To prove the existence of chaos, we apply the technique based on the Li–Yorke definition [24], and make use of *chaotic external inputs* in the networks. We say that the external inputs are chaotic if they belong to a collection of functions which satisfy the ingredients of chaos. That is, we consider members of a chaotic set as external input terms, and, as a result, we obtain solutions which display chaotic behavior.

Existence of a chaotic attractor in SICNNs with impulses was numerically observed in [25] without a theoretical support, as well it is the case for the paper [26]. The presence of chaos in SICNNs with impulsive effects [27] by means of chaotic external inputs will be considered in Sect. 8.3.

8.1.2 Preliminaries

Throughout the section, $\mathbb{R}$ will stand for the set of real numbers, and the norm

$$\|u\| = \max_{(i,j)} |u_{ij}|$$

will be used, where

$$u = \{u_{ij}\} = (u_{11}, \ldots, u_{1n}, \ldots, u_{m1}, \ldots, u_{mn}) \in \mathbb{R}^{m\times n}$$

and m, n are natural numbers.

Suppose that $\mathscr{B}$ is a collection of continuous functions $\psi(t) = \{\psi_{ij}(t)\}$, $i = 1, 2, \ldots, m$, $j = 1, 2, \ldots, n$, such that $\sup_{t\in\mathbb{R}} \|\psi(t)\| \le M$, where M is a positive number. We start by describing the ingredients of Li–Yorke chaos for the collection $\mathscr{B}$.

We say that a couple $(\psi(t), \widetilde{\psi}(t)) \in \mathscr{B} \times \mathscr{B}$ is proximal if for arbitrary small $\varepsilon > 0$ and arbitrary large $E > 0$, there exist infinitely many disjoint intervals of length not less than E such that $\|\psi(t) - \widetilde{\psi}(t)\| < \varepsilon$, for each t from these intervals. On the other hand, a couple $(\psi(t), \widetilde{\psi}(t)) \in \mathscr{B} \times \mathscr{B}$ is called frequently (ε_0, Δ)-separated if there exist positive numbers ε_0, Δ and infinitely many disjoint intervals of length not less than Δ, such that $\|\psi(t) - \widetilde{\psi}(t)\| > \varepsilon_0$, for each t from these intervals. It is worth saying that the numbers ε_0 and Δ depend on the functions $\psi(t)$ and $\widetilde{\psi}(t)$.

A couple $(\psi(t), \widetilde{\psi}(t)) \in \mathscr{B} \times \mathscr{B}$ is a Li–Yorke pair if it is proximal and frequently (ε_0, Δ)-separated for some positive numbers ε_0 and Δ. Moreover, an uncountable set $\mathscr{C} \subset \mathscr{B}$ is called a scrambled set if $\mathscr{C}$ does not contain any periodic functions and each couple of different functions inside $\mathscr{C} \times \mathscr{C}$ is a Li–Yorke pair.

$\mathscr{B}$ is called a Li–Yorke chaotic set if (i) There exists a positive number T_0 such that $\mathscr{B}$ possesses a periodic function of period kT_0, for any $k \in \mathbb{N}$; (ii) $\mathscr{B}$ possesses a scrambled set $\mathscr{C}$; (iii) For any function $\psi(t) \in \mathscr{C}$ and any periodic function $\widetilde{\psi}(t) \in \mathscr{B}$, the couple $\left(\psi(t), \widetilde{\psi}(t)\right)$ is frequently (ε_0, Δ)-separated for some positive numbers ε_0 and Δ.

One can obtain a new Li–Yorke chaotic set from a given one as follows. Suppose that $h : \mathbb{R}^{m\times n} \to \mathbb{R}^{\overline{m}\times\overline{n}}$ is a function which satisfies for all $u_1, u_2 \in \mathbb{R}^{m\times n}$ that

$$L_1 \|u_1 - u_2\| \le \|h(u_1) - h(u_2)\| \le L_2 \|u_1 - u_2\|, \tag{8.1.2}$$

where L_1 and L_2 are positive numbers. One can verify that if the collection $\mathscr{B}$ is Li–Yorke chaotic then the collection $\mathscr{B}_h$ whose elements are of the form $h(\psi(t))$, $\psi(t) \in \mathscr{B}$, is also Li–Yorke chaotic.

The following conditions are needed:

(C1) $\gamma = \min\limits_{(i,j)} a_{ij} > 0$;

(C2) There exist positive numbers M_{ij} such that $\sup\limits_{t\in\mathbb{R}} \left|L_{ij}(t)\right| \le M_{ij}$;

(C3) There exists a positive number M_f such that $\sup\limits_{s\in\mathbb{R}} |f(s)| \le M_f$;

(C4) There exists a positive number L_f such that $|f(s_1) - f(s_2)| \le L_f |s_1 - s_2|$ for all $s_1, s_2 \in \mathbb{R}$;

(C5) $M_f \max\limits_{(i,j)} \dfrac{\sum_{C_{kl}\in N_r(i,j)} C_{ij}^{kl}}{a_{ij}} < 1$;

(C6) $\dfrac{\overline{c}(L_f K_0 + M_f)}{\gamma} < 1$, where $\overline{c}$ and K_0 are defined as $\overline{c} = \max\limits_{(i,j)} \sum\limits_{C_{kl}\in N_r(i,j)} C_{ij}^{kl}$

and $K_0 = \dfrac{\max\limits_{(i,j)} \dfrac{M_{ij}}{a_{ij}}}{1 - M_f \max\limits_{(i,j)} \dfrac{\sum_{C_{kl}\in N_r(i,j)} C_{ij}^{kl}}{a_{ij}}}$.

Using the theory of quasilinear equations [28], one can verify that a bounded on $\mathbb{R}$ function $x(t) = \left\{x_{ij}(t)\right\}$ is a solution of the network (8.1.1) if and only if the following integral equation is satisfied

$$x_{ij}(t) = -\int_{-\infty}^{t} \mathrm{e}^{-a_{ij}(t-s)}\Big[\sum_{C_{kl}\in N_r(i,j)} C_{ij}^{kl} f(x_{kl}(s))x_{ij}(s) - L_{ij}(s)\Big]ds. \tag{8.1.3}$$

A result about existence of bounded on $\mathbb{R}$ solutions is as follows.

Lemma 8.1 *For any $L(t) = \left\{L_{ij}(t)\right\}$, $i = 1, 2, \ldots, m$, $j = 1, 2, \ldots, n$, there exists a unique bounded on $\mathbb{R}$ solution $\phi_L(t) = \left\{\phi_L^{ij}(t)\right\}$ of the network (8.1.1) such that $\sup_{t\in\mathbb{R}} \|\phi_L(t)\| \le K_0$.*

Proof Consider the set C_0 of continuous functions $u(t) = \{u_{ij}(t)\}, i = 1, 2, \ldots, m$, $j = 1, 2, \ldots, n$, such that $\|u\|_1 \le K_0$, where $\|u\|_1 = \sup_{t\in\mathbb{R}} \|u(t)\|$. Define on C_0 the operator Π as

$$(\Pi u)_{ij}(t) \equiv -\int_{-\infty}^{t} \mathrm{e}^{-a_{ij}(t-s)}\Big[\sum_{C_{kl}\in N_r(i,j)} C_{ij}^{kl} f(u_{kl}(s))u_{ij}(s) - L_{ij}(s)\Big]ds,$$

where $u(t) = \{u_{ij}(t)\}$ and $\Pi u(t) = \{(\Pi u)_{ij}(t)\}$. If $u(t)$ belongs to C_0 then

$$\begin{aligned}\big|(\Pi u)_{ij}(t)\big| &\le \int_{-\infty}^{t} e^{-a_{ij}(t-s)}\Big[\sum_{C_{kl}\in N_r(i,j)} C_{ij}^{kl}\,|f(u_{kl})(s)|\,\big|u_{ij}(s)\big| + \big|L_{ij}(s)\big|\Big]ds \\ &\le \frac{1}{a_{ij}}\Big(M_{ij} + M_f K_0 \sum_{C_{kl}\in N_r(i,j)} C_{ij}^{kl}\Big).\end{aligned}$$

Accordingly, we have $\|\Pi u\|_1 \le \max\limits_{(i,j)} \dfrac{M_{ij}}{a_{ij}} + M_f K_0 \max\limits_{(i,j)} \dfrac{\sum_{C_{kl}\in N_r(i,j)} C_{ij}^{kl}}{a_{ij}} = K_0$. Therefore, $\Pi(C_0) \subseteq C_0$.

On the other hand, for any $u, v \in C_0$,

$$\begin{aligned}\big|(\Pi u)_{ij}(t) - (\Pi v)_{ij}(t)\big| &\le \int_{-\infty}^{t} e^{-a_{ij}(t-s)} \sum_{C_{kl}\in N_r(i,j)} C_{ij}^{kl}\Big|f(u_{kl}(s))u_{ij}(s) \\ &-f(u_{kl}(s))v_{ij}(s)\Big|ds + \int_{-\infty}^{t} e^{-a_{ij}(t-s)} \sum_{C_{kl}\in N_r(i,j)} C_{ij}^{kl}\Big|f(u_{kl}(s))v_{ij}(s) \\ &-f(v_{kl}(s))v_{ij}(s)\Big|ds \\ &\le (L_f K_0 + M_f) \max_{(i,j)} \frac{\sum_{C_{kl}\in N_r(i,j)} C_{ij}^{kl}}{a_{ij}} \|u-v\|_1 .\end{aligned}$$

Thus, $\|\Pi u - \Pi v\|_1 \le (L_f K_0 + M_f) \max\limits_{(i,j)} \dfrac{\sum_{C_{kl}\in N_r(i,j)} C_{ij}^{kl}}{a_{ij}} \|u-v\|_1$, and condition $(C6)$ implies that the operator Π is contractive.

Consequently, for any $L(t)$, there exists a unique bounded on $\mathbb{R}$ solution $\phi_L(t)$ of the network (8.1.1) such that $\sup\limits_{t\in\mathbb{R}} \|\phi_L(t)\| \le K_0$. □

For a given $L(t) = \{L_{ij}(t)\}$, $i = 1, 2, \ldots, m$, $j = 1, 2, \ldots, n$, let us denote by $x_L(t, x_0) = \Big\{x_L^{ij}(t, x_0)\Big\}$ the unique solution of the SICNN (8.1.1) with $x_L(0, x_0) = x_0$. We note that the solution $x_L(t, x_0)$ is not necessarily bounded on $\mathbb{R}$.

Consider the set $\mathscr{L}$ whose elements are functions of the form $L(t) = \{L_{ij}(t)\}$, $i = 1, 2, \ldots, m$, $j = 1, 2, \ldots, n$, such that $\sup\limits_{t\in\mathbb{R}} \big|L_{ij}(t)\big| \le M_{ij}$ for each i and j.

Suppose that $\mathscr{A}$ is the collection of functions consisting of the bounded on $\mathbb{R}$ solutions $\phi_L(t)$ of system (8.1.1), where $L(t) \in \mathscr{L}$. In the present section, we assume that $\mathscr{L}$ is an equicontinuous family on $\mathbb{R}$.

The following assertion confirms the attractiveness of the set $\mathscr{A}$.

Lemma 8.2 *For any $x_0 \in \mathbb{R}^{m\times n}$ and $L(t) = \{L_{ij}(t)\}$, $i = 1, 2, \ldots, m$, $j = 1, 2, \ldots, n$, we have $\|x_L(t, x_0) - \phi_L(t)\| \to 0$ as $t \to \infty$.*

Proof Making use of the relation

$$x_L^{ij}(t, x_0) - \phi_L^{ij}(t) = e^{-a_{ij}t}\left(x_L^{ij}(0, x_0) - \phi_L^{ij}(0)\right)$$
$$-\int_0^t e^{-a_{ij}(t-s)}\Big[\sum_{C_{kl}\in N_r(i,j)} C_{ij}^{kl} f(x_L^{kl}(s, x_0))x_L^{ij}(s, x_0)$$
$$-\sum_{C_{kl}\in N_r(i,j)} C_{ij}^{kl} f(\phi_L^{kl}(s))\phi_L^{ij}(s)\Big]ds$$

we obtain for $t \geq 0$ that

$$\left|x_L^{ij}(t, x_0) - \phi_L^{ij}(t)\right| \leq e^{-a_{ij}t}\left|x_L^{ij}(0, x_0) - \phi_L^{ij}(0)\right|$$
$$+M_f \sum_{C_{kl}\in N_r(i,j)} C_{ij}^{kl} \int_0^t e^{-a_{ij}(t-s)}\left|x_L^{ij}(s, x_0) - \phi_L^{ij}(s)\right|ds$$
$$+L_f K_0 \int_0^t e^{-a_{ij}(t-s)} \sum_{C_{kl}\in N_r(i,j)} C_{ij}^{kl}\left|x_L^{kl}(s, x_0) - \phi_L^{kl}(s)\right|ds.$$

The last inequality implies for $t \geq 0$ that

$$e^{\gamma t}\left\|x_L(t, x_0) - \phi_L(t)\right\|$$
$$\leq \|x_0 - \phi_L(0)\| + \overline{c}(L_f K_0 + M_f)\int_0^t e^{\gamma s}\left\|x_L(s, x_0) - \phi_L(s)\right\| ds.$$

Applying Gronwall–Bellman Lemma, one can attain that

$$\|x_L(t, x_0) - \phi_L(t)\| \leq \|x_0 - \phi_L(0)\|\, e^{[\overline{c}(L_f K_0 + M_f) - \gamma]t},\ t \geq 0.$$

Consequently, $\|x_L(t, x_0) - \phi_L(t)\| \to 0$ as $t \to \infty$, in accordance with condition $(C6)$. □

Our purpose in the next part is to prove rigorously that if the collection $\mathscr{L}$ is chaotic in the sense of Li–Yorke then the same is true for $\mathscr{A}$. In other words, if the external input terms $L_{ij}(t)$ behave chaotically, then the dynamics of the SICNNs are also chaotic.

8.1.3 Chaotic Dynamics

The replication of the ingredients of Li–Yorke chaos from the collection $\mathscr{L}$ to the collection $\mathscr{A}$ will be affirmed in the following two lemmas, and the main conclusion will be stated in Theorem 8.1. We start with the following lemma, which indicates existence of proximality in the collection $\mathscr{A}$.

Lemma 8.3 *If a couple of functions* $\left(L(t), \widetilde{L}(t)\right) \in \mathscr{L} \times \mathscr{L}$ *is proximal, then the same is true for the couple* $\left(\phi_L(t), \phi_{\widetilde{L}}(t)\right) \in \mathscr{A} \times \mathscr{A}$.

Proof Fix an arbitrary small positive number ε and an arbitrary large positive number E. Set

$$R = 2\left(M_f K_0 \max_{(i,j)} \frac{\sum_{C_{kl} \in N_r(i,j)} C_{ij}^{kl}}{a_{ij}} + \max_{(i,j)} \frac{M_{ij}}{a_{ij}} \right)$$

and

$$0 < \alpha \leq \frac{\gamma - \overline{c}(L_f K_0 + M_f)}{1 + \gamma - \overline{c}(L_f K_0 + M_f)}.$$

Suppose that a given pair $\left(L(t), \widetilde{L}(t)\right) \in \mathscr{L} \times \mathscr{L}$ is proximal. There exist a sequence of real numbers $\left\{E_q\right\}$ satisfying $E_q \geq E$ for each $q \in \mathbb{N}$ and a sequence $\left\{t_q\right\}$, $t_q \to \infty$ as $q \to \infty$, such that $\left\|L(t) - \widetilde{L}(t)\right\| < \alpha\varepsilon$ for each t from the disjoint intervals $J_q = [t_q, t_q + E_q]$, $q \in \mathbb{N}$. Let us denote $\phi_L(t) = \left\{\phi_L^{ij}(t)\right\}$ and $\phi_{\widetilde{L}}(t) = \left\{\phi_{\widetilde{L}}^{ij}(t)\right\}$.

Fix $q \in \mathbb{N}$. For $t \in J_q$, using the relation (8.1.3), one can reach up for any i and j that

$$\phi_L^{ij}(t) - \phi_{\widetilde{L}}^{ij}(t) = -\int_{-\infty}^{t} e^{-a_{ij}(t-s)}\Big[\sum_{C_{kl} \in N_r(i,j)} C_{ij}^{kl} f(\phi_L^{kl}(s))\phi_L^{ij}(s) - L_{ij}(s)$$
$$- \sum_{C_{kl} \in N_r(i,j)} C_{ij}^{kl} f(\phi_{\widetilde{L}}^{kl}(s))\phi_{\widetilde{L}}^{ij}(s) + \widetilde{L}_{ij}(s)\Big] ds.$$

By means of the last equation, one can obtain that

$$\left|\phi_L^{ij}(t) - \phi_{\widetilde{L}}^{ij}(t)\right| \leq 2\left(M_f K_0 \frac{\sum_{C_{kl} \in N_r(i,j)} C_{ij}^{kl}}{a_{ij}} + \frac{M_{ij}}{a_{ij}} \right) e^{-a_{ij}(t-t_q)}$$
$$+ \frac{\alpha\varepsilon}{a_{ij}} \left(1 - e^{-a_{ij}(t-t_q)}\right)$$
$$+ \overline{c}(L_f K_0 + M_f) \int_{t_q}^{t} e^{-a_{ij}(t-s)} \left\|\phi_L(s) - \phi_{\widetilde{L}}(s)\right\| ds.$$

Accordingly, for $t \in J_q$ we have that

$$e^{\gamma t} \left\| \phi_L(t) - \phi_{\widetilde{L}}(t) \right\| \leq Re^{\gamma t_q} + \frac{\alpha \varepsilon}{\gamma} \left(e^{\gamma t} - e^{\gamma t_q} \right)$$
$$+ \overline{c}(L_f K_0 + M_f) \int_{t_q}^{t} e^{\gamma s} \left\| \phi_L(s) - \phi_{\widetilde{L}}(s) \right\| ds.$$

Application of Gronwall's Lemma to the last inequality implies for $t \in J_q$ that

$$\left\| \phi_L(t) - \phi_{\widetilde{L}}(t) \right\| \leq \frac{\alpha \varepsilon}{\gamma - \overline{c}(L_f K_0 + M_f)} \left(1 - e^{[\overline{c}(L_f K_0 + M_f) - \gamma](t - t_q)} \right)$$
$$+ Re^{[\overline{c}(L_f K_0 + M_f) - \gamma](t - t_q)}.$$

Suppose that the number E is sufficiently large such that

$$E > \frac{2}{\gamma - \overline{c}(L_f K_0 + M_f)} \ln \left(\frac{R}{\alpha \varepsilon} \right).$$

In this case, if t belongs to the interval $[t_q + E/2, t_q + E_q]$, then

$$Re^{[\overline{c}(L_f K_0 + M_f) - \gamma](t - t_q)} < \alpha \varepsilon.$$

Thus, for $t \in [t_q + E/2, t_q + E_q]$. the inequality

$$\left\| \phi_L(t) - \phi_{\widetilde{L}}(t) \right\| < \left(1 + \frac{1}{\gamma - \overline{c}(L_f K_0 + M_f)} \right) \alpha \varepsilon \leq \varepsilon.$$

is valid. Consequently, since the last inequality holds for each t from the disjoint intervals $J_q^1 = [t_q + E/2, t_q + E_q]$, $q \in \mathbb{N}$, the couple $\left(\phi_L(t), \phi_{\widetilde{L}}(t) \right) \in \mathscr{A} \times \mathscr{A}$ is proximal. □

Now, let us continue with the replication the second main ingredient of Li–Yorke chaos in the next lemma.

Lemma 8.4 *If a couple $\left(L(t), \widetilde{L}(t) \right) \in \mathscr{L} \times \mathscr{L}$ is frequently (ε_0, Δ)-separated for some positive numbers ε_0 and Δ, then there exist positive numbers ε_1 and $\overline{\Delta}$ such that the couple $\left(\phi_L(t), \phi_{\widetilde{L}}(t) \right) \in \mathscr{A} \times \mathscr{A}$ is frequently $(\varepsilon_1, \overline{\Delta})$-separated.*

Proof Suppose that a given couple $\left(L(t), \widetilde{L}(t) \right) \in \mathscr{L} \times \mathscr{L}$ is frequently (ε_0, Δ) separated, for some $\varepsilon_0 > 0$ and $\Delta > 0$. In this case, there exist infinitely many disjoint intervals J_q, $q \in \mathbb{N}$, each with length not less than Δ, such that $\left\| L(t) - \widetilde{L}(t) \right\| > \varepsilon_0$ for each t from these intervals. In the proof, we will verify the existence of positive numbers ε_1, $\overline{\Delta}$ and infinitely many disjoint intervals $J_q^1 \subset J_q$, $q \in \mathbb{N}$, each with length $\overline{\Delta}$, such that the inequality $\left\| \phi_L(t) - \phi_{\widetilde{L}}(t) \right\| > \varepsilon_1$ holds for each t from the intervals J_q^1, $q \in \mathbb{N}$.

According to the equicontinuity of $\mathscr{L}$, one can find a positive number $\tau < \Delta$, such that for any $t_1, t_2 \in \mathbb{R}$ with $|t_1 - t_2| < \tau$, the inequality

$$\left|\left(L_{ij}(t_1) - \widetilde{L}_{ij}(t_1)\right) - \left(L_{ij}(t_2) - \widetilde{L}_{ij}(t_2)\right)\right| < \frac{\varepsilon_0}{2} \tag{8.1.4}$$

holds for all $1 \le i \le m$, $1 \le j \le n$.

Suppose that for each $q \in \mathbb{N}$, the number s_q denotes the midpoint of the interval J_q. Let us define a sequence $\{\theta_q\}$ through the equation $\theta_q = s_q - \tau/2$.

Let us fix an arbitrary $q \in \mathbb{N}$. One can find integers i_0, j_0, such that

$$\left|L_{i_0 j_0}(s_q) - \widetilde{L}_{i_0 j_0}(s_q)\right| = \left\|L(s_q) - \widetilde{L}(s_q)\right\| > \varepsilon_0. \tag{8.1.5}$$

Making use of the inequality (8.1.4), for all $t \in [\theta_q, \theta_q + \tau]$ we have

$$\begin{aligned}
&\left|L_{i_0 j_0}(s_q) - \widetilde{L}_{i_0 j_0}(s_q)\right| - \left|L_{i_0 j_0}(t) - \widetilde{L}_{i_0 j_0}(t)\right| \\
&\le \left|\left(L_{i_0 j_0}(t) - \widetilde{L}_{i_0 j_0}(t)\right) - \left(L_{i_0 j_0}(s_q) - \widetilde{L}_{i_0 j_0}(s_q)\right)\right| \\
&< \frac{\varepsilon_0}{2}
\end{aligned}$$

and therefore, by means of (8.1.5), we achieve that the inequality

$$\left|L_{i_0 j_0}(t) - \widetilde{L}_{i_0 j_0}(t)\right| > \left|L_{i_0 j_0}(s_q) - \widetilde{L}_{i_0 j_0}(s_q)\right| - \frac{\varepsilon_0}{2} > \frac{\varepsilon_0}{2} \tag{8.1.6}$$

is valid for all $t \in [\theta_q, \theta_q + \tau]$.

For each i and j, one can find numbers $\zeta_{ij}^q \in [\theta_q, \theta_q + \tau]$ such that

$$\int_{\theta_q}^{\theta_q+\tau} \left(L(s) - \widetilde{L}(s)\right) ds = \tau \left(L_{11}(\zeta_{11}^q) - \widetilde{L}_{11}(\zeta_{11}^q), \ldots, L_{mn}(\zeta_{mn}^q) - \widetilde{L}_{mn}(\zeta_{mn}^q)\right).$$

Thus, according to the inequality (8.1.6), we have that

$$\left\|\int_{\theta_q}^{\theta_q+\tau} \left(L(s) - \widetilde{L}(s)\right) ds\right\| \ge \tau \left|L_{i_0 j_0}(\zeta_{i_0 j_0}^q) - \widetilde{L}_{i_0 j_0}(\zeta_{i_0 j_0}^q)\right| > \frac{\tau \varepsilon_0}{2}. \tag{8.1.7}$$

For $t \in [\theta_q, \theta_q + \tau]$, using the couple of relations

$$\phi_L^{ij}(t) = \phi_L^{ij}(\theta_q) - \int_{\theta_q}^{t} \Big[a_{ij} + \sum_{C_{kl} \in N_r(i,j)} C_{ij}^{kl} f(\phi_L^{kl}(s))\Big] \phi_L^{ij}(s) ds + \int_{\theta_q}^{t} L_{ij}(s) ds,$$

and

$$\phi_{\widetilde{L}}^{ij}(t) = \phi_{\widetilde{L}}^{ij}(\theta_q) - \int_{\theta_q}^{t} \Big[a_{ij} + \sum_{C_{kl} \in N_r(i,j)} C_{ij}^{kl} f(\phi_{\widetilde{L}}^{kl}(s))\Big]\phi_{\widetilde{L}}^{ij}(s)ds + \int_{\theta_q}^{t} \widetilde{L}_{ij}(s)ds,$$

it can be verified that

$$\phi_{L}^{ij}(\theta_q+\tau) - \phi_{\widetilde{L}}^{ij}(\theta_q+\tau) = \int_{\theta_q}^{\theta_q+\tau} \left(L_{ij}(s) - \widetilde{L}_{ij}(s)\right) ds$$
$$+(\phi_{L}^{ij}(\theta_q) - \phi_{\widetilde{L}}^{ij}(\theta_q)) - \int_{\theta_q}^{\theta_q+\tau} a_{ij}\left(\phi_{L}^{ij}(s) - \phi_{\widetilde{L}}^{ij}(s)\right) ds$$
$$-\int_{\theta_q}^{\theta_q+\tau} \Big[\sum_{C_{kl} \in N_r(i,j)} C_{ij}^{kl} f(\phi_{L}^{kl}(s))\phi_{L}^{ij}(s) - \sum_{C_{kl} \in N_r(i,j)} C_{ij}^{kl} f(\phi_{\widetilde{L}}^{kl}(s))\phi_{\widetilde{L}}^{ij}(s)\Big]ds.$$

Hence we achieve that

$$\left\|\phi_L(\theta_q+\tau) - \phi_{\widetilde{L}}(\theta_q+\tau)\right\| \geq \left\|\int_{\theta_q}^{\theta_q+\tau} \left(L(s) - \widetilde{L}(s)\right)ds\right\|$$
$$-\left\|\phi_L(\theta_q) - \phi_{\widetilde{L}}(\theta_q)\right\| - \max_{(i,j)}\left|\int_{\theta_q}^{\theta_q+\tau} a_{ij}\left(\phi_{L}^{ij}(s) - \phi_{\widetilde{L}}^{ij}(s)\right) ds\right|$$
$$-\max_{(i,j)}\Big|\int_{\theta_q}^{\theta_q+\tau} \Big[\sum_{C_{kl} \in N_r(i,j)} C_{ij}^{kl} f(\phi_{L}^{kl}(s))\phi_{L}^{ij}(s)$$
$$-\sum_{C_{kl} \in N_r(i,j)} C_{ij}^{kl} f(\phi_{\widetilde{L}}^{kl}(s))\phi_{\widetilde{L}}^{ij}(s)\Big]ds\Big|. \tag{8.1.8}$$

Let us denote $\overline{\gamma} = \max_{(i,j)} a_{ij}$ and $H_0 = \max_{(i,j)} M_{ij}$. The inequalities (8.1.7) and (8.1.8) together imply that

$$\max_{t\in[\theta_q,\theta_q+\tau]} \left\|\phi_L(t) - \phi_{\widetilde{L}}(t)\right\| \geq \left\|\phi_L(\theta_q+\tau) - \phi_{\widetilde{L}}(\theta_q+\tau)\right\|$$
$$> \frac{\tau\varepsilon_0}{2} - [1+\tau\overline{\gamma}+\tau\overline{c}(L_f K_0 + M_f)] \max_{t\in[\theta_q,\theta_q+\tau]} \left\|\phi_L(t) - \phi_{\widetilde{L}}(t)\right\|.$$

Therefore, we have $\max_{t\in[\theta_q,\theta_q+\tau]} \left\|\phi_L(t) - \phi_{\widetilde{L}}(t)\right\| > \overline{\varepsilon}$, where

$$\overline{\varepsilon} = \frac{\tau\varepsilon_0}{2[2+\tau\overline{\gamma}+\tau\overline{c}(L_f K_0 + M_f)]}.$$

Suppose that $\max_{t\in[\theta_q,\theta_q+\tau]} \left\|\phi_L(t) - \phi_{\widetilde{L}}(t)\right\| = \left\|\phi_L(\xi_q) - \phi_{\widetilde{L}}(\xi_q)\right\|$, for some $\xi_q \in [\theta_q, \theta_q+\tau]$. Define

$$\overline{\Delta} = \min\left\{\frac{\tau}{2}, \frac{\overline{\varepsilon}}{4\left(H_0 + K_0\overline{\gamma} + M_f K_0\overline{c}\right)}\right\}$$

and let

$$\theta_q^1 = \begin{cases} \xi_q, & \text{if } \xi_q \le \theta_q + \tau/2 \\ \xi_q - \overline{\Delta}, & \text{if } \xi_q > \theta_q + \tau/2 \end{cases}.$$

For $t \in \left[\theta_q^1, \theta_q^1 + \overline{\Delta}\right]$, by favor of the integral equation

$$\begin{aligned}
&\phi_L^{ij}(t) - \phi_{\widetilde{L}}^{ij}(t) = (\phi_L^{ij}(\xi_q) - \phi_{\widetilde{L}}^{ij}(\xi_q)) \\
&+ \int_{\xi_q}^t \left(L_{ij}(s) - \widetilde{L}_{ij}(s)\right) ds - \int_{\xi_q}^t a_{ij}\left(\phi_L^{ij}(s) - \phi_{\widetilde{L}}^{ij}(s)\right) ds \\
&- \int_{\xi_q}^t \Big[\sum_{C_{kl}\in N_r(i,j)} C_{ij}^{kl} f(\phi_L^{kl}(s))\phi_L^{ij}(s) - \sum_{C_{kl}\in N_r(i,j)} C_{ij}^{kl} f(\phi_{\widetilde{L}}^{kl}(s))\phi_{\widetilde{L}}^{ij}(s)\Big] ds,
\end{aligned}$$

we have

$$\begin{aligned}
&\left\|\phi_L(t) - \phi_{\widetilde{L}}(t)\right\| \ge \left\|\phi_L(\xi_q) - \phi_{\widetilde{L}}(\xi_q)\right\| \\
&- \max_{(i,j)}\left|\int_{\xi_q}^t \left(L_{ij}(s) - \widetilde{L}_{ij}(s)\right) ds\right| - \max_{(i,j)}\left|\int_{\xi_q}^t a_{ij}\left(\phi_L^{ij}(s) - \phi_{\widetilde{L}}^{ij}(s)\right) ds\right| \\
&- \max_{(i,j)}\left|\int_{\xi_q}^t \Big[\sum_{C_{kl}\in N_r(i,j)} C_{ij}^{kl} f(\phi_L^{kl}(s))\phi_L^{ij}(s) - \sum_{C_{kl}\in N_r(i,j)} C_{ij}^{kl} f(\phi_{\widetilde{L}}^{kl}(s))\phi_{\widetilde{L}}^{ij}(s)\Big] ds\right| \\
&> \overline{\varepsilon} - 2\overline{\Delta}\left(H_0 + K_0\overline{\gamma} + M_f K_0\overline{c}\right) \\
&\ge \frac{\overline{\varepsilon}}{2}.
\end{aligned}$$

Consequently, for each t from the disjoint intervals $J_q^1 = \left[\theta_q^1, \theta_q^1 + \overline{\Delta}\right]$, $q \in \mathbb{N}$, the inequality $\left\|\phi_L(t) - \phi_{\widetilde{L}}(t)\right\| > \varepsilon_1$ holds, where $\varepsilon_1 = \overline{\varepsilon}/2$. □

The following theorem, which is the main result of the present section, indicates that the network (8.1.1) is chaotic, provided that the external inputs are chaotic.

Theorem 8.1 *If $\mathscr{L}$ is a Li–Yorke chaotic set, then the same is true for $\mathscr{A}$.*

Proof Assume that the set $\mathscr{L}$ is Li–Yorke chaotic. Under the circumstances, there exists a positive number T_0 such that for any natural number k, $\mathscr{L}$ possesses a periodic function of period kT_0. One can confirm that $L(t) \in \mathscr{L}$ is kT_0-periodic if and only if $\phi_L(t) \in \mathscr{A}$ is kT_0-periodic. Therefore, the set $\mathscr{A}$ contains a kT_0-periodic function for any natural number k.

Next, suppose that $\mathscr{L}_S$ is a scrambled set inside $\mathscr{L}$ and take into account the collection $\mathscr{A}_S$ with elements of the form $\phi_L(t)$, where $L(t) \in \mathscr{L}_S$. Since $\mathscr{L}_S$ is uncountable, the set $\mathscr{A}_S$ is also uncountable. Due to the one-to-one correspondence between the periodic functions inside $\mathscr{L}$ and $\mathscr{A}$, no periodic functions exist inside $\mathscr{A}_S$.

According to Lemmas 8.3 and 8.4, $\mathscr{A}_S$ is a scrambled set. Moreover, Lemma 8.4 implies that each couple of functions inside $\mathscr{A}_S \times \mathscr{A}_P$ is frequently $(\varepsilon_1, \overline{\Delta})$-separated for some positive numbers ε_1 and $\overline{\Delta}$, where $\mathscr{A}_P$ denotes the set of all periodic functions inside $\mathscr{A}$. Consequently, the set $\mathscr{A}$ is Li–Yorke chaotic. □

Remark 8.1 Combining the result of Theorem 8.1 with the one of Lemma 8.2, we conclude that a *chaotic attractor* takes place in the dynamics of system (8.1.1).

8.1.4 Examples

To actualize the results of Sect. 8.1, one needs a source of external inputs, $L_{ij}(t)$, which are ensured to be chaotic in the Li–Yorke sense. For this reason, in the first example, we will take into account SICNNs whose external inputs are relay functions with chaotically changing switching moments. Then, to support our new theoretical results, we will make use of the solutions of this network as external inputs for another SICNNs, which is the main illustrative object for the results of Sect. 8.1. To increase the flexibility of our method for applications, we will also take advantage of nonlinear functions to build chaotic inputs.

Example 8.1 Let us introduce the SICNN

$$\frac{dz_{ij}}{dt} = -b_{ij}z_{ij} - \sum_{D_{kl}\in N_1(i,j)} D_{ij}^{kl} g(z_{kl}(t))z_{ij} + \nu_{ij}(t, t_0), \tag{8.1.9}$$

in which $i, j = 1, 2, 3,$

$$\begin{pmatrix} b_{11} & b_{12} & b_{13} \\ b_{21} & b_{22} & b_{23} \\ b_{31} & b_{32} & b_{33} \end{pmatrix} = \begin{pmatrix} 8 & 4 & 7 \\ 10 & 6 & 5 \\ 6 & 4 & 1 \end{pmatrix},$$

$$\begin{pmatrix} D_{11} & D_{12} & D_{13} \\ D_{21} & D_{22} & D_{23} \\ D_{31} & D_{32} & D_{33} \end{pmatrix} = \begin{pmatrix} 0.006 & 0 & 0.001 \\ 0.009 & 0.002 & 0.003 \\ 0 & 0.005 & 0.004 \end{pmatrix}.$$

In Eq. (8.1.9), D_{ij} denotes the cell at the (i, j) position of the lattice, and for each i, j, the relay function $\nu_{ij}(t, t_0)$ is defined by the equation

$$\nu_{ij}(t, t_0) = \begin{cases} \alpha_{ij}, & \text{if } \zeta_{2q}(t_0) < t \le \zeta_{2q+1}(t_0), \\ \beta_{ij}, & \text{if } \zeta_{2q-1}(t_0) < t \le \zeta_{2q}(t_0), \end{cases}$$

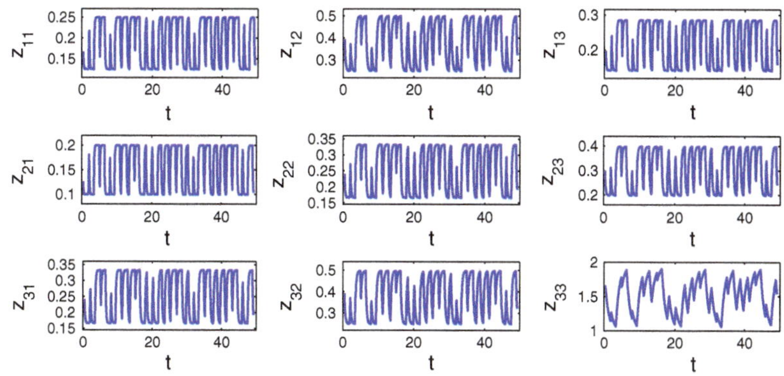

Fig. 8.1 The chaotic behavior of the SICNN (8.1.9)

where $t_0 \in [0, 1]$ and the numbers $\zeta_q(t_0)$, $q \in \mathbb{Z}$, denote the switching moments, which are the same for all i, j. The switching moments are defined through the formula $\zeta_q(t_0) = q + \kappa_q(t_0)$, $q \in \mathbb{Z}$, where the sequence $\{\kappa_q(t_0)\}$, $\kappa_0(t_0) = t_0$, is generated by the logistic equation $\kappa_{q+1}(t_0) = 3.9\kappa_q(t_0)(1-\kappa_q(t_0))$, which is chaotic in the Li–Yorke sense [24]. More information about the dynamics of relay systems and replication of chaos can be found in papers [29–34].

In system (8.1.9), let $g(s) = s^2$ and $\alpha_{ij} = 1$, $\beta_{ij} = 2$ for all i, j. By results of the paper [29], the family $\{\nu_{ij}(t, t_0)\}$, $t_0 \in [0, 1]$, is chaotic in the sense of Li–Yorke, and the collection $\mathscr{L}$ consisting of elements of the form $z(t) = \{z_{ij}(t)\}$, where $z(t)$ are bounded on $\mathbb{R}$ solutions of (8.1.9), is a Li–Yorke chaotic set.

Next, we consider the simulations of the network (8.1.9). Figure 8.1 represents the chaotic solution $z(t) = \{z_{ij}(t)\}$ of (8.1.9) with $z_{11}(t_0) = 0.1678$, $z_{12}(t_0) = 0.3956$, $z_{13}(t_0) = 0.1987$, $z_{21}(t_0) = 0.1261$, $z_{22}(t_0) = 0.2405$, $z_{23}(t_0) = 0.3012$, $z_{31}(t_0) = 0.2412$, $z_{32}(t_0) = 0.3942$, $z_{33}(t_0) = 1.6692$, where $t_0 = 0.45$.

In Example 8.1, to procure a Li–Yorke chaotic set, we used an SICNN in the form of (8.1.1), where the terms $L_{ij}(t)$ are replaced by relay functions $\nu_{ij}(t, t_0)$, whose switching moments change chaotically. Now, to support the results of the present section, we will construct another SICNN, but this time we will use external inputs of the form $L_{ij}(t) = h_{ij}(z(t))$, where $z(t)$ are the chaotic solutions of the network (8.1.9) and $h(v) = \{h_{ij}(v)\}$ is a nonlinear function which satisfies the inequality (8.1.2).

Example 8.2 Consider the following SICNN,

$$\frac{dx_{ij}}{dt} = -a_{ij}x_{ij} - \sum_{C_{kl}\in N_1(i,j)} C_{ij}^{kl} f(x_{kl}(t))x_{ij} + L_{ij}(t), \qquad (8.1.10)$$

in which $i, j = 1, 2, 3,$

$$\begin{pmatrix} a_{11} & a_{12} & a_{13} \\ a_{21} & a_{22} & a_{23} \\ a_{31} & a_{32} & a_{33} \end{pmatrix} = \begin{pmatrix} 5 & 12 & 2 \\ 6 & 4 & 8 \\ 2 & 9 & 3 \end{pmatrix},$$

$$\begin{pmatrix} C_{11} & C_{12} & C_{13} \\ C_{21} & C_{22} & C_{23} \\ C_{31} & C_{32} & C_{33} \end{pmatrix} = \begin{pmatrix} 0.02 & 0.04 & 0.06 \\ 0.04 & 0.07 & 0.09 \\ 0.03 & 0.04 & 0.08 \end{pmatrix},$$

and $f(s) = \frac{1}{2}s^3$. One can calculate that

$$\sum_{C_{kl}\in N_1(1,1)} C_{11}^{kl} = 0.17, \quad \sum_{C_{kl}\in N_1(1,2)} C_{12}^{kl} = 0.32, \quad \sum_{C_{kl}\in N_1(1,3)} C_{13}^{kl} = 0.26,$$

$$\sum_{C_{kl}\in N_1(2,1)} C_{21}^{kl} = 0.24, \quad \sum_{C_{kl}\in N_1(2,2)} C_{22}^{kl} = 0.47, \quad \sum_{C_{kl}\in N_1(2,3)} C_{23}^{kl} = 0.38,$$

$$\sum_{C_{kl}\in N_1(3,1)} C_{31}^{kl} = 0.18, \quad \sum_{C_{kl}\in N_1(3,2)} C_{32}^{kl} = 0.35, \quad \sum_{C_{kl}\in N_1(3,3)} C_{33}^{kl} = 0.28.$$

In the previous example, we obtained a network whose solutions behave chaotically. Now, we will make these solutions as external inputs for (8.1.10), with the help of a nonlinear function h.

Define a function $h(v) = \{h_{ij}(v)\}$, where $v = \{v_{ij}\}$, $i, j = 1, 2, 3$, through the equations $h_{11}(v) = 2v_{11} + \sin(v_{11})$, $h_{12}(v) = \frac{3}{2}v_{12}^2$, $h_{13}(v) = e^{v_{13}}$, $h_{21}(v) = \tan\left(\frac{v_{21}}{2}\right)$, $h_{22}(v) = v_{22} + \arctan v_{22}$, $h_{23}(v) = \frac{v_{23}^2 - v_{23} - 1}{v_{23} - 1}$, $h_{31}(v) = \frac{2}{3}(2 + v_{31})^{3/2}$, $h_{32}(v) = \tanh(v_{32})$, $h_{33}(v) = \frac{1}{4}v_{33}^3 + \frac{1}{5}v_{33}$. We note that the inequality (8.1.2) can be verified using the bounded regions where each component function $z_{ij}(t)$ lies in. Accordingly, the set $\mathscr{L}_h$ whose elements are of the form $h(z(t))$, $z(t) \in \mathscr{L}$, where $\mathscr{L}$ is the set of bounded on $\mathbb{R}$ solutions of (8.1.9), is Li–Yorke chaotic. Moreover, for each $z(t) \in \mathscr{L}$ we have $\left|h_{ij}(z(t))\right| \le M_{ij}$, where $M_{11} = 0.78$, $M_{12} = 0.54$, $M_{13} = 1.35$, $M_{21} = 0.11$, $M_{22} = 0.69$, $M_{23} = 2.11$, $M_{31} = 2.41$, $M_{32} = 0.51$, and $M_{33} = 2.4$.

Consider the network (8.1.10) with $L_{ij}(t) = h_{ij}(z(t))$, where $h(z(t)) = \{h_{ij}(z(t))\}$ belongs to $\mathscr{L}_h$. In that case, the condition $(C6)$ holds for (8.1.10) with $M_f = 0.864$, $L_f = 2.16$, $K_0 = 1.36$, $\gamma = 2$, and $\overline{c} = 0.47$. The results of Theorem 8.1 ensure us to say that the collection $\mathscr{A}$ with elements $\phi_z(t)$, $z(t) \in \mathscr{L}$, is Li–Yorke chaotic.

In the SICNN (8.1.10), we use the chaotically behaving solution $z(t) = \{z_{ij}(t)\}$ which is simulated in Example 8.1, and depict in Fig. 8.2 the solution of (8.1.10) with the initial data $x_{11}(t_0) = 0.1341$, $x_{12}(t_0) = 0.0247$, $x_{13}(t_0) = 0.6493$, $x_{21}(t_0) = 0.0143$, $x_{22}(t_0) = 0.1503$, $x_{23}(t_0) = 0.2394$, $x_{31}(t_0) = 1.1574$, $x_{32}(t_0) = 0.0467$, and $x_{33}(t_0) = 0.5145$, where $t_0 = 0.45$. Figure 8.2 reveals that each cell

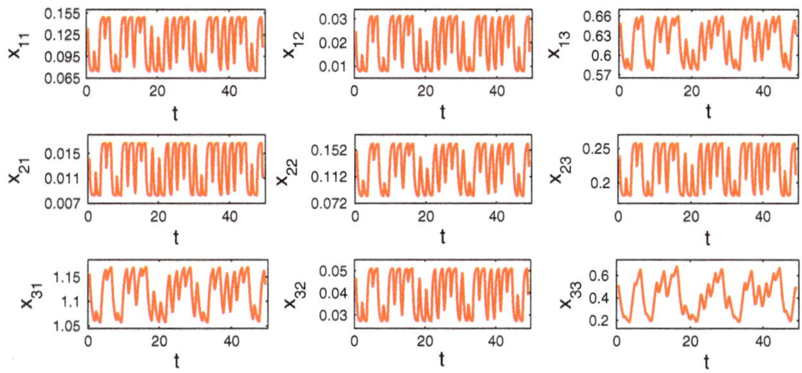

Fig. 8.2 The chaotic behavior of the SICNN (8.1.10)

Fig. 8.3 The projection of the chaotic attractor of the network (8.1.10) on the $x_{22} - x_{31} - x_{33}$ space

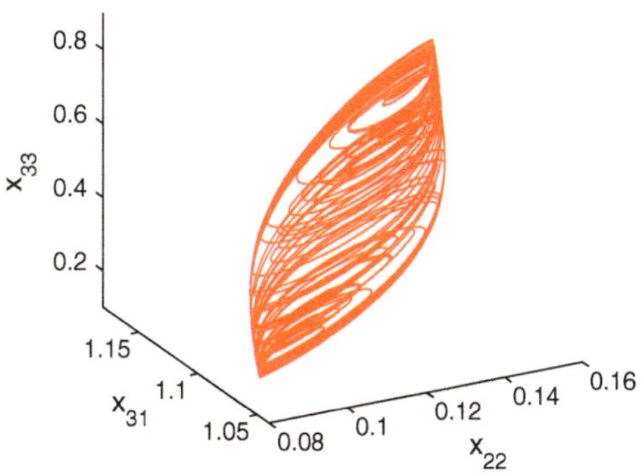

C_{ij}, $i, j = 1, 2, 3$, behave chaotically, and this supports the result mentioned in Theorem 8.1. Moreover, Fig. 8.3 shows the projection of the same trajectory on the $x_{22} - x_{31} - x_{33}$ space, and this figure also confirms the results of the present section.

8.2 Attraction of Chaos by Retarded SICNNs

In the present section, dynamics of retarded shunting inhibitory cellular neural networks (SICNNs) is investigated with Li–Yorke chaotic external inputs and outputs. Within the scope of our results, we prove the presence of generalized synchronization in coupled retarded SICNNs, and confirm it by means of the auxiliary system approach. We have obtained more than just synchronization, as it is proved that the Li–Yorke chaos is extended with its ingredients, proximality, and frequent separation, which have not been considered in the theory of synchronization at all. Our procedure is used to synchronize chains of unidirectionally coupled neural networks.

The results may explain the high performance of brain functioning and can be extended by specific stability analysis methods. Illustrations supporting the results are depicted. For the first time in the literature, proximality, and frequent separation features are demonstrated numerically for continuous-time dynamics.

8.2.1 Introduction

Cellular neural networks (CNNs) have been paid much attention due to their local connectivity and easy hardware implementation. Time delays occur during the hardware implementation of neural networks because of the finite switching speed of the amplifiers. The occurrence of time delays may lead to an oscillation and instability of the networks [35, 36]. Moreover, the introduction of delay in the signals transmitted among the cells of CNNs is required by the process of moving images [37]. Therefore, the consideration of neural networks with time delay is important for applications.

Chaotic dynamics has been widely investigated in neural networks [38–58]. In their study, King et al. [45] observed chaotic behavior in a model of the central dopaminergic neuronal system. It is shown in paper [41] that chaos can be expected in mathematical models of neural systems possessing time delays. In order to study the dynamical properties of a neural network in chaotic wandering state, Kuroiwa et al. [46] utilized a model which was proposed by Aihara et al. [39]. The same model was also used in [52] to investigate the synchronization characteristics in response to external inputs in a coupled lattice based on a Newman–Watts model. The existence of a period-doubling cascade was demonstrated by Wang [55] in a discrete time neural network. Ke and Oommen [44] considered the chaotic and pattern recognition properties of a neural network, which is based on the logistic map. In the paper [43], the existence of chaos was demonstrated in the dynamics of fractional-order Hopfield type neural networks. The presence of chaos in the Hodgkin-Huxley model with its original parameters was revealed in [40], where the solutions were found by displaying rectangles in a cross section whose images under the return map produce a Smale horseshoe. Moreover, the verification of chaotic behavior in Hopfield neural networks was provided by virtue of the horseshoes in the studies [42, 47]. The problem of creating a robust chaotic neural network was studied by Potapov and Ali [51]. Furthermore, chaotic dynamics in CNNs were studied in the papers [48, 56–58].

The presence of chaos in neural networks is useful for separating image segments [52], information processing [49, 50], and synchronization of neural networks [59–64]. Besides, the synchronization phenomenon is also observable in the dynamics of coupled chaotic CNNs [65, 66]. The detection and characterization of synchronization in neural networks is of great interest, since they may provide the opportunity to understand how the brain and nervous system works [67]. Chaotic dynamics can improve the performance of CNNs on problems that have local minima in energy (cost) functions, since chaotic behavior of CNNs can help the network avoid local minima and reach the global optimum [68]. Moreover, chaotic dynamics in CNNs is an important tool for the studies of chaotic communication [69–71] and combinatorial optimization problems [72].

The term chaos, as a mathematical notion, has first been used by Li and Yorke [24] for one-dimensional difference equations. The concept of snap-back repellers for high-dimensional maps was introduced by Marotto [73]. According to the results of the paper [73], if a multidimensional continuously differentiable map has a snap-back repeller, then it is Li–Yorke chaotic. Li et al. [74] used Marotto's Theorem to prove the existence of Li–Yorke chaos in a spatiotemporal chaotic system. Li–Yorke sensitivity, which links the Li–Yorke chaos with the notion of sensitivity, was studied in [75], and generalizations of Li–Yorke chaos to mappings in Banach spaces and complete metric spaces were provided in [76]. In the present section, we develop the concept of Li–Yorke chaos to the multidimensional dynamics of retarded shunting inhibitory cellular neural networks, and prove its existence rigorously.

Marotto's Theorem is also useful in the theory of neural networks to prove the presence of chaos rigorously. It was used by Lin and Ruan [77] to determine chaotic dynamics in a pacemaker neuron type integrate-and-fire circuit having two states with a periodic pulse-train input. Moreover, in the paper [78], the chaos was approved by virtue of the Marotto's Theorem in discrete time-delayed Hopfield neural networks.

A class of CNNs which was introduced by Bouzerdoum and Pinter [6] is shunting inhibitory cellular neural networks (SICNNs). SICNNs have been extensively applied in psychophysics, speech, perception, robotics, adaptive pattern recognition, vision and image processing [7–13, 79].

The model of SICNNs in the most original formulation [6] is as follows. Consider a two-dimensional grid of processing cells, and let C_{ij}, $i = 1, 2, \ldots, m$, $j = 1, 2, \ldots, n$, denote the cell at the (i, j) position of the lattice. The r-neighborhood of C_{ij} is defined as

$$N_r(i, j) = \{C_{kl} : \max\{|k - i|, |l - j|\} \leq r,\ 1 \leq k \leq m,\ 1 \leq l \leq n\}.$$

In SICNNs, neighboring cells exert mutual inhibitory interactions of the shunting type. The dynamics of the cell C_{ij} is described by the nonlinear ordinary differential equation

$$\frac{dx_{ij}(t)}{dt} = -a_{ij}x_{ij}(t) - \sum_{C_{kl} \in N_r(i,j)} C_{ij}^{kl} f(x_{kl}(t))x_{ij}(t) + L_{ij}(t),$$

where x_{ij} is the activity of the cell C_{ij}; $L_{ij}(t)$ is the external input to C_{ij}; the constant $a_{ij} > 0$ represents the passive decay rate of the cell activity; $C_{ij}^{kl} \geq 0$ is the connection or coupling strength of the postsynaptic activity of the cell C_{kl} transmitted to the cell C_{ij}; and the activation function $f(x_{kl})$ is a positive continuous function representing the output or firing rate of the cell C_{kl}.

In the present section, we consider SICNNs with delay in the form

$$\frac{dx_{ij}(t)}{dt} = -a_{ij}x_{ij}(t) - \sum_{C_{kl}\in N_r(i,j)} C_{ij}^{kl} f(x_{kl}(t-\tau))x_{ij}(t) + L_{ij}(t), \quad (8.2.11)$$

where τ is a positive number.

To give an explanation of the title of this section, let us start with the chaos to be attracted. It is a set of bounded functions, $\mathscr{L}$, chaotic in the Li–Yorke sense. In the next section, the set $\mathscr{L}$ will be described in a detailed form. We apply the elements of the chaotic set, $L(t) = \{L_{ij}(t)\}$, as external inputs in the SICNN (8.2.11). Next, we verify that the network (8.2.11) outputs a set of solutions of the same nature as the set $\mathscr{L}$, which are bounded functions chaotic in the Li–Yorke sense. We denote the set of the outputs of (8.2.11) by $\widetilde{\mathscr{L}}$. Thus, we say that the SICNN (8.2.11) "attracts" a chaotic set $\mathscr{L}$ if it produces the chaotic output $\widetilde{\mathscr{L}}$. It is worth noting that under the conditions that will be introduced in the next section, the SICNN (8.2.11) does not possess chaos provided that the external inputs are not chaotic, but regular or absent. In the papers [16, 17], retarded SICNNs were considered with periodic/almost periodic inputs, and it was demonstrated that the same regular dynamics appear in the outputs.

Li–Yorke chaos is chosen in this section mainly for two reasons. First, the presence of Li–Yorke chaos can be obtained through the reduction to scalar discrete equations, for instance, to the logistic map. This reduction can be done in the multidimensional case. Second, Marotto's Theorem allows to study the chaos by reduction to *multidimensional* discrete equations. In the parametric sense, the chaos is generic, for example, the logistic map $x_{n+1} = \mu x_n(1 - x_n)$ is chaotic for the parameter value μ between 3.84 and 4 [24].

In their studies, Freeman and his collaborators [80–85] revealed that chaotic dynamics is an inevitable attribution of brain activities. Considering the brain as a collection of neural networks, one may suppose that the chaos appearance can happen in two ways. The first one is the "endogenous chaos," which is generated by an individual neural network itself without an influence from outside. This type of chaos appearance was widely investigated in the literature [40, 42–44, 47, 48, 51, 55–58]. The second way is the extension of chaos from one network to another. One can consider the synchronization of chaos [86–97] within the scope of the latter way. However, synchronization of chaos relies deeply on its description as well as on the verification of asymptotic closeness between the outputs. Therefore, this type of chaos extension brings us far from the effectiveness of chaos for the brain activities. It brings us to the comprehension of schizophrenia, insomnia, and epilepsy [41] rather than regular brain functioning. Nevertheless, Breakspear and Terry [98] reported that synchronization plays an important role for activities of healthy brain. That is why it is important to find mathematical methods for the chaos extension between neural networks, where the asymptotic closeness is fully removed or its presence is weakened in some sense.

In the case of identical synchronization [94], one requires the condition

$$\lim_{t\to\infty} \|y(t) - x(t)\| = 0,$$

where x and y denote the states of the drive and response systems, respectively. This type of asymptotic relation is strong and to weaken it, one should consider the theory of generalized synchronization [67, 86, 88, 91, 92, 96, 97]. In this theory, the previous relation is replaced by

$$\lim_{t\to\infty} \|y(t) - \phi(x(t))\| = 0,$$

where ϕ is a transformation. The presence of the synchronization manifold $y = \phi(x)$ in the drive-response systems is mostly investigated by numerical analyses [67, 86, 96]. The concept of generalized synchronization for coupled systems with delay was considered in [97]. In the present section, we suggest an easy theoretical approach to verify the presence of synchronization based on the exponential convergence of outputs. Moreover, by the traditional simulation methods [67, 86], we will check that generalized synchronization takes place in the attraction of chaos. It is worth noting that we verify the ingredients of Li–Yorke chaos, which cannot be realized by the concept of synchronization at all, and this is one of the principal novelties of our results. The ingredients, *proximality and frequent separation, may play an essential role in the brain dynamics*. This idea can be supported if one follows the experimental analyses of Freeman and his collaborators [80–85], and develop researches in this direction.

Investigations of neural networks will not be adequate for application problems unless delay is not introduced in the models. Therefore, a large number of papers paid special attention to the presence of delay in SICNNs [14–17, 19–23, 99–102]. In these papers, the existence and stability of periodic, almost periodic and antiperiodic solutions of SICNNs were studied. Despite the fact that SICNNs and chaos are important in neoroscience, there are still very few papers which consider chaos in this type of neural networks. As far as we know, the subject was considered only in the studies [25, 26], and the analyses were made only *numerically* without a theoretical support. That is the reason why even *a type of chaos* was not indicated in these studies. The theoretical approach for SICNNs based on the rigorous definition of chaos presented by Li and Yorke [24] was started in our paper [103], where we discussed the chaotification of SICNNs without delay. The way of chaos expansion in continuous-time dynamics was also considered in the paper [34] without time delay by taking into account chaos in the sense of both Li–Yorke [24] and Devaney [104] as well as for period-doubling cascade [105, 106] and intermittency [107].

The novelty of the present section is the discussion of chaotic dynamics in SICNNs with time delay. The investigation of chaotic outputs in retarded SICNNs is much more sophisticated than the one without delay [103]. The method of adaptation of the Li–Yorke chaos for differential equations with retardation considered in the present section is new not only for neural networks, but also for the theory as a whole. This also provides a contribution to the chaos theory. Moreover, we analyze the relation between generalized synchronization [67, 86, 88, 91, 92, 96, 97] and our approach about the chaotification of neural systems in a detailed form, and such discussions have never been reported before for SICNNs in the literature. In Sect. 8.2.4, we take into account retarded SICNNs with external inputs in the form of relay functions. However, in the studies [29–31, 33], relay systems were considered without time delay. Our results are also applicable to other kinds of recurrent networks such as Hopfield and Cohen–Grossberg neural networks [5, 108–114].

Motivated by the deficiency of mathematical methods for chaos recognition in neural networks and the importance of irregular behavior for effective brain activities, we suggest the results of the present section. It is the first time in the literature that rigorous mathematical methods are used to prove not only the presence of chaos in retarded SICNNs, but also how chaos can be exported between neural networks. Another novelty is the precise achievement of chaos in the sense of Li–Yorke with its ingredients, proximality and frequent separation, which may play an important role for the working principle of a nervous system. To the best of our knowledge, the numerical demonstration of the proximality and frequent separation features for continuous-time dynamics have never been reported before (see Fig. 8.8 and the related text). Our results can provide further research areas in neuroscience, in particular, by the consideration of experiments of Freeman and other neurobiologists [41, 54, 80–85].

The primary contributions of the present section are summarized below:

(i) We give a mathematical description of the Li–Yorke chaos for continuous-time neural networks with delay. Moreover, simulations of the ingredients of Li–Yorke chaos, proximality and frequent separation, have been performed for continuous-time dynamics for the first time.
(ii) By means of external inputs, we theoretically prove the presence of chaos in retarded SICNNs with arbitrary high number of cells, and we provide a way of chaos extension among coupled neural networks with delay.
(iii) We make use of the exponential convergence of solutions (see Lemma 8.6) to prove the presence of generalized synchronization in coupled retarded SICNNs, and confirm its presence by means of the auxiliary system approach [67, 86]. Our procedure can be easily extended to synchronize chains of unidirectionally coupled neural networks with delay. This may be important in neuroscience to explain the high performance of brain functioning [54, 84]. The proposed approach cannot be reduced to generalized synchronization, since we have obtained more than just synchronization. We prove that the Li–Yorke chaos is extended with its ingredients, proximality, and frequent separation, which have not been considered in the theory of synchronization at all.
(iv) Our results can be extended in neuroscience by specific stability analysis methods, for example, by the linear matrix inequality technique [62, 115–118].

The rest of the section is organized as follows. In Sect. 8.2.2, the description of Li–Yorke chaos is presented and two lemmas about the existence of unique bounded on $\mathbb{R}$ solutions of SICNNs and their stability are provided. In Sect. 8.2.3, the presence of Li–Yorke chaos is theoretically proved for retarded SICNNs of the form (8.2.11). Section 8.2.4 is devoted for an example. In this part, a chain of SICNNs is used to show the effectiveness of the proposed results. Moreover, the ingredients of Li–Yorke chaos are demonstrated numerically. Finally, we compared our method with generalized synchronization both theoretically and numerically in Sect. 8.2.5.

8.2.2 Preliminaries

Throughout the section, $\mathbb{R}$ and $\mathbb{N}$ will denote the sets of real numbers and natural numbers, respectively. Moreover, the norm $\|u\| = \max_{(i,j)} |u_{ij}|$ will be used, where $u = \{u_{ij}\} = (u_{11}, \ldots, u_{1n}, \ldots, u_{m1} \ldots, u_{mn}) \in \mathbb{R}^{m\times n}$.

The description of Li–Yorke chaos that will be utilized in the present section is as follows. Suppose that $\mathscr{L}$ is a collection of continuous functions $L(t) = \{L_{ij}(t)\}$, $i = 1, 2, \ldots, m$, $j = 1, 2, \ldots, n$, such that $\sup_{t\in\mathbb{R}} \|L(t)\| \le M$, where M is a positive number.

We say that a couple $(L(t), \overline{L}(t)) \in \mathscr{L} \times \mathscr{L}$ is proximal if for arbitrary small $\varepsilon > 0$ and arbitrary large $E > 0$, there exists an interval J with a length no less than E such that $\|L(t) - \overline{L}(t)\| < \varepsilon$ for $t \in J$. On the other hand, a couple $(L(t), \overline{L}(t)) \in \mathscr{L} \times \mathscr{L}$ is called frequently (ε_0, Δ)-separated if there exist positive numbers ε_0, Δ and infinitely many intervals $J_q = [\alpha_q, \beta_q]$, $q \in \mathbb{N}$, each with a length no less than Δ, such that $\beta_q \to \infty$ as $q \to \infty$ and $\|L(t) - \overline{L}(t)\| > \varepsilon_0$ for each t from these intervals. It is worth noting that the numbers ε_0 and Δ depend on the functions $L(t)$ and $\overline{L}(t)$.

A couple $(L(t), \overline{L}(t)) \in \mathscr{L} \times \mathscr{L}$ is a Li–Yorke pair if it is proximal and frequently (ε_0, Δ)-separated for some positive numbers ε_0 and Δ. Moreover, an uncountable set $\mathscr{L}_S \subset \mathscr{L}$ is called a scrambled set if $\mathscr{L}_S$ does not contain any periodic functions and each couple of different functions inside $\mathscr{L}_S \times \mathscr{L}_S$ is a Li–Yorke pair.

The collection $\mathscr{L}$ is called a Li–Yorke chaotic set if: (i) there exists a positive number T_0 such that $\mathscr{L}$ possesses a periodic function of period kT_0 for any $k \in \mathbb{N}$; (ii) $\mathscr{L}$ possesses a scrambled set $\mathscr{L}_S$; (iii) for any function $L(t) \in \mathscr{L}_S$ and any periodic function $\overline{L}(t) \in \mathscr{L}$, the couple $(L(t), \overline{L}(t))$ is frequently (ε_0, Δ)-separated for some positive numbers ε_0 and Δ.

The following conditions are required:

(C1) There exist positive numbers M_{ij} such that $\sup_{t\in\mathbb{R}} |L_{ij}(t)| \le M_{ij}$;

(C2) There exists a positive number M_f such that $\sup_{s\in\mathbb{R}} |f(s)| \le M_f$;

(C3) There exists a positive number L_f such that $|f(s_1) - f(s_2)| \le L_f |s_1 - s_2|$ for all $s_1, s_2 \in \mathbb{R}$;

(C4) $M_f\delta < 1$, where $\delta = \max\limits_{(i,j)} \dfrac{\sum_{C_{kl}\in N_r(i,j)} C_{ij}^{kl}}{a_{ij}}$;

(C5) $(M_f + K_0 L_f)\delta < 1$, where $K_0 = \dfrac{\overline{M}}{1 - M_f\delta}$ and $\overline{M} = \max\limits_{(i,j)} \dfrac{M_{ij}}{a_{ij}}$.

One can confirm that a bounded on $\mathbb{R}$ function $x(t) = \{x_{ij}(t)\}$, $i = 1, 2, \ldots, m$, $j = 1, 2, \ldots, n$, is a solution of the network (8.2.11) if and only if the following integral equation is satisfied

$$x_{ij}(t) = -\int_{-\infty}^{t} \mathrm{e}^{-a_{ij}(t-s)} \Big[\sum_{C_{kl}\in N_r(i,j)} C_{ij}^{kl} f(x_{kl}(s-\tau))x_{ij}(s) - L_{ij}(s)\Big] ds. \tag{8.2.12}$$

The following assertion is about the existence and uniqueness of bounded on $\mathbb{R}$ solutions of system (8.2.11).

Lemma 8.5 *Suppose that the conditions $(C1)$–$(C5)$ are valid. Then, for any $L(t) = \{L_{ij}(t)\}$, $i = 1, 2, \ldots, m$, $j = 1, 2, \ldots, n$, there exists a unique bounded on $\mathbb{R}$ solution $\phi_L(t)$ of the network (8.2.11) such that* $\sup\limits_{t\in\mathbb{R}} \|\phi_L(t)\| \le K_0$.

Proof Consider the set C_0 of continuous functions $u(t) = \{u_{ij}(t)\}$, $i = 1, 2, \ldots, m$, $j = 1, 2, \ldots, n$, which are defined on $\mathbb{R}$, such that $\|u\|_0 \le K_0$, where $\|u\|_0 = \sup\limits_{t\in\mathbb{R}} \|u(t)\|$. Define the operator Π on C_0 as

$$(\Pi u(t))_{ij} = -\int_{-\infty}^{t} \mathrm{e}^{-a_{ij}(t-s)} \Big[\sum_{C_{kl}\in N_r(i,j)} C_{ij}^{kl} f(u_{kl}(s-\tau))u_{ij}(s) - L_{ij}(s)\Big] ds,$$

where $\Pi u(t) = \{(\Pi u)_{ij}(t)\}$. If $u(t)$ belongs to C_0, then we have

$$\begin{aligned} \big|(\Pi u(t))_{ij}\big| &\le \int_{-\infty}^{t} e^{-a_{ij}(t-s)} \Big[\sum_{C_{kl}\in N_r(i,j)} C_{ij}^{kl} \,|f(u_{kl}(s-\tau))|\, \big|u_{ij}(s)\big| + \big|L_{ij}(s)\big| \Big] ds \\ &\le \frac{1}{a_{ij}} \Big(M_{ij} + M_f K_0 \sum_{C_{kl}\in N_r(i,j)} C_{ij}^{kl} \Big). \end{aligned}$$

Accordingly, the inequality $\|\Pi u\|_0 \le \overline{M} + M_f K_0 \delta = K_0$ holds. Therefore, $\Pi(C_0) \subseteq C_0$.

On the other hand, for any $u(t), v(t) \in C_0$, one can verify that

$$\left|(\Pi u(t))_{ij} - (\Pi v(t))_{ij}\right| \le \int_{-\infty}^{t} e^{-a_{ij}(t-s)} \sum_{C_{kl}\in N_r(i,j)} C_{ij}^{kl} \Big| f(u_{kl}(s-\tau))u_{ij}(s)$$
$$- f(u_{kl}(s-\tau))v_{ij}(s)\Big| ds$$
$$+ \int_{-\infty}^{t} e^{-a_{ij}(t-s)} \sum_{C_{kl}\in N_r(i,j)} C_{ij}^{kl} \Big| f(u_{kl}(s-\tau))v_{ij}(s) - f(v_{kl}(s-\tau))v_{ij}(s)\Big| ds$$
$$\le (M_f + K_0 L_f)\frac{\sum_{C_{kl}\in N_r(i,j)} C_{ij}^{kl}}{a_{ij}} \|u - v\|_0 .$$

Thus, $\|\Pi u - \Pi v\|_0 \le (M_f + K_0 L_f)\delta \, \|u - v\|_0$, and the operator Π is contractive according to the condition $(C5)$. Consequently, for any $L(t)$, there exists a unique bounded on $\mathbb{R}$ solution $\phi_L(t)$ of system (8.2.11) such that $\sup_{t\in\mathbb{R}} \|\phi_L(t)\| \le K_0$. □

Consider the collection $\mathscr{L}$ whose elements are functions of the form $L(t) = \{L_{ij}(t)\}$, $i = 1, 2, \ldots, m$, $j = 1, 2, \ldots, n$, such that $\sup_{t\in\mathbb{R}} |L_{ij}(t)| \le M_{ij}$ for each i and j. Suppose that $\widetilde{\mathscr{L}}$ denotes the set of bounded on $\mathbb{R}$ solutions $\phi_L(t)$ of the network (8.2.11), where $L(t) = \{L_{ij}(t)\}$ belongs to $\mathscr{L}$. In the present section, we assume that $\widetilde{\mathscr{L}}$ is an equicontinuous family on $\mathbb{R}$.

Making use of the technique indicated in the proof of Theorem 2 [16], one can prove the following assertion, which confirms the attractiveness of the set $\widetilde{\mathscr{L}}$. A similar result for systems without delay was obtained in the paper [34].

Lemma 8.6 *If the conditions* $(C1)$*–*$(C5)$ *are fulfilled, then for a fixed* $L(t) = \{L_{ij}(t)\}$, $i = 1, 2, \ldots, m$, $j = 1, 2, \ldots, n$, *all solutions of system (8.2.11) converge exponentially to the unique bounded on* $\mathbb{R}$ *solution* $\phi_L(t)$.

8.2.3 Li–Yorke Chaos

Our purpose in the present section is to demonstrate that the network (8.2.11) behaves chaotically provided that the external inputs are chaotic. In the following lemmas, we will take advantage of the sets $\mathscr{L}$ and $\widetilde{\mathscr{L}}$, which are defined in Sect. 8.2.2. The main result will be mentioned in Theorem 8.2.

Let us denote $K_1 = \dfrac{2\overline{M}}{1 - (M_f + K_0 L_f)\delta}$, $\gamma = \min_{(i,j)} a_{ij}$ and $\overline{\delta} = \max_{(i,j)} \dfrac{\sum_{C_{kl}\in N_r(i,j)} C_{ij}^{kl}}{2a_{ij} - \gamma}$. We note that the number γ is positive since each a_{ij}, $i = 1, 2, \ldots, m$, $j = 1, 2, \ldots, n$, are positive.

The following conditions are needed:

(C6) $\left[M_f + (K_0 + K_1)L_f\right]\delta < 1$;
(C7) $2\left(M_f + K_0 L_f e^{\gamma\tau/2}\right)\overline{\delta} < 1$.

The next lemma is about the proximality feature of system (8.2.11).

Lemma 8.7 *Under the conditions (C1)–(C7), if a pair* $\left(L(t), \overline{L}(t)\right) \in \mathscr{L} \times \mathscr{L}$ *is proximal, then the same is true for the pair* $\left(\phi_L(t), \phi_{\overline{L}}(t)\right) \in \widetilde{\mathscr{L}} \times \widetilde{\mathscr{L}}$.

Proof Set $R_0 = \dfrac{2K_0}{1 - 2(M_f + K_0 L_f e^{\gamma\tau/2})\overline{\delta}}$, $R_1 = \dfrac{1}{\gamma[1 - \delta(M_f + K_0 L_f)]}$ and take a positive number η such that $\eta \leq 1/(R_0 + R_1)$. Fix an arbitrary small number $\varepsilon > 0$ and a positive number E such that $E > \dfrac{4}{\gamma} \ln\left(\dfrac{1}{\eta\varepsilon}\right)$. Because the pair $\left(L(t), \overline{L}(t)\right) \in \mathscr{L} \times \mathscr{L}$ is proximal, there exists an interval $J = [\sigma, \sigma + E_0]$, where $E_0 \geq E$, such that $\left\| L(t) - \overline{L}(t) \right\| < \eta\varepsilon$ for $t \in J$.

The bounded on $\mathbb{R}$ solutions $\phi_L(t) = \left\{\phi_L^{ij}(t)\right\}$ and $\phi_{\overline{L}}(t) = \left\{\phi_{\overline{L}}^{ij}(t)\right\}$, $i = 1, 2, \ldots, m$, $j = 1, 2, \ldots, n$, satisfy the relation

$$\phi_L^{ij}(t) - \phi_{\overline{L}}^{ij}(t) = e^{-a_{ij}(t-\sigma)}\left(\phi_L^{ij}(\sigma) - \phi_{\overline{L}}^{ij}(\sigma)\right) + \int_\sigma^t e^{-a_{ij}(t-s)}\left(L_{ij}(s) - \overline{L}_{ij}(s)\right)ds$$
$$- \int_\sigma^t e^{-a_{ij}(t-s)} \sum_{C_{kl} \in N_r(i,j)} C_{ij}^{kl}\left[f(\phi_L^{kl}(s-\tau))\phi_L^{ij}(s) - f(\phi_{\overline{L}}^{kl}(s-\tau))\phi_{\overline{L}}^{ij}(s)\right]ds.$$

Denote by $w(t) = \left\{w_{ij}(t)\right\}$ the difference $\phi_L(t) - \phi_{\overline{L}}(t)$. Then for each i and j, we have that

$$w_{ij}(t) = e^{-a_{ij}(t-\sigma)}\left(\phi_L^{ij}(\sigma) - \phi_{\overline{L}}^{ij}(\sigma)\right) + \int_\sigma^t e^{-a_{ij}(t-s)}\left(L_{ij}(s) - \overline{L}_{ij}(s)\right)ds$$
$$- \int_\sigma^t e^{-a_{ij}(t-s)} \sum_{C_{kl} \in N_r(i,j)} C_{ij}^{kl}\left[f(w_{kl}(s-\tau) + \phi_{\overline{L}}^{kl}(s-\tau))\left(w_{ij}(s) + \phi_{\overline{L}}^{ij}(s)\right)\right.$$
$$\left. - f(\phi_{\overline{L}}^{kl}(s-\tau))\phi_{\overline{L}}^{ij}(s)\right]ds.$$

Let Ψ be the set of continuous functions $w(t) = \left\{w_{ij}(t)\right\}$, $i = 1, 2, \ldots, m$, $j = 1, 2, \ldots, n$, which are defined on $\mathbb{R}$, such that $\|w(t)\| \leq R_0 e^{-\gamma(t-\sigma)/2} + R_1\eta\varepsilon$ for $\sigma - \tau \leq t \leq \sigma + E_0$ and $\|w\|_0 \leq K_1$, where $\|w\|_0 = \sup_{t \in \mathbb{R}} \|w(t)\|$.

Define on Ψ the operator $\widetilde{\Pi}$ as follows:

$$(\widetilde{\Pi} w(t))_{ij} = \begin{cases} \phi_L^{ij}(t) - \phi_{\overline{L}}^{ij}(t), \ t < \sigma, \\ e^{-a_{ij}(t-\sigma)}\left(\phi_L^{ij}(\sigma) - \phi_{\overline{L}}^{ij}(\sigma)\right) + \displaystyle\int_\sigma^t e^{-a_{ij}(t-s)}(L_{ij}(s) - \overline{L}_{ij}(s))ds \\ - \displaystyle\int_\sigma^t e^{-a_{ij}(t-s)} \sum_{C_{kl} \in N_r(i,j)} C_{ij}^{kl}\Big[f(w_{kl}(s-\tau) + \phi_{\overline{L}}^{kl}(s-\tau)) \\ \times(w_{ij}(s) + \phi_{\overline{L}}^{ij}(s)) - f(\phi_{\overline{L}}^{kl}(s-\tau))\phi_{\overline{L}}^{ij}(s)\Big]ds, \ t \geq \sigma. \end{cases}$$

First, we will show that $\widetilde{\Pi} : \Psi \to \Psi$. Indeed, if $w(t)$ belongs to Ψ, then for $t \in [\sigma, \sigma + E_0]$ it is true that

$$
\begin{aligned}
&\left|(\widetilde{\Pi} w(t))_{ij}\right| \le e^{-a_{ij}(t-\sigma)} \left|\phi_L^{ij}(\sigma) - \phi_{\overline{L}}^{ij}(\sigma)\right| + \int_\sigma^t e^{-a_{ij}(t-s)} \left|L_{ij}(s) - \overline{L}_{ij}(s)\right| ds \\
&+ \int_\sigma^t e^{-a_{ij}(t-s)} \sum_{C_{kl} \in N_r(i,j)} C_{ij}^{kl} \left| f(w_{kl}(s-\tau) + \phi_{\overline{L}}^{kl}(s-\tau))\right| \left|w_{ij}(s)\right| ds \\
&+ \int_\sigma^t e^{-a_{ij}(t-s)} \sum_{C_{kl} \in N_r(i,j)} C_{ij}^{kl} \\
&\times \left| f(w_{kl}(s-\tau) + \phi_{\overline{L}}^{kl}(s-\tau)) - f(\phi_{\overline{L}}^{kl}(s-\tau))\right| \left|\phi_{\overline{L}}^{kl}(s)\right| ds \\
&\le 2K_0 e^{-\gamma(t-\sigma)} + \int_\sigma^t e^{-a_{ij}(t-s)} \eta\varepsilon ds \\
&+ \int_\sigma^t e^{-a_{ij}(t-s)} \sum_{C_{kl} \in N_r(i,j)} C_{ij}^{kl} M_f \Big(R_0 e^{-\gamma(s-\sigma)/2} + R_1 \eta\varepsilon\Big) ds \\
&+ \int_\sigma^t e^{-a_{ij}(t-s)} \sum_{C_{kl} \in N_r(i,j)} C_{ij}^{kl} K_0 L_f \Big(R_0 e^{-\gamma(s-\tau-\sigma)/2} + R_1 \eta\varepsilon\Big) ds \\
&= 2K_0 e^{-\gamma(t-\sigma)} + \frac{\eta\varepsilon}{a_{ij}} \left(1 - e^{-a_{ij}(t-\sigma)}\right) \\
&+ 2R_0 \left(M_f + K_0 L_f e^{\gamma\tau/2}\right) \frac{\sum_{C_{kl} \in N_r(i,j)} C_{ij}^{kl}}{2a_{ij} - \gamma} e^{-\gamma(t-\sigma)/2} \left(1 - e^{-(a_{ij} - \gamma/2)(t-\sigma)}\right) \\
&+ R_1 \eta\varepsilon (M_f + K_0 L_f) \frac{\sum_{C_{kl} \in N_r(i,j)} C_{ij}^{kl}}{a_{ij}} \left(1 - e^{-a_{ij}(t-\sigma)}\right).
\end{aligned}
$$

Hence, for $t \in [\sigma, \sigma + E_0]$, it can be verified that

$$
\begin{aligned}
&\left\|\widetilde{\Pi} w(t)\right\| \le 2K_0 e^{-\gamma(t-\sigma)} + \frac{\eta\varepsilon}{\gamma} + 2R_0 \left(M_f + K_0 L_f e^{\gamma\tau/2}\right) \overline{\delta} e^{-\gamma(t-\sigma)/2} \\
&+ R_1 \eta\varepsilon (M_f + K_0 L_f) \delta \\
&\le 2\left[K_0 + R_0 (M_f + K_0 L_f e^{\gamma\tau/2}) \overline{\delta}\right] e^{-\gamma(t-\sigma)/2} + \eta\varepsilon \left[\frac{1}{\gamma} + R_1 (M_f + K_0 L_f) \delta\right] \\
&= R_0 e^{-\gamma(t-\sigma)/2} + R_1 \eta\varepsilon.
\end{aligned}
$$

Since $R_0 > 2K_0$, the inequality $\left\|\widetilde{\Pi} w(t)\right\| \le R_0 e^{-\gamma(t-\sigma)/2} + R_1 \eta\varepsilon$ holds also for $\sigma - \tau \le t < \sigma$.

On the other hand, if $w(t)$ belongs to Ψ, then making benefit of the inequality $K_1 \geq 2K_0$ one can confirm for $t \geq \sigma$ that

$$\begin{aligned}
&\left|(\widetilde{\Pi} w(t))_{ij}\right| \leq e^{-a_{ij}(t-\sigma)} \left|\phi_L^{ij}(\sigma) - \phi_{\overline{L}}^{ij}(\sigma)\right| + \int_\sigma^t 2M_{ij} e^{-a_{ij}(t-s)} ds \\
&+ \int_\sigma^t e^{-a_{ij}(t-s)} \sum_{C_{kl}\in N_r(i,j)} C_{ij}^{kl} \left|f(w_{kl}(s-\tau) + \phi_{\overline{L}}^{kl}(s-\tau))\right| \left|w_{ij}(s)\right| ds \\
&+ \int_\sigma^t e^{-a_{ij}(t-s)} \sum_{C_{kl}\in N_r(i,j)} C_{ij}^{kl} L_f \left|w_{kl}(s-\tau)\right| \left|\phi_{\overline{L}}^{ij}(s)\right| ds \\
&\leq 2K_0 e^{-a_{ij}(t-\sigma)} + \left(\frac{2M_{ij}}{a_{ij}} + K_1(M_f + K_0 L_f)\frac{\sum_{C_{kl}\in N_r(i,j)} C_{ij}^{kl}}{a_{ij}}\right)\left(1 - e^{-a_{ij}(t-\sigma)}\right) \\
&\leq e^{-a_{ij}(t-\sigma)} \left[2K_0 - 2\overline{M} - K_1(M_f + K_0 L_f)\delta\right] + 2\overline{M} + K_1(M_f + K_0 L_f)\delta \\
&\leq K_1.
\end{aligned}$$

Therefore, the inequality $\left\|\widetilde{\Pi} w\right\|_0 \leq K_1$ is valid. Thus, $\widetilde{\Pi}(\Psi) \subseteq \Psi$.

Now, we shall verify that the operator $\widetilde{\Pi}$ is a contraction. Suppose that $w(t)$, $\overline{w}(t) \in \Psi$. For $t \geq \sigma$, we have that

$$\begin{aligned}
&\left|(\widetilde{\Pi} w(t))_{ij} - (\widetilde{\Pi}\overline{w}(t))_{ij}\right| \leq \int_\sigma^t e^{-a_{ij}(t-s)} \sum_{C_{kl}\in N_r(i,j)} C_{ij}^{kl} \Big| f(w_{kl}(s-\tau) + \phi_{\overline{L}}^{kl}(s-\tau)) \\
&\times (w_{ij}(s) + \phi_{\overline{L}}^{ij}(s)) - f(\overline{w}_{kl}(s-\tau) + \phi_{\overline{L}}^{kl}(s-\tau))(\overline{w}_{ij}(s) + \phi_{\overline{L}}^{ij}(s))\Big| ds \\
&\leq \int_\sigma^t e^{-a_{ij}(t-s)} \sum_{C_{kl}\in N_r(i,j)} C_{ij}^{kl} \left|f(w_{kl}(s-\tau) + \phi_{\overline{L}}^{kl}(s-\tau))\right| \left|w_{ij}(s) - \overline{w}_{ij}(s)\right| ds \\
&+ \int_\sigma^t e^{-a_{ij}(t-s)} \sum_{C_{kl}\in N_r(i,j)} C_{ij}^{kl} L_f \left|w_{kl}(s-\tau) - \overline{w}_{kl}(s-\tau)\right| \left|\overline{w}_{ij}(s) + \phi_{\overline{L}}^{ij}(s)\right| ds \\
&\leq \left[M_f + (K_0 + K_1)L_f\right] \frac{\sum_{C_{kl}\in N_r(i,j)} C_{ij}^{kl}}{a_{ij}} \sup_{t\geq\sigma-\tau} \left\|w(t) - \overline{w}(t)\right\|.
\end{aligned}$$

In view of the equation $\left\|\widetilde{\Pi} w(t) - \widetilde{\Pi}\overline{w}(t)\right\| = 0$ for $t < \sigma$, the last inequality implies that

$$\left\|\widetilde{\Pi} w - \widetilde{\Pi}\overline{w}\right\|_0 \leq \left[M_f + (K_0 + K_1)L_f\right]\delta \left\|w - \overline{w}\right\|_0,$$

and the operator $\widetilde{\Pi}$ is contractive according to the condition $(C6)$.

By means of the uniqueness of solutions, one can conclude that $w(t) = \phi_L(t) - \phi_{\overline{L}}(t)$ is the unique fixed point of the operator $\widetilde{\Pi}$.

Since the number E satisfies the inequality $E > \dfrac{4}{\gamma} \ln\left(\dfrac{1}{\eta\varepsilon}\right)$, we have

$$e^{-\gamma(t-\sigma)/2} < \eta\varepsilon,$$

provided that $t \geq \sigma + E/2$. Therefore, the inequality $\|\phi_L(t) - \phi_{\overline{L}}(t)\| < (R_0 + R_1)\eta\varepsilon \leq \varepsilon$ holds for $t \in [\sigma + E/2, \sigma + E_0]$. Consequently, the pair $\big(\phi_L(t), \phi_{\overline{L}}(t)\big) \in \widetilde{\mathscr{L}} \times \widetilde{\mathscr{L}}$ is proximal. □

Lemma 8.8 *Suppose that the conditions (C1)–(C5) are fulfilled. If a pair* $\big(L(t), \overline{L}(t)\big) \in \mathscr{L} \times \mathscr{L}$ *is frequently* (ε_0, Δ)*-separated for some positive numbers* ε_0 *and* Δ*, then there exist positive numbers* ε_1 *and* $\overline{\Delta}$ *such that the pair* $\big(\phi_L(t), \phi_{\overline{L}}(t)\big) \in \widetilde{\mathscr{L}} \times \widetilde{\mathscr{L}}$ *is frequently* $(\varepsilon_1, \overline{\Delta})$*-separated.*

Proof Since the pair $\big(L(t), \overline{L}(t)\big) \in \mathscr{L} \times \mathscr{L}$ is frequently (ε_0, Δ) separated for some numbers $\varepsilon_0 > 0$, $\Delta > 0$, there exist infinitely many intervals $J_q = [\alpha_q, \beta_q]$, $q \in \mathbb{N}$, each with a length no less than Δ, such that $\beta_q \to \infty$ as $q \to \infty$, and $\|L(t) - \overline{L}(t)\| > \varepsilon_0$ for each t from these intervals. The essence of the proof is to determine numbers $\varepsilon_1 > 0$, $\overline{\Delta} > 0$ and infinitely many intervals $\overline{J}_q = [\overline{\alpha}_q, \overline{\beta}_q]$, $q \in \mathbb{N}$, each with length $\overline{\Delta}$, such that $\overline{\beta}_q \to \infty$ as $q \to \infty$, and $\|\phi_L(t) - \phi_{\overline{L}}(t)\| > \varepsilon_1$ for each t from the intervals $\overline{J}_q$, $q \in \mathbb{N}$.

Since $\mathscr{L}$ is an equicontinuous family on $\mathbb{R}$, there exists a positive number κ such that for any $t_1, t_2 \in \mathbb{R}$ with $|t_1 - t_2| < \kappa$, the inequality

$$\big|\big(L_{ij}(t_1) - \overline{L}_{ij}(t_1)\big) - \big(L_{ij}(t_2) - \overline{L}_{ij}(t_2)\big)\big| < \frac{\varepsilon_0}{2} \tag{8.2.13}$$

holds for all $i = 1, 2, \ldots, m$ and $j = 1, 2, \ldots, n$. For each $q \in \mathbb{N}$, set $\theta_q = \beta_q - \kappa/2$.

Let us fix $q \in \mathbb{N}$. There exist integers i_0, j_0 such that

$$\big|L_{i_0 j_0}(\beta_q) - \overline{L}_{i_0 j_0}(\beta_q)\big| = \big\|L(\beta_q) - \overline{L}(\beta_q)\big\| > \varepsilon_0. \tag{8.2.14}$$

By virtue of the inequality (8.2.13), it can be verified for each $t \in \big[\theta_q, \theta_q + \kappa\big]$ that

$$\begin{aligned}
&\big|L_{i_0 j_0}(\beta_q) - \overline{L}_{i_0 j_0}(\beta_q)\big| - \big|L_{i_0 j_0}(t) - \overline{L}_{i_0 j_0}(t)\big| \\
&\leq \big|\big(L_{i_0 j_0}(t) - \overline{L}_{i_0 j_0}(t)\big) - \big(L_{i_0 j_0}(\beta_q) - \overline{L}_{i_0 j_0}(\beta_q)\big)\big| \\
&< \frac{\varepsilon_0}{2}.
\end{aligned}$$

Therefore, making use of (8.2.14), one can confirm for $\theta_q \leq t \leq \theta_q + \kappa$ that

$$\big|L_{i_0 j_0}(t) - \overline{L}_{i_0 j_0}(t)\big| > \big|L_{i_0 j_0}(\beta_q) - \overline{L}_{i_0 j_0}(\beta_q)\big| - \frac{\varepsilon_0}{2} > \frac{\varepsilon_0}{2}. \tag{8.2.15}$$

For each i and j, there exist numbers $\zeta_{ij}^q \in [\theta_q, \theta_q + \kappa]$ such that

$$\int_{\theta_q}^{\theta_q+\kappa} \big(L(s) - \overline{L}(s)\big)\, ds = \kappa \left(L_{11}(\zeta_{11}^q) - \overline{L}_{11}(\zeta_{11}^q), \ldots, L_{mn}(\zeta_{mn}^q) - \overline{L}_{mn}(\zeta_{mn}^q)\right).$$

Thus, by means of the inequality (8.2.15), we obtain that

$$\left\| \int_{\theta_q}^{\theta_q+\kappa} \left(L(s) - \overline{L}(s)\right) ds \right\| \geq \kappa \left| L_{i_0 j_0}(\zeta^q_{i_0 j_0}) - \overline{L}_{i_0 j_0}(\zeta^q_{i_0 j_0}) \right| > \frac{\kappa \varepsilon_0}{2}. \quad (8.2.16)$$

For $t \in [\theta_q, \theta_q + \kappa]$, by the help of the relations

$$\phi_L^{ij}(t) = \phi_L^{ij}(\theta_q) - \int_{\theta_q}^{t} \Big[a_{ij} + \sum_{C_{kl} \in N_r(i,j)} C_{ij}^{kl} f(\phi_L^{kl}(s-\tau))\Big] \phi_L^{ij}(s) ds + \int_{\theta_q}^{t} L_{ij}(s) ds$$

and

$$\phi_{\overline{L}}^{ij}(t) = \phi_{\overline{L}}^{ij}(\theta_q) - \int_{\theta_q}^{t} \Big[a_{ij} + \sum_{C_{kl} \in N_r(i,j)} C_{ij}^{kl} f(\phi_{\overline{L}}^{kl}(s-\tau))\Big] \phi_{\overline{L}}^{ij}(s) ds + \int_{\theta_q}^{t} \overline{L}_{ij}(s) ds,$$

we attain that

$$\begin{aligned}
\phi_L^{ij}(\theta_q+\kappa) - \phi_{\overline{L}}^{ij}(\theta_q+\kappa) &= \int_{\theta_q}^{\theta_q+\kappa} \left(L_{ij}(s) - \overline{L}_{ij}(s)\right) ds \\
&+ (\phi_L^{ij}(\theta_q) - \phi_{\overline{L}}^{ij}(\theta_q)) - \int_{\theta_q}^{\theta_q+\kappa} a_{ij} \left(\phi_L^{ij}(s) - \phi_{\overline{L}}^{ij}(s)\right) ds \\
&- \int_{\theta_q}^{\theta_q+\kappa} \sum_{C_{kl} \in N_r(i,j)} C_{ij}^{kl} \Big[f(\phi_L^{kl}(s-\tau))\phi_L^{ij}(s) - f(\phi_{\overline{L}}^{kl}(s-\tau))\phi_{\overline{L}}^{ij}(s)\Big] ds.
\end{aligned}$$

Hence, we have that

$$\begin{aligned}
\|\phi_L(\theta_q+\kappa) - \phi_{\overline{L}}(\theta_q+\kappa)\| &\geq \left\| \int_{\theta_q}^{\theta_q+\kappa} \left(L(s) - \overline{L}(s)\right) ds \right\| \\
&- \|\phi_L(\theta_q) - \phi_{\overline{L}}(\theta_q)\| - \max_{(i,j)} \left| \int_{\theta_q}^{\theta_q+\kappa} a_{ij} \left(\phi_L^{ij}(s) - \phi_{\overline{L}}^{ij}(s)\right) ds \right| \\
&- \max_{(i,j)} \Big| \int_{\theta_q}^{\theta_q+\kappa} \sum_{C_{kl} \in N_r(i,j)} C_{ij}^{kl} \Big[f(\phi_L^{kl}(s-\tau))\phi_L^{ij}(s) \\
&- f(\phi_{\overline{L}}^{kl}(s-\tau))\phi_{\overline{L}}^{ij}(s)\Big] ds \Big|. \qquad (8.2.17)
\end{aligned}$$

Set $\overline{a} = \max_{(i,j)} a_{ij}, \overline{c} = \max_{(i,j)} \sum_{C_{kl} \in N_r(i,j)} C_{ij}^{kl}, H_0 = \max_{(i,j)} M_{ij}$. The inequalities (8.2.16) and (8.2.17) together imply that

$$\max_{t\in[\theta_q-\tau,\theta_q+\kappa]} \left\|\phi_L(t)-\phi_{\overline{L}}(t)\right\| \geq \left\|\phi_L(\theta_q+\kappa)-\phi_{\overline{L}}(\theta_q+\kappa)\right\|$$
$$> \frac{\kappa\varepsilon_0}{2} - [1+\kappa\overline{a}+\kappa\overline{c}(L_f K_0+M_f)] \max_{t\in[\theta_q-\tau,\theta_q+\kappa]} \left\|\phi_L(t)-\phi_{\overline{L}}(t)\right\|.$$

Therefore,

$$\max_{t\in[\theta_q-\tau,\theta_q+\kappa]} \left\|\phi_L(t)-\phi_{\overline{L}}(t)\right\| > \overline{\varepsilon},$$

where $\overline{\varepsilon} = \dfrac{\kappa\varepsilon_0}{2[2+\kappa\overline{a}+\kappa\overline{c}(L_f K_0+M_f)]}$.

Now, suppose that $\max_{t\in[\theta_q-\tau,\theta_q+\kappa]} \left\|\phi_L(t)-\phi_{\overline{L}}(t)\right\| = \left\|\phi_L(\xi_q)-\phi_{\overline{L}}(\xi_q)\right\|$, for some $\xi_q \in [\theta_q-\tau, \theta_q+\kappa]$. Take a positive number Δ_0 such that

$$\Delta_0 \leq \frac{\overline{\varepsilon}}{4\left(H_0+K_0\overline{a}+M_f K_0\overline{c}\right)}.$$

For $t \in \left[\xi_q-\Delta_0, \xi_q+\Delta_0\right]$, with the aid of the relation

$$\phi_L^{ij}(t)-\phi_{\overline{L}}^{ij}(t) = (\phi_L^{ij}(\xi_q)-\phi_{\overline{L}}^{ij}(\xi_q))$$
$$+\int_{\xi_q}^t \left(L_{ij}(s)-\overline{L}_{ij}(s)\right)ds - \int_{\xi_q}^t a_{ij}\left(\phi_L^{ij}(s)-\phi_{\overline{L}}^{ij}(s)\right)ds$$
$$-\int_{\xi_q}^t \sum_{C_{kl}\in N_r(i,j)} C_{ij}^{kl}\Big[f(\phi_L^{kl}(s-\tau))\phi_L^{ij}(s)-f(\phi_{\overline{L}}^{kl}(s-\tau))\phi_{\overline{L}}^{ij}(s)\Big]ds,$$

we obtain the inequality

$$\left\|\phi_L(t)-\phi_{\overline{L}}(t)\right\| \geq \left\|\phi_L(\xi_q)-\phi_{\overline{L}}(\xi_q)\right\|$$
$$-\max_{(i,j)}\left|\int_{\xi_q}^t \left(L_{ij}(s)-\overline{L}_{ij}(s)\right)ds\right| - \max_{(i,j)}\left|\int_{\xi_q}^t a_{ij}\left(\phi_L^{ij}(s)-\phi_{\overline{L}}^{ij}(s)\right)ds\right|$$
$$-\max_{(i,j)}\left|\int_{\xi_q}^t \sum_{C_{kl}\in N_r(i,j)} C_{ij}^{kl}\Big[f(\phi_L^{kl}(s-\tau))\phi_L^{ij}(s)-f(\phi_{\overline{L}}^{kl}(s-\tau))\phi_{\overline{L}}^{ij}(s)\Big]ds\right|$$
$$> \overline{\varepsilon} - 2\Delta_0\left(H_0+K_0\overline{a}+M_f K_0\overline{c}\right)$$
$$\geq \frac{\overline{\varepsilon}}{2}.$$

Hence, we have $\left\|\phi_L(t)-\phi_{\overline{L}}(t)\right\| > \overline{\varepsilon}/2$ for each t from the intervals $\overline{J}_q = \left[\overline{\alpha}_q, \overline{\beta}_q\right]$, $q \in \mathbb{N}$, where $\overline{\alpha}_q = \xi_q - \Delta_0$ and $\overline{\beta}_q = \xi_q + \Delta_0$. One can confirm that $\overline{\beta}_q \to \infty$ as $q \to \infty$.

Consequently, the pair $\left(\phi_L(t), \phi_{\overline{L}}(t)\right) \in \widetilde{\mathscr{L}} \times \widetilde{\mathscr{L}}$ is frequently $(\varepsilon_1, \overline{\Delta})$-separated, where $\varepsilon_1 = \overline{\varepsilon}/2$ and $\overline{\Delta} = 2\Delta_0$. □

The main result of the present section is as follows.

Theorem 8.2 *Under the conditions* $(C1)$–$(C7)$*, the set* $\widetilde{\mathscr{L}}$ *is Li–Yorke chaotic, provided that the same is true for the set* $\mathscr{L}$*.*

Proof Since the set $\mathscr{L}$ is Li–Yorke chaotic, there exists a positive number T_0 such that for any $k \in \mathbb{N}$, $\mathscr{L}$ possesses a periodic function with period kT_0. One can use the integral equation (8.2.12) together with condition $(C5)$ to verify that $L(t) \in \mathscr{L}$ is kT_0-periodic if and only if $\phi_L(t) \in \widetilde{\mathscr{L}}$ is kT_0-periodic. Thus, for each $k \in \mathbb{N}$, the set $\widetilde{\mathscr{L}}$ contains a kT_0-periodic function.

Suppose that $\mathscr{L}_S$ is a scrambled set inside $\mathscr{L}$. Consider the collection $\widetilde{\mathscr{L}}_S$ with elements of the form $\phi_L(t)$, where $L(t) \in \mathscr{L}_S$. Because of the one-to-one correspondence between the elements of $\mathscr{L}_S$ and $\widetilde{\mathscr{L}}_S$, the set $\widetilde{\mathscr{L}}_S$ is uncountable. Moreover, no periodic functions exist inside $\widetilde{\mathscr{L}}_S$, since no such functions take place inside $\mathscr{L}_S$.

Lemmas 8.7 and 8.8 together ensure that $\widetilde{\mathscr{L}}_S$ is a scrambled set. Additionally, Lemma 8.8 implies that any pair of functions inside $\widetilde{\mathscr{L}}_S \times \widetilde{\mathscr{L}}_P$ is frequently $(\varepsilon_1, \overline{\Delta})$-separated for some positive numbers ε_1 and $\overline{\Delta}$, where $\widetilde{\mathscr{L}}_P$ denotes the set of all periodic functions inside $\widetilde{\mathscr{L}}$. As a consequence, the set $\widetilde{\mathscr{L}}$ is Li–Yorke chaotic. □

Since time delay is an inevitable feature of neural networks, the result presented in Theorem 8.2 is much more realistic then the one obtained in [103]. That is, unless retardation is not introduced in the models, investigations of neural networks will not be adequate for application problems. Introducing delay requests a more sophisticated mathematical analysis, and this is the first time in the literature that the approach developed in [34, 103] is applied to functional differential equations.

Suppose that $F : \mathbb{R}^{m \times n} \to \mathbb{R}^{\overline{m} \times \overline{n}}$ is a function such that for all $s_1, s_2 \in \mathbb{R}^{m \times n}$ the inequality

$$L_1 \left\| s_1 - s_2 \right\| \leq \left\| F(s_1) - F(s_2) \right\| \leq L_2 \left\| s_1 - s_2 \right\|, \qquad (8.2.18)$$

is valid, where L_1 and L_2 are positive numbers. One can verify that if a collection $\mathscr{L}$ of functions is Li–Yorke chaotic, then the collection with elements of the form $F(L(t))$, where $L(t) \in \mathscr{L}$, is also Li–Yorke chaotic.

In the next section, we will focus on a neural system consisting of three layers such that each layer is a SICNN, and the connections between the layers are provided through nonlinear functions that satisfy the inequality (8.2.18).

8.2.4 An Example

In the theory of neural networks, one can consider interconnected collections of neurons, called layers. Additionally, a neural system is a collection of neural networks, which can be considered as single layers. Each neuron in a neural network is capable of receiving input signals, processing them, and sending an output signal. Neural signals consist of short electrical pulses, which are called action potentials or spikes. That is why the discontinuity phenomena is a natural property of neural networks. A chain of action potentials emitted by a single neuron is called a spike train. Action potentials in a spike train are usually well separated, and it is impossible to excite a second spike during or immediately after a first one [1]. In this section, we take into account an example of a neural system consisting of three layers, where each layer is a SICNN. Discontinuous external inputs are used in the first layer to provide the chaos.

Consider the retarded SICNNs

$$\frac{dx_{ij}}{dt} = -a_{ij}x_{ij} - \sum_{C_{kl}\in N_1(i,j)} C_{ij}^{kl} f(x_{kl}(t-\tau_1))x_{ij} + L_{ij}(t), \tag{8.2.19}$$

$$\frac{dy_{ij}}{dt} = -b_{ij}y_{ij} - \sum_{\overline{C}_{kl}\in N_1(i,j)} \overline{C}_{ij}^{kl} g(y_{kl}(t-\tau_2))y_{ij} + \overline{L}_{ij}(t), \tag{8.2.20}$$

$$\frac{dz_{ij}}{dt} = -c_{ij}z_{ij} - \sum_{\overline{\overline{C}}_{kl}\in N_1(i,j)} \overline{\overline{C}}_{ij}^{kl} h(z_{kl}(t-\tau_3))z_{ij} + \overline{\overline{L}}_{ij}(t), \tag{8.2.21}$$

in which $i, j = 1, 2, 3,$

$$\begin{pmatrix} a_{11} & a_{12} & a_{13} \\ a_{21} & a_{22} & a_{23} \\ a_{31} & a_{32} & a_{33} \end{pmatrix} = \begin{pmatrix} 3 & 7 & 2 \\ 9 & 4 & 5 \\ 1 & 3 & 6 \end{pmatrix},$$

$$\begin{pmatrix} C_{11} & C_{12} & C_{13} \\ C_{21} & C_{22} & C_{23} \\ C_{31} & C_{32} & C_{33} \end{pmatrix} = \begin{pmatrix} 0 & 0.008 & 0.002 \\ 0.001 & 0.003 & 0.007 \\ 0.004 & 0 & 0.006 \end{pmatrix},$$

$$\begin{pmatrix} b_{11} & b_{12} & b_{13} \\ b_{21} & b_{22} & b_{23} \\ b_{31} & b_{32} & b_{33} \end{pmatrix} = \begin{pmatrix} 4 & 7 & 5 \\ 3 & 6 & 8 \\ 10 & 9 & 4 \end{pmatrix},$$

$$\begin{pmatrix} \overline{C}_{11} & \overline{C}_{12} & \overline{C}_{13} \\ \overline{C}_{21} & \overline{C}_{22} & \overline{C}_{23} \\ \overline{C}_{31} & \overline{C}_{32} & \overline{C}_{33} \end{pmatrix} = \begin{pmatrix} 0.004 & 0.007 & 0.002 \\ 0 & 0.006 & 0.003 \\ 0.005 & 0.009 & 0.008 \end{pmatrix},$$

$$\begin{pmatrix} c_{11} & c_{12} & c_{13} \\ c_{21} & c_{22} & c_{23} \\ c_{31} & c_{32} & c_{33} \end{pmatrix} = \begin{pmatrix} 2 & 8 & 4 \\ 1 & 1 & 3 \\ 6 & 2 & 5 \end{pmatrix},$$

$$\begin{pmatrix} \overline{\overline{C}}_{11} & \overline{\overline{C}}_{12} & \overline{\overline{C}}_{13} \\ \overline{\overline{C}}_{21} & \overline{\overline{C}}_{22} & \overline{\overline{C}}_{23} \\ \overline{\overline{C}}_{31} & \overline{\overline{C}}_{32} & \overline{\overline{C}}_{33} \end{pmatrix} = \begin{pmatrix} 0.006 & 0 & 0.002 \\ 0.004 & 0.001 & 0.008 \\ 0 & 0.007 & 0.002 \end{pmatrix},$$

$f(s) = \frac{1}{2}s^2$, $g(s) = \frac{1}{3}s^3$, $h(s) = \sqrt{s}$, $\tau_1 = 0.5$, $\tau_2 = 3$ and $\tau_3 = 2.5$.

To obtain chaotic SICNNs with delay by means of the presented method, one needs a collection of external inputs which are known to be chaotic in the sense of Li–Yorke. For that reason, in system (8.2.19), the external inputs, $L_{ij}(t)$, will be considered as relay functions with chaotically changing switching moments [29–31, 33]. More precisely, we set $L_{ij}(t) = \nu_{ij}(t, \zeta)$, where

$$\nu_{ij}(t, \zeta) = \begin{cases} \alpha_{ij}, & \text{if } \zeta_{2q} < t \le \zeta_{2q+1}, \\ \beta_{ij}, & \text{if } \zeta_{2q-1} < t \le \zeta_{2q}, \end{cases} . \tag{8.2.22}$$

In the relay function (8.2.22), $\alpha = \{\alpha_{ij}\}$, and $\beta = \{\beta_{ij}\}$ are different from each other and the sequence $\zeta = \{\zeta_q\}$, $q \in \mathbb{Z}$, of switching moments are the same for each i and j. The sequence ζ is defined by the formula $\zeta_q = q + \kappa_q$, $q \in \mathbb{Z}$, where the sequence $\{\kappa_q\}$, $\kappa_0 \in [0, 1]$, is generated through the equation $\kappa_{q+1} = \lambda(\kappa_q)$, and $\lambda(s) = 3.9s(1 - s)$ is the logistic map, which is chaotic in the Li–Yorke sense [24]. The interval $[0, 1]$ is invariant under the iterations of the map $\lambda(s)$ [119]. The presence of chaos in the SICNN (8.2.19) can be proved in a similar manner to those mentioned in Sect. 8.2.3. It is worth noting that external inputs of the form (8.2.22) are Li–Yorke chaotic and this type of inputs have never been considered before in the literature for SICNNs with delay.

Let us use $\alpha_{ij} = 1.2$ and $\beta_{ij} = 2.5$ in (8.2.19). Clearly,

$$\sum_{C_{kl} \in N_1(1,1)} C_{11}^{kl} = 0.012, \quad \sum_{C_{kl} \in N_1(1,2)} C_{12}^{kl} = 0.021, \quad \sum_{C_{kl} \in N_1(1,3)} C_{13}^{kl} = 0.020,$$

$$\sum_{C_{kl} \in N_1(2,1)} C_{21}^{kl} = 0.016, \quad \sum_{C_{kl} \in N_1(2,2)} C_{22}^{kl} = 0.031, \quad \sum_{C_{kl} \in N_1(2,3)} C_{23}^{kl} = 0.026,$$

$$\sum_{C_{kl} \in N_1(3,1)} C_{31}^{kl} = 0.008, \quad \sum_{C_{kl} \in N_1(3,2)} C_{32}^{kl} = 0.021, \quad \sum_{C_{kl} \in N_1(3,3)} C_{33}^{kl} = 0.016.$$

One can confirm that the conditions $(C1)$–$(C7)$ hold for system (8.2.19) with $\gamma = 1$, $L_f = 2.5$, $M_f = 3.125$, $\overline{M} = 2.5$, $\delta = 0.01$, $\overline{\delta} = 0.008$, $K_0 = 2.581$, and

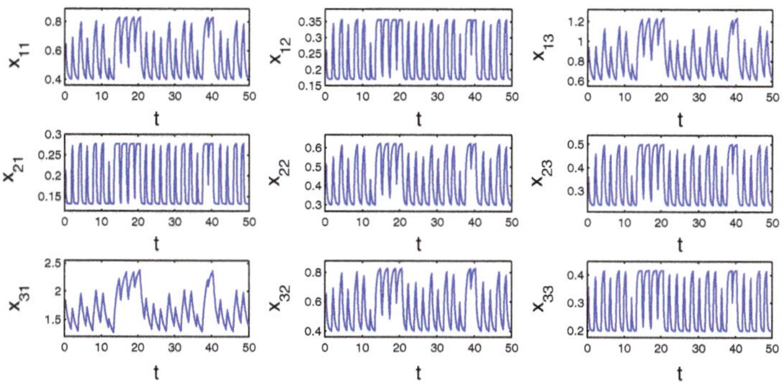

Fig. 8.4 The irregular behavior in each cell of the SICNN (8.2.19)

$K_1 = 5.53$. Therefore, the collection $\mathscr{L}_x$ of bounded on $\mathbb{R}$ solutions of (8.2.19) with different ζ is a Li–Yorke chaotic set.

Consider the constant function $u_1(t) = \left\{u_1^{ij}(t)\right\}$ with $u_1^{11}(t) = 0.652$, $u_1^{12}(t) = 0.263$, $u_1^{13}(t) = 0.942$, $u_1^{21}(t) = 0.215$, $u_1^{22}(t) = 0.517$, $u_1^{23}(t) = 0.364$, $u_1^{31}(t) = 1.846$, $u_1^{32}(t) = 0.658$, and $u_1^{33}(t) = 0.361$. We use the sequence ζ with $\zeta_0 = 0.38$, and represent in Fig. 8.4 the solution $x(t) = \left\{x_{ij}(t)\right\}$ of (8.2.19) satisfying $x(t) = u_1(t)$ for $t_0 - \tau_1 \le t \le t_0$, where $t_0 = 0.38$. Figure 8.4 reveals that each coordinate of the solution behaves chaotically.

Now, we shall focus on the SICNN (8.2.20). Consider the function $\varphi(v) = \left\{\varphi_{ij}(v)\right\}$, where $v = \left\{v_{ij}\right\}$ and $\varphi_{11}(v) = 0.5\tanh(v_{11})$, $\varphi_{12}(v) = 2v_{12} + \arctan v_{12}$, $\varphi_{13}(v) = 2v_{13}^2$, $\varphi_{21}(v) = \sqrt{v_{31}}$, $\varphi_{22}(v) = 0.4e^{v_{32}}$, $\varphi_{23}(v) = v_{33} + 0.7\cos v_{33}$, $\varphi_{31}(v) = \dfrac{1}{v_{21}^2+1}$, $\varphi_{32}(v) = \dfrac{v_{22}^2 + 2v_{22} + 2}{v_{22}+1}$, $\varphi_{33}(v) = (0.9 + v_{23})^3$. In system (8.2.20), we set $\overline{L}_{ij}(t) = \varphi_{ij}(x(t))$ for each $i, j = 1, 2, 3$. That is, the external inputs $\overline{L}_{ij}(t)$ of the network (8.2.20) are provided through the outputs of (8.2.19).

The function φ satisfies the inequality (8.2.18) inside the compact region where the chaotic attractor of system (8.2.19) takes place. Therefore, the collection which consists of elements of the form $\varphi(x(t))$, $x(t) \in \mathscr{L}_x$, is a Li–Yorke chaotic set.

One can evaluate that

$$\sum_{\overline{C}_{kl}\in N_1(1,1)} \overline{C}_{11}^{kl} = 0.017, \quad \sum_{\overline{C}_{kl}\in N_1(1,2)} \overline{C}_{12}^{kl} = 0.022, \quad \sum_{\overline{C}_{kl}\in N_1(1,3)} \overline{C}_{13}^{kl} = 0.018,$$

$$\sum_{\overline{C}_{kl}\in N_1(2,1)} \overline{C}_{21}^{kl} = 0.031, \quad \sum_{\overline{C}_{kl}\in N_1(2,2)} \overline{C}_{22}^{kl} = 0.044, \quad \sum_{\overline{C}_{kl}\in N_1(2,3)} \overline{C}_{23}^{kl} = 0.035,$$

$$\sum_{\overline{C}_{kl}\in N_1(3,1)} \overline{C}_{31}^{kl} = 0.020, \quad \sum_{\overline{C}_{kl}\in N_1(3,2)} \overline{C}_{32}^{kl} = 0.031, \quad \sum_{\overline{C}_{kl}\in N_1(3,3)} \overline{C}_{33}^{kl} = 0.026,$$

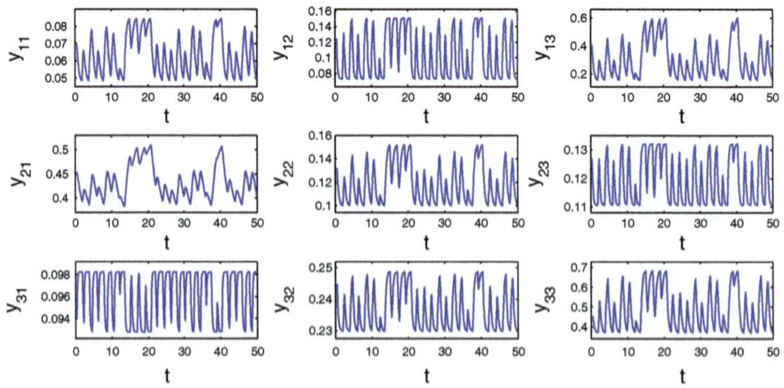

Fig. 8.5 The motions that appear in the cells of the SICNN (8.2.20). Our theoretical discussions are supported such that each cell behaves chaotically

and the conditions $(C1)$–$(C7)$ hold for system (8.2.20) with $\gamma = 3$, $L_g = 0.49$, $M_g = 0.1145$, $\overline{M} = 0.762175$, $\delta = \overline{\delta} = 0.031/3$, $K_0 = 0.7631$, and $K_1 = 1.5322$. Consequently, the set $\mathscr{L}_y$ of bounded on $\mathbb{R}$ solutions of system (8.2.20) is Li–Yorke chaotic in accordance with Theorem 8.2.

We represent in Fig. 8.5 the solution of system (8.2.20) with $y(t) = u_2(t)$ for $t_0 - \tau_2 \le t \le t_0$, where $u_2(t) = \left\{u_2^{ij}\right\}$ is a constant function defined as $u_2^{11}(t) = 0.071$, $u_2^{12}(t) = 0.125$, $u_2^{13}(t) = 0.412$, $u_2^{21}(t) = 0.454$, $u_2^{22}(t) = 0.132$, $u_2^{23}(t) = 0.127$, $u_2^{31}(t) = 0.094$, $u_2^{32}(t) = 0.245$, $u_2^{33}(t) = 0.442$ and $t_0 = 0.38$. Figure 8.5 supports the theoretical results such that the dynamics of the SICNN (8.2.20) is chaotic.

In a similar way, in system (8.2.21), we take $\overline{\overline{L}}_{ij}(t) = \psi_{ij}(y(t))$, $i, j = 1, 2, 3$, where the function $\psi(v) = \{\psi_{ij}(v)\}$ is defined through the equations $\psi_{11}(v) = \frac{1}{3}v_{33}$, $\psi_{12}(v) = 2v_{32} + \sin v_{32}$, $\psi_{13}(v) = 10v_{31}^3$, $\psi_{21}(v) = (1 + v_{11})^{1/3}$, $\psi_{22}(v) = 0.5 \arctan v_{12}$, $\psi_{23}(v) = 2v_{13}$, $\psi_{31}(v) = \tanh v_{21}$, $\psi_{32}(v) = \dfrac{1}{v_{22} + 2}$ and $\psi_{33}(v) = 1.5\sqrt{1 + v_{23}}$. It can be verified that the inequality (8.2.18) holds for the function ψ, and the collection with elements of the form $\psi(y(t))$, $y(t) \in \mathscr{L}_y$, is Li–Yorke chaotic.

In system (8.2.21), we have that

$$\sum_{\overline{\overline{C}}_{kl} \in N_1(1,1)} \overline{\overline{C}}_{11}^{kl} = 0.011, \quad \sum_{\overline{\overline{C}}_{kl} \in N_1(1,2)} \overline{\overline{C}}_{12}^{kl} = 0.021, \quad \sum_{\overline{\overline{C}}_{kl} \in N_1(1,3)} \overline{\overline{C}}_{13}^{kl} = 0.011,$$

$$\sum_{\overline{\overline{C}}_{kl} \in N_1(2,1)} \overline{\overline{C}}_{21}^{kl} = 0.018, \quad \sum_{\overline{\overline{C}}_{kl} \in N_1(2,2)} \overline{\overline{C}}_{22}^{kl} = 0.03, \quad \sum_{\overline{\overline{C}}_{kl} \in N_1(2,3)} \overline{\overline{C}}_{23}^{kl} = 0.02,$$

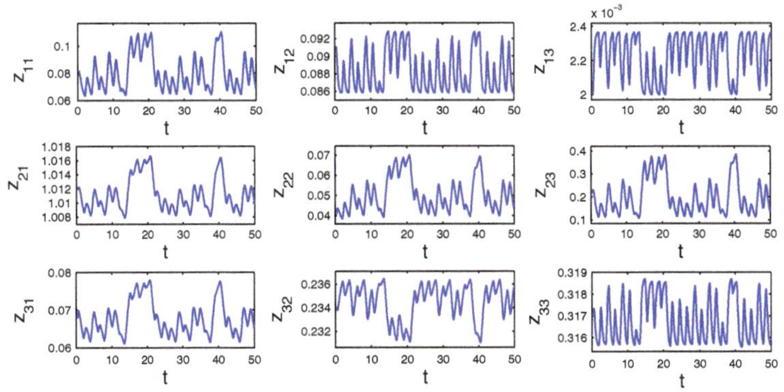

Fig. 8.6 The chaotic motions in each cell of the SICNN (8.2.21)

$$\sum_{\overline{\overline{C}}_{kl}\in N_1(3,1)} \overline{\overline{C}}_{31}^{kl} = 0.012, \quad \sum_{\overline{\overline{C}}_{kl}\in N_1(3,2)} \overline{\overline{C}}_{32}^{kl} = 0.022, \quad \sum_{\overline{\overline{C}}_{kl}\in N_1(3,3)} \overline{\overline{C}}_{33}^{kl} = 0.018.$$

Moreover, the conditions $(C1)$–$(C7)$ are valid with $\gamma = 1, L_h = 0.4951, M_h = 1.01$, $\overline{M} = 1.0292$, $\delta = \overline{\delta} = 0.03$, $K_0 = 1.0614$, and $K_1 = 2.1579$. Thus, in compliance with Theorem 8.2, the dynamics of system (8.2.21) is Li–Yorke chaotic. That is, the set $\mathscr{L}_z$ of bounded on $\mathbb{R}$ solutions of (8.2.21) is chaotic in the sense of Li–Yorke.

The behavior of the SICNN (8.2.21) is observable in Fig. 8.6, which depicts the solution with $z(t) = u_3(t)$ for $t_0 - \tau_3 \le t \le t_0$, where $t_0 = 0.38$ and the constant function $u_3(t) = \left\{u_3^{ij}\right\}$ is defined as $u_3^{11}(t) = 0.082$, $u_3^{12}(t) = 0.091$, $u_3^{13}(t) = 0.002$, $u_3^{21}(t) = 1.012$, $u_3^{22}(t) = 0.041$, $u_3^{23}(t) = 0.217$, $u_3^{31}(t) = 0.068$, $u_3^{32}(t) = 0.234$, $u_3^{33}(t) = 0.317$. The illustration supports our results such that the SICNN (8.2.21) exhibits chaos.

Even if we consider constant initial functions in the simulations, the illustrated outputs in Figs. 8.4, 8.5 and 8.6 converge to bounded on $\mathbb{R}$ solutions, which are known to be chaotic, and that is the reason why chaotic behavior is observable. Moreover, it is possible to use other values of the delays τ_1, τ_2, and τ_3 in the neural system (8.2.19) – (8.2.20) – (8.2.21) provided that the condition $(C7)$ is fulfilled.

To confirm one more time that the neural system (8.2.19) – (8.2.20) – (8.2.21) exhibits chaotic motions, we illustrate in Fig. 8.7 the projection of the trajectory with $x(t) = u_1(t)$, $t_0 - \tau_1 \le t \le t_0$, $y(t) = u_2(t)$, $t_0 - \tau_2 \le t \le t_0$, $z(t) = u_3(t)$, $t_0 - \tau_3 \le t \le t_0$, on the $x_{22} - y_{21} - z_{33}$ space, where $t_0 = 0.38$. Figure 8.7 supports our results such that a chaotic attractor takes place in the dynamics of the neural system. The obtained chaos for the neural system (8.2.19) – (8.2.20) – (8.2.21) is in the sense of Li–Yorke, and it is remarkable that the presence of chaos with a precise type in neural systems consisting of retarded SICNNs has never been reported before.

In order to illustrate the proximality and frequent separation features in the neural system (8.2.19) – (8.2.20) – (8.2.21), we represent in Fig. 8.8 the x_{22}, y_{22}, and z_{22} coordinates of the solutions corresponding to the sequence ζ with $\zeta_0 = 0.38$ and

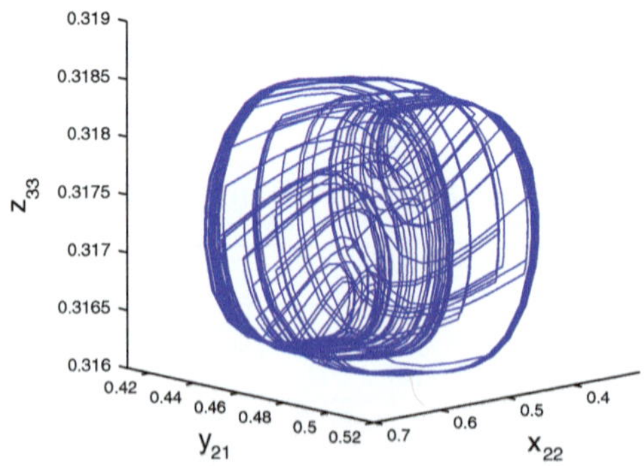

Fig. 8.7 The projection of the chaotic trajectory of the neural system (8.2.19) – (8.2.20) – (8.2.21) on the x_{22}–y_{21}–z_{33} space

$\zeta_0 = 0.39$ in blue and red colors, respectively. In the former, we use the initial conditions $x(t) = u_1(t)$, $t_0 - \tau_1 \leq t \leq t_0$, $y(t) = u_2(t)$, $t_0 - \tau_2 \leq t \leq t_0$, $z(t) = u_3(t)$, $t_0 - \tau_3 \leq t \leq t_0$, where $t_0 = 0.38$. For the solution shown in red color, the initial conditions $x(t) = \bar{u}_1(t)$, $t_1 - \tau_1 \leq t \leq t_1$, $y(t) = \bar{u}_2(t)$, $t_1 - \tau_2 \leq t \leq t_1$, $z(t) = \bar{u}_3(t)$, $t_1 - \tau_3 \leq t \leq t_1$, where $t_1 = 0.39$, are used. Here, $\bar{u}_1(t) = \left\{u_1^{ij}(t)\right\}$, $\bar{u}_2(t) = \left\{u_2^{ij}(t)\right\}$ and $\bar{u}_3(t) = \left\{u_3^{ij}(t)\right\}$ are constant functions defined as $\bar{u}_1^{11}(t) = 0.428$, $\bar{u}_1^{12}(t) = 0.351$, $\bar{u}_1^{13}(t) = 0.745$, $\bar{u}_1^{21}(t) = 0.623$, $\bar{u}_1^{22}(t) = 0.553$, $\bar{u}_1^{23}(t) = 0.254$, $\bar{u}_1^{31}(t) = 1.725$, $\bar{u}_1^{32}(t) = 0.742$, $\bar{u}_1^{33}(t) = 0.249$, $\bar{u}_2^{11}(t) = 0.086$, $\bar{u}_2^{12}(t) = 0.234$, $\bar{u}_2^{13}(t) = 0.321$, $\bar{u}_2^{21}(t) = 0.253$, $\bar{u}_2^{22}(t) = 0.201$, $\bar{u}_2^{23}(t) = 0.113$, $\bar{u}_2^{31}(t) = 0.105$, $\bar{u}_2^{32}(t) = 0.194$, $\bar{u}_2^{33}(t) = 0.454$, $\bar{u}_3^{11}(t) = 0.095$, $\bar{u}_3^{12}(t) = 0.094$, $\bar{u}_3^{13}(t) = 0.001$, $\bar{u}_3^{21}(t) = 1.145$, $\bar{u}_3^{22}(t) = 0.038$, $\bar{u}_3^{23}(t) = 0.332$, $\bar{u}_3^{31}(t) = 0.089$, $\bar{u}_3^{32}(t) = 0.251$, and $\bar{u}_3^{33}(t) = 0.212$.

It is seen in Fig. 8.8 that the x_{22}, y_{22} and z_{22} coordinates of the represented outputs are separated from each other by a positive number approximately for $1105 \leq t \leq 1110.3$, $1105 \leq t \leq 1111.4$, and $1105 \leq t \leq 1112.3$, respectively. On the other hand, one can observe the presence of the proximality feature in each of the coordinates such that the solutions are almost identical approximately for $1120 \leq t \leq 1133$. In addition to this, the solutions are again separated from each other by a positive number approximately for $1133 \leq t \leq 1140$.

8.2.5 *Synchronization of Chaos*

In the example presented in Sect. 8.2.4, a neural system consisting of three layers is considered. Each layer of the neural system (8.2.19) – (8.2.20) – (8.2.21) is, in fact, a retarded SICNN. The layers are connected in a unidirectional way such that between the layers we have feed-forward connections. The schematic diagram of the neural

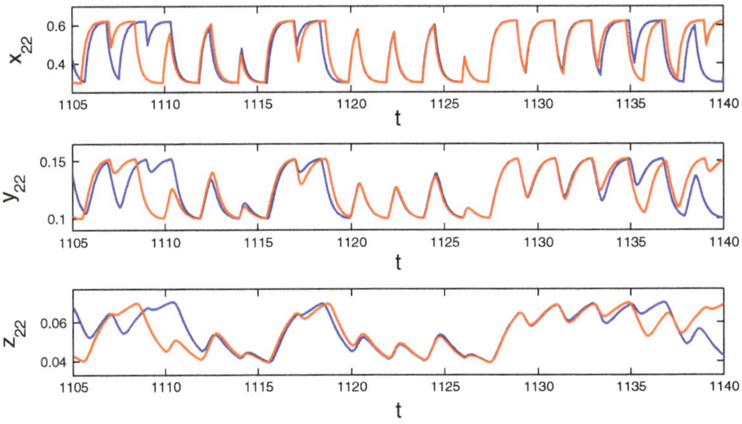

Fig. 8.8 The presence of the proximality and frequent separation features in the neural system (8.2.19) – (8.2.20) – (8.2.21)

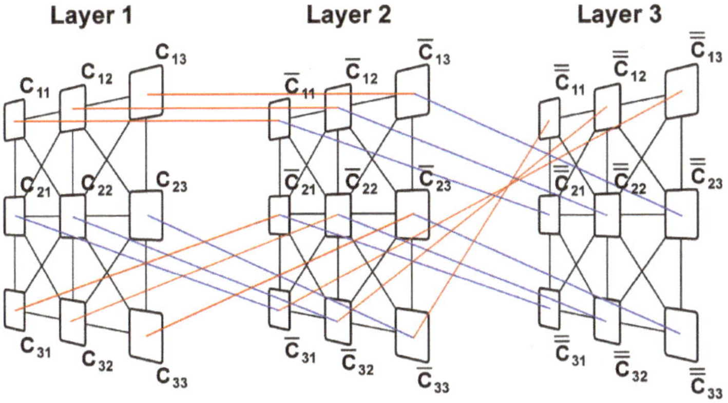

Fig. 8.9 The schematic diagram of the neural system (8.2.19) – (8.2.20) – (8.2.21). The layers are unidirectionally coupled, and each layer of the neural system is a SICNN. The couplings between the cells of different layers are presented in *blue* and *red colors*, while the connections within each SICNN are shown in *black color*

system is shown in Fig. 8.9, where the unidirectional connections between the cells of different layers are presented in blue and red colors. It is worth noting that feed-backward connections exist within the layers, and black color is used to depict them in the figure. The first layer admits the chaos due to the external inputs in the form of chaotic relay functions. The chaotic outputs of the first layer are used as external inputs for the second one; and therefore, the latter also possesses chaotic motions in accordance with our theoretical results. Besides, being affected by the outputs of the second layer, the SICNN (8.2.21) exhibits chaos too. As a result, the system (8.2.19) – (8.2.20) – (8.2.21) admits chaotic motions, and we call this process as the *chaotification* of the neural system.

The first notions of chaotic synchronization were introduced and developed in the papers [87, 89, 94, 95]. Afraimovich et al. [87] proposed the synchronization of chaotic systems that are different and not restricted in coupling. To realize this proposal, Rulkov et al. [96] considered the concept of generalized synchronization (GS) for unidirectionally coupled systems with a skew product structure in the form

$$x'(t) = F(x(t)) \tag{8.2.23}$$

and

$$y'(t) = G(x(t), y(t)). \tag{8.2.24}$$

The systems (8.2.23) and (8.2.24) are called the drive and response systems, respectively. GS [67, 86, 88, 91, 92, 96] is said to occur if there exist sets B_x, B_y of initial conditions and a transformation ϕ, defined on the chaotic attractor of (8.2.23), such that for all $x(0) \in B_x$, $y(0) \in B_y$ the relation $\lim_{t\to\infty} \|y(t) - \phi(x(t))\| = 0$ holds. In the case of GS, a motion starting on $B_x \times B_y$ collapses onto a manifold $M \subset B_x \times B_y$ of synchronized motions. The transformation ϕ is not required to exist for the transient trajectories. If ϕ is the identity transformation, then identical synchronization takes place [94].

It is formulated in paper [92] that GS occurs in the coupled system (8.2.23) – (8.2.24) if and only if for all $x_0 \in B_x, y_1, y_2 \in B_y$, the asymptotic stability criterion

$$\lim_{t\to\infty} \|y(t, x_0, y_1) - y(t, x_0, y_2)\| = 0$$

holds, where $y(t, x_0, y_1)$ and $y(t, x_0, y_2)$ are the solutions of (8.2.24) with the same $x(t)$ such that $y(0, x_0, y_1) = y_1$, $y(0, x_0, y_2) = y_2$, and $x(0) = x_0$.

Now, let us discuss the concept of GS for the neural system (8.2.19) – (8.2.20) – (8.2.21). Lemma 8.6 implies that for a fixed output $x(t) = \{x_{ij}(t)\}$ of (8.2.19), the criterion

$$\lim_{t\to\infty} \|y(t, x(t), \varphi_1(t)) - y(t, x(t), \varphi_2(t))\| = 0$$

holds for arbitrary initial functions $\varphi_1(t)$ and $\varphi_2(t)$, where $y(t, x(t), \varphi_1(t))$, and $y(t, x(t), \varphi_2(t))$ denote the solutions of the network (8.2.20) with $y(t, x(t), \varphi_1(t)) = \varphi_1(t)$ and $y(t, x(t), \varphi_2(t)) = \varphi_2(t)$ for $t \in [-\tau_2, 0]$. Therefore, one can conclude that GS occurs in the dynamics of the coupled SICNNs (8.2.19) – (8.2.20). It is worth noting that a similar discussion is valid for the SICNNs (8.2.20) and (8.2.21), and they are also synchronized in the generalized sense. Since different coefficients and different external inputs are used in the networks (8.2.20) and (8.2.21), one can confirm that different synchronization manifolds take place for the couples (8.2.19) – (8.2.20), (8.2.20) – (8.2.21), and (8.2.19) – (8.2.21).

We apply Lemma 8.6 to prove the presence of GS and we need just to verify the conditions $(C1)$–$(C7)$ to confirm the presence of synchronization in a couple of neural networks. It is shown for the first time in the literature that the technique is used to synchronize a chain of SICNNs. Since the synchronization manifolds are not the same for different pairs of SICNNs from the chain, the complexity of chaos in the family increases. The results may be used to explain the high performance of brain functioning [54, 84].

A numerical method that can be used to investigate coupled systems for GS is the auxiliary system approach [67, 86]. We will use this approach to support the theoretical discussions about the presence of GS in the neural system (8.2.19) – (8.2.20) – (8.2.21). First, we start with the coupled SICNNs (8.2.19) – (8.2.20).

Let us take into account the auxiliary system

$$\frac{d\zeta_{ij}}{dt} = -b_{ij}\zeta_{ij} - \sum_{\overline{C}_{kl}\in N_1(i,j)} \overline{C}_{ij}^{kl} g(\zeta_{kl}(t-\tau_2))\zeta_{ij} + \overline{L}_{ij}(t), \qquad (8.2.25)$$

which is an identical copy of (8.2.20).

In the networks (8.2.20) and (8.2.25), we use the external inputs $x_{ij}(t), i = 1, 2, 3,$ $j = 1, 2, 3,$ which are depicted in Fig. 8.4, and represent in Fig. 8.10 the projection of the stroboscopic plot of the network (8.2.19) – (8.2.20) – (8.2.25) on the $y_{22} - \zeta_{22}$ plane using the initial functions $y(t) = u_2(t)$, which was described in the previous section, and $\zeta(t) = v(t)$ for $t_0 - \tau_2 \leq t \leq t_0$, where $t_0 = 0.38$ and $v(t) = \{v_{ij}\}$ is the constant function defined as $v_{11}(t) = 0.114$, $v_{12}(t) = 0.191$, $v_{13}(t) = 0.302$, $v_{21}(t) = 0.512$, $v_{22}(t) = 0.041$, $v_{23}(t) = 0.215$, $v_{31}(t) = 0.287$, $v_{32}(t) = 0.158$, $v_{33}(t) = 0.294$. In the simulation the first 50 iterations are omitted. One can see in Fig. 8.10 that the plot is on the line $\zeta_{22} = y_{22}$, and this result supports our theoretical discussions about the presence of GS for the coupled SICNNs (8.2.19) – (8.2.20). A similar simulation can be performed for the coupled SICNNs (8.2.20) – (8.2.21).

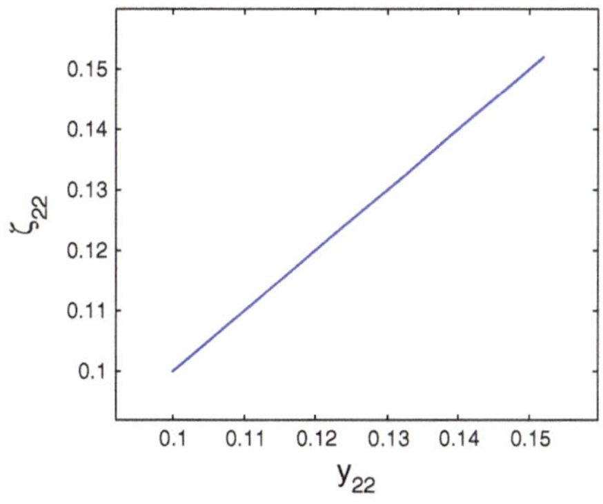

Fig. 8.10 Application of the auxiliary system approach to the coupled SICNNs (8.2.19) – (8.2.20) indicates that GS exists for the couple

8.3 Impulsive SICNNs with Chaotic Postsynaptic Currents

In the present section, we investigate the dynamics of shunting inhibitory cellular neural networks (SICNNs) with impulsive effects. We give a mathematical description of the chaos for the multidimensional dynamics of impulsive SICNNs, and prove its existence rigorously by taking advantage of the external inputs. The Li–Yorke definition of chaos is used in our theoretical discussions. In the considered model, the impacts satisfy the cell and shunting principles. This strengthens the application role of the results, and makes the analysis of impulsive neural networks deeper. The technique is exceptionally useful for SICNNs with arbitrary number of cells. We make benefit of unidirectionally coupled SICNNs to exemplify our results. Moreover, the appearance of cyclic irregular behavior observed in neuroscience is numerically demonstrated for discontinuous dynamics of impulsive SICNNs.

8.3.1 Introduction

Bouzerdoum and Pinter [6] introduced and analyzed a class of cellular neural networks (CNNs), namely the shunting inhibitory cellular neural networks (SICNNs), which have been extensively applied in psychophysics, speech, perception, robotics, adaptive pattern recognition, vision and image processing [7–13, 79]. The layers in SICNNs are arranged into two-dimensional arrays of processing units, called cells, where each cell is coupled to its neighboring units only. The interactions among cells within a single layer is mediated via the biophysical mechanism of recurrent shunting inhibition, where the shunting conductance of each cell is modulated by voltages of neighboring cells [6].

In the most original formulation [6], the model of SICNNs is as follows. Consider a two-dimensional grid of processing cells, and let C_{ij}, $i = 1, 2, \ldots, m$, $j = 1, 2, \ldots, n$, denote the cell at the (i, j) position of the lattice. The r-neighborhood of C_{ij} is defined as

$$N_r(i, j) = \{C_{hl} : \max\{|h - i|, |l - j|\} \leq r,\ 1 \leq h \leq m,\ 1 \leq l \leq n\}.$$

In SICNNs, neighboring cells exert mutual inhibitory interactions of the shunting type. The dynamics of the cell C_{ij} is described by the nonlinear ordinary differential equation

$$\frac{dx_{ij}(t)}{dt} = -a_{ij}x_{ij}(t) - \sum_{C_{hl} \in N_r(i,j)} C_{ij}^{hl} f(x_{hl}(t))x_{ij}(t) + L_{ij}(t), \qquad (8.3.26)$$

where x_{ij} is the activity of the cell C_{ij}; $L_{ij}(t)$ is the external input to C_{ij}; the constant $a_{ij} > 0$ represents the passive decay rate of the cell activity; $C_{ij}^{hl} \geq 0$ is the connection or coupling strength of the postsynaptic activity of the cell C_{hl} transmitted to the cell

C_{ij}; and the activation function $f(x_{hl})$ is a positive continuous function representing the output or firing rate of the cell C_{hl}.

According to the switching phenomenon, frequency changes or other sudden noises, the states of the electronic networks are often subject to instantaneous perturbations and experience abrupt changes at certain instants [2–4, 108]. In other words, they exhibit impulsive effects. Therefore, neural network models with impulsive effects are more accurate to describe the evolutionary processes of the systems.

In the present section, we will consider impulsive SICNNs in the form

$$\begin{aligned} \frac{dx_{ij}(t)}{dt} &= -a_{ij}x_{ij}(t) - \sum_{C_{hl}\in N_r(i,j)} C_{ij}^{hl} f(x_{hl}(t))x_{ij}(t) + L_{ij}(t),\ t \neq \theta_k, \\ \Delta x_{ij}|_{t=\theta_k} &= b_{ij}x_{ij}(\theta_k) + \sum_{C_{hl}\in N_r(i,j)} D_{ij}^{hl} g(x_{hl}(\theta_k))x_{ij}(\theta_k) + I_{ij}^k, \end{aligned} \tag{8.3.27}$$

where $b_{ij} \neq -1$ for each $i = 1, 2, \ldots, m$ and $j = 1, 2, \ldots, n$, the sequence $\{\theta_k\}$, $k = 0, \pm 1, \pm 2, \ldots$, of impact moments is strictly increasing, $\Delta x_{ij}|_{t=\theta_k} = x_{ij}(\theta_k+) - x_{ij}(\theta_k)$ and $x_{ij}(\theta_k+) = \lim_{t\to\theta_k+} x_{ij}(t)$. In SICNN (8.3.27), the couple $\left(L_{ij}(t), I_{ij}^k\right)$ is the external input to the cell C_{ij}. Similarly to the continuous interactions of neural networks through synapses, one can say about impact type of interactions [2–4]. We will say that the impact in the SICNN (8.3.27) is subject to the cell and shunting principles since of the term $\sum_{C_{hl}\in N_r(i,j)} D_{ij}^{hl} g(x_{hl}(\theta_k))x_{ij}(\theta_k)$, where $D_{ij}^{hl} \geq 0$ is the impact coupling strength of the postsynaptic activity of the cell C_{hl} transmitted to the cell C_{ij} and the impact activation function $g(x_{hl})$ represents the output localized at a moment of impact of the cell C_{hl}. In the theoretical discussions, the functions $L_{ij}(t)$ will be assumed to be continuous. Our main objective is to verify chaos in the dynamics of the network (8.3.27), provided that the external inputs $L_{ij}(t)$ behave chaotically.

According to Chua and Yang [120], the cellular structure makes cells of a CNN communicate with each other directly only through its neighbors, and because of the local interconnection feature, CNNs are much more amenable to VLSI implementation than general neural networks. In previous analyses of impulsive SICNNs [22, 25, 26, 121], the cell principle [120] and shunting phenomenon [6] have not been applied to the impacts of neurons. Contrarily, in the present section, we make the benefit of the cell and shunting principles. In other words, the term $\sum_{C_{hl}\in N_r(i,j)} D_{ij}^{hl} g(x_{hl}(\theta_k))x_{ij}(\theta_k)$ is inserted in the SICNN (8.3.27). We suppose that this novelty strengthens the theoretical results, and it is important for applications due to the suggestions mentioned in the papers [6, 120].

The dynamics of CNNs with impulsive effects have been widely investigated in the literature [22, 122–129]. The problem of global exponential stability for CNNs with time-varying delays and fixed moments of impulses was considered in the studies [122, 128] by means of the Lyapunov functions and the Razumikhin technique.

Wang and Liu [127] used the method of variation of parameters and Lyapunov functionals to obtain sufficient conditions for the exponential stability of impulsive CNNs with time delays. Besides, Song et al. [125] dealed with the exponential stability of distributed delayed and impulsive CNNs with partially Lipschitz continuous activation functions. Li et al. [123] investigated impulsive CNNs with time-varying and distributed delays, and obtained some sufficient conditions that ensure the existence, uniqueness, and global exponential stability of the equilibrium point. Taking advantage of piecewise continuous Lyapunov functions and the Razumikhin technique combined with Young's inequality, the stability of impulsive CNNs were analyzed by Stamova and Ilarionov [126]. On the other hand, contraction mapping principle and Krasnoselski's fixed point theorem were utilized by Pan and Cao [124] to verify the existence of antiperiodic solutions of delayed cellular neural networks with impulsive effects. Moreover, Yang and Cao [129] considered the global exponential stability as well as the existence of a periodic solution for delayed cellular neural networks with impulsive effects based on the Halanay inequality, mathematical induction, and fixed point theorem.

Chaotic dynamics is an object of great interest in the theory of neural networks [38–40, 44, 46, 49–53, 55, 82, 84], and CNNs are not excluded [48, 56–58]. The presence of chaotic attractors was observed in two-cell nonautonomous and three-cell autonomous CNNs in the studies [57, 58]. Moreover, Yan et al. [56] proposed algebraic conditions for the control of multiple time-delayed chaotic CNNs, and Liu and Wang [48] investigated the effect of variable thresholds in chaotic CNNs.

The presence of chaos in neural networks is useful for separating image segments [52], information processing [49, 50], and synchronization of neural networks [59–61, 63]. Besides, the synchronization phenomenon is also observable in the dynamics of coupled chaotic CNNs [65, 66]. Chaotic dynamics can improve the performance of CNNs on problems that have local minima in energy (cost) functions, since chaotic behavior of CNNs can help the network avoid local minima and reach the global optimum [68]. Furthermore, chaotic dynamics in CNNs is an important tool for the studies of chaotic communication [69–71] and combinatorial optimization problems [72].

As a mathematical notion, the term chaos has first been used by Li and Yorke [24] for one-dimensional discrete equations. According to Marotto [73], a multidimensional continuously differentiable map exhibits chaos in the sense of Li–Yorke, provided that it has a snap-back repeller. Marotto's Theorem was utilized in [74] to prove the existence of Li–Yorke chaos in a spatiotemporal chaotic system. This theorem is also a powerful tool in the theory of neural networks. For instance, it was used by Lin and Ruan [77] to determine the existence of chaos in a pacemaker neuron type integrate-and-fire circuit having two states with a periodic pulse-train input. Moreover, in the study [78], the chaos was approved by virtue of the Marotto's Theorem in discrete time-delayed Hopfield neural networks. Li–Yorke sensitivity, which links the Li–Yorke chaos with the notion of sensitivity, was studied in [75], and generalizations of Li–Yorke chaos to mappings in Banach spaces and complete metric spaces were provided in [76]. Impulsive systems can be used as an appropriate source of chaotic motions and there are several studies on the subject [25, 27,

130–132]. In the present section, we develop the concept of Li–Yorke chaos to the multidimensional dynamics of impulsive SICNNs, and prove its presence rigorously.

Many results concerning the dynamics of SICNNs have been published in the last decade. The existence and stability of periodic, almost periodic and antiperiodic outputs for SICNNs with delay have been studied in the papers [14–17, 19–21, 23, 99, 100, 102] using external inputs with the same type of regularity. Besides, the existence and stability of periodic and almost periodic solutions for impulsive SICNNs without the cell and shunting principles in the impacts were considered by Sun [26] and Xia et al. [22], respectively. However, in this section, we make use of chaotic external inputs and obtain chaos in the outputs of impulsive SICNNs with impacts subject to the cell and shunting principles. In the paper [25], the existence of a chaotic attractor in SICNNs with impulses was numerically demonstrated without a theoretical support, and the chaos type was not indicated. Contrarily, in the present section, we rigorously prove the presence of chaos in impulsive SICNNs with a precise type of chaos. Our approach has already been discussed in [103] for SICNNs without impulses. The main novelty of the present section is the discussion of the problem with impulsive effects, which request a more sophisticated analysis and a new approach of the proofs. In the paper [34], the chaos extension in continuous-time dynamics was considered without impulsive effects. The technique presented in this section was also approved for attraction of chaos in differential equations with impulses and applied to mechanical problems (by means of Duffing oscillators) in the study [132].

One of the advantages of our results is the suitability to obtain high-dimensional neural systems. A possible chaos extension mechanism is represented in Fig. 8.11. N pieces of SICNNs are shown in Fig. 8.11 such that SICNN 1 is the source of chaotic motions, and the other networks, SICNN 2, SICNN 3, ..., SICNN N, are influenced by the outputs of SICNN 1. The couplings between SICNN 1 and the remaining networks are all unidirectional. According to our theoretical results, the neural system consisting of N pieces of SICNNs possess chaos under the conditions that will be presented in the next section. We call this type of chaos extension process

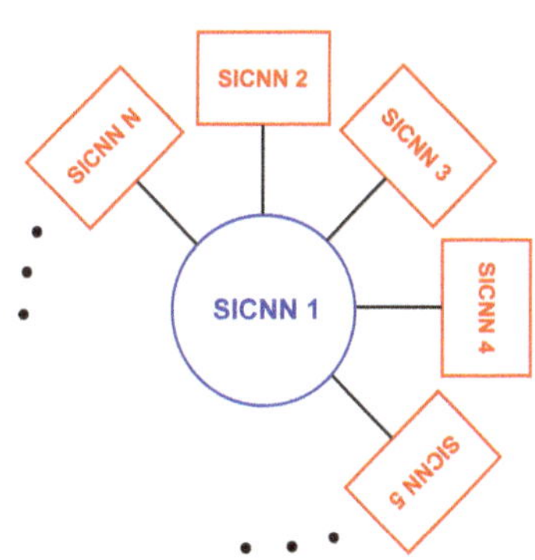

Fig. 8.11 The core mechanism of chaos extension

Fig. 8.12 The chain mechanism of chaos extension

as the "core" mechanism. In Sect. 8.3.4, we will make use of the core mechanism to demonstrate how to obtain high-dimensional neural systems.

Another possible mechanism that can be used to obtain high-dimensional chaotic neural systems is the "chain" mechanism, which is shown in Fig. 8.12. The figure represents consecutively connected N pieces of SICNNs such that the couplings between the networks are unidirectional. In the first coupling, we take into account SICNN 1 as the source of chaos. The chaotic outputs of SICNN 1, which are used as inputs for SICNN 2, gives rise to the appearance of chaos in the latter. In the next coupling, SICNN 2 is considered as the source of chaotic motions with respect to the third network. That is, SICNN 2 changes its role in the process. Similarly, in the remaining couplings, the role of the previously chaotically influenced SICNN changes and we start to use it as the source of chaotic external inputs for the next network. As a result, all of the networks become chaotic as well as the whole neural system consisting of N pieces of SICNNs. It is worth noting that the type of the chaos is preserved in this procedure.

In the mechanisms, one can take the number of networks, N, arbitrarily large, even with the possibility of infinite number of networks in the core mechanism. Other mechanisms are also possible, for example, by means of the "composition" of the proposed ones.

In their study, Skarda and Freeman [84] reported the formation of periodic and chaotic EEG signals when a rabbit was given known and unknown odorants, respectively. Additionally, Yao and Freeman [133] observed the presence of chaotic behavior near-periodic motions in a model of the olfactory system. The emergence of near-periodic chaos in continuous-time systems without impulses was considered in [134] by means of weak chaotic perturbations applied to systems that possess stable periodic solutions. In the study [135], the brain units such as neurons, cortical columns, and neuronal modules were supposed to be weakly connected. The presence of weak synaptic connections in the hippocampals cells and between neurons in the cortex was experimentally observed by McNaughton et al. [136] and Abeles [137], respectively. In the present section, by establishing weak connections between SICNNs, we numerically demonstrate the appearance of near-periodic discontinuous chaos.

The rest of the section is organized as follows. In Sect. 8.3.2, we introduce the description of Li–Yorke chaos for impulsive SICNNs, and prove the existence, uniqueness, and attractiveness feature of the bounded solutions. The main result of the present section is indicated in Sect. 8.3.3, where we prove the presence of chaos in the dynamics of the impulsive SICNNs (8.3.27). Illustrative examples are presented in Sect. 8.3.4.

8.3.2 Preliminaries

Throughout the section, $\mathbb{R}$, $\mathbb{Z}$ and $\mathbb{N}$ will stand for the sets of real numbers, integers, and natural numbers, respectively. We will use the norm $\|w\| = \max_{(i,j)} |w_{ij}|$, where $w = \{w_{ij}\} = (w_{11}, \ldots, w_{1n}, \ldots, w_{m1} \ldots, w_{mn}) \in \mathbb{R}^{m\times n}$.

We say that a function $\psi(t) = \{\psi_{ij}(t)\}$, $i = 1, 2, \ldots, m$, $j = 1, 2, \ldots, n$, which is defined on $\mathbb{R}$, belongs to the set $\mathscr{PC}(\mathbb{R})$ if it is left-continuous and continuous except, possibly, at the points where it has discontinuities of the first kind. The definition of a Li–Yorke chaotic set of piecewise continuous functions that will be used in the present section is as follows.

Suppose that $\mathscr{B}$ is a set of uniformly bounded functions $\psi(t) = \{\psi_{ij}(t)\}$, $i = 1, 2, \ldots, m$, $j = 1, 2, \ldots, n$, which belong to $\mathscr{PC}(\mathbb{R})$ and have common points of discontinuity.

We say that a couple $(\psi(t), \widetilde{\psi}(t)) \in \mathscr{B}\times\mathscr{B}$ is proximal if for arbitrary small $\varepsilon > 0$ and arbitrary large $E > 0$, there exists an interval J with a length no less than E such that $\|\psi(t) - \widetilde{\psi}(t)\| < \varepsilon$ for $t \in J$. Besides, a couple $(\psi(t), \widetilde{\psi}(t)) \in \mathscr{B}\times\mathscr{B}$ is called frequently (ε_0, Δ)-separated if there exist positive numbers ε_0, Δ and infinitely many disjoint intervals, each with a length no less than Δ, such that $\|\psi(t) - \widetilde{\psi}(t)\| > \varepsilon_0$ for each t from these intervals, and each of these intervals contains at most one discontinuity point of both $\psi(t)$ and $\widetilde{\psi}(t)$. It is worth noting that the numbers ε_0 and Δ depend on the functions $\psi(t)$ and $\widetilde{\psi}(t)$.

A couple $(\psi(t), \widetilde{\psi}(t)) \in \mathscr{B}\times\mathscr{B}$ is a Li–Yorke pair if it is proximal and frequently (ε_0, Δ)-separated for some positive numbers ε_0 and Δ. Moreover, an uncountable set $\mathscr{C} \subset \mathscr{B}$ is called a scrambled set if $\mathscr{C}$ does not contain any periodic functions and each couple of different functions inside $\mathscr{C}\times\mathscr{C}$ is a Li–Yorke pair.

We say that the collection $\mathscr{B}$ is a Li–Yorke chaotic set if: (i) There exists a positive number T_0 such that $\mathscr{B}$ admits a periodic function of period mT_0, for any $m \in \mathbb{N}$; (ii) $\mathscr{B}$ possesses a scrambled set $\mathscr{C}$; (iii) For any function $\psi(t) \in \mathscr{C}$ and any periodic function $\widetilde{\psi}(t) \in \mathscr{B}$, the couple $(\psi(t), \widetilde{\psi}(t))$ is frequently (ε_0, Δ)-separated for some positive numbers ε_0 and Δ.

Let us describe a method for obtaining a new Li–Yorke chaotic set of piecewise continuous functions from a given one. Suppose that $\varphi : \mathbb{R}^{m\times n} \to \mathbb{R}^{\overline{m}\times\overline{n}}$ is a function which satisfies for all $u, v \in \mathbb{R}^{m\times n}$ that

$$L_1 \|u - v\| \le \|\varphi(u) - \varphi(v)\| \le L_2 \|u - v\|, \qquad (8.3.28)$$

where L_1 and L_2 are some positive numbers. In this case, if $\mathscr{B}$ is a Li–Yorke chaotic set, then the collection $\mathscr{B}_\varphi$ whose elements are of the form $\varphi(\psi(t))$, $\psi(t) \in \mathscr{B}$, is also Li–Yorke chaotic.

For any interval I_0, we will denote by $i(I_0)$ the number of elements of the sequence $\{\theta_k\}$, $k \in \mathbb{Z}$, that belong to I_0. Let us denote $u_{ij}(t, s) = e^{-a_{ij}(t-s)}(1 + b_{ij})^{i([s,t))}$, where $t \ge s$ [131].

The following conditions are required:

(C1) There exist a positive number T and a natural number p such that $\theta_{k+p} = \theta_k + T$ for all $k \in \mathbb{Z}$;

(C2) $\lambda = \min_{(i,j)} \lambda_{ij} > 0$, where $\lambda_{ij} = a_{ij} - \frac{p}{T} \ln |1 + b_{ij}|$;

(C3) There exist positive numbers M and M_{ij} such that $\sup_{t\in\mathbb{R}} |f(t)| + \sup_{t\in\mathbb{R}} |g(t)| \le M$ and $\sup_{t\in\mathbb{R}} \left|L_{ij}(t)\right| + \sup_{k\in\mathbb{Z}} \left|I_{ij}^k\right| \le M_{ij}$;

(C4) There exists a positive number L_0 such that $|f(s_1) - f(s_2)| + |g(s_1) - g(s_2)| \le L_0 |s_1 - s_2|$ for all $s_1, s_2 \in \mathbb{R}$;

(C5) $KM\delta < 1$, where $\delta = \max_{(i,j)} \left(\frac{\sum_{C_{hl}\in N_r(i,j)} C_{ij}^{hl}}{\lambda_{ij}} + \frac{p \sum_{C_{hl}\in N_r(i,j)} D_{ij}^{hl}}{1 - e^{-\lambda_{ij} T}} \right)$.

In view of the condition $(C1)$, we have $\left| i([s,t)) - \frac{p}{T}(t-s) \right| \le p$ for all $t \ge s$. Moreover, the conditions $(C1)$ and $(C2)$ imply the existence of a positive number K such that for each $i = 1, 2, \ldots, m$ and $j = 1, 2, \ldots, n$, the inequality $\left|u_{ij}(t,s)\right| \le Ke^{-\lambda_{ij}(t-s)}$ holds for all $t \ge s$.

Throughout the section, the notations

$$H_0 = \frac{K}{1 - KM\delta} \max_{(i,j)} \left(\frac{M_{ij}}{\lambda_{ij}} + \frac{pM_{ij}}{1 - e^{-\lambda_{ij} T}} \right),$$

$$\overline{c} = \max_{(i,j)} \sum_{C_{hl}\in N_r(i,j)} C_{ij}^{hl},$$

$$\overline{d} = \max_{(i,j)} \sum_{C_{hl}\in N_r(i,j)} D_{ij}^{hl}$$

and

$$b_0 = \min_{(i,j)} \left(\left|1 + b_{ij}\right| - M \sum_{C_{hl}\in N_r(i,j)} D_{ij}^{hl} \right)$$

will be used.

The following conditions are also needed:

(C6) $K(M + L_0 H_0)\delta < 1$;

(C7) $-\lambda + K\overline{c}(M + L_0 H_0) + \frac{p}{T} \ln(1 + K\overline{d}(M + L_0 H_0)) < 0$;

(C8) $b_0 - L_0 H_0 \overline{d} > 0$;

(C9) $I_{ij}^{k+p} = I_{ij}^k$ for each $i = 1, 2, \ldots, m$, $j = 1, 2, \ldots, n$ and $k \in \mathbb{Z}$.

The conditions $(C1)$ and $(C9)$ are essentially required for the existence of infinitely many periodic solutions, which is one of the features of the Li–Yorke chaos. Similar conditions were also used in the studies [25, 26]. In SICNNs of the form (8.3.26), for each i and j, the passive decay rate of the cell activity a_{ij} is assumed to be positive, since it represents the ratio of the resting conductance to the membrane capacitance, which are connected in parallel to each other, in the electrical equivalent circuit of a cell [6]. The condition $(C2)$ is natural for impulsive SICNNs (8.3.27), since it is the counterpart of the aforementioned feature of SICNNs without impulses, and a similar condition was required by Xia et al. [22]. The Lipschitz continuity and boundedness of the activation functions were used in the studies [16, 17, 20, 25], and such conditions are specified in $(C3)$ and $(C4)$ for the functions f and g. The conditions $(C5)$–$(C7)$ can be achieved by means of the smallness of the coupling strengths C_{ij}^{hl} and D_{ij}^{hl}. Similar conditions were used in the paper [16]. The condition $(C8)$ can also be attained by the smallness of the constants D_{ij}^{hl} and it is needed for the frequent separation feature.

We say that a left-continuous function $x(t) = \{x_{ij}(t)\}$ is a solution of (8.3.27) if: (i) It has discontinuities only at the points θ_k, $k \in \mathbb{Z}$, and these discontinuities are of the first kind; (ii) The derivatives $\dfrac{dx_{ij}(t)}{dt}$ exist at each point $t \in \mathbb{R} \setminus \{\theta_k\}$, and the left-sided derivatives exist at the points θ_k, $k \in \mathbb{Z}$; (iii) The differential equations are satisfied by $x_{ij}(t)$ on $\mathbb{R} \setminus \{\theta_k\}$, and they hold for the left derivatives of $x_{ij}(t)$ at every point θ_k, $k \in \mathbb{Z}$; (iv) The jump equations are satisfied by $x_{ij}(t)$ for every $k \in \mathbb{Z}$.

According to the results of [131, 138], if conditions $(C1)$–$(C3)$ hold, then a bounded on $\mathbb{R}$ function $x(t) = \{x_{ij}(t)\}$ is a solution of the network (8.3.27) if and only if the relation

$$\begin{aligned} x_{ij}(t) = &-\int_{-\infty}^{t} u_{ij}(t,s)\Big[\sum_{C_{hl}\in N_r(i,j)} C_{ij}^{hl} f(x_{hl}(s))x_{ij}(s) - L_{ij}(s)\Big]ds \\ &+ \sum_{-\infty<\theta_k<t} u_{ij}(t,\theta_k+)\Big[\sum_{C_{hl}\in N_r(i,j)} D_{ij}^{hl} f(x_{hl}(s))x_{ij}(s) + I_{ij}^{k}\Big] \end{aligned} \tag{8.3.29}$$

is satisfied for each i and j.

The next lemma is about the existence and uniqueness of the bounded on $\mathbb{R}$ solutions of network (8.3.27).

Lemma 8.9 *If the conditions $(C1)$–$(C6)$ are fulfilled, then for any $L(t) = \{L_{ij}(t)\}$, $i = 1, 2, \ldots, m$, $j = 1, 2, \ldots, n$, there exists a unique bounded on $\mathbb{R}$ solution $\phi_L(t) = \{\phi_L^{ij}(t)\}$ of the network (8.3.27) such that $\sup_{t\in\mathbb{R}} \|\phi_L(t)\| \le H_0$.*

Proof Fix an arbitrary function $L(t) = \{L_{ij}(t)\}$, and consider the set $\mathscr{S}_0$ of functions $w(t) = \{w_{ij}(t)\} \in \mathscr{PC}(\mathbb{R})$ which have discontinuities at the points θ_k, $k \in \mathbb{Z}$, such that $\|w\|_1 \le H_0$, where $\|w\|_1 = \sup_{t\in\mathbb{R}} \|w(t)\|$. The set $\mathscr{S}_0$ is complete [131].

Define on $\mathscr{S}_0$ the operator Π as

$$\begin{aligned} (\Pi w(t))_{ij} \equiv &-\int_{-\infty}^{t} u_{ij}(t,s)\Big[\sum_{C_{hl}\in N_r(i,j)} C_{ij}^{hl} f(w_{hl}(s))w_{ij}(s) - L_{ij}(s)\Big]ds \\ &+ \sum_{-\infty<\theta_k<t} u_{ij}(t,\theta_k+)\Big[\sum_{C_{hl}\in N_r(i,j)} D_{ij}^{hl} g(w_{hl}(\theta_k))w_{ij}(\theta_k) + I_{ij}^{k}\Big], \end{aligned}$$

where $\Pi w(t) = \{(\Pi w(t))_{ij}\}$.

First, we shall show that $\Pi(\mathscr{S}_0) \subseteq \mathscr{S}_0$. If $w(t)$ belongs to $\mathscr{S}_0$, then it is easy to verify that

$$\begin{aligned}\left|(\Pi w(t))_{ij}\right| \leq & \int_{-\infty}^{t} K e^{-\lambda_{ij}(t-s)} \Big(M H_0 \sum_{C_{hl} \in N_r(i,j)} C_{ij}^{hl} + M_{ij} \Big) ds \\ & + \sum_{-\infty<\theta_k<t} K e^{-\lambda_{ij}(t-\theta_k)} \Big(M H_0 \sum_{C_{hl} \in N_r(i,j)} D_{ij}^{hl} + M_{ij} \Big).\end{aligned}$$

Making use of the inequality $\sum_{-\infty<\theta_k<t} e^{-\lambda_{ij}(t-\theta_k)} \leq \frac{p}{1-e^{-\lambda_{ij}T}}$, one can obtain that

$$\begin{aligned}\left|(\Pi w(t))_{ij}\right| \leq & \frac{K}{\lambda_{ij}} \Big(M H_0 \sum_{C_{hl} \in N_r(i,j)} C_{ij}^{hl} + M_{ij} \Big) \\ & + \frac{Kp}{1-e^{-\lambda_{ij}T}} \Big(M H_0 \sum_{C_{hl} \in N_r(i,j)} D_{ij}^{hl} + M_{ij} \Big).\end{aligned}$$

Accordingly, the inequality $\|\Pi w\|_1 \leq K M H_0 \delta + K \max_{(i,j)} \left(\frac{M_{ij}}{\lambda_{ij}} + \frac{p M_{ij}}{1-e^{-\lambda_{ij}T}} \right) = H_0$ holds. Therefore, $\Pi(\mathscr{S}_0) \subseteq \mathscr{S}_0$.

Next, we will verify that the operator Π is a contraction. For any $w(t), \overline{w}(t) \in \mathscr{S}_0$, one can attain that

$$\begin{aligned}(\Pi w(t))_{ij} - (\Pi \overline{w}(t))_{ij} = & - \int_{-\infty}^{t} u_{ij}(t,s) \sum_{C_{hl} \in N_r(i,j)} C_{ij}^{hl} \Big[f(w_{hl}(s)) w_{ij}(s) - \\ & f(\overline{w}_{hl}(s)) \overline{w}_{ij}(s) \Big] ds \\ & + \sum_{-\infty<\theta_k<t} u_{ij}(t,\theta_k+) \sum_{C_{hl} \in N_r(i,j)} D_{ij}^{hl} \Big[g(w_{hl}(\theta_k)) w_{ij}(\theta_k) - g(\overline{w}_{hl}(\theta_k)) \overline{w}_{ij}(\theta_k) \Big].\end{aligned}$$

Therefore, we have

$$\begin{aligned}\left|(\Pi w(t))_{ij} - (\Pi \overline{w}(t))_{ij}\right| \leq & \int_{-\infty}^{t} K e^{-\lambda_{ij}(t-s)} \sum_{C_{hl} \in N_r(i,j)} C_{ij}^{hl} \\ & \times \Big(|f(w_{hl}(s))| \left|w_{ij}(s) - \overline{w}_{ij}(s)\right| \\ & + \left|\overline{w}_{ij}(s)\right| |f(w_{hl}(s)) - f(\overline{w}_{hl}(s))| \Big) ds \\ & + \sum_{-\infty<\theta_k<t} K e^{-\lambda_{ij}(t-\theta_k)} \sum_{C_{hl} \in N_r(i,j)} D_{ij}^{hl} \Big(|g(w_{hl}(\theta_k))| \left|w_{ij}(\theta_k) - \overline{w}_{ij}(\theta_k)\right|\end{aligned}$$

$$+ \left|\overline{w}_{ij}(\theta_k)\right| |g(w_{hl}(\theta_k)) - g(\overline{w}_{hl}(\theta_k))| \Big)ds$$

$$\leq K(M + L_0 H_0)\left(\frac{\sum_{C_{hl}\in N_r(i,j)} C_{ij}^{hl}}{\lambda_{ij}} + \frac{p\sum_{C_{hl}\in N_r(i,j)} D_{ij}^{hl}}{1 - e^{-\lambda_{ij}T}}\right)\|w - \overline{w}\|_1 .$$

The last inequality yields $\|\Pi w - \Pi\overline{w}\|_1 \leq K(M + L_0H_0)\delta\,\|w - \overline{w}\|_1$. Hence, in accordance with condition $(C6)$, the operator Π is contractive. Consequently, there exists a unique bounded on $\mathbb{R}$ solution $\phi_L(t) = \left\{\phi_L^{ij}(t)\right\}$ of the network (8.3.27) such that $\sup_{t\in\mathbb{R}} \|\phi_L(t)\| \leq H_0$. □

As mentioned in Lemma 8.9, in the remaining parts of this section, we will denote by $\phi_L(t) = \left\{\phi_L^{ij}(t)\right\}$ the unique bounded on $\mathbb{R}$ solution of the impulsive SICNN (8.3.27). Moreover, for a given external input $L(t) = \left\{L_{ij}(t)\right\}$ and initial data $x_0 \in \mathbb{R}^{m\times n}$, let us denote by $x_L(t, x_0) = \left\{x_L^{ij}(t, x_0)\right\}$ the unique solution of (8.3.27) with $x_L(0, x_0) = x_0$. We note that the solution $x_L(t, x_0)$ is not necessarily bounded on $\mathbb{R}$.

Consider the collection $\mathscr{L}$ whose elements are equicontinuous functions of the form $L(t) = \left\{L_{ij}(t)\right\}$ such that $\sup_{t\in\mathbb{R}}\left|L_{ij}(t)\right| + \sup_{k\in\mathbb{Z}}\left|I_{ij}^k\right| \leq M_{ij}$ for each $i = 1, 2, \ldots, m$ and $j = 1, 2, \ldots, n$. Suppose that $\mathscr{A}$ denotes the set of bounded on $\mathbb{R}$ solutions $\phi_L(t)$ of the network (8.3.27), where $L(t)$ belongs to $\mathscr{L}$.

The following lemma is about the attractiveness of the set $\mathscr{A}$.

Lemma 8.10 *Suppose that the conditions* $(C1)$–$(C7)$ *are valid. Then for any* $x_0 \in \mathbb{R}^{m\times n}$ *and* $L(t) = \left\{L_{ij}(t)\right\}$, *we have* $\|x_L(t, x_0) - \phi_L(t)\| \to 0$ *as* $t \to \infty$.

Proof Fix an arbitrary $x_0 \in \mathbb{R}^{m\times n}$ and an arbitrary function $L(t) = \left\{L_{ij}(t)\right\}$. For $t \geq 0$, making use of the relations

$$\begin{aligned} x_L^{ij}(t, x_0) &= u_{ij}(t, 0)x_L^{ij}(0, x_0) \\ &- \int_0^t u_{ij}(t, s)\Big[\sum_{C_{hl}\in N_r(i,j)} C_{ij}^{hl} f(x_L^{hl}(s, x_0))x_L^{ij}(s, x_0) - L_{ij}(s)\Big]ds \\ &+ \sum_{0\leq\theta_k<t} u_{ij}(t, \theta_k+)\Big[\sum_{C_{hl}\in N_r(i,j)} D_{ij}^{hl} g(x_L^{hl}(\theta_k, x_0))x_L^{ij}(\theta_k, x_0) + I_{ij}^k\Big] \end{aligned}$$

and

$$\begin{aligned} \phi_L^{ij}(t) &= u_{ij}(t, 0)\phi_L^{ij}(0) \\ &- \int_0^t u_{ij}(t, s)\Big[\sum_{C_{hl}\in N_r(i,j)} C_{ij}^{hl} f(\phi_L^{hl}(s))\phi_L^{ij}(s) - L_{ij}(s)\Big]ds \\ &+ \sum_{0\leq\theta_k<t} u_{ij}(t, \theta_k+)\Big[\sum_{C_{hl}\in N_r(i,j)} D_{ij}^{hl} g(\phi_L^{hl}(\theta_k))\phi_L^{ij}(\theta_k) + I_{ij}^k\Big], \end{aligned}$$

we obtain that

$$\begin{aligned}
&\left|x_L^{ij}(t,x_0)-\phi_L^{ij}(t)\right| \le Ke^{-\lambda_{ij}t}\left|x_L^{ij}(0,x_0)-\phi_L^{ij}(0)\right| \\
&+\int_0^t Ke^{-\lambda_{ij}(t-s)}\sum_{C_{hl}\in N_r(i,j)} C_{ij}^{hl}\Big(\left|f(x_L^{hl}(s,x_0))\right|\left|x_L^{ij}(s,x_0)-\phi_L^{ij}(s)\right| \\
&+\left|\phi_L^{ij}(s)\right|\left|f(x_L^{hl}(s,x_0))-f(\phi_L^{hl}(s))\right|\Big)ds \\
&+\sum_{0\le\theta_k<t} Ke^{-\lambda_{ij}(t-\theta_k)}\sum_{C_{hl}\in N_r(i,j)} D_{ij}^{hl}\Big(\left|g(x_L^{hl}(\theta_k,x_0))\right|\left|x_L^{ij}(\theta_k,x_0)-\phi_L^{ij}(\theta_k)\right| \\
&+\left|\phi_L^{ij}(\theta_k)\right|\left|g(x_L^{hl}(\theta_k,x_0))-g(\phi_L^{hl}(\theta_k))\right|\Big) \\
&\le Ke^{-\lambda_{ij}t}\left\|x_0-\phi_L(0)\right\| \\
&+\int_0^t K(M+L_0H_0)e^{-\lambda_{ij}(t-s)}\sum_{C_{hl}\in N_r(i,j)} C_{ij}^{hl}\left\|x_L(s,x_0)-\phi_L(s)\right\|ds \\
&+\sum_{0\le\theta_k<t} K(M+L_0H_0)e^{-\lambda_{ij}(t-\theta_k)}\sum_{C_{hl}\in N_r(i,j)} D_{ij}^{hl}\left\|x_L(\theta_k,x_0)-\phi_L(\theta_k)\right\|.
\end{aligned}$$

The last inequality implies for $t\ge 0$ that

$$\begin{aligned}
&\left\|x_L(t,x_0)-\phi_L(t)\right\| \le Ke^{-\lambda t}\left\|x_0-\phi_L(0)\right\| \\
&+K\overline{c}(M+L_0H_0)\int_0^t e^{-\lambda(t-s)}\left\|x_L(s,x_0)-\phi_L(s)\right\|ds \\
&+K\overline{d}(M+L_0H_0)\sum_{0\le\theta_k<t} e^{-\lambda(t-\theta_k)}\left\|x_L(\theta_k,x_0)-\phi_L(\theta_k)\right\|.
\end{aligned}$$

Let us define the function $u(t)=e^{\lambda t}\left\|x_L(t,x_0)-\phi_L(t)\right\|$. Then we have that

$$u(t)\le K\left\|x_0-\phi_L(0)\right\|+K\overline{c}(M+L_0H_0)\int_0^t u(s)ds+K\overline{d}(M+L_0H_0)\sum_{0\le\theta_k<t}u(\theta_k).$$

With the aid of the Gronwall–Bellman Lemma for piecewise continuous functions, one can verify that

$$\begin{aligned}
&u(t)\le K\left\|x_0-\phi_L(0)\right\|e^{K\overline{c}(M+L_0H_0)t}[1+K\overline{d}(M+L_0H_0)]^{i([0,t))} \\
&\le K[1+K\overline{d}(M+L_0H_0)]^p\left\|x_0-\phi_L(0)\right\|e^{[K\overline{c}(M+L_0H_0)+(p/T)\ln(1+K\overline{d}(M+L_0H_0))]t}.
\end{aligned}$$

Thus, the inequality

$$\begin{aligned}
&\left\|x_L(t,x_0)-\phi_L(t)\right\|\le K[1+K\overline{d}(M+L_0H_0)]^p\left\|x_0-\phi_L(0)\right\| \\
&\times e^{[-\lambda+K\overline{c}(M+L_0H_0)+(p/T)\ln(1+K\overline{d}(M+L_0H_0))]t}
\end{aligned}$$

holds for all $t \geq 0$. Consequently, in accordance with condition $(C7)$, we have that $\|x_L(t, x_0) - \phi_L(t)\| \to 0$ as $t \to \infty$. □

The next section is devoted for the chaotic dynamics of the network (8.3.27).

8.3.3 The Existence of Chaos

In this subsection, we will rigorously prove that if the collection $\mathscr{L}$ is chaotic in the sense of Li–Yorke, then the same is true for the collection $\mathscr{A}$. Before the main result of the present section that will be stated in Theorem 8.3, we will mention about the main ingredients of Li–Yorke chaos, proximality, and frequent separation features, in Lemmas 8.11 and 8.12, respectively. The lemmas are as follows.

Lemma 8.11 *Suppose that the conditions* $(C1)$–$(C7)$ *hold. If a couple of functions* $\left(L(t), \overline{L}(t)\right) \in \mathscr{L} \times \mathscr{L}$ *is proximal, then the same is true for the couple* $\left(\phi_L(t), \phi_{\overline{L}}(t)\right) \in \mathscr{A} \times \mathscr{A}$.

Proof Set $R = 2K \max\limits_{(i,j)} \left[\dfrac{M H_0 \sum_{C_{hl} \in N_r(i,j)} C_{ij}^{hl} + M_{ij}}{\lambda_{ij}} + \dfrac{p M H_0 \sum_{C_{hl} \in N_r(i,j)} D_{ij}^{hl}}{1 - e^{-\lambda_{ij} T}} \right]$ and $\alpha = \lambda - K\overline{c}(M + L_0 H_0) - (p/T)\ln(1 + K\overline{d}(M + L_0 H_0))$. The number α is positive by condition $(C7)$. Fix an arbitrary small positive number ε and an arbitrary large positive number E which satisfies the inequality

$$E \geq \frac{2}{\alpha} \ln\left(\frac{\gamma R[1 + K\overline{d}(M + L_0 H_0)]^p}{\varepsilon} \right).$$

Let γ be a number such that $\gamma \geq 1 + \dfrac{K}{\lambda} + \dfrac{K^2\overline{c}(M + L_0 H_0)}{\lambda\alpha}[1 + K\overline{d}(M + L_0 H_0)]^p + K^2\overline{d}(M + L_0 H_0)[1 + K\overline{d}(M + L_0 H_0)]^p \dfrac{p e^{2\alpha T}}{\lambda(e^{\alpha T} - 1)}$. Since the pair $\left(L(t), \overline{L}(t)\right) \in \mathscr{L} \times \mathscr{L}$ is proximal, there exists an interval $J = [\sigma, \sigma + E_1]$ with $E_1 \geq E$ such that $\left\|L(t) - \overline{L}(t)\right\| < \varepsilon/\gamma$ for all $t \in J$.

By means of the relations

$$\begin{aligned}
\phi_L^{ij}(t) = &- \int_{-\infty}^{t} u_{ij}(t, s)\Big[\sum_{C_{hl} \in N_r(i,j)} C_{ij}^{hl} f(\phi_L^{hl}(s))\phi_L^{ij}(s) - L_{ij}(s) \Big] ds \\
&+ \sum_{-\infty < \theta_k < t} u_{ij}(t, \theta_k +)\Big[\sum_{C_{hl} \in N_r(i,j)} D_{ij}^{hl} g(\phi_L^{hl}(\theta_k))\phi_L^{ij}(\theta_k) + I_{ij}^k \Big]
\end{aligned}$$

and

$$\phi_{\overline{L}}^{ij}(t) = -\int_{-\infty}^{t} u_{ij}(t,s)\Big[\sum_{C_{hl}\in N_r(i,j)} C_{ij}^{hl} f(\phi_{\overline{L}}^{hl}(s))\phi_{\overline{L}}^{ij}(s) - \overline{L}_{ij}(s)\Big]ds$$
$$+ \sum_{-\infty<\theta_k<t} u_{ij}(t,\theta_k+)\Big[\sum_{C_{hl}\in N_r(i,j)} D_{ij}^{hl} g(\phi_{\overline{L}}^{hl}(\theta_k))\phi_{\overline{L}}^{ij}(\theta_k) + I_{ij}^{k}\Big],$$

one can obtain for $t \geq \sigma$ that

$$\phi_{L}^{ij}(t) - \phi_{\overline{L}}^{ij}(t) = -\int_{-\infty}^{\sigma} u_{ij}(t,s)\Big[\sum_{C_{hl}\in N_r(i,j)} C_{ij}^{hl} f(\phi_{L}^{hl}(s))\phi_{L}^{ij}(s) - L_{ij}(s)$$
$$- \sum_{C_{hl}\in N_r(i,j)} C_{ij}^{hl} f(\phi_{\overline{L}}^{hl}(s))\phi_{\overline{L}}^{ij}(s) + \overline{L}_{ij}(s)\Big]ds$$
$$-\int_{\sigma}^{t} u_{ij}(t,s)\Big[\sum_{C_{hl}\in N_r(i,j)} C_{ij}^{hl} f(\phi_{L}^{hl}(s))\phi_{L}^{ij}(s) - L_{ij}(s)$$
$$- \sum_{C_{hl}\in N_r(i,j)} C_{ij}^{hl} f(\phi_{\overline{L}}^{hl}(s))\phi_{\overline{L}}^{ij}(s) + \overline{L}_{ij}(s)\Big]ds$$
$$+ \sum_{-\infty<\theta_k\leq\sigma} u_{ij}(t,\theta_k+) \sum_{C_{hl}\in N_r(i,j)} D_{ij}^{hl}\Big[g(\phi_{L}^{hl}(\theta_k))\phi_{L}^{ij}(\theta_k) - g(\phi_{\overline{L}}^{hl}(\theta_k))\phi_{\overline{L}}^{ij}(\theta_k)\Big]$$
$$+ \sum_{\sigma<\theta_k<t} u_{ij}(t,\theta_k+) \sum_{C_{hl}\in N_r(i,j)} D_{ij}^{hl}\Big[g(\phi_{L}^{hl}(\theta_k))\phi_{L}^{ij}(\theta_k) - g(\phi_{\overline{L}}^{hl}(\theta_k))\phi_{\overline{L}}^{ij}(\theta_k)\Big].$$

If t belongs to the interval J, then making use of the inequalities

$$\Big| -\int_{-\infty}^{\sigma} u_{ij}(t,s)\Big[\sum_{C_{hl}\in N_r(i,j)} C_{ij}^{hl} f(\phi_{L}^{hl}(s))\phi_{L}^{ij}(s) - L_{ij}(s)$$
$$- \sum_{C_{hl}\in N_r(i,j)} C_{ij}^{hl} f(\phi_{\overline{L}}^{hl}(s))\phi_{\overline{L}}^{ij}(s) + \overline{L}_{ij}(s)\Big]ds\Big|$$
$$\leq \int_{-\infty}^{\sigma} 2Ke^{-\lambda_{ij}(t-s)}\Big(MH_0 \sum_{C_{hl}\in N_r(i,j)} C_{ij}^{hl} + M_{ij}\Big)ds$$
$$= \frac{2K(MH_0 \sum_{C_{hl}\in N_r(i,j)} C_{ij}^{hl} + M_{ij})}{\lambda_{ij}} e^{-\lambda_{ij}(t-\sigma)}$$

and

$$\begin{aligned}
&\Big| \sum_{-\infty<\theta_k\leq\sigma} u_{ij}(t,\theta_k+) \sum_{C_{hl}\in N_r(i,j)} D_{ij}^{hl} \Big[g(\phi_L^{hl}(\theta_k))\phi_L^{ij}(\theta_k) - g(\phi_{\overline{L}}^{hl}(\theta_k))\phi_{\overline{L}}^{ij}(\theta_k) \Big] \Big| \\
&\leq \sum_{-\infty<\theta_k\leq\sigma} 2KMH_0 e^{-\lambda_{ij}(t-\theta_k)} \sum_{C_{hl}\in N_r(i,j)} D_{ij}^{hl} \\
&\leq 2pKMH_0 \frac{\sum_{C_{hl}\in N_r(i,j)} D_{ij}^{hl}}{1-e^{-\lambda_{ij}T}} e^{-\lambda_{ij}(t-\sigma)},
\end{aligned}$$

we attain that

$$\begin{aligned}
&\left|\phi_L^{ij}(t) - \phi_{\overline{L}}^{ij}(t)\right| \\
&\leq 2K\Big[\frac{MH_0 \sum_{C_{hl}\in N_r(i,j)} C_{ij}^{hl} + M_{ij}}{\lambda_{ij}} + \frac{pMH_0 \sum_{C_{hl}\in N_r(i,j)} D_{ij}^{hl}}{1-e^{-\lambda_{ij}T}} \Big] e^{-\lambda_{ij}(t-\sigma)} \\
&+ \int_\sigma^t Ke^{-\lambda_{ij}(t-s)} \sum_{C_{hl}\in N_r(i,j)} C_{ij}^{hl} \Big(\left|f(\phi_L^{hl}(s))\right| \left|\phi_L^{ij}(s) - \phi_{\overline{L}}^{ij}(s)\right| \\
&+ \left|\phi_{\overline{L}}^{ij}(s)\right| \left|f(\phi_L^{hl}(s)) - f(\phi_{\overline{L}}^{hl}(s))\right| \Big) ds \\
&+ \sum_{\sigma<\theta_k<t} Ke^{-\lambda_{ij}(t-\theta_k)} \sum_{C_{hl}\in N_r(i,j)} D_{ij}^{hl} \Big(\left|g(\phi_L^{hl}(\theta_k))\right| \left|\phi_L^{ij}(\theta_k) - \phi_{\overline{L}}^{ij}(\theta_k)\right| \\
&+ \left|\phi_{\overline{L}}^{ij}(\theta_k)\right| \left|g(\phi_L^{hl}(\theta_k)) - g(\phi_{\overline{L}}^{hl}(\theta_k))\right| \Big) + \int_\sigma^t Ke^{-\lambda_{ij}(t-s)} \left|L_{ij}(s) - \overline{L}_{ij}(s)\right| ds.
\end{aligned}$$

The last inequality implies for $t \in J$ that

$$\begin{aligned}
\left\|\phi_L(t) - \phi_{\overline{L}}(t)\right\| &\leq Re^{-\lambda(t-\sigma)} + \frac{K\varepsilon}{\gamma\lambda}(1-e^{-\lambda(t-\sigma)}) \\
&+ \int_\sigma^t K\overline{c}(M+L_0H_0)e^{-\lambda(t-s)} \left\|\phi_L(s) - \phi_{\overline{L}}(s)\right\| ds \\
&+ \sum_{\sigma<\theta_k<t} K\overline{d}(M+L_0H_0)e^{-\lambda(t-\theta_k)} \left\|\phi_L(\theta_k) - \phi_{\overline{L}}(\theta_k)\right\|.
\end{aligned}$$

Define the functions

$$u(t) = e^{\lambda t} \left\|\phi_L(t) - \phi_{\overline{L}}(t)\right\|$$

and

$$\psi(t) = \left(R - \frac{K\varepsilon}{\gamma\lambda}\right) e^{\lambda\sigma} + \frac{K\varepsilon}{\gamma\lambda} e^{\lambda t}.$$

In that case, we have

$$u(t) \leq \psi(t) + \int_{\sigma}^{t} K\overline{c}(M + L_0H_0)u(s)ds + \sum_{\sigma<\theta_k<t} K\overline{d}(M + L_0H_0)u(\theta_k), \; t \in J.$$

The application of the Gronwall's Lemma for piecewise continuous functions to the last inequality yields

$$\begin{aligned} u(t) &\leq \psi(t) + \int_{\sigma}^{t} K\overline{c}(M + L_0H_0) \\ &\times[1 + K\overline{d}(M + L_0H_0)]^{i((s,t))}\psi(s)e^{K\overline{c}(M+L_0H_0)(t-s)}ds \\ &+ \sum_{\sigma<\theta_k<t} K\overline{d}(M + L_0H_0)[1 + K\overline{d}(M + L_0H_0)]^{i((\theta_k,t))}\psi(\theta_k)e^{K\overline{c}(M+L_0H_0)(t-\theta_k)}. \end{aligned}$$

By virtue of the equation

$$\begin{aligned} &1 + \int_{\sigma}^{t} K\overline{c}(M + L_0H_0)[1 + K\overline{d}(M + L_0H_0)]^{i((s,t))}e^{K\overline{c}(M+L_0H_0)(t-s)}ds \\ &+ \sum_{\sigma<\theta_k<t} K\overline{d}(M + L_0H_0)[1 + K\overline{d}(M + L_0H_0)]^{i((\theta_k,t))}e^{K\overline{c}(M+L_0H_0)(t-\theta_k)} \\ &= [1 + K\overline{d}(M + L_0H_0)]^{i((\sigma,t))}e^{K\overline{c}(M+L_0H_0)(t-\sigma)}, \end{aligned}$$

one can obtain that

$$\begin{aligned} u(t) &\leq [1 + K\overline{d}(M + L_0H_0)]^{i((\sigma,t))}e^{K\overline{c}(M+L_0H_0)(t-\sigma)}\left(R - \frac{K\varepsilon}{\gamma\lambda}\right)e^{\lambda\sigma} + \frac{K\varepsilon}{\gamma\lambda}e^{\lambda t} \\ &+K\overline{c}(M + L_0H_0)\frac{K\varepsilon}{\gamma\lambda}\int_{\sigma}^{t} e^{\lambda s}[1 + K\overline{d}(M + L_0H_0)]^{i((s,t))}e^{K\overline{c}(M+L_0H_0)(t-s)}ds \\ &+K\overline{d}(M + L_0H_0)\frac{K\varepsilon}{\gamma\lambda}\sum_{\sigma<\theta_k<t} e^{\lambda\theta_k}[1 + K\overline{d}(M + L_0H_0)]^{i((\theta_k,t))}e^{K\overline{c}(M+L_0H_0)(t-\theta_k)}. \end{aligned}$$

Since the inequality

$$[1 + K\overline{d}(M + L_0H_0)]^{i((s,t))}e^{K\overline{c}(M+L_0H_0)(t-s)} \leq [1 + K\overline{d}(M + L_0H_0)]^{p}e^{(\lambda-\alpha)(t-s)}$$

holds for all $t \geq s$, we have for $t \in J$ that

$$u(t) \le [1 + K\overline{d}(M + L_0H_0)]^p \left(R - \frac{K\varepsilon}{\gamma\lambda}\right) e^{\lambda\sigma} e^{(\lambda-\alpha)(t-\sigma)} + \frac{K\varepsilon}{\gamma\lambda} e^{\lambda t}$$
$$+K\overline{c}(M + L_0H_0)[1 + K\overline{d}(M + L_0H_0)]^p \frac{K\varepsilon}{\gamma\lambda} \int_\sigma^t e^{\lambda s} e^{(\lambda-\alpha)(t-s)} ds$$
$$+K\overline{d}(M + L_0H_0)[1 + K\overline{d}(M + L_0H_0)]^p \frac{K\varepsilon}{\gamma\lambda} \sum_{\sigma<\theta_k<t} e^{\lambda\theta_k} e^{(\lambda-\alpha)(t-\theta_k)}.$$

Now, let $q = \left\lfloor \frac{t-\sigma}{T} \right\rfloor$. That is, q is the greatest integer which not larger than $\frac{t-\sigma}{T}$. One can verify that

$$\sum_{\sigma<\theta_k<t} e^{\alpha\theta_k} \le \sum_{\sigma<\theta_k<\sigma+(q+1)T} e^{\alpha\theta_k} \le \sum_{l=0}^{q} \sum_{\sigma+lT\le\theta_k<\sigma+(l+1)T} e^{\alpha\theta_k} < \sum_{l=0}^{q} p e^{\alpha[\sigma+(l+1)T]}$$
$$= p e^{(\sigma+T)\alpha} \frac{e^{(q+1)\alpha T} - 1}{e^{\alpha T} - 1} \le \frac{p e^{(\sigma+T)\alpha}}{e^{\alpha T} - 1} \left(e^{\alpha(t-\sigma+T)} - 1\right) < \frac{p e^{2\alpha T}}{e^{\alpha T} - 1} e^{\alpha t}.$$

Therefore,

$$u(t) < R[1 + K\overline{d}(M + L_0H_0)]^p e^{\lambda t} e^{-\alpha(t-\sigma)}$$
$$+\frac{K\varepsilon}{\gamma\lambda} e^{\lambda t} \left[1 - (1 + K\overline{d}(M + L_0H_0))^p e^{-\alpha(t-\sigma)}\right]$$
$$+K^2\overline{c}(M + L_0H_0)[1 + K\overline{d}(M + L_0H_0)]^p \frac{\varepsilon}{\gamma\lambda\alpha} \left(1 - e^{-\alpha(t-\sigma)}\right) e^{\lambda t}$$
$$+K^2\overline{d}(M + L_0H_0)[1 + K\overline{d}(M + L_0H_0)]^p \frac{\varepsilon p e^{2\alpha T}}{\gamma\lambda(e^{\alpha T} - 1)} e^{\lambda t}.$$

If we multiply both sides of the last inequality by $e^{-\lambda t}$, then we get

$$\left\|\phi_L(t) - \phi_{\overline{L}}(t)\right\| < R[1 + K\overline{d}(M + L_0H_0)]^p e^{-\alpha(t-\sigma)}$$
$$+\frac{K\varepsilon}{\gamma\lambda} \left[1 - (1 + K\overline{d}(M + L_0H_0))^p e^{-\alpha(t-\sigma)}\right]$$
$$+K^2\overline{c}(M + L_0H_0)[1 + K\overline{d}(M + L_0H_0)]^p \frac{\varepsilon}{\gamma\lambda\alpha} \left(1 - e^{-\alpha(t-\sigma)}\right)$$
$$+K^2\overline{d}(M + L_0H_0)[1 + K\overline{d}(M + L_0H_0)]^p \frac{\varepsilon p e^{2\alpha T}}{\gamma\lambda(e^{\alpha T} - 1)}.$$

Thus, one can obtain for $t \in J$ that

$$\begin{aligned}\left\|\phi_L(t)-\phi_{\overline{L}}(t)\right\| &< R[1+K\overline{d}(M+L_0H_0)]^p e^{-\alpha(t-\sigma)}\\ &+\frac{K\varepsilon}{\gamma\lambda}\Big[1+\frac{K\overline{c}}{\alpha}(M+L_0H_0)(1+K\overline{d}(M+L_0H_0))^p\\ &+K\overline{d}(M+L_0H_0)(1+K\overline{d}(M+L_0H_0))^p\frac{pe^{2\alpha T}}{e^{\alpha T}-1}\Big].\end{aligned}$$

Since the number E is sufficiently large such that

$$E \geq \frac{2}{\alpha}\ln\left(\frac{\gamma R[1+K\overline{d}(M+L_0H_0)]^p}{\varepsilon}\right),$$

we have that

$$R[1+K\overline{d}(M+L_0H_0)]^p e^{-\alpha(t-\sigma)} \leq \frac{\varepsilon}{\gamma},\ t\in[\sigma+E/2,\sigma+E_1].$$

Hence, the inequality

$$\begin{aligned}&\left\|\phi_L(t)-\phi_{\overline{L}}(t)\right\| < \frac{\varepsilon}{\gamma}\Big[1+\frac{K}{\lambda}+\frac{K^2\overline{c}(M+L_0H_0)}{\lambda\alpha}(1+K\overline{d}(M+L_0H_0))^p\\ &+K^2\overline{d}(M+L_0H_0)(1+K\overline{d}(M+L_0H_0))^p\frac{pe^{2\alpha T}}{\lambda(e^{\alpha T}-1)}\Big]\\ &\leq \varepsilon\end{aligned}$$

holds for all $t \in J^1$, where $J^1 = [\sigma + E/2, \sigma + E_1]$. We note that the interval J^1 has a length no less than $E/2$. Consequently, the couple $\left(\phi_L(t), \phi_{\overline{L}}(t)\right) \in \mathscr{A} \times \mathscr{A}$ is proximal. □

The next assertion is about the frequent separation feature.

Lemma 8.12 *Suppose that the conditions* $(C1)$*–*$(C6)$*,* $(C8)$ *are fulfilled. If a couple* $\left(L(t), \overline{L}(t)\right) \in \mathscr{L} \times \mathscr{L}$ *is frequently* (ε_0, Δ)*-separated for some positive numbers* ε_0 *and* Δ*, then there exist positive numbers* ε_1 *and* $\overline{\Delta}$ *such that the couple* $\left(\phi_L(t), \phi_{\overline{L}}(t)\right) \in \mathscr{A} \times \mathscr{A}$ *is frequently* $(\varepsilon_1, \overline{\Delta})$*-separated.*

Proof Because the couple $\left(L(t), \overline{L}(t)\right) \in \mathscr{L} \times \mathscr{L}$ is frequently (ε_0, Δ) separated for some $\varepsilon_0 > 0$ and $\Delta > 0$, there exist infinitely many disjoint intervals $J_q, q \in \mathbb{N}$, each with a length no less than Δ, such that $\left\|L(t) - \overline{L}(t)\right\| > \varepsilon_0$ for each t from these intervals. In the proof, we will verify the existence of numbers $\varepsilon_1 > 0$, $\overline{\Delta} > 0$ and infinitely many disjoint intervals $J_q^1 \subset J_q, q \in \mathbb{N}$, each with length $\overline{\Delta}$, such that the inequality $\left\|\phi_L(t) - \phi_{\overline{L}}(t)\right\| > \varepsilon_1$ holds for each t from the intervals $J_q^1, q \in \mathbb{N}$.

According to the equicontinuity of $\mathscr{L}$, one can find a positive number $\tau < \Delta$, such that for any $t_1, t_2 \in \mathbb{R}$ with $|t_1 - t_2| < \tau$, the inequality

$$\left|\left(L_{ij}(t_1) - \overline{L}_{ij}(t_1)\right) - \left(L_{ij}(t_2) - \overline{L}_{ij}(t_2)\right)\right| < \frac{\varepsilon_0}{2} \tag{8.3.30}$$

holds for all $1 \leq i \leq m$, $1 \leq j \leq n$.

Suppose that for each $q \in \mathbb{N}$, the number s_q denotes the midpoint of the interval J_q. Let us define a sequence $\{\kappa_q\}$ through the equation $\kappa_q = s_q - \tau/2$.

Let us fix an arbitrary $q \in \mathbb{N}$. One can find integers i_0, j_0, such that

$$\left|L_{i_0 j_0}(s_q) - \overline{L}_{i_0 j_0}(s_q)\right| = \left\|L(s_q) - \overline{L}(s_q)\right\| > \varepsilon_0. \tag{8.3.31}$$

Making use of the inequality (8.3.30), for all $t \in \left[\kappa_q, \kappa_q + \tau\right]$ we have

$$\begin{aligned}
&\left|L_{i_0 j_0}(s_q) - \overline{L}_{i_0 j_0}(s_q)\right| - \left|L_{i_0 j_0}(t) - \overline{L}_{i_0 j_0}(t)\right| \\
&\leq \left|\left(L_{i_0 j_0}(t) - \overline{L}_{i_0 j_0}(t)\right) - \left(L_{i_0 j_0}(s_q) - \overline{L}_{i_0 j_0}(s_q)\right)\right| \\
&< \frac{\varepsilon_0}{2},
\end{aligned}$$

and therefore, by means of (8.3.31), we attain that the inequality

$$\left|L_{i_0 j_0}(t) - \overline{L}_{i_0 j_0}(t)\right| > \left|L_{i_0 j_0}(s_q) - \overline{L}_{i_0 j_0}(s_q)\right| - \frac{\varepsilon_0}{2} > \frac{\varepsilon_0}{2} \tag{8.3.32}$$

is valid for all $t \in \left[\kappa_q, \kappa_q + \tau\right]$.

For each i and j, one can find numbers $\zeta_{ij}^q \in [\kappa_q, \kappa_q + \tau]$ such that

$$\int_{\kappa_q}^{\kappa_q+\tau} \left(L(s) - \overline{L}(s)\right) ds = \tau \left(L_{11}(\zeta_{11}^q) - \overline{L}_{11}(\zeta_{11}^q), \ldots, L_{mn}(\zeta_{mn}^q) - \overline{L}_{mn}(\zeta_{mn}^q)\right).$$

Thus, according to the inequality (8.3.32), we have that

$$\left\|\int_{\kappa_q}^{\kappa_q+\tau} \left(L(s) - \overline{L}(s)\right) ds\right\| \geq \tau \left|L_{i_0 j_0}(\zeta_{i_0 j_0}^q) - \overline{L}_{i_0 j_0}(\zeta_{i_0 j_0}^q)\right| > \frac{\tau \varepsilon_0}{2}. \tag{8.3.33}$$

Making use of the relations

$$\begin{aligned}
\phi_L^{ij}(t) = \phi_L^{ij}(\kappa_q) &- \int_{\kappa_q}^{t} \Big(a_{ij} + \sum_{C_{hl} \in N_r(i,j)} C_{ij}^{hl} f(\phi_L^{hl}(s))\Big) \phi_L^{ij}(s) ds + \int_{\kappa_q}^{t} L_{ij}(s) ds \\
&+ \sum_{\kappa_q \leq \theta_k < t} \Big(b_{ij} + \sum_{C_{hl} \in N_r(i,j)} D_{ij}^{hl} g(\phi_L^{hl}(\theta_k))\Big) \phi_L^{ij}(\theta_k) + \sum_{\kappa_q \leq \theta_k < t} I_{ij}^k
\end{aligned}$$

and

$$\phi_{\overline{L}}^{ij}(t) = \phi_{\overline{L}}^{ij}(\kappa_q) - \int_{\kappa_q}^{t} \Big(a_{ij} + \sum_{C_{hl}\in N_r(i,j)} C_{ij}^{hl} f(\phi_{\overline{L}}^{hl}(s))\Big)\phi_{\overline{L}}^{ij}(s)ds + \int_{\kappa_q}^{t} \overline{L}_{ij}(s)ds$$
$$+ \sum_{\kappa_q\le\theta_k<t} \Big(b_{ij} + \sum_{C_{hl}\in N_r(i,j)} D_{ij}^{hl} g(\phi_{\overline{L}}^{hl}(\theta_k))\Big)\phi_{\overline{L}}^{ij}(\theta_k) + \sum_{\kappa_q\le\theta_k<t} I_{ij}^{k}$$

we obtain that

$$\Big|\phi_{L}^{ij}(\kappa_q+\tau) - \phi_{\overline{L}}^{ij}(\kappa_q+\tau)\Big| \ge \Big|\int_{\kappa_q}^{\kappa_q+\tau} \big(L_{ij}(s) - \overline{L}_{ij}(s)\big)\,ds\Big| - \Big|\phi_{L}^{ij}(\kappa_q) - \phi_{\overline{L}}^{ij}(\kappa_q)\Big|$$
$$- \int_{\kappa_q}^{\kappa_q+\tau} a_{ij}\Big|\phi_{L}^{ij}(s) - \phi_{\overline{L}}^{ij}(s)\Big|ds - \sum_{\kappa_q\le\theta_k<\kappa_q+\tau} |b_{ij}|\,\Big|\phi_{L}^{ij}(\theta_k) - \phi_{\overline{L}}^{ij}(\theta_k)\Big|$$
$$- \int_{\kappa_q}^{\kappa_q+\tau} \sum_{C_{hl}\in N_r(i,j)} C_{ij}^{hl}\Big|f(\phi_{L}^{hl}(s))\phi_{L}^{ij}(s) - f(\phi_{\overline{L}}^{hl}(s))\phi_{\overline{L}}^{ij}(s)\Big|ds$$
$$- \sum_{\kappa_q\le\theta_k<\kappa_q+\tau} \sum_{C_{hl}\in N_r(i,j)} D_{ij}^{hl}\Big|g(\phi_{L}^{hl}(\theta_k))\phi_{L}^{ij}(\theta_k) - g(\phi_{\overline{L}}^{hl}(\theta_k))\phi_{\overline{L}}^{ij}(\theta_k)\Big|$$
$$\ge \Big|\int_{\kappa_q}^{\kappa_q+\tau} \big(L_{ij}(s) - \overline{L}_{ij}(s)\big)\,ds\Big| - \sup_{t\in[\kappa_q,\kappa_q+\tau]} \big\|\phi_L(t) - \phi_{\overline{L}}(t)\big\|$$
$$-\tau a_{ij} \sup_{t\in[\kappa_q,\kappa_q+\tau]} \big\|\phi_L(t) - \phi_{\overline{L}}(t)\big\| - \frac{p}{T}(T+\tau)|b_{ij}| \sup_{t\in[\kappa_q,\kappa_q+\tau]} \big\|\phi_L(t) - \phi_{\overline{L}}(t)\big\|$$
$$-\tau(M + L_0H_0) \sum_{C_{hl}\in N_r(i,j)} C_{ij}^{hl} \sup_{t\in[\kappa_q,\kappa_q+\tau]} \big\|\phi_L(t) - \phi_{\overline{L}}(t)\big\|$$
$$-\frac{p}{T}(T+\tau)(M + L_0H_0) \sum_{C_{hl}\in N_r(i,j)} D_{ij}^{hl} \sup_{t\in[\kappa_q,\kappa_q+\tau]} \big\|\phi_L(t) - \phi_{\overline{L}}(t)\big\|.$$

By means of the inequality (8.3.33), one can show that

$$\sup_{t\in[\kappa_q,\kappa_q+\tau]} \big\|\phi_L(t) - \phi_{\overline{L}}(t)\big\| \ge \big\|\phi_L(\kappa_q+\tau) - \phi_{\overline{L}}(\kappa_q+\tau)\big\|$$
$$> \frac{\tau\varepsilon_0}{2} - (1+P_0) \sup_{t\in[\kappa_q,\kappa_q+\tau]} \big\|\phi_L(t) - \phi_{\overline{L}}(t)\big\|,$$

where

$$P_0 = \max_{(i,j)}\Big[\tau a_{ij} + \tau(M + L_0H_0) \sum_{C_{hl}\in N_r(i,j)} C_{ij}^{hl} + \frac{p}{T}(T+\tau)|b_{ij}|$$
$$+(M + L_0H_0)\frac{p}{T}(T+\tau) \sum_{C_{hl}\in N_r(i,j)} D_{ij}^{hl}\Big].$$

Hence, we have that $\sup_{t\in[\kappa_q,\kappa_q+\tau]} \left\|\phi_L(t)-\phi_{\overline{L}}(t)\right\| > M_0$, where $M_0=\dfrac{\tau\varepsilon_0}{2\,(2+P_0)}$.

Set

$$M_1=2\max_{(i,j)}\Big(H_0a_{ij}+MH_0\sum_{C_{hl}\in N_r(i,j)}C_{ij}^{hl}+M_{ij}\Big),$$

$$b_1=\max_{(i,j)}\Big(|b_{ij}|+(M+H_0L_0)\sum_{C_{hl}\in N_r(i,j)}D_{ij}^{hl}\Big)$$

and

$$\underline{\theta}=\min_{1\le k\le p}(\theta_{k+1}-\theta_k).$$

Define the numbers

$$\varepsilon_1=\frac{M_0}{2}\min\Big\{b_0-L_0H_0\overline{d},\ \frac{1}{1+b_1}\Big\}$$

and

$$\overline{\Delta}=\min\Big\{\underline{\theta},\ \frac{M_0}{2M_1(2+b_1)},\ \frac{M_0\left(b_0-L_0H_0\overline{d}\right)}{2M_1(1+b_0-L_0H_0\overline{d})}\Big\}.$$

Suppose that there exists $\xi_q\in[\kappa_q,\kappa_q+\tau]$ such that

$$\sup_{t\in[\kappa_q,\kappa_q+\tau]}\left\|\phi_L(t)-\phi_{\overline{L}}(t)\right\|=\left\|\phi_L(\xi_q)-\phi_{\overline{L}}(\xi_q)\right\|.$$

Let $\kappa_q^1=\begin{cases}\xi_q, & \text{if } \xi_q\le\kappa_q+\tau/2\\ \xi_q-\overline{\Delta}, & \text{if } \xi_q>\kappa_q+\tau/2\end{cases}$. Since $\overline{\Delta}\le\underline{\theta}$, there exists at most one impulsive moment on the interval $(\kappa_q^1,\kappa_q^1+\overline{\Delta})$.

We shall start by considering the case $\xi_q>\kappa_q+\dfrac{\tau}{2}$. Assume that there exists an impulsive moment $\theta_{k_0}\in(\kappa_q^1,\kappa_q^1+\overline{\Delta})$. For $t\in(\theta_{k_0},\kappa_q^1+\overline{\Delta})$, making use of the equation

$$\begin{aligned}\phi_L^{ij}(t)-\phi_{\overline{L}}^{ij}(t)&=\left(\phi_L^{ij}(\xi_q)-\phi_{\overline{L}}^{ij}(\xi_q)\right)-\int_{\xi_q}^t a_{ij}\left(\phi_L^{ij}(s)-\phi_{\overline{L}}^{ij}(s)\right)ds\\&-\int_{\xi_q}^t\sum_{C_{hl}\in N_r(i,j)}C_{ij}^{hl}\left(f(\phi_L^{hl}(s))\phi_L^{ij}(s)-f(\phi_{\overline{L}}^{hl}(s))\phi_{\overline{L}}^{ij}(s)\right)ds\\&+\int_{\xi_q}^t\left(L_{ij}(s)-\overline{L}_{ij}(s)\right)ds,\end{aligned}$$

one can verify that $\left\|\phi_L(t) - \phi_{\overline{L}}(t)\right\| > M_0 - \overline{\Delta} M_1 > \dfrac{M_0}{2} > \varepsilon_1$. In particular, we have that

$$\left\|\phi_L(\theta_{k_0}+) - \phi_{\overline{L}}(\theta_{k_0}+)\right\| > M_0 - \overline{\Delta} M_1.$$

Because the inequality

$$\begin{aligned}
&\left|\phi_L^{ij}(\theta_{k_0}+) - \phi_{\overline{L}}^{ij}(\theta_{k_0}+)\right| \le |1 + b_{ij}| \left|\phi_L^{ij}(\theta_{k_0}) - \phi_{\overline{L}}^{ij}(\theta_{k_0})\right| \\
&+ M \sum_{C_{hl} \in N_r(i,j)} D_{ij}^{hl} \left|\phi_L^{ij}(\theta_{k_0}) - \phi_{\overline{L}}^{ij}(\theta_{k_0})\right| \\
&+ H_0 L_0 \sum_{C_{hl} \in N_r(i,j)} D_{ij}^{hl} \left|\phi_L^{hl}(\theta_{k_0}) - \phi_{\overline{L}}^{hl}(\theta_{k_0})\right|
\end{aligned}$$

is valid for each i and j, it is easy to obtain that

$$\left\|\phi_L(\theta_{k_0}) - \phi_{\overline{L}}(\theta_{k_0})\right\| \ge \frac{\left\|\phi_L(\theta_{k_0}+) - \phi_{\overline{L}}(\theta_{k_0}+)\right\|}{1 + b_1} > \frac{M_0 - \overline{\Delta} M_1}{1 + b_1}.$$

For $t \in (\kappa_q^1, \theta_{k_0})$, the relation

$$\begin{aligned}
&\phi_L^{ij}(t) - \phi_{\overline{L}}^{ij}(t) = \left(\phi_L^{ij}(\theta_{k_0}) - \phi_{\overline{L}}^{ij}(\theta_{k_0})\right) - \int_{\theta_{k_0}}^{t} a_{ij} \left(\phi_L^{ij}(s) - \phi_{\overline{L}}^{ij}(s)\right) ds \\
&- \int_{\theta_{k_0}}^{t} \sum_{C_{hl} \in N_r(i,j)} C_{ij}^{hl} \left(f(\phi_L^{hl}(s))\phi_L^{ij}(s) - f(\phi_{\overline{L}}^{hl}(s))\phi_{\overline{L}}^{ij}(s)\right) ds \\
&+ \int_{\theta_{k_0}}^{t} \left(L_{ij}(s) - \overline{L}_{ij}(s)\right) ds
\end{aligned}$$

implies that

$$\begin{aligned}
&\left\|\phi_L(t) - \phi_{\overline{L}}(t)\right\| > \frac{M_0 - \overline{\Delta} M_1}{1 + b_1} - \overline{\Delta} M_1 \\
&= \frac{1}{1 + b_1}[M_0 - \overline{\Delta} M_1(2 + b_1)] \\
&\ge \frac{M_0}{2(1 + b_1)} \\
&\ge \varepsilon_1.
\end{aligned}$$

On the other hand, if none of the impulsive moments belong to $(\kappa_q^1, \kappa_q^1 + \overline{\Delta})$, then for each t from this interval we have that $\left\|\phi_L(t) - \phi_{\overline{L}}(t)\right\| > M_0 - \overline{\Delta} M_1 > \varepsilon_1$. Therefore, in the case of $\xi_q > \kappa_q + \tau/2$, the inequality $\left\|\phi_L(t) - \phi_{\overline{L}}(t)\right\| > \varepsilon_1$ holds for all $t \in (\kappa_q^1, \kappa_q^1 + \overline{\Delta})$, regardless of the existence of an impulsive moment in this interval.

Next, we consider the case $\xi_q \leq \kappa_q + \dfrac{\tau}{2}$. If there exists an impulsive moment $\theta_{k_0} \in (\kappa_q^1, \kappa_q^1 + \overline{\Delta})$, then one can use a similar evaluation as in the case discussed above to show for $t \in (\kappa_q^1, \theta_{k_0}]$ that $\left\|\phi_L(t) - \phi_{\overline{L}}(t)\right\| > M_0 - \overline{\Delta} M_1 > \varepsilon_1$. Moreover, the inequality

$$\left|\phi_L^{ij}(\theta_{k_0}+) - \phi_{\overline{L}}^{ij}(\theta_{k_0}+)\right| \geq \left(\left|1 + b_{ij}\right| - M \sum_{C_{hl} \in N_r(i,j)} D_{ij}^{hl}\right) \left|\phi_L^{ij}(\theta_{k_0}) - \phi_{\overline{L}}^{ij}(\theta_{k_0})\right|$$
$$- \sum_{C_{hl} \in N_r(i,j)} D_{ij}^{hl} L_0 H_0 \left|\phi_L^{hl}(\theta_{k_0}) - \phi_{\overline{L}}^{hl}(\theta_{k_0})\right|$$

yields

$$\begin{aligned} &\left\|\phi_L(\theta_{k_0}+) - \phi_{\overline{L}}(\theta_{k_0}+)\right\| \geq \left(b_0 - L_0 H_0 \overline{d}\right) \left\|\phi_L(\theta_{k_0}) - \phi_{\overline{L}}(\theta_{k_0})\right\| \\ &> \left(b_0 - L_0 H_0 \overline{d}\right) (M_0 - \overline{\Delta} M_1). \end{aligned}$$

Thus, for $t \in (\theta_{k_0}, \kappa_q^1 + \overline{\Delta})$, the relation

$$\phi_L^{ij}(t) - \phi_{\overline{L}}^{ij}(t) = \left(\phi_L^{ij}(\theta_{k_0}+) - \phi_{\overline{L}}^{ij}(\theta_{k_0}+)\right) - \int_{\theta_{k_0}}^t a_{ij} \left(\phi_L^{ij}(s) - \phi_{\overline{L}}^{ij}(s)\right) ds$$
$$- \int_{\theta_{k_0}}^t \sum_{C_{hl} \in N_r(i,j)} C_{ij}^{hl} \left(f(\phi_L^{hl}(s))\phi_L^{ij}(s) - f(\phi_{\overline{L}}^{kl}(s))\phi_{\overline{L}}^{ij}(s)\right) ds$$
$$+ \int_{\theta_{k_0}}^t \left(L_{ij}(s) - \overline{L}_{ij}(s)\right) ds$$

implies that

$$\begin{aligned} &\left\|\phi_L(t) - \phi_{\overline{L}}(t)\right\| > \left(b_0 - L_0 H_0 \overline{d}\right) (M_0 - \overline{\Delta} M_1) - \overline{\Delta} M_1 \\ &\geq \frac{\left(b_0 - L_0 H_0 \overline{d}\right) M_0}{2} \\ &\geq \varepsilon_1. \end{aligned}$$

Therefore, for all $t \in (\kappa_q^1, \kappa_q^1 + \overline{\Delta})$ we have that $\left\|\phi_L(t) - \phi_{\overline{L}}(t)\right\| > \varepsilon_1$. One can also show that the same inequality holds even if no impulsive moments exist inside the interval $(\kappa_q^1, \kappa_q^1 + \overline{\Delta})$.

Now, suppose that there exists an impulsive moment $\theta_{l_0} \in [\kappa_q, \kappa_q + \tau]$ such that

$$\sup_{t \in [\kappa_q, \kappa_q + \tau]} \left\|\phi_L(t) - \phi_{\overline{L}}(t)\right\| = \left\|\phi_L(\theta_{l_0}+) - \phi_{\overline{L}}(\theta_{l_0}+)\right\|.$$

Let us define $\kappa_q^1 = \begin{cases} \theta_{l_0}, & \text{if } \theta_{l_0} \leq \kappa_q + \tau/2 \\ \theta_{l_0} - \overline{\Delta}, & \text{if } \theta_{l_0} > \kappa_q + \tau/2 \end{cases}$. It is worth noting that the interval $(\kappa_q^1, \kappa_q^1 + \overline{\Delta})$ does not contain any impulsive moments. If $\theta_{l_0} > \kappa_q + \frac{\tau}{2}$, then using the inequality

$$\left\|\phi_L(\theta_{l_0}) - \phi_{\overline{L}}(\theta_{l_0})\right\| \geq \frac{\left\|\phi_L(\theta_{l_0}+) - \phi_{\overline{L}}(\theta_{l_0}+)\right\|}{1+b_1} > \frac{M_0}{1+b_1},$$

one can verify for $t \in (\kappa_q^1, \kappa_q^1 + \overline{\Delta})$ that

$$\left\|\phi_L(t) - \phi_{\overline{L}}(t)\right\| \geq \left\|\phi_L(\theta_{l_0}) - \phi_{\overline{L}}(\theta_{l_0})\right\| - \overline{\Delta} M_1 > \frac{M_0}{1+b_1} - \overline{\Delta} M_1 > \frac{M_0}{2(1+b_1)} \geq \varepsilon_1.$$

In a similar way, if $\theta_{l_0} \leq \kappa_q + \frac{\tau}{2}$, then we have for $t \in (\kappa_q^1, \kappa_q^1 + \overline{\Delta})$ that

$$\left\|\phi_L(t) - \phi_{\overline{L}}(t)\right\| \geq \left\|\phi_L(\theta_{l_0}+) - \phi_{\overline{L}}(\theta_{l_0}+)\right\| - \overline{\Delta} M_1 > M_0 - \overline{\Delta} M_1 > \frac{M_0}{2} > \varepsilon_1.$$

Hence, on each of the intervals $J_q^1 = (\kappa_q^1, \kappa_q^1 + \overline{\Delta})$, $q \in \mathbb{N}$, the inequality

$$\left\|\phi_L(t) - \phi_{\overline{L}}(t)\right\| > \varepsilon_1$$

holds.

Consequently, the pair $\left(\phi_L(t), \phi_{\overline{L}}(t)\right) \in \mathscr{A} \times \mathscr{A}$ is frequently $(\varepsilon_1, \overline{\Delta})$-separated. □

The main result of the present section is given in the following theorem.

Theorem 8.3 *Suppose that the conditions* $(C1)$–$(C9)$ *are valid. If* $\mathscr{L}$ *is a Li–Yorke chaotic set which possesses a* ρT*-periodic function for each natural number* ρ*, then the set* $\mathscr{A}$ *is also Li–Yorke chaotic.*

Proof Using the conditions $(C1)$–$(C6)$ and $(C9)$, one can show that if $L(t) \in \mathscr{L}$ is an ρT-periodic function for some natural number ρ, then the bounded on $\mathbb{R}$ solution $\phi_L(t) \in \mathscr{A}$ is also a periodic function with the same period, and vice versa. Therefore, the collection $\mathscr{A}$ contains ρT-periodic functions for each natural number ρ.

Suppose that the set $\mathscr{C}_{\mathscr{L}}$ is a scrambled set inside $\mathscr{L}$. Define the set $\mathscr{C}_{\mathscr{A}} = \{\phi_L(t) \mid L(t) \in \mathscr{C}_{\mathscr{L}}\}$. There is a one-to-one correspondence between the sets $\mathscr{C}_{\mathscr{L}}$ and $\mathscr{C}_{\mathscr{A}}$. Because the set $\mathscr{C}_{\mathscr{L}}$ is uncountable, the same is true for $\mathscr{C}_{\mathscr{A}}$. Moreover, no periodic functions exist inside $\mathscr{C}_{\mathscr{A}}$, since there are no such functions inside $\mathscr{C}_{\mathscr{L}}$.

Lemmas 8.11 and 8.12 together imply that the set $\mathscr{C}_{\mathscr{A}}$ is a scrambled set. On the other hand, according to Lemma 8.12, for any function $\phi_L(t) \in \mathscr{C}_{\mathscr{A}}$ and any periodic function $\phi_{\overline{L}}(t) \in \mathscr{A}$, there exist positive numbers ε_1 and $\overline{\Delta}$ such that the pair $(\phi_L(t), \phi_{\overline{L}}(t))$ is frequently $(\varepsilon_1, \overline{\Delta})$-separated. Consequently, the set $\mathscr{A}$ is Li–Yorke chaotic. □

The results of the present section reveal that if the external inputs $L_{ij}(t)$ are chaotic, then the impulsive SICNNs (8.3.27) behave chaotically. Accordingly, to illustrate our results, we need external inputs which are ensured to be chaotic in the Li–Yorke sense. In the next section, to obtain such external inputs, we will consider SICNNs in the form of (8.3.26) whose external inputs are relay functions with chaotically changing switching moments [29–31, 33, 34, 103, 131, 132]. Moreover, we will take advantage of the core mechanism, which is represented in Fig. 8.11, in order to set up a neural system consisting of three SICNNs.

8.3.4 Examples

Each neuron in a neural network is capable of receiving input signals, processing them and sending an output signal. Neural signals consist of short electrical pulses called action potentials or spikes. A chain of action potentials emitted by a single neuron is called a spike train. Action potentials in a spike train are usually well separated, and it is impossible to excite a second spike during or immediately after a first one [1]. That is why the discontinuity phenomenon is a natural property of neural networks. In this section, we take into account an example of a neural system consisting of three SICNNs. Discontinuous external inputs in a rectangular form are used in the first SICNN to provide the chaos.

Let us consider the SICNN

$$\frac{dx_{ij}(t)}{dt} = -a_{ij}x_{ij}(t) - \sum_{C_{hl}\in N_1(i,j)} C_{ij}^{hl} f(x_{hl}(t))x_{ij}(t) + L_{ij}(t), \qquad (8.3.34)$$

in which $i, j = 1, 2, 3,$

$$\begin{pmatrix} a_{11} & a_{12} & a_{13} \\ a_{21} & a_{22} & a_{23} \\ a_{31} & a_{32} & a_{33} \end{pmatrix} = \begin{pmatrix} 6 & 8 & 10 \\ 1 & 9 & 4 \\ 12 & 7 & 5 \end{pmatrix},$$

$$\begin{pmatrix} C_{11} & C_{12} & C_{13} \\ C_{21} & C_{22} & C_{23} \\ C_{31} & C_{32} & C_{33} \end{pmatrix} = \begin{pmatrix} 0.004 & 0.002 & 0 \\ 0.006 & 0.008 & 0.005 \\ 0.009 & 0.007 & 0.003 \end{pmatrix}.$$

In the network (8.3.34), we set $L_{ij}(t) = R_{ij}(t, \zeta)$, where the relay function $R_{ij}(t, \zeta)$ is defined by the equation

$$R_{ij}(t, \zeta) = \begin{cases} \alpha_{ij}, & \text{if } \zeta_{2q} < t \le \zeta_{2q+1}, \\ \beta_{ij}, & \text{if } \zeta_{2q-1} < t \le \zeta_{2q}, \end{cases}.$$

Here, the numbers ζ_q, $q \in \mathbb{Z}$, denote the switching moments and they are the same for all i and j. The sequence $\zeta = \{\zeta_q\}$ is defined through the formula $\zeta_q = q + \vartheta_q$, $q \in \mathbb{Z}$, where the sequence $\{\vartheta_q\}$, $\vartheta_0 \in [0,1]$, is generated by the logistic map $\vartheta_{q+1} = 3.9\vartheta_q(1-\vartheta_q)$, which is chaotic in the Li–Yorke sense [24]. We note that the interval $[0,1]$ is invariant under the iterations of the map [119]. More information about the dynamics of relay systems can be found in the studies [29–31, 33, 34, 103, 131, 132].

We consider the SICNN (8.3.34) with $f(s) = 0.6\sqrt{s}$ and $\alpha_{ij} = 1.5$, $\beta_{ij} = 0.4$ for all i, j. According to the results of [29, 103], the SICNN (8.3.34) exhibits chaotic motions for $\zeta_0 \in [0,1]$, and the collection $\mathscr{L}$ consisting of the bounded on $\mathbb{R}$ solutions of (8.3.34) corresponding to different values of ζ_0 is a Li–Yorke chaotic set, which admits infinitely many periodic solutions with periods 2ρ for each natural number ρ.

Figure 8.13 shows the solution $x(t) = \{x_{ij}(t)\}$ of the SICNN (8.3.34) with $\zeta_0 = 0.192$ corresponding to the initial data $x_{11}(t_0) = 0.1407, x_{12}(t_0) = 0.1548, x_{13}(t_0) = 0.1092$, $x_{21}(t_0) = 0.9168$, $x_{22}(t_0) = 0.1451$, $x_{23}(t_0) = 0.3276$, $x_{31}(t_0) = 0.1046$, $x_{32}(t_0) = 0.0992$, $x_{33}(t_0) = 0.2518$, where $t_0 = 0.192$. It is seen in Fig. 8.13 that each cell of the SICNN (8.3.34) possesses chaos.

Next, we consider the impulsive SICNN

$$\begin{aligned}
\frac{dy_{ij}(t)}{dt} &= -\overline{a}_{ij}y_{ij}(t) - \sum_{\overline{C}_{hl}\in N_1(i,j)} \overline{C}_{ij}^{hl} f_1(y_{hl}(t))y_{ij}(t) + \overline{L}_{ij}(t),\ t \neq \theta_k, \\
\Delta y_{ij}|_{t=\theta_k} &= \overline{b}_{ij}y_{ij}(\theta_k) + \sum_{\overline{C}_{hl}\in N_1(i,j)} \overline{D}_{ij}^{hl} g_1(y_{hl}(\theta_k))y_{ij}(\theta_k) + \overline{I}_{ij}^k,
\end{aligned} \tag{8.3.35}$$

where $i, j = 1, 2, 3$, $f_1(s) = 0.4s^{7/2}$, $g_1(s) = 0.2s^2$, $\theta_k = 2k$, $k \in \mathbb{Z}$, $\overline{I}_{ij}^k = 0.001$ for each i, j and k,

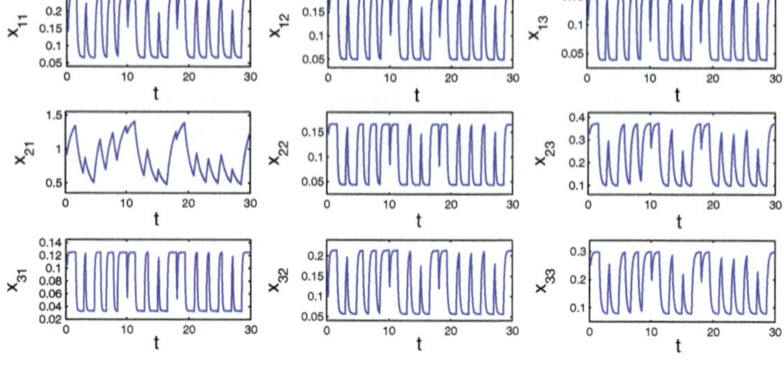

Fig. 8.13 The chaotic behavior of the SICNNs (8.3.34)

$$\begin{pmatrix} \overline{a}_{11} & \overline{a}_{12} & \overline{a}_{13} \\ \overline{a}_{21} & \overline{a}_{22} & \overline{a}_{23} \\ \overline{a}_{31} & \overline{a}_{32} & \overline{a}_{33} \end{pmatrix} = \begin{pmatrix} 7 & 4 & 8 \\ 5 & 9 & 6 \\ 10 & 7 & 5 \end{pmatrix},$$

$$\begin{pmatrix} \overline{b}_{11} & \overline{b}_{12} & \overline{b}_{13} \\ \overline{b}_{21} & \overline{b}_{22} & \overline{b}_{23} \\ \overline{b}_{31} & \overline{b}_{32} & \overline{b}_{33} \end{pmatrix} = \begin{pmatrix} -0.5 & 0.6 & -0.4 \\ 0.4 & -0.5 & -0.3 \\ 0.7 & 0.5 & -0.2 \end{pmatrix},$$

$$\begin{pmatrix} \overline{C}_{11} & \overline{C}_{12} & \overline{C}_{13} \\ \overline{C}_{21} & \overline{C}_{22} & \overline{C}_{23} \\ \overline{C}_{31} & \overline{C}_{32} & \overline{C}_{33} \end{pmatrix} = \begin{pmatrix} 0.006 & 0.001 & 0.004 \\ 0 & 0.003 & 0.009 \\ 0.012 & 0.005 & 0 \end{pmatrix},$$

$$\begin{pmatrix} \overline{D}_{11} & \overline{D}_{12} & \overline{D}_{13} \\ \overline{D}_{21} & \overline{D}_{22} & \overline{D}_{23} \\ \overline{D}_{31} & \overline{D}_{32} & \overline{D}_{33} \end{pmatrix} = \begin{pmatrix} 0.001 & 0.005 & 0.007 \\ 0.009 & 0 & 0.008 \\ 0.006 & 0.003 & 0.002 \end{pmatrix}.$$

Define the function $\varphi(v) = \{\varphi_{ij}(v)\}$, where $v = \{v_{ij}\}$, $i, j = 1, 2, 3$, through the equations $\varphi_{11}(v) = 3v_{11}^2$, $\varphi_{12}(v) = \arctan(v_{12})$, $\varphi_{13}(v) = 4(0.3 + 2v_{13})^3$, $\varphi_{21}(v) = v_{21}$, $\varphi_{22}(v) = \tanh(10v_{22})$, $\varphi_{23}(v) = v_{23}^{3/2}$, $\varphi_{31}(v) = 3v_{31} + \sin(v_{31})$, $\varphi_{32}(v) = 2v_{32} + 0.1v_{32}^3$, $\varphi_{33}(v) = \dfrac{5v_{33}}{1 + v_{33}}$. In system (8.3.35), we set $L_{ij}(t) = \varphi_{ij}(x(t))$. That is, we make use of the outputs of the SICNN (8.3.34) as external inputs for the impulsive SICNN (8.3.35).

It is worth noting that the nonlinear function φ satisfies the inequality (8.3.28) on the compact region in which the chaotic attractor of system (8.3.34) takes place. Accordingly, the set $\mathscr{L}_\varphi$ whose elements are of the form $\varphi(x(t))$, $x(t) \in \mathscr{L}$, is Li–Yorke chaotic.

One can verify that the conditions $(C1)$–$(C9)$ hold for the network (8.3.35) with $p = 1$, $T = 2$, $K = 4$, $\overline{c} = 0.04$, $\overline{d} = 0.041$, $M = 0.0482$, $L_0 = 0.3017$, $b_0 = 0.4980238$, $H_0 = 7.3109$, $\lambda \approx 3.764998$, and $\delta \approx 0.045279$, where the approximations for λ and δ are given with accuracy of six digits in the decimal part. Therefore, the dynamics of the network (8.3.35) is Li–Yorke chaotic according to Theorem 8.3.

In the impulsive SICNN (8.3.35), let us use the solution of (8.3.34) that is represented in Fig. 8.13. Figure 8.14 depicts the output of (8.3.35) with the initial data $y_{11}(t_0) = 0.0119$, $y_{12}(t_0) = 0.0306$, $y_{13}(t_0) = 0.0541$, $y_{21}(t_0) = 0.1591$, $y_{22}(t_0) = 0.0635$, $y_{23}(t_0) = 0.0203$, $y_{31}(t_0) = 0.0339$, $y_{32}(t_0) = 0.0346$, $y_{33}(t_0) = 0.0419$, where $t_0 = 0.192$. Figure 8.14 supports our theoretical results such that the SICNN (8.3.35) exhibits chaotic motions. The 3-dimensional projection of the same solution on the $y_{13} - y_{21} - y_{33}$ space is shown in Fig. 8.15, which confirms one more time the presence of chaos in the dynamics of the network.

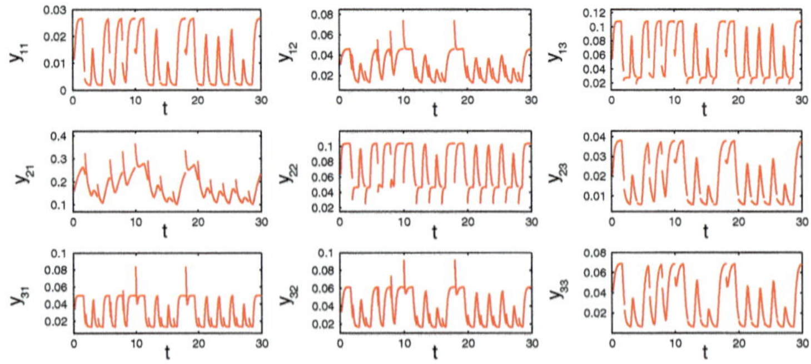

Fig. 8.14 The chaotic behavior of the SICNNs (8.3.35)

Fig. 8.15 The three-dimensional projection of the chaotic trajectory of the impulsive SICNN (8.3.35) on the $y_{13} - y_{21} - y_{33}$ space

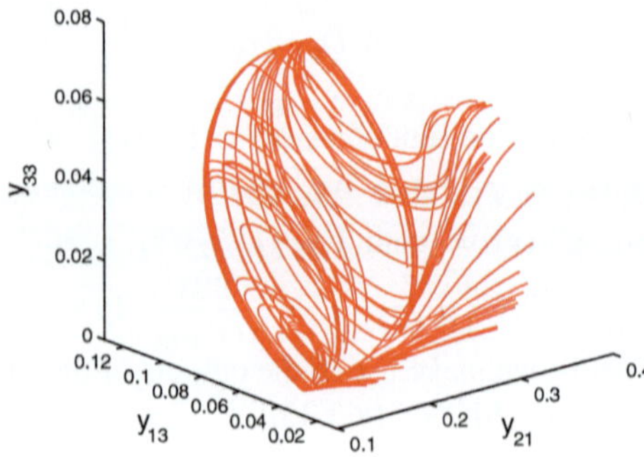

Now, let us take into account the impulsive SICNN

$$\begin{aligned} \frac{dz_{ij}(t)}{dt} &= -\widetilde{a}_{ij} z_{ij}(t) - \sum_{\widetilde{C}_{hl} \in N_1(i,j)} \widetilde{C}_{ij}^{hl} f_2(z_{hl}(t)) z_{ij}(t) + \widetilde{L}_{ij}(t), \ t \neq \eta_k, \\ \Delta z_{ij}|_{t=\eta_k} &= \widetilde{b}_{ij} z_{ij}(\eta_k) + \sum_{\widetilde{C}_{hl} \in N_1(i,j)} \widetilde{D}_{ij}^{hl} g_2(z_{hl}(\eta_k)) z_{ij}(\eta_k) + \widetilde{I}_{ij}^k, \end{aligned} \tag{8.3.36}$$

where $i, j = 1, 2, 3$, $f_2(s) = 0.6s^3$, $g_2(s) = 0.1s^4$, $\eta_k = 4k, k \in \mathbb{Z}$, $\widetilde{I}_{ij}^k = 0.02(-1)^k$ for each i, j and k,

$$\begin{pmatrix} \widetilde{a}_{11} \ \widetilde{a}_{12} \ \widetilde{a}_{13} \\ \widetilde{a}_{21} \ \widetilde{a}_{22} \ \widetilde{a}_{23} \\ \widetilde{a}_{31} \ \widetilde{a}_{32} \ \widetilde{a}_{33} \end{pmatrix} = \begin{pmatrix} 0.3 \ 0.5 \ 0.2 \\ 0.4 \ 0.9 \ 0.2 \\ 0.6 \ 0.7 \ 0.4 \end{pmatrix},$$

$$\begin{pmatrix} \widetilde{b}_{11} \ \widetilde{b}_{12} \ \widetilde{b}_{13} \\ \widetilde{b}_{21} \ \widetilde{b}_{22} \ \widetilde{b}_{23} \\ \widetilde{b}_{31} \ \widetilde{b}_{32} \ \widetilde{b}_{33} \end{pmatrix} = \begin{pmatrix} 0.12 \ 0.21 \ 0.24 \\ 0.16 \ 0.18 \ 0.01 \\ 0.32 \ 0.15 \ 0.35 \end{pmatrix},$$

$$\begin{pmatrix} \widetilde{C}_{11} & \widetilde{C}_{12} & \widetilde{C}_{13} \\ \widetilde{C}_{21} & \widetilde{C}_{22} & \widetilde{C}_{23} \\ \widetilde{C}_{31} & \widetilde{C}_{32} & \widetilde{C}_{33} \end{pmatrix} = \begin{pmatrix} 0.002 & 0 & 0.007 \\ 0.001 & 0.004 & 0.003 \\ 0.005 & 0.009 & 0.008 \end{pmatrix},$$

$$\begin{pmatrix} \widetilde{D}_{11} & \widetilde{D}_{12} & \widetilde{D}_{13} \\ \widetilde{D}_{21} & \widetilde{D}_{22} & \widetilde{D}_{23} \\ \widetilde{D}_{31} & \widetilde{D}_{32} & \widetilde{D}_{33} \end{pmatrix} = \begin{pmatrix} 0.005 & 0.008 & 0.003 \\ 0.006 & 0.004 & 0.003 \\ 0.009 & 0.002 & 0.007 \end{pmatrix}.$$

We will demonstrate numerically the appearance of near-periodic discontinuous chaos in the dynamics of the SICNN (8.3.36) when the networks (8.3.34) and (8.3.36) are weakly connected.

First of all, let us consider the network (8.3.36) with the external inputs $L_{11}(t) = 0.008$, $L_{12}(t) = 0.007$, $L_{13}(t) = 0.002$, $L_{21}(t) = 0.017$, $L_{22}(t) = 0.032$, $L_{23}(t) = 0.018$, $L_{31}(t) = 0.013$, $L_{32}(t) = 0.016$, $L_{33}(t) = 0.007$, which are constant functions, such that the network admits a unique periodic solution.

The output of SICNN (8.3.36) corresponding to the initial data $z_{11}(0.5) = 0.0332$, $z_{12}(0.5) = 0.0372$, $z_{13}(0.5) = 0.0257$, $z_{21}(0.5) = 0.0445$, $z_{22}(0.5) = 0.0361$, $z_{23}(0.5) = 0.0973$, $z_{31}(0.5) = 0.0245$, $z_{32}(0.5) = 0.0246$, $z_{33}(0.5) = 0.0232$ is shown in Fig. 8.16, where it is seen that the represented output approaches to the periodic solution of the network (8.3.36).

In order to obtain motions that behave chaotically around the discontinuous periodic solution shown in Fig. 8.16, we make use of the external inputs

$$L_{11}(t) = 0.008 + 0.035x_{11}(t), \quad L_{12}(t) = 0.007 + 0.037x_{12}(t),$$

$$L_{13}(t) = 0.002 + 0.032x_{13}(t), \quad L_{21}(t) = 0.017 + 0.009x_{21}(t),$$

$$L_{22}(t) = 0.032 + 0.084x_{22}(t), \quad L_{23}(t) = 0.018 + 0.029x_{23}(t),$$

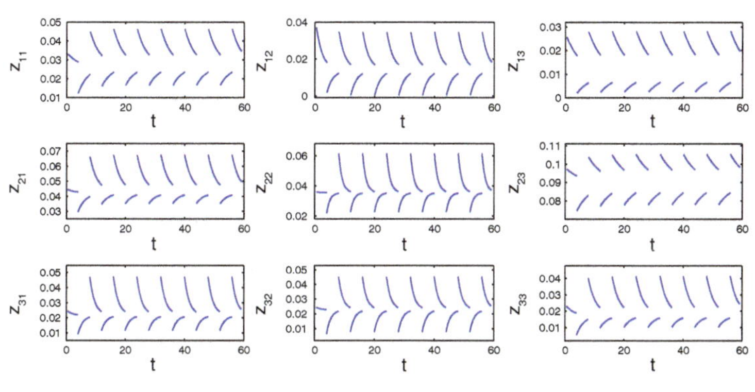

Fig. 8.16 The periodic solution of the SICNN (8.3.36) with the external inputs $L_{11}(t) = 0.008$, $L_{12}(t) = 0.007$, $L_{13}(t) = 0.002$, $L_{21}(t) = 0.017$, $L_{22}(t) = 0.032$, $L_{23}(t) = 0.018$, $L_{31}(t) = 0.013$, $L_{32}(t) = 0.016$, $L_{33}(t) = 0.007$

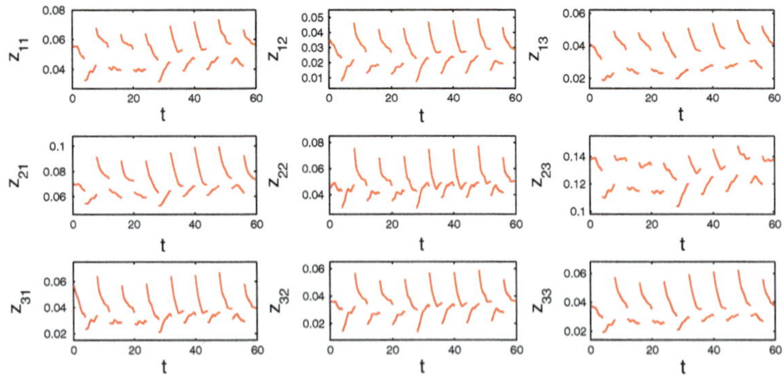

Fig. 8.17 The appearance of near-periodic discontinuous chaos in the SICNN (8.3.36)

$$L_{31}(t) = 0.013 + 0.083x_{31}(t), \quad L_{32}(t) = 0.016 + 0.045x_{32}(t),$$

$$L_{33}(t) = 0.007 + 0.023x_{33}(t)$$

in the SICNN (8.3.36), where $x(t) = \{x_{ij}(t)\}$ are the outputs of the SICNN (8.3.34).

The conditions $(C1)$–$(C9)$ are valid for the SICNN (8.3.36) with $p = 2$, $T = 8$, $K = 1.8225$, $\overline{c} = 0.039$, $\overline{d} = 0.047$, $M = 0.0026$, $L_0 = 0.0478$, $b_0 = 1.0099298$, $H_0 = 0.6861$, $\lambda \approx 0.146222$ and $\delta \approx 0.224958$, where the approximations for λ and δ are given with accuracy of six digits in the decimal part. According to Theorem 8.3, the network (8.3.36) possesses chaos in the sense of Li–Yorke.

Making use of the solution of (8.3.34) that is depicted in Fig. 8.13, we represent in Fig. 8.17 the output of the impulsive SICNN (8.3.36) corresponding to the initial data $z_{11}(t_0) = 0.0561$, $z_{12}(t_0) = 0.0353$, $z_{13}(t_0) = 0.0412$, $z_{21}(t_0) = 0.0706$, $z_{22}(t_0) = 0.0451$, $z_{23}(t_0) = 0.1381$, $z_{31}(t_0) = 0.0572$, $z_{32}(t_0) = 0.0367$, $z_{33}(t_0) = 0.0379$, where $t_0 = 0.192$. One can observe that the represented motion behaves chaotically near the periodic solution shown in Fig. 8.16.

8.4 Cyclic/Toroidal Chaos in Hopfield Neural Networks

In this section, we discuss the appearance of cyclic and toroidal chaos in Hopfield neural networks. The theoretical results may strongly relate to investigations of brain activities performed by neurobiologists. As new phenomena, extension of chaos by entrainment of several limit cycles as well as the attraction of cyclic chaos by an equilibrium are discussed. Appropriate simulations that support the theoretical results are depicted. Stabilization of tori in a chaotic attractor is realized not only for neural networks, but also for differential equations theory, and this phenomenon has never been reported before in the literature. It is demonstrated that the proposed chaos generation technique cannot be considered as generalized synchronization.

8.4.1 Introduction

There is a certain adequacy for the real world and its reflection by brain activities. The presence of chaos in neural networks is useful for separating image segments [52], information processing [49, 50] and synchronization [59–61, 63]. The chaos can be generated either by a neural network itself (endogenous chaos), or a chaotic influence outside of the neural network can be realized in its output (exogenous chaos). The endogenous chaos in neural networks has been widely investigated in the literature [38–40, 42, 44, 46, 47, 51, 53, 55, 82, 139–144], but the latter has not been effectively discussed yet. This is not because the problem is not natural, but the absence of a rigorously developed input/output mechanism for the phenomenon seems to be the reason. This is why we were attracted by the problem of chaos generation.

In their experiments, Skarda and Freeman [84] obtained different kinds of electroencephalogram (EEG) signals when known and unknown odorants were given to a rabbit. For known odorants, the signals were in the form of a *limit cycle*, but for unknown ones, they were *chaotic*. According to the experimental results, it was proposed that deterministic chaos is utilized in neural activities for learning new sensory patterns as well as ensuring continual access to previously learned sensory patterns. The roles of chaos for brain behavior have been investigated in many papers. For example, Watanabe et al. [54] demonstrated that the chaotic dynamics works as means to learn new patterns and increases the *memory capacity* of neural networks. The group of theorists, Guevara et al. [41], suggested that chaotic behavior may be responsible for dynamical diseases such as schizophrenia, insomnia, epilepsy, and dyskinesia. It was shown in the paper [41] that the periodic forcing of neural oscillator models can lead to chaos. This is similar to the case that was primarily observed in electrical devices through Van der Pol and Duffing oscillators in pioneer papers [145–149]. Actually, this is not surprising since brain activities can be mostly considered as electrical processes.

A sensory cortex is conceived in [83] as a global attractor with many "wings." When the cortex is at rest, the wings are shut. When a known stimulus arrives, the system moves to an appropriate wing and a burst of oscillation is observed. In paper [133], it was revealed that "each of the wings are either a near-limit cycle (a narrow band chaos) or a broad band chaos." One should emphasize that the near-limit cycle chaos can result from the entrainment of limit cycles by chaos, which is theoretically proved in [134] and considered as one of the main ways of chaos generation in the present research.

Discussing wings as neural networks, one can suppose that there is the opportunity of chaos production by a wing itself and extension of chaos from one wing to another. The latter case has not been considered in the literature yet, at least mathematically. That is why we decided to investigate the problem in the present section. More precisely, for the first time in the literature, we consider the extension of chaos in the following ways: (i) Entrainment of a limit cycle by chaos, (ii) Attraction of a cyclic chaos by an equilibrium, (iii) Entrainment of two and more limit cycles by chaos, (iv) Attractions of chaotic cycles by an equilibrium. These chaos extension types provide us with the mathematical support for Freeman's "wings" of a sensory cortex. The extension of chaos has not been considered in the previous investigations, and we hope that the rigorous mathematical background for the phenomenon may give a positive effect for the researches of neurobiologists. This is also true for our results concerning quasiperiodic and almost periodic motions in the basis of chaotic attractors. Since the appearance of limit cycle and near-limit cycle chaos were experimentally observed in the studies [84, 133], one can hypothesize that both limit tori and near-torus chaos can be dynamical representatives of brain processes too.

The present section is suggested as an application of our investigations about chaos extension developed in the papers [34, 103, 134] to give an additional mathematical light on the ideas developed by neurobiologists, primarily, Freeman and his collaborators [80–83, 85, 133]. The main dynamical result considered in this section is the entrainment by chaos, which is understood as *the deformation of limit cycles to chaotic cycles.* Our results are useful for analyzing chaos extension among collectives of neural networks based on generation of chaos by input/output mechanisms built through differential equations. According to Skarda and Freeman [84], limit cycles and chaotic dynamics are of prime importance in odor recognition. Moreover, it was observed in [84] that the brain's EEG activity changes from limit cyclical to near-cycle chaotic if a familiar odor was replaced by an unknown one. This can be interpreted through our paper [134] as the chaotification of limit cycles. Additionally, if we accept that complexity of chaos is important for the memory capacity, then one can suppose that to increase a memory we need to do the same with the complexity of chaos. From this point of view, it is interesting to say about regular unstable motions which constitute a basis (skeleton) of chaotic attractors. These are usually assumed to be periodic motions [24, 104, 105]. Beside the periodic motions, quasiperiodic, almost periodic, and recurrent motions can also be considered as a basis of chaos [150–152]. As chaos increases the capacity of memorizing [54, 84], one can suppose that chaos with the basis of quasiperiodic motions provides a memory with a larger capacity than that with periodic motions. This is true if we compare chaos with quasiperiodic unstable motions with a one having a skeleton of almost periodic motions. That is why the problem of chaos generation by neural networks which is based on unstable quasiperiodic or almost periodic unstable solutions is of strong importance. In this section, it is shown that one can create quasiperiodic motions in chaotic attractors as well as join different quasiperiodic motions to obtain quasiperiodic motions with a larger number of incommensurate frequencies. Moreover, we discuss the problem of chaos control, which can also be considered as theoretical basis of learning and recognition, if one accepts the ideas in the papers [80, 81, 84, 85, 133]. We suggest that the appearance of limit cycles in experiments with brain behavior [54, 84, 133] results from the stabilization of one of the unstable periodic solutions, which are already present in a wing. This stabilization can be done either by external perturbation or by control (of Pyragas type [153]), which is triggered by stimuli.

To have a unity and uniform delivering in the section, the discussions are developed using Hopfield neural networks (HNNs) [5, 154–156], but they can also be realized for other types of neural networks [111, 112, 120, 157–161].

HNNs [5, 154–156] are continuous-time dynamical systems described by the following nonlinear ordinary differential equations:

$$C_i \frac{dp_i}{dt} = -\frac{p_i}{R_i} + \sum_{j=1}^{N} w_{ij} f_j(p_j) + I_i, \quad i = 1, 2, \ldots, N, \tag{8.4.37}$$

where N is the number of neurons, p_i is the total input to neuron i, the bounded monotonic differentiable function f_j is the activation function acted on neuron j, C_i and R_i are the parameters corresponding, respectively, to a capacitance and a resistance, I_i is the external input of neuron i and w_{ij} is the synaptic connection value between neuron i and neuron j.

In an equivalent form, the HNN (8.4.37) can be represented as

$$\dot{p} = -Cp + Wf(p) + I,$$

where $p = (p_1, p_2, \ldots, p_N)^T$, the diagonal matrix $C = diag\{c_1, c_2, \ldots, c_N\}$, which is associated with C_i and R_i, has positive diagonal entries, $W = (w_{ij})_{N\times N}$ is the connection matrix, $f(p) = (f_1(p_1), f_2(p_2), \ldots, f_N(p_N))^T$ and $I = (I_1, I_2, \ldots, I_N)^T$ is the external input vector.

Weak synaptic connections between neurons are observable in the dynamics of brain, and a method to characterize the weakness of synaptic connections is to consider amplitudes of postsynaptic potentials measured in the soma of neurons while the neuron membrane potential is far below the threshold value [162]. The brain units such as neurons, cortical columns, and neuronal modules are supposed to be weakly connected and modeled as autonomous quasiperiodic oscillators in the paper [135]. McNaughton et al. [136] revealed weak synaptic connections in the hippocampal cells by means of the investigation of excitatory postsynaptic potentials. Moreover, weak interactions between neurons in the cortex are observed by Abeles [137] as a result of the analysis of cross correlograms obtained from pairs of neurons. On the other hand, according to Pasemann et al. [163], periodic and quasiperiodic solutions in biological and artificial systems are of fundamental importance as they are associated with central pattern generators. Therefore, the investigations of coupled neural networks that possess periodic or quasiperiodic solutions with weak connections are of prime importance.

In the present section, we establish weak connections between two HNNs, one with a chaotic attractor and another with an attracting limit cycle or attracting torus. As a result we obtain a chaotic cycle/torus, that is, motions that behave chaotically around the limit cycle or torus.

Stability is one of the main properties which are suggested to be started with pioneer papers [5]. It attracts the attention of other authors nowadays [164–171].

However, starting with results on chaos, the role of papers on unstable motions and their stabilization has been increased significantly.

In the literature, the generation of chaos is considered within the scope of synchronization theory [67, 86, 87, 91, 92, 94, 96]. For two coupled systems to be synchronized, the chaos of the response system has to be asymptotically close to that of the driver. We do not use this proximity in our results, and we demonstrate that chaos generation around limit cycles and tori are not reducible to synchronization, in general.

8.4.2 Entrainment by Chaos in HNNs

Let us consider the HNN

$$\dot{x} = -Cx + Wf(x) + I, \tag{8.4.38}$$

where $x \in \mathbb{R}^m$, $C = diag\{c_1, c_2, \ldots, c_m\}$, $c_i > 0$ for each $i = 1, 2, \ldots, m$, $W = (w_{ij})_{m\times m}$ is the connection matrix and I is the external input vector.

Next, we take into account the HNN

$$\dot{y} = -Dy + \overline{W}g(y) + \varepsilon h(x(t)), \tag{8.4.39}$$

where $y \in \mathbb{R}^n$, $x(t)$ are solutions of (8.4.38), ε is a nonzero constant, $h : \mathbb{R}^m \to \mathbb{R}^n$ is a continuous function, $D = diag\{d_1, d_2, \ldots, d_n\}$, $d_i > 0$ for each $i = 1, 2, \ldots, n$ and $\overline{W} = (\overline{w}_{ij})_{n\times n}$ is the connection matrix. It is worth noting that the unidirectionally coupled networks (8.4.38)+(8.4.39) have a skew product structure.

We mainly assume that the HNN

$$\dot{u} = -Du + \overline{W}g(u) \tag{8.4.40}$$

possesses an orbitally stable limit cycle.

On the other hand, we also assume that the network (8.4.38) admits a chaotic attractor, let us say a set in $\mathbb{R}^m$. Fix x_0 from the attractor and take a solution $x(t)$ of (8.4.38) with $x(0) = x_0$. Since we use the solution $x(t)$ as an external input in the network (8.4.39), we call it as *chaotic function*. The chaotic functions may be irregular as well as regular (periodic and unstable) [104–106, 172].

The network (8.4.38) is called sensitive if there exist positive numbers ε_0 and Δ such that for an arbitrary positive number δ_0 and for each chaotic solution $x(t)$ of (8.4.38), there exist a chaotic solution $\overline{x}(t)$ of the same network and an interval $J \subset [0, \infty)$, with a length no less than Δ, such that $\|x(0) - \overline{x}(0)\| < \delta_0$ and $\|x(t) - \overline{x}(t)\| > \varepsilon_0$ for all $t \in J$.

For a given chaotic solution $x(t)$ of (8.4.38), let us denote by $\phi_{x(t)}(t, y_0)$, $y_0 \in \mathbb{R}^n$, the solution of (8.4.39) with $\phi_{x(t)}(0, y_0) = y_0$. The network (8.4.39) replicates the sensitivity of (8.4.38) if there exist positive numbers ε_1 and $\overline{\Delta}$ such that for an arbitrary positive number δ_1 and for each solution $\phi_{x(t)}(t, y_0)$, there exist an interval $J^1 \subset [0, \infty)$, with a length no less than $\overline{\Delta}$, and a solution $\phi_{\overline{x}(t)}(t, y_1)$ such that $\|y_0 - y_1\| < \delta_1$ and $\left\|\phi_{x(t)}(t, y_0) - \phi_{\overline{x}(t)}(t, y_1)\right\| > \varepsilon_1$ for all $t \in J^1$. Moreover, we say that the network (8.4.39) is chaotic if it replicates the sensitivity of (8.4.38) and the system (8.4.38)+(8.4.39) possesses infinitely many unstable periodic solutions in a bounded region.

The following theorem is based on the entrainment of limit cycles by chaos considered in the paper [134], where the replication of sensitivity and the existence of infinitely many unstable periodic solutions were rigorously proved.

Theorem 8.4 *If there exists a number $L > 0$ such that $\|h(s_1) - h(s_2)\| \geq L\|s_1 - s_2\|$ for all $s_1, s_2 \in \mathbb{R}^m$ and the number $|\varepsilon|$ is sufficiently small, then there exists a neighborhood $\mathscr{N}$ of the orbitally stable limit cycle of (8.4.40) such that solutions of (8.4.39) which start inside $\mathscr{N}$ behave chaotically around the limit cycle. That is, the solutions are sensitive and there are infinitely many unstable periodic solutions.*

To illustrate the result of Theorem 8.4, let us consider the HNN [42]

$$\begin{aligned}\dot{u}_1 &= -u_1 + 3.4\tanh(u_1) - 1.6\tanh(u_2) + 0.7\tanh(u_3)\\ \dot{u}_2 &= -u_2 + 2.5\tanh(u_1) + 0.95\tanh(u_3)\\ \dot{u}_3 &= -u_3 - 3.5\tanh(u_1) + 0.5\tanh(u_2),\end{aligned} \tag{8.4.41}$$

which is in the form of (8.4.40). It is mentioned in [42] that the network (8.4.41) possesses a limit cycle with the Lyapunov exponents 0, −0.1356 and −0.1466. Therefore, 1 is a simple characteristic multiplier of the corresponding variational system, and the remaining characteristic multipliers are in modulus less than 1. According to the Andronov–Witt Theorem [173], the limit cycle of (8.4.41) is orbitally stable.

Next, we take into account the following HNN,

$$\begin{aligned}\dot{x}_1 &= -x_1 + 2\tanh(x_1) - 1.2\tanh(x_2)\\ \dot{x}_2 &= -x_2 + 2\tanh(x_1) + 1.71\tanh(x_2) + 1.15\tanh(x_3)\\ \dot{x}_3 &= -x_3 - 4.75\tanh(x_1) + 1.1\tanh(x_3).\end{aligned} \tag{8.4.42}$$

In the paper [143], it is shown that the network (8.4.42) admits a positive Lyapunov exponent and possesses chaotic motions. We will use it as a system of the form (8.4.38), which entrains the limit cycle of (8.4.41) by the chaos.

Making use of the solutions of (8.4.42) as external inputs for (8.4.41), we set up the following HNN:

$$\begin{aligned}&\dot{y}_1 = -y_1 + 3.4\tanh(y_1) - 1.6\tanh(y_2) + 0.7\tanh(y_3)\\ &+0.0136\tanh(x_1(t)) - 0.0015\tanh(x_2(t)) + 0.0025\tanh(x_3(t))\\ &\dot{y}_2 = -y_2 + 2.5\tanh(y_1) + 0.95\tanh(y_3)\\ &+0.0004\tanh(x_1(t)) + 0.0212\tanh(x_2(t)) - 0.0005\tanh(x_3(t))\\ &\dot{y}_3 = -y_3 - 3.5\tanh(y_1) + 0.5\tanh(y_2)\\ &+0.0012\tanh(x_1(t)) + 0.0023\tanh(x_2(t)) + 0.0145\tanh(x_3(t)).\end{aligned} \tag{8.4.43}$$

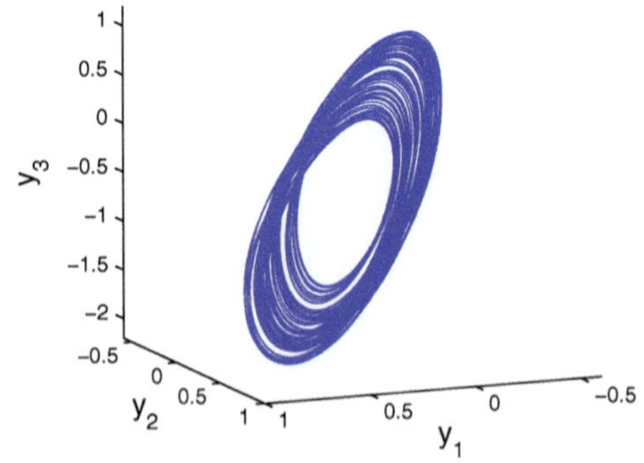

Fig. 8.18 The chaotic trajectory of HNN (8.4.43). The figure supports Theorem 8.4 such that the trajectory behaves chaotically around the limit cycle of HNN (8.4.41)

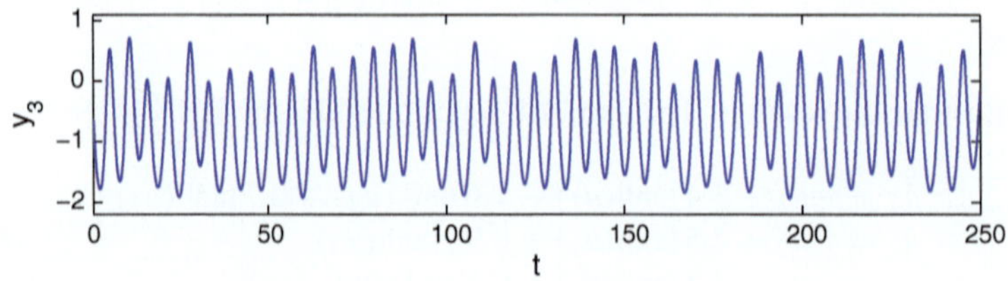

Fig. 8.19 The chaotic behavior of the y_3 coordinate of HNN (8.4.43)

The network (8.4.43) is in the form of (8.4.39), and according to Theorem 8.4, it possesses chaotic motions around the limit cycle of (8.4.41).

To simulate the result, let us use in HNN (8.4.43) the chaotic solution $x(t)$ of (8.4.42) with $x_1(0) = -0.109$, $x_2(0) = -0.832$, $x_3(0) = 1.721$, and represent the trajectory of (8.4.43) with $y_1(0) = 0.645$, $y_2(0) = 0.243$, $y_3(0) = -0.628$ in Fig. 8.18. The figure supports the result of Theorem 8.4 such that the limit cycle of (8.4.41) is entrained by the chaos. Moreover, the irregular behavior of the y_3 coordinate over time is illustrated in Fig. 8.19.

8.4.2.1 Sensitivity Analysis

The replication of sensitivity in more general coupled systems is rigorously proved in the paper [134]. Here, we will show through simulations the replication of sensitivity by HNNs.

Li et al. [47] theoretically verified the existence of horseshoe chaos in the HNN

$$\begin{aligned}\dot{x}_1 &= -x_1 + 2\tanh(x_1) - \tanh(x_2)\\ \dot{x}_2 &= -x_2 + 1.7\tanh(x_1) + 1.71\tanh(x_2) + 1.1\tanh(x_3)\\ \dot{x}_3 &= -2x_3 - 2.5\tanh(x_1) - 2.9\tanh(x_2) + 0.56\tanh(x_3).\end{aligned} \tag{8.4.44}$$

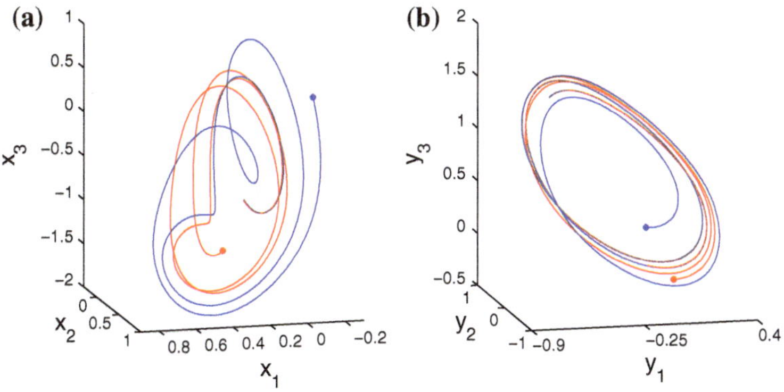

Fig. 8.20 Extension of sensitivity in the coupled HNNs (8.4.44)+(8.4.45)

Additionally, we consider the HNN

$$\begin{aligned}\dot{y}_1 &= -y_1 + 3.4\tanh(y_1) - 1.6\tanh(y_2) + 0.7\tanh(y_3)\\&+0.02\tanh(x_1(t)) + 0.035\tanh(x_3(t))\\\dot{y}_2 &= -y_2 + 2.5\tanh(y_1) + 0.95\tanh(y_3) + 0.025\tanh(x_2(t))\\\dot{y}_3 &= -y_3 - 3.5\tanh(y_1) + 0.5\tanh(y_2) + 0.004\tanh(x_1(t))\\&-0.01\tanh(x_2(t)) + 0.05\tanh(x_3(t)),\end{aligned} \quad (8.4.45)$$

which is obtained using of the solutions of (8.4.44) as external inputs in (8.4.41).

To demonstrate numerically the replication of sensitivity, we illustrate in Fig. 8.20 two initially nearby trajectories of the coupled network (8.4.44)+(8.4.45), one with the initial data $x_1(0) = 0.236$, $x_2(0) = 0.543$, $x_3(0) = -0.745$, $y_1(0) = -0.751$, $y_2(0) = -0.672$, $y_3(0) = 1.641$, represented in blue, and another with the initial data $x_1(0) = 0.237$, $x_2(0) = 0.541$, $x_3(0) = -0.752$, $y_1(0) = -0.749$, $y_2(0) = -0.674$, $y_3(0) = 1.643$, pictured in red. Figure 8.20a, b, show the projections of these trajectories on the $x_1 - x_2 - x_3$ and $y_1 - y_2 - y_3$ spaces, respectively. It is seen in Fig. 8.20a, that the sensitivity feature is present in the HNN (8.4.44) such that the initially nearby solutions eventually diverge. On the other hand, it is seen in Fig. 8.20b, that the trajectories are initially close to each other and are then separated, that is, the sensitivity is replicated by the network (8.4.45). The simulations are performed for $t \in [0, 21]$.

8.4.2.2 Chaos Around Tori

Verification of the entrainment of limit tori by chaos is a theoretically difficult task. Nevertheless, let us show that near-torus chaos is possible for HNNs. For these needs, similarly to the near-limit cycle chaos, we will use the following neural networks.

According to the simulation results of the study [141], the HNN

$$\begin{aligned} \dot{x}_1 &= -x_1 + \tanh(x_1) + 0.5\tanh(x_2) - 3\tanh(x_3) - \tanh(x_4) \\ \dot{x}_2 &= -x_2 + 2.3\tanh(x_2) + 3\tanh(x_3) \\ \dot{x}_3 &= -x_3 + 3\tanh(x_1) - 3\tanh(x_2) + \tanh(x_3) \\ \dot{x}_4 &= -100x_4 + 100\tanh(x_1) + 170\tanh(x_4) \end{aligned} \tag{8.4.46}$$

is hyperchaotic such that it possesses two positive Lyapunov exponents. The chaos will be applied as an input for the following Hopfield neural network,

$$\begin{aligned} \dot{u}_1 &= -u_1 + \tanh(u_1) + 0.5\tanh(u_2) - 3\tanh(u_3) - \tanh(u_4) \\ \dot{u}_2 &= -u_2 - 0.1\tanh(u_1) + 2\tanh(u_2) + 3\tanh(u_3) \\ \dot{u}_3 &= -u_3 + 3\tanh(u_1) - 3\tanh(u_2) + \tanh(u_3) \\ \dot{u}_4 &= -100u_4 + 100\tanh(u_1) + 170\tanh(u_4). \end{aligned} \tag{8.4.47}$$

It is shown in paper [139] that the HNN (8.4.47) admits the Lyapunov exponents 0, 0, -0.2092, and -46.8691 such that the network possesses a regular torus, which attracts near solutions.

Now, let us perturb the last HNN by solutions of (8.4.46) as external inputs to obtain

$$\begin{aligned} \dot{y}_1 &= -y_1 + \tanh(y_1) + 0.5\tanh(y_2) - 3\tanh(y_3) \\ &- \tanh(y_4) + 0.0257\tanh(x_1(t)) \\ \dot{y}_2 &= -y_2 - 0.1\tanh(y_1) + 2\tanh(y_2) + 3\tanh(y_3) \\ &+ 0.0223\tanh(x_2(t)) \\ \dot{y}_3 &= -y_3 + 3\tanh(y_1) - 3\tanh(y_2) + \tanh(y_3) + 0.0159\tanh(x_3(t)) \\ \dot{y}_4 &= -100y_4 + 100\tanh(y_1) + 170\tanh(y_4) + 0.0334\tanh(x_4(t)). \end{aligned} \tag{8.4.48}$$

Figure 8.21 shows the trajectory of (8.4.48) with $x_1(0) = -0.1321$, $x_2(0) = -0.3589$, $x_3(0) = 0.3914$, $x_4(0) = -1.7219$, $y_1(0) = 0.0259$, $y_2(0) = -0.0096$, $y_3(0) = -0.2383$, $y_4(0) = -1.5493$. One can see that the motion is chaotic and surrounds the torus. Furthermore, the y_2 coordinate of the solution is represented in Fig. 8.22. The simulation results reveal that the HNN (8.4.48) possesses motions that behave chaotically around the torus of (8.4.47).

8.4.2.3 Comparison with Synchronization of Chaos

The main role of synchronization [87, 94, 96] is to predict the properties of the response system or the drive system. Thus, our results may be considered as an indicative of a type of synchronization, if one accepts the following properties to be predicted: the existence of infinitely many unstable periodic solutions with the same periods as those for the drive system, ingredients of chaos, strange attractors, the possibility of controlling chaos, etc.

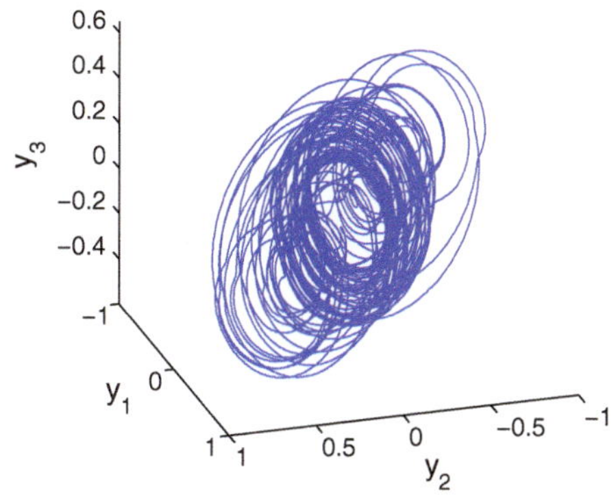

Fig. 8.21 The chaotic motion around the torus of HNN (8.4.47)

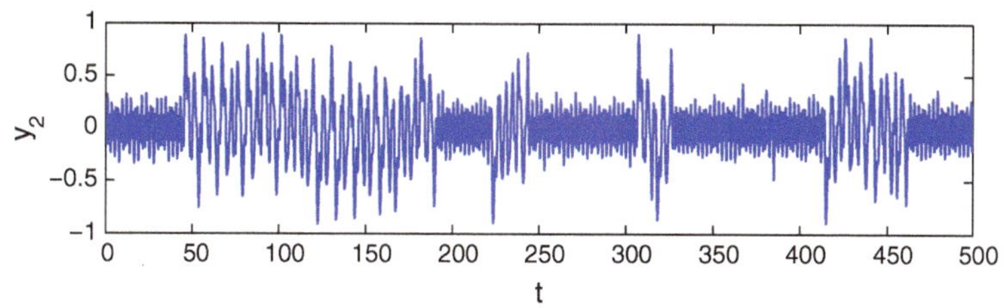

Fig. 8.22 The graph of the y_2 coordinate of HNN (8.4.48)

To analyze our results for generalized synchronization [67, 86, 91, 92, 96], we will consider the chaos produced by the networks (8.4.46) and (8.4.48). The auxiliary system approach [67, 86] as well as the method of conditional Lyapunov exponents [92, 94] will be applied to indicate the presence or absence of generalized synchronization in the couple (8.4.46) + (8.4.48) considered this time as drive-response systems (as it is accepted in the synchronization theory).

Let us take into account the auxiliary system

$$\begin{aligned}
\dot{z}_1 &= -z_1 + \tanh(z_1) + 0.5\tanh(z_2) - 3\tanh(z_3) \\
&- \tanh(z_4) + 0.0257\tanh(x_1(t)) \\
\dot{z}_2 &= -z_2 - 0.1\tanh(z_1) + 2\tanh(z_2) + 3\tanh(z_3) \\
&+ 0.0223\tanh(x_2(t)) \\
\dot{z}_3 &= -z_3 + 3\tanh(z_1) - 3\tanh(z_2) + \tanh(z_3) + 0.0159\tanh(x_3(t)) \\
\dot{z}_4 &= -100z_4 + 100\tanh(z_1) + 170\tanh(z_4) + 0.0334\tanh(x_4(t)).
\end{aligned} \tag{8.4.49}$$

Making use of the initial data $x_1(0) = -0.1321$, $x_2(0) = -0.3589$, $x_3(0) = 0.3914$, $x_4(0) = -1.7219$, $y_1(0) = 0.0259$, $y_2(0) = -0.0096$, $y_3(0) = -0.2383$, $y_4(0) = -1.5493$, $z_1(0) = 0.1376$, $z_2(0) = -0.0469$, $z_3(0) = 0.2524$, $z_4(0) = 1.7589$, and omitting the first 1000 iterations, we obtain the stroboscopic plot of system (8.4.46) + (8.4.48) + (8.4.49) whose projection on the $y_2 - z_2$ plane is shown

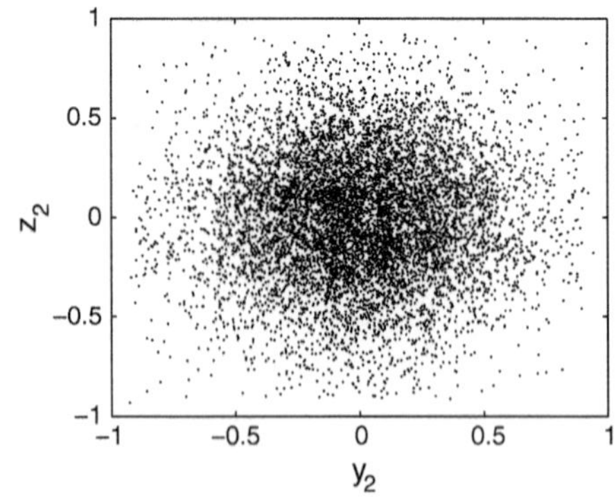

Fig. 8.23 The auxiliary system approach applied to the coupled HNNs (8.4.46) + (8.4.48)

in Fig. 8.23. Since the plot is not on the line $z_2 = y_2$, we conclude that generalized synchronization does not occur.

Next, to determine the conditional Lyapunov exponents, we consider the following variational equations for the HNN (8.4.48),

$$\begin{aligned}
\dot\eta_1 &= [-1+\text{sech}^2(y_1(t))]\eta_1 + 0.5\ \text{sech}^2(y_2(t))\eta_2 - 3\ \text{sech}^2(y_3(t))\eta_3\\
&- \text{sech}^2(y_4(t))\eta_4\\
\dot\eta_2 &= -0.1\ \text{sech}^2(y_1(t))\eta_1 + [-1+2\ \text{sech}^2(y_2(t))]\eta_2\\
&+3\ \text{sech}^2(y_3(t))\eta_3\\
\dot\eta_3 &= 3\ \text{sech}^2(y_1(t))\eta_1 - 3\ \text{sech}^2(y_2(t))\eta_2 + [-1+\text{sech}^2(y_3(t))]\eta_3\\
\dot\eta_4 &= 100\ \text{sech}^2(y_1(t))\eta_1 + [-100+170\ \text{sech}^2(y_4(t))]\eta_4.
\end{aligned} \tag{8.4.50}$$

Taking into account the solution $y(t)$ of (8.4.48) corresponding to the initial data $x_1(0) = -0.1321$, $x_2(0) = -0.3589$, $x_3(0) = 0.3914$, $x_4(0) = -1.7219$, $y_1(0) = 0.0259$, $y_2(0) = -0.0096$, $y_3(0) = -0.2383$, $y_4(0) = -1.5493$, we evaluated the largest Lyapunov exponent of system (8.4.50) as 0.105747. That is, the network (8.4.48) admits a positive conditional Lyapunov exponent, and this result reveals one more time the absence of generalized synchronization in the coupled HNNs (8.4.46) + (8.4.48).

We have shown that the method of extension of chaos by entrainment of tori is not generalized synchronization. This was also affirmed in several other simulations for limit cycles and tori in the paper [134].

8.4.3 Control of Cyclic/Toroidal Chaos in Neural Networks

In the present section, we will apply the instrument of chaos extension to obtain and control chaos in collectives of neural networks. New phenomena of the entrainment of two limit cycles by chaos and attraction of two chaotic cycles by an equilibrium will be demonstrated. Moreover, we will exhibit that the OGY (Ott, Grebogi, Yorke) control method [174] can be applied to stabilize not only periodic motions, but also tori. The control applied to the chaos generating HNNs also affects chaos of the perturbed HNNs. The results of the section may provide new ideas on brain activities, if one takes into account the experimental results in [54, 83, 84].

8.4.3.1 Entrainment of Two Limit Cycles by Chaos

We will use one more time the HNN (8.4.42) as the source of chaotic inputs, but this time the following HNN is to be perturbed by the inputs,

$$\begin{aligned} \dot{u}_1 &= -u_1 + 1.5\tanh(u_1) + 2.9\tanh(u_2) + 0.8\tanh(u_3) \\ \dot{u}_2 &= -u_2 - 3.5\tanh(u_1) + 1.18\tanh(u_2) \\ \dot{u}_3 &= -u_3 + 2.977\tanh(u_1) - 22\tanh(u_2) + 0.47\tanh(u_3). \end{aligned} \tag{8.4.51}$$

According to the results of the study [140], the network (8.4.51) admits two limit cycles with the Lyapunov exponents 0, -0.1792 and -0.7083 such that the cycles are orbitally stable by the Andronov–Witt Theorem [173].

Beside the last equations, consider the following HNN,

$$\begin{aligned} &\dot{y}_1 = -y_1 + 1.5\tanh(y_1) + 2.9\tanh(y_2) + 0.8\tanh(y_3) + 0.04\tanh(x_1(t)) \\ &+0.03\tanh(x_2(t)) + 0.006\tanh(x_3(t)) \\ &\dot{y}_2 = -y_2 - 3.5\tanh(y_1) + 1.18\tanh(y_2) - 0.002\tanh(x_1(t)) \\ &+0.06\tanh(x_2(t)) \\ &\dot{y}_3 = -y_3 + 2.977\tanh(y_1) - 22\tanh(y_2) + 0.47\tanh(y_3) \\ &-0.001\tanh(x_2(t)) + 0.04\tanh(x_3(t)). \end{aligned} \tag{8.4.52}$$

By localizing the result of Theorem 8.4 near the two limit cycles, one can conclude that (8.4.52) admits two chaotic cycles. Figure 8.24 represents the trajectories of (8.4.52) corresponding to the initial data $x_1(0) = 1.903$, $x_2(0) = 0.221$, $x_3(0) = -4.011$, $y_1(0) = 0.713$, $y_2(0) = 0.273$, $y_3(0) = -10.001$, and $x_1(0) = -0.532$, $x_2(0) = -1.647$, $x_3(0) = 2.607$, $y_1(0) = 0.571$, $y_2(0) = 0.117$, $y_3(0) = -0.079$ shown in blue and red colors, respectively. One can see in Fig. 8.24 that two chaotic cycles appear in the dynamics of the network (8.4.52). We call this phenomenon as the entrainment of two limit cycles by chaos.

One can predict that the appearance of cyclic chaos can be implemented for HNNs with not only two cycles, but also several ones. Moreover, the chaos extension by the entrainment procedure can be realized for different types of neural networks.

8.4.3.2 Attraction of Two Chaotic Cycles by an Equilibrium

In our paper [34], we considered extension of chaos in neighborhoods of attracting equilibria. From the simple observation for a dynamical system that a periodic solution used as a perturbation may cause a new cycle under certain conditions, one can

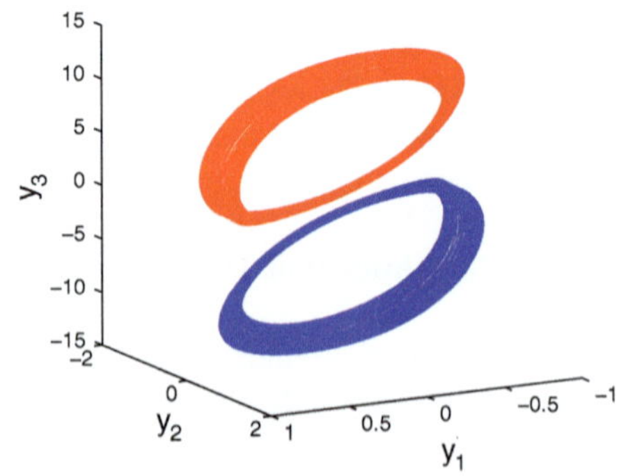

Fig. 8.24 Entrainment of two limit cycles by chaos

conclude that near-limit cycle chaotic inputs can lead to similar outputs for systems with stable equilibria. To verify this hypothesis numerically, let us apply the double cyclic chaos obtained for system (8.4.52) as an input to the HNN

$$\begin{aligned}\dot{u}_1 &= -u_1 + 0.005\tanh(u_1)) + 0.009\tanh(u_2) - 0.008\tanh(u_3)\\ \dot{u}_2 &= -u_2 - 0.001\tanh(u_1) + 0.007\tanh(u_2) - 0.003\tanh(u_3)\\ \dot{u}_3 &= -u_3 + 0.009\tanh(u_1) - 0.002\tanh(u_2) + 0.004\tanh(u_3),\end{aligned} \tag{8.4.53}$$

which admits the primitive asymptotically stable solution, to set up the HNN

$$\begin{aligned}\dot{z}_1 &= -z_1 + 0.005\tanh(z_1)) + 0.009\tanh(z_2) - 0.008\tanh(z_3)\\ &+4\tanh(y_1(t)) + \tanh(y_2(t))\\ \dot{z}_2 &= -z_2 - 0.001\tanh(z_1) + 0.007\tanh(z_2) - 0.003\tanh(z_3)\\ &+0.5\tanh(y_1(t)) + 2\tanh(y_2(t)) + 0.5\tanh(y_3(t))\\ \dot{z}_3 &= -z_3 + 0.009\tanh(z_1) - 0.002\tanh(z_2) + 0.004\tanh(z_3)\\ &+2\tanh(y_1(t)) - \tanh(y_3(t)).\end{aligned} \tag{8.4.54}$$

Figure 8.25 shows the simulation results such that two chaotic cycles appear in the dynamics of the network (8.4.54). In the simulation, we used the solutions of (8.4.52) represented in Fig. 8.24, and depict with the corresponding same colors in Fig. 8.25 the trajectories of (8.4.54) with the initial data $z_1(0) = 0.705$, $z_2(0) = 0.487$, $z_3(0) = 0.997$, and $z_1(0) = -0.142$, $z_2(0) = 0.408$, $z_3(0) = -0.873$ in blue and red, respectively.

8.4.3.3 OGY Control of a Torus

The control of chaos in neural networks is supposed to be the reason for the appearance of limit cycles in the experiments of neurobiologists [54, 84]. We showed in [134] how the Pyragas control method [153] stabilizes entrained limit cycles. It is easy to see that the simulations can be adapted for neural networks in the form of

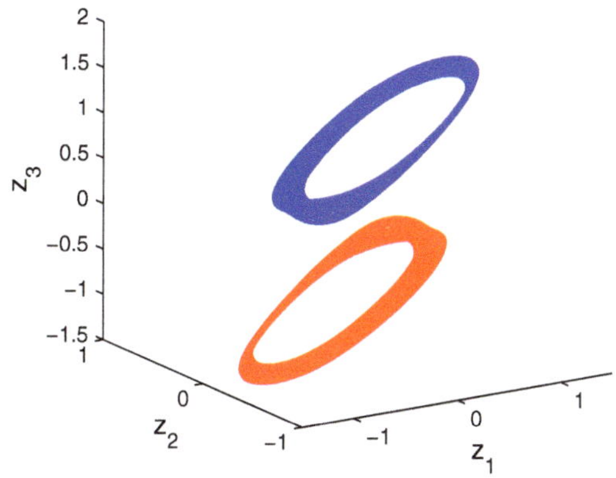

Fig. 8.25 Attraction of two chaotic cycles by an equilibrium

(8.4.42) + (8.4.43). In this subsection, we will demonstrate a novel application of the OGY control [174, 175] to stabilize tori. Since the OGY control is for discrete equations, we will start with the description of piecewise constant perturbations, which will be controlled by the method.

Consider the function

$$P(t,\theta) = \begin{cases} 1.6, & \text{if } \theta_{2i} < t \le \theta_{2i+1}, \\ 0.2, & \text{if } \theta_{2i+1} < t \le \theta_{2i+2}, \end{cases} \tag{8.4.55}$$

where i is a nonnegative integer, the sequence $\theta = \{\theta_i\}$ is defined through the equation $\theta_i = i + \zeta_i$ with $\zeta_{i+1} = F_\lambda(\zeta_i)$, $\zeta_0 \in [0, 1]$, and $F_\lambda(u) = \lambda u(1-u)$ is the logistic map. The map $F_\lambda(u)$ is chaotic through period-doubling cascade for $\lambda = 3.8$, and the interval $[0, 1]$ is invariant under its iterations [175].

Let us describe the OGY control method for the logistic map [175]. Suppose that the parameter λ in the map $F_\lambda(u)$ is allowed to vary in the range $[3.8-\varepsilon, 3.8+\varepsilon]$, where ε is a given small positive number. Consider an arbitrary solution $\{\zeta_i\}$, $\zeta_0 \in [0, 1]$, of the map and denote by $\zeta^{(j)}$, $j = 1, 2, \ldots, p$, the target p-periodic orbit to be stabilized. In the control procedure [174, 175], at each iteration step i after the control mechanism is switched on, we consider the logistic map with the parameter value $\lambda = \bar{\lambda}_i$, where

$$\bar{\lambda}_i = 3.8\left(1 + \frac{(2\zeta^{(j)}-1)(\zeta_i - \zeta^{(j)})}{\zeta^{(j)}(1-\zeta^{(j)})}\right), \tag{8.4.56}$$

provided that the number on the right-hand side of the formula (8.4.56) belongs to the interval $[3.8-\varepsilon, 3.8+\varepsilon]$. In other words, formula (8.4.56) is valid if the trajectory $\{\zeta_i\}$ is sufficiently close to the target periodic orbit. Otherwise, we take $\bar{\lambda}_i = 3.8$ so that the system evolves at its original parameter value, and wait until the trajectory $\{\zeta_i\}$ enters in a sufficiently small neighborhood of the periodic orbit $\zeta^{(j)}$,

$j = 1, 2, \ldots, p$, such that the inequality $-\varepsilon \leq 3.8\dfrac{(2\zeta^{(j)} - 1)(\zeta_i - \zeta^{(j)})}{\zeta^{(j)}(1 - \zeta^{(j)})} \leq \varepsilon$ holds. If this is the case, the chaos control is not achieved immediately after switching on the control mechanism. Instead, there is a transition time before the desired periodic orbit is stabilized. The transition time increases if the number ε decreases [67].

Let us introduce the Hopfield neural network

$$\begin{aligned}
\dot{x}_1 &= -7x_1 + 0.012\tanh(x_1) - 0.016\tanh(x_2) + 0.003\tanh(x_3)\\
\dot{x}_2 &= -4x_2 - 0.004\tanh(x_1) + 0.013\tanh(x_2) + 0.005\tanh(x_3)\\
&+P(t,\theta)\\
\dot{x}_3 &= -6x_3 + 0.008\tanh(x_1) + 0.005\tanh(x_2) + 0.009\tanh(x_3)\\
&+\sin(4t) + P(t,\theta),
\end{aligned} \tag{8.4.57}$$

where the piecewise constant function $P(t, \theta)$ described by (8.4.55) is used as a chaotic input. Since the functions $P(t, \theta)$ and $\sin(4t)$ lead to the presence of infinitely many quasiperiodic inputs with incommensurate periods, multiples of 2 and $\pi/2$, respectively, one can use the results of [29–31, 33] to conclude that the HNN (8.4.57) with $\lambda = 3.8$ possesses a chaotic attractor with infinitely many unstable quasiperiodic solutions.

Using the solutions of (8.4.57) as inputs for the HNN

$$\begin{aligned}
\dot{u}_1 &= -3u_1 + 0.003\tanh(u_1) - 0.005\tanh(u_2) - 0.013\tanh(u_3)\\
\dot{u}_2 &= -8u_2 + 0.007\tanh(u_1) + 0.008\tanh(u_2) + 0.007\tanh(u_3)\\
\dot{u}_3 &= -6u_3 - 0.004\tanh(u_1) - 0.006\tanh(u_2) + 0.002\tanh(u_3),
\end{aligned} \tag{8.4.58}$$

we set up the network

$$\begin{aligned}
\dot{y}_1 &= -3y_1 + 0.003\tanh(y_1) - 0.005\tanh(y_2) - 0.013\tanh(y_3)\\
&+1.5x_1(t)\\
\dot{y}_2 &= -8y_2 + 0.007\tanh(y_1) + 0.008\tanh(y_2) + 0.007\tanh(y_3)\\
&+1.8x_2(t)\\
\dot{y}_3 &= -6y_3 - 0.004\tanh(y_1) - 0.006\tanh(y_2) + 0.002\tanh(y_3)\\
&+1.2x_3(t).
\end{aligned} \tag{8.4.59}$$

It is worth noting that the origin is the asymptotically stable equilibrium point of (8.4.58). According to the results of the study [34], the network (8.4.59) possesses a chaotic attractor with infinitely many unstable quasiperiodic solutions, provided that the value $\lambda = 3.8$ is used in (8.4.57).

The trajectories of (8.4.57) and (8.4.59) with $\lambda = 3.8$ corresponding to the initial data $x_1(t_0) = -0.0007$, $x_2(t_0) = 0.3983$, $x_3(t_0) = 0.2061$, $y_1(t_0) = -0.0004$, $y_2(t_0) = 0.0801$, $y_3(t_0) = 0.0487$, where $t_0 = 0.281$, are represented in Fig. 8.26, (a) and (b), respectively. Moreover, the graphs of the x_3 and y_3 coordinates of the same trajectories are depicted in Fig. 8.27. The simulations reveal that both of the HNNs (8.4.57) and (8.4.59) exhibit chaotic motions.

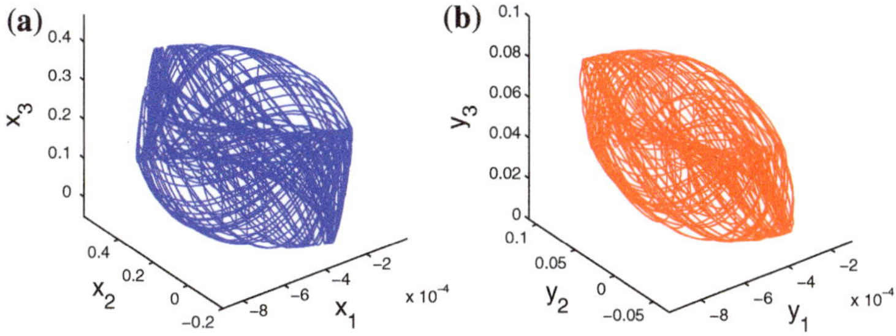

Fig. 8.26 The chaotic trajectories of (8.4.57) and (8.4.59) are represented in (**a**) and (**b**), respectively

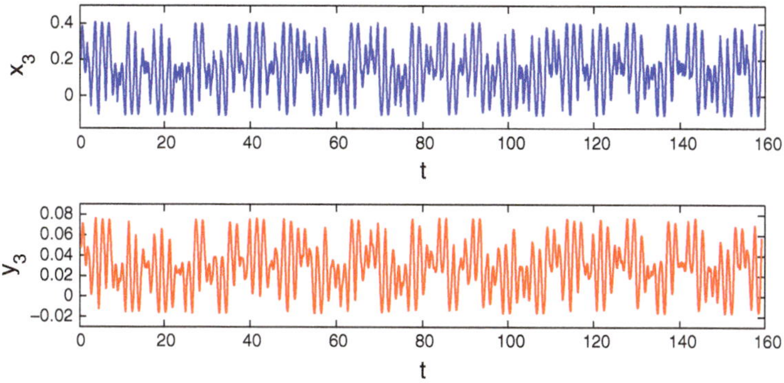

Fig. 8.27 The chaotic behavior in the networks (8.4.57) and (8.4.59)

Next, we consider the solution of the coupled networks (8.4.57) + (8.4.59) with the same initial data as considered in Figs. 8.26 and 8.27, and apply the OGY control method around the fixed point 2.8/3.8 of the logistic map $F_{3.8}(u)$. Figure 8.28 shows the simulation results for the x_3 and y_3 coordinates. The value $\varepsilon = 0.06$ is used in the simulation. The control mechanism is switched on at $t = \theta_{30}$ and switched off at $t = \theta_{60}$. The control becomes dominant approximately at $t = 45$ and its effect lasts approximately until $t = 115$, after which the instability becomes dominant and irregular behavior develops again. It is seen that a quasiperiodic solution of the HNN (8.4.57) is stabilized, and accordingly, the chaos of the HNN (8.4.59) is controlled by the stabilization of the corresponding quasiperiodic solution. On the other hand, Fig. 8.29a, b, represent the stabilized tori of the networks (8.4.57) and (8.4.59), respectively.

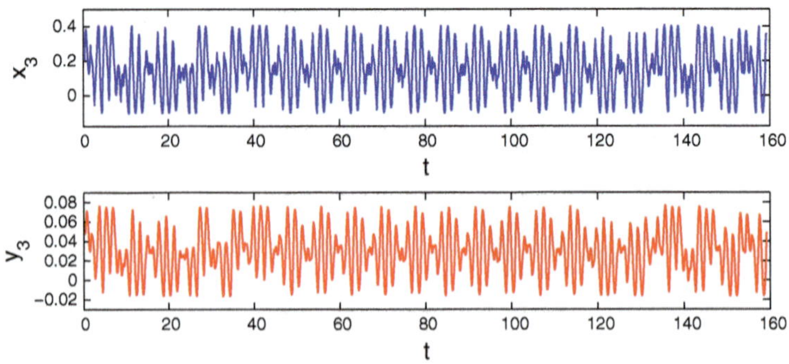

Fig. 8.28 The application of the OGY control method to stabilize the quasiperiodic solutions of (8.4.57) and (8.4.59). The control is switched on at $t = \theta_{30}$ and switched off at $t = \theta_{60}$. The value $\varepsilon = 0.06$ is used in the control procedure

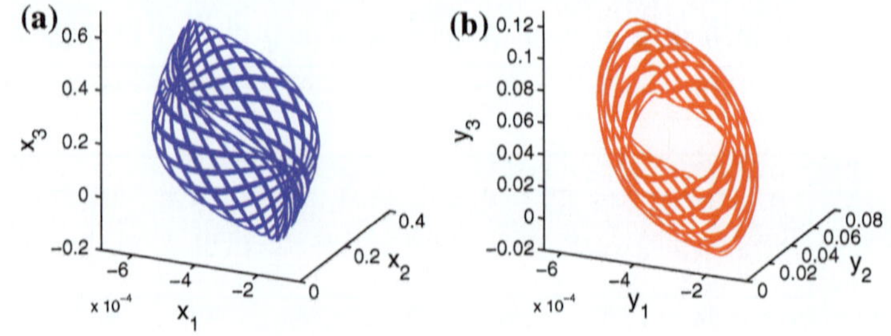

Fig. 8.29 The stabilized tori of HNNs (8.4.57) and (8.4.59) are shown in (**a**) and (**b**), respectively

8.5 Notes

In Sect. 8.1, it is shown that SICNNs with chaotic external inputs admit a chaotic attractor. Considering this phenomenon through the input–output analysis, one can say about chaos expansion among nonlinearly coupled SICNNs. The presented two examples considered together illustrate the possibility. Our method can be applied to other types of chaos, for example, that one analyzed through period-doubling cascade. The approach is suitable for the control of unstable periodic motions. The results of Sect. 8.1 were published in the paper [103] and they can be applied to the studies of chaotic communication, combinatorial optimization problems and on problems that have local minima in energy (cost) functions.

Delayed neural networks have applications in many areas such as signal and image processing, associative memories, combinatorial optimization and automatic control. Because of the finite switching speed of the amplifiers, time delays occur during the hardware implementation of neural networks. Therefore, it is of prime importance to study neural networks with time delays.

Chaotic dynamics is useful in neural networks for separating image segments and information processing. The presence of synchronization in neural networks provides a criterion for the existence of a dynamical correspondence between the systems, and helps for a better understanding of neural processes. Moreover, chaos can improve the performance of CNNs on problems that have local minima in energy (cost) functions and it is an important tool for the studies of chaotic communication and combinatorial optimization problems.

In Sect. 8.2, SICNNs with delay are considered with chaotic external inputs, and this is the first time that a theoretically approved chaos is obtained in such networks. As an example, we have considered a neural system consisting of three layers such that each layer is a retarded SICNN. Piecewise constant external inputs are utilized in the first layer of this neural system to ensure the presence of chaos in the sense of Li–Yorke. The results of the obtained chaotification process is discussed through the generalized synchronization point of view, and the proximality and frequent separation features are demonstrated numerically. The results of Sect. 8.2 can be extended easily if the delay is variable and also for the case of advanced argument. Our approach can be applied to other types of chaos such as the one analyzed through period-doubling cascade.

Freeman and his collaborators [80–85] achieved remarkable observations and conclusions that reveal the essentialness of deterministic chaos for the brain functioning. Another hypothesis is that chaos is undesirable and it occurs in brains subject to pathological malfunctions [41]. This also provides an interesting and considerable direction to the analysis of neural network problems in the chaos theory. We suppose that Sect. 8.2 can give some contributions in both directions. The proposed chaotification procedure indicates not only the advantage of the deterministic chaos over random noise for the analysis, but also significant properties of self-organization [176, 177]. Our results may be useful for the investigation of environmental inputs of the brain both on low and high levels of organization as well as learning by considering it as the creation of new structures (motions) in neural networks.

The brain comprises functionally specialized areas, which perform specific tasks and have differentiated parts or structures within. These different structures have to work together for a cerebral activity to occur. In the papers [178–180], the authors proposed the presence of synchronization as the underlying reason for such processes. Breakspear and Terry [98] reported the detection of generalized synchronization between different brain regions by means of electroencephalogram signals. In Sect. 8.2, we have demonstrated the presence of generalized synchronization by means of interconnected SICNNs with delay, and our results may provide an opportunity to understand the complex structure of the brain and the rest of the nervous system. The results of Sect. 8.2 were published in the paper [181].

In Sect. 8.3, SICNNs with fixed moments of impulses under the influence of chaotic external inputs are considered. The description of Li–Yorke chaos for the multidimensional dynamics of impulsive SICNNs is given. This is the first time in the literature that discontinuous Li–Yorke chaos is rigorously approved not only for SICNNs, but also in neuroscience. The presence of the ingredients of Li–Yorke chaos, proximality and frequent separation, are mathematically verified. The presented technique is appropriate for impulsive SICNNs with arbitrary number of cells.

Another novelty in Sect. 8.3 is the consideration of the impacts with the cell and shunting principles. The advantage of the novelty is grounded to the arguments of the studies [6, 120].

According to the presented results, it is possible to obtain arbitrarily high-dimensional neural systems by means of the core or chain mechanisms (see Figs. 8.11 and 8.12), as well as their combinations. We illustrated the usefulness of our results by taking into account a neural system consisting of three SICNNs in Sect. 8.3.4. It is worth noting that the obtained chaos in the neural system (8.3.34) – (8.3.35) – (8.3.36) is controllable [33, 34, 132], and a way to control the chaos of the neural system is to stabilize an unstable periodic solution of the SICNN (8.3.34). For instance, the OGY [174] and Pyragas [153] control methods applied to the logistic map can be used for this purpose. The problem of period-doubling route to chaos [105, 106] and extension of intermittency [107] by impulsive SICNNs can also be considered through the presented method. Moreover, our approach can be useful for modeling secure communication systems [182–186].

The appearance of cyclic irregular behavior in neural systems was observed by Freeman and his collaborators [84, 133]. We numerically demonstrated the presence of near-periodic discontinuous chaotic motions of SICNNs. The obtained result can be useful for investigations of weakly coupled impulsive neural networks. The results of Sect. 8.3 can be found in [187].

In Sect. 8.4, we have provided theoretical arguments for the entrainment of limit cycles and tori by chaos in neural networks by applying basic Hopfield neural networks. In Sect. 8.4.3, several opportunities of the chaos extension are considered when the number of limit cycles vary, and we performed the attraction of chaotic cycles by equilibria. These all demonstrate the potentials of our approach, which can be realized in the theory of neural networks.

It is natural to suppose that instead of a unique limit cycle or near-limit cycle chaos as it was the case in the experiments [80, 81, 83, 85, 133], one and the same stimulus may cause to the presence of several such behaviors if the experiments are performed intentionally. Our dynamical results support this idea, and they can be developed easily in the mathematical sense (and hopefully in brain behavior researches) for various numbers and types of stimuli as well as chaotic and regular outputs. We suggest that not only limit cycles and near-limit cycle chaos, but also limit tori and near-limit tori chaos can be investigated in experiments. Another possible experimental program concerning our results is to follow the papers [49, 50, 54, 84], where it was claimed that the memory capacity depends strongly on chaos. Loosely speaking, complexity of behavior, its degree of irregularity, is proportional to the memory capacity. It is obvious that to have a larger memory, we have to make chaos more "complex". For example, it is known that periodic solutions (unstable) are in the basis of Li–Yorke and Devaney chaos [24, 104]. By replacing the periodic motions with quasiperiodic or even almost periodic ones [150], we have more complex chaos.

In papers [54, 84, 133], the limit cycle appearance in the chaotic set of motions was mentioned without an explicit indication of the reason for the phenomenon. One can suspect that this is because of the chaos control [153, 174], that is, the stabilization of periodic solutions. However, the control procedure uses a special mechanism in which the solutions are involved [153, 174]. One can suggest that external stimuli are not controlling the cycle, but just trigger the control mechanism in neural networks. From this point of view, we have to say that our results extend the comprehension of the mechanism from limit cycles to tori. Moreover, we develop the idea that the control of chaos applied to a certain neural network can be extended to those which are adjoint with the controlled one. The extension of chaos control may give some positive information for the synchronized behavior in large society of neural networks to govern a motion of human body.

Another process in brain behavior that our results concern with is the synchronization of neural networks. Since chaos is an attribute of neural networks and synchronization is necessary for the effective brain work, one should say about synchronization of chaos in neural networks. For the moment, the most developed one is the generalized synchronization [67, 86, 91, 92, 96], which requests asymptotic closeness of drive and response systems. In Sect. 8.4.2.3, it is proved by the method of auxiliary system approach and conditional Lyapunov exponents applied to coupled networks (8.4.46) + (8.4.48) that the presented method is not generalized synchronization. In other words, there are not necessarily asymptotic relations between the two networks. Moreover, our method reveals that the systems participating in the extension of chaos are synchronized in the sense that chaos may admit similar properties such as presence of motions with the same periods, similarity of chaotic attractors and bifurcation diagrams, property to be controlled simultaneously, Shilnikov orbits, intermittency and so on [29–31, 33, 34, 103, 134]. The results of Sect. 8.4 were published in the paper [188] and they may be useful for neurobiologists to give more directions as well as mathematical apparatus for the future joint investigations.

Chaotic itinerancy [189] is a universal dynamics in high-dimensional systems, showing itinerant motion among varieties of low-dimensional ordered states through high-dimensional chaos. This phenomenon occurs in nonequilibrium neural networks [53] and analysis of brain activities [83]. In its degenerated form, chaotic itinerancy relates to intermittency [107] since both of them represent dynamical interchange of irregularity and regularity. Likewise the itinerant chaos observed in brain activities, low-dimensional chaos occurs in our results, and high-dimensional chaos takes place when all subsystems are considered as a whole. The main difference between our technique and chaotic itinerancy is in the elapsed time for the occurrence of the processes. No itinerant motion is observable in our discussions and all resultant chaotic subsystems process simultaneously, whereas the low-dimensional chaotic motions take place as time elapses in the case of chaotic itinerancy. The knowledge of the chaos type is another difference between chaotic itinerancy and our approach [29, 30, 33, 34, 103].

References

1. W. Gerstner, W.M. Kistler, *Spiking Neuron Models: Single Neurons, Populations, Plasticity* (Cambridge University Press, Cambridge, 2002)

2. J. Lu, D.W.C. Ho, J. Cao, J. Kurths, Exponential synchronization of linearly coupled neural networks with impulsive disturbances. IEEE Trans. Neural Netw. **22**, 329–335 (2011)
3. T. Yang, *Impulsive Systems and Control: Theory and Applications* (Nova Science, New York, 2001)
4. Z. Yang, D. Xu, Stability analysis of delay neural networks with impulsive effects. IEEE Trans. Circuits Syst.-II Express Br. **52**, 517–521 (2005)
5. J.J. Hopfield, Neurons with graded response have collective computational properties like those of two-state neurons. Proc. Natl. Acad. Sci. USA **81**, 3088–3092 (1984)
6. A. Bouzerdoum, R.B. Pinter, Shunting inhibitory cellular neural networks: derivation and stability analysis. IEEE Trans. Circuits Syst.-I: Fund. Theory Appl. **40**, 215–221 (1993)
7. A. Bouzerdoum, R.B. Pinter, Nonlinear lateral inhibition applied to motion detection in the fly visual system, in *Nonlinear Vision*, ed. by R.B. Pinter, B. Nabet (CRC Press, Boca Raton, 1992), pp. 423–450
8. A. Bouzerdoum, B. Nabet, R.B. Pinter, Analysis and analog implementation of directionally sensitive shunting inhibitory cellular neural networks, in *Visual Information Processing: From Neurons to Chips, Proceeding SPIE 1473* (1991), pp. 29–38
9. A. Bouzerdoum, R.B. Pinter, A shunting inhibitory motion detector that can account for the functional characteristics of fly motion sensitive interneurons, in *Proceedings of IJCNN International Joint Conference on Neural Networks* (1990), pp. 149–153
10. G.A. Carpenter, S. Grossberg, The ART of adaptive pattern recognition by a self-organizing neural network. Computer **21**, 77–88 (1988)
11. K. Fukushima, Analysis of the process of visual pattern recognition by the neocognitron. Neural Netw. **2**, 413–420 (1989)
12. M.E. Jernigan, G.F. McLean, Lateral inhibition and image processing, in *Nonlinear Vision*, ed. by R.B. Pinter, B. Nabet (CRC Press, Boca Raton, 1992), pp. 451–462
13. R.B. Pinter, R.M. Olberg, E. Warrant, Luminance adaptation of preferred object size in identified dragonfly movement detectors, in *Proceedings of IEEE International Conference on Systems, Man and Cybernetics* (1989), pp. 682–686
14. M. Cai, W. Xiong, Almost periodic solutions for shunting inhibitory cellular neural networks without global Lipschitz and bounded activation functions. Phys. Lett. A **362**, 417–423 (2007)
15. H.-S. Ding, J. Liang, T.-J. Xiao, Existence of almost periodic solutions for SICNNs with time-varying delays. Phys. Lett. A **372**, 5411–5416 (2008)
16. X. Huang, J. Cao, Almost periodic solution of shunting inhibitory cellular neural networks with time-varying delay. Phys. Lett. A **314**, 222–231 (2003)
17. Y. Li, C. Liu, L. Zhu, Global exponential stability of periodic solution for shunting inhibitory CNNs with delays. Phys. Lett. A **337**, 46–54 (2005)
18. L. Li, Z. Fang, Y. Yang, A shunting inhibitory cellular neural network with continuously distributed delays of neutral type. Nonlinear Anal. Real World Appl. **13**, 1186–1196 (2012)
19. C. Ou, Almost periodic solutions for shunting inhibitory cellular neural networks. Nonlinear Anal. Real World Appl. **10**, 2652–2658 (2009)
20. G. Peng, L. Huang, Anti-periodic solutions for shunting inhibitory cellular neural networks with continuously distributed delays. Nonlinear Anal. Real World Appl. **10**, 2434–2440 (2009)
21. J. Shao, Anti-periodic solutions for shunting inhibitory cellular neural networks with time-varying delays. Phys. Lett. A **372**, 5011–5016 (2008)
22. Y. Xia, J. Cao, Z. Huang, Existence and exponential stability of almost periodic solution for shunting inhibitory cellular neural networks with impulses. Chaos Solitons Fractals **34**, 1599–1607 (2007)
23. W. Zhao, H. Zhang, On almost periodic solution of shunting inhibitory cellular neural networks with variable coefficients and time-varying delays. Nonlinear Anal. Real World Appl. **9**, 2326–2336 (2008)
24. T.Y. Li, J.A. Yorke, Period three implies chaos. Am. Math. Mon. **82**, 985–992 (1975)
25. Z. Gui, W. Ge, Periodic solution and chaotic strange attractor for shunting inhibitory cellular neural networks with impulses. Chaos **16**(3), 033116 (2006)

26. J. Sun, Stationary oscillation for chaotic shunting inhibitory cellular neural networks with impulses. Chaos **17**, 043123 (2007)
27. M.U. Akhmet, Li-Yorke chaos in the system with impacts. J. Math. Anal. Appl. **351**, 804–810 (2009)
28. J.K. Hale, *Ordinary Differential Equations* (Krieger Publishing Company, Malabar, 1980)
29. M.U. Akhmet, Creating a chaos in a system with relay. Int. J. Qual. Theory Differ. Equ. Appl. **3**, 3–7 (2009)
30. M.U. Akhmet, Devaney's chaos of a relay system. Commun. Nonlinear Sci. Numer. Simul. **14**, 1486–1493 (2009)
31. M.U. Akhmet, Homoclinical structure of the chaotic attractor. Commun. Nonlinear Sci. Numer. Simul. **15**, 819–822 (2010)
32. M.U. Akhmet, M.O. Fen, Chaos generation in hyperbolic systems. Interdiscip. J. Discon. Nonlinear. Complex. **1**, 367–386 (2012)
33. M.U. Akhmet, M.O. Fen, Chaotic period-doubling and OGY control for the forced Duffing equation. Commun. Nonlinear Sci. Numer. Simul. **17**, 1929–1946 (2012)
34. M.U. Akhmet, M.O. Fen, Replication of chaos. Commun. Nonlinear Sci. Numer. Simul. **18**, 2626–2666 (2013)
35. J. Cao, Global asymptotic stability of neural networks with transmission delays. Int. J. Syst. Sci. **31**, 1313–1316 (2000)
36. C.M. Marcus, R.M. Westervelt, Stability of analog neural networks with delay. Phys. Rev. A **39**, 347–359 (1989)
37. T. Roska, L.O. Chua, Cellular neural networks with non-linear and delay-type template elements and non-uniform grids. Int. J. Circuit Theory Appl. **20**, 469–481 (1992)
38. K. Aihara, G. Matsumoto, Chaotic oscillations and bifurcations in squid giant axons, in *Chaos*, ed. by A. Holden (Manchester University Press, Manchester, 1986), pp. 257–269
39. K. Aihara, T. Takabe, M. Toyoda, Chaotic neural networks. Phys. Lett. A **144**, 333–340 (1990)
40. J. Guckenheimer, R.A. Oliva, Chaos in the Hodgkin-Huxley model. SIAM J. Appl. Dyn. Syst. **1**(1), 105–114 (2002)
41. M.R. Guevara, L. Glass, M.C. Mackey, A. Shrier, Chaos in neurobiology. IEEE Trans. Syst. Man Cybern. SMC **13**(5), 790–798 (1983)
42. W.-Z. Huang, Y. Huang, Chaos of a new class of Hopfield neural networks. Appl. Math. Comput. **206**, 1–11 (2008)
43. E. Kaslik, S. Sivasundaram, Nonlinear dynamics and chaos in fractional-order neural networks. Neural Netw. **32**, 245–256 (2012)
44. Q. Ke, B.J. Oommen, Logistic neural networks: Their chaotic and pattern recognition properties. Neurocomputing **125**, 184–194 (2014)
45. R. King, J.D. Barchas, B.A. Huberman, Chaotic behavior in dopamine neurodynamics. Proc. Natl. Acad. Sci. USA **81**, 1244–1247 (1984)
46. J. Kuroiwa, N. Masutani, S. Nara, K. Aihara, Chaotic wandering and its sensitivity to external input in a chaotic neural network, in *Proceedings of the 9th International Conference on Neural Information Processing (ICONIP'O2)*, ed. by L. Wang, J.C. Rajapakse, K. Fukushima, S.Y. Lee, X. Yao (Orchid Country Club, Singapore, 2002), pp. 353–357
47. J. Li, F. Liu, Z.-H. Guan, T. Li, A new chaotic Hopfield neural network and its synthesis via parameter switchings. Neurocomputing **117**, 33–39 (2013)
48. W. Liu, L. Wang, Variable thresholds in the chaotic cellular neural network, in *Proceedings of International Joint Conference on Neural Networks*, Orlando, Florida, USA, 12–17 August, 2007
49. S. Nara, P. Davis, Chaotic wandering and search in a cycle-memory neural network. Prog. Theor. Phys. **88**(5), 845–855 (1992)
50. S. Nara, P. Davis, M. Kawachi, H. Totsuji, Chaotic memory dynamics in a recurrent neural network with cycle memories embedded by pseudo-inverse method. Int. J. Bifurc. Chaos **5**(4), 1205–1212 (1995)
51. A. Potapov, M.K. Ali, Robust chaos in neural networks. Phys. Lett. A **277**(6), 310–322 (2000)

52. M. Shibasaki, M. Adachi, Response to external input of chaotic neural networks based on Newman-Watts model, in *The 2012 International Joint Conference on Neural Networks*, ed. by J. Liu, C. Alippi, B. Bouchon-Meunier, G.W. Greenwood, H.A. Abbass (Brisbane, Australia, 2012), pp. 1–7
53. I. Tsuda, Chaotic itinerancy as a dynamical basis of hermeneutics in brain and mind. World Future **32**, 167–184 (1991)
54. M. Watanabe, K. Aihara, S. Kondo, Self-organization dynamics in chaotic neural networks. Control Chaos Math. Model. **8**, 320–333 (1997)
55. X. Wang, Period-doublings to chaos in a simple neural network: an analytical proof. Complex Syst. **5**, 425–441 (1991)
56. L. Yan, H. He, P. Xiong, Algebraic condition of control for multiple time-delayed chaotic cellular neural networks, in *Fourth International Workshop on Advanced Computational Intelligence Wuhan* (Hubei, China, October 19–21, 2011), pp. 596–600
57. F. Zou, J.A. Nossek, A chaotic attractor with cellular neural networks. IEEE Trans. Circuits Syst. **38**, 811–812 (1991)
58. F. Zou, J.A. Nossek, Bifurcation and chaos in cellular neural networks. IEEE Trans. Circuits Syst.-I: Fundam. Theory Appl. **40**, 166–173 (1993)
59. J. Cao, J. Lu, Adaptive synchronization of neural networks with or without time-varying delay. Chaos **16**, 013133 (2006)
60. Q. Liu, S. Zhang, Adaptive lag synchronization of chaotic Cohen-Grossberg neural networks with discrete delays. Chaos **22**(3), 033123 (2012)
61. W. Lu, T. Chen, Synchronization of coupled connected neural networks with delays. IEEE Trans. Circuits Syst.-I: Regul. Pap. **51**(12), 2491–2503 (2004)
62. Y. Shi, P. Zhu, K. Qin, Projective synchronization of different chaotic neural networks with mixed time delays based on an integral sliding mode controller. Neurocomputing **123**, 443–449 (2014)
63. W. Yu, J. Cao, W. Lu, Synchronization control of switched linearly coupled neural networks with delay. Neurocomputing **73**(4–6), 858–866 (2010)
64. F. Yu, H. Jiang, Global exponential synchronization of fuzzy cellular neural networks with delays and reaction-diffusion terms. Neurocomputing **74**, 509–515 (2011)
65. S. Jankowski, A. Londei, C. Mazur, A. Lozowski, Synchronization phenomena in 2D chaotic CNN, in *CNNA-94 Third IEEE International Workshop on Cellular Neural Networks and their Applications*, Rome, Italy, 18–21 December (1994), pp. 339–344
66. D.J. Rijlaarsdam, V.M. Mladenov, Synchronization of chaotic cellular neural networks based on Rössler cells, in *8th Seminar on Neural Network Applications in Electrical Engineering, NEUREL-2006 Faculty of Electrical Engineering*, University of Belgrade, Serbia, 25–27 September (2006), pp. 41–43
67. J.M. Gonzáles-Miranda, *Synchronization and Control of Chaos* (Imperial College Press, London, 2004)
68. J.A.K. Suykens, M.E. Yalcin, J. Vandewalle, Coupled chaotic simulated annealing processes, in *IEEE ISCAS*, Bangkok, Thailand, May (2003), pp. 582–585
69. R. Caponetto, M. Lavorgna, L. Occhipinti, Cellular neural networks in secure transmission applications, in *CNNA96: Fourth IEEE International Workshop on Cellular Neural Networks and Their Applications*, Seville, Spain, 24–26 June (1996) pp. 411–416
70. J. Lei, Z. Lei, The chaotic cipher based on CNNs and its application in network, in *International Symposium on Intelligence Information Processing and Trusted Computing* (2011), pp. 184–187
71. Z. Yifeng, H. Zhengya, A secure communication scheme based on cellular neural network, in *IEEE International Conference on Intelligent Processing Systems* (1997), pp. 521–524
72. M. Ohta, K. Yamashita, A chaotic neural network for reducing the peak-to-average power ratio of multicarrier modulation, in *International Joint Conference on Neural Networks* (2003) pp. 864–868
73. F.R. Marotto, Snap-back repellers imply chaos in $\mathbb{R}^n$. J. Math. Anal. Appl. **63**, 199–223 (1978)

74. P. Li, Z. Li, W.A. Halang, G. Chen, Li-Yorke chaos in a spatiotemporal chaotic system. Chaos Solitons Fractals **33**(2), 335–341 (2007)
75. E. Akin, S. Kolyada, Li-Yorke sensitivity. Nonlinearity **16**, 1421–1433 (2003)
76. P. Kloeden, Z. Li, Li-Yorke chaos in higher dimensions: a review. J. Differ. Equ. Appl. **12**, 247–269 (2006)
77. W. Lin, J. Ruan, Chaotic dynamics of an integrate-and-fire circuit with periodic pulse-train input. IEEE Trans. Circuits Syst.-I: Fundam. Theory Appl. **50**, 686–693 (2003)
78. E. Kaslik, S. Balint, Complex and chaotic dynamics in a discrete-time-delayed Hopfield neural network with ring architecture. Neural Netw. **22**, 1411–1418 (2009)
79. H.N. Cheung, A. Bouzerdoum, W. Newland, Properties of shunting inhibitory cellular neural networks for colour image enhancement, in *Proceedings of 6th International Conference on Neural Information Processing Perth*, vol. 3 (1999), pp. 1219–1223
80. J. Eisenberg, W.J. Freeman, B. Burke, Hardware architecture of a neural network model simulating pattern recognition by the olfactory bulb. Neural Netw. **2**(4), 315–325 (1989)
81. W.J. Freeman, Y. Yao, B. Burke, Central pattern generating and recognizing in olfactory bulb: a correlation learning rule. Neural Netw. **1**(4), 277–288 (1988)
82. W.J. Freeman, Tutorial on neurobiology: from single neurons to brain chaos. Int. J. Bifurc. Chaos **2**(3), 451–482 (1992)
83. W.J. Freeman, J.M. Barrie, Chaotic oscillations and the genesis of meaning in cerebral cortex, in *Temporal Coding in the Brain*, ed. by G. Buzsáki, R. Llinás, W. Singer, A. Berthoz, Y. Christen (Springer, Berlin, 1994), pp. 13–37
84. C.A. Skarda, W.J. Freeman, How brains make chaos in order to make sense of the world. Behav. Brain Sci. **10**(2), 161–173 (1987)
85. C.A. Skarda, W.J. Freeman, Chaos and the new science of the brain. Concepts Neurosci. **1**(2), 275–285 (1990)
86. H.D.I. Abarbanel, N.F. Rulkov, M.M. Sushchik, Generalized synchronization of chaos: the auxiliary system approach. Phys. Rev. E **53**, 4528–4535 (1996)
87. V.S. Afraimovich, N.N. Verichev, M.I. Rabinovich, Stochastic synchronization of oscillation in dissipative systems. Radiophys. Quantum Electron. **29**, 795–803 (1986)
88. V. Afraimovich, J.R. Chazottes, A. Cordonet, Nonsmooth functions in generalized synchronization of chaos. Phys. Lett. A **283**, 109–112 (2001)
89. H. Fujisaka, T. Yamada, Stability theory of synchronized motion in coupled-oscillator systems. Prog. Theor. Phys. **69**, 32–47 (1983)
90. J.M.V. Grzybowski, E.E.N. Macau, T. Yoneyama, Isochronal synchronization of time delay and delay-coupled chaotic systems. J. Phys. A: Math. Theor. **44**, 175103 (2011)
91. B.R. Hunt, E. Ott, J.A. Yorke, Differentiable generalized synchronization of chaos. Phys. Rev. E **55**(4), 4029–4034 (1997)
92. L. Kocarev, U. Parlitz, Generalized synchronization, predictability, and equivalence of unidirectionally coupled dynamical systems. Phys. Rev. Lett. **76**(11), 1816–1819 (1996)
93. A.A. Koronovskii, O.I. Moskalenko, S.A. Shurygina, A.E. Hramov, Generalized synchronization in discrete maps. New point of view on weak and strong synchronization. Chaos Solitons Fractals **46**, 12–18 (2013)
94. L.M. Pecora, T.L. Carroll, Synchronization in chaotic systems. Phys. Rev. Lett. **64**, 821–825 (1990)
95. A.S. Pikovsky, On the interaction of strange attractors. Z. Phys. B **55**, 149–154 (1984)
96. N.F. Rulkov, M.M. Sushchik, L.S. Tsimring, H.D.I. Abarbanel, Generalized synchronization of chaos in directionally coupled chaotic systems. Phys. Rev. E **51**(2), 980–994 (1995)
97. D.V. Senthilkumar, R. Suresh, M. Lakshmanan, J. Kurths, Global generalized synchronization in networks of different time-delay systems. EPL **103**, 50010 (2013)
98. M. Breakspear, J.R. Terry, Detection and description of non-linear interdependence in normal multichannel human EEG data. Clin. Neurophysiol. **113**, 735–753 (2002)
99. L. Chen, H. Zhao, Global stability of almost periodic solution of shunting inhibitory cellular neural networks with variable coefficients. Chaos Solitons Fractals **35**, 351–357 (2008)

100. B. Liu, L. Huang, Existence and stability of almost periodic solutions for shunting inhibitory cellular neural networks with time-varying delays. Chaos Solitons Fractals **31**, 211–217 (2007)
101. L. Peng, W. Wang, Anti-periodic solutions for shunting inhibitory cellular neural networks with time-varying delays in leakage terms. Neurocomputing **111**, 27–33 (2013)
102. Q. Zhou, B. Xiao, Y. Yu, L. Peng, Existence and exponential stability of almost periodic solutions for shunting inhibitory cellular neural networks with continuously distributed delays. Chaos Solitons Fractals **34**, 860–866 (2007)
103. M.U. Akhmet, M.O. Fen, Shunting inhibitory cellular neural networks with chaotic external inputs. Chaos **23**, 023112 (2013)
104. R.L. Devaney, *An Introduction to Chaotic Dynamical Systems* (Addison-Wesley, Menlo Park, 1989)
105. M.J. Feigenbaum, Universal behavior in nonlinear systems. Los Alamos Sci. **1**/Summer, 4–27 (1980)
106. E. Sander, J.A. Yorke, Period-doubling cascades galore. Ergod. Theory Dyn. Syst. **31**, 1249–1267 (2011)
107. Y. Pomeau, P. Manneville, Intermittent transition to turbulence in dissipative dynamical systems. Commun. Math. Phys. **74**, 189–197 (1980)
108. M. Akhmet, E. Yılmaz, *Neural Networks with Discontinuous/Impact Activations* (Springer, New York, 2014)
109. M. Atencia, G. Joya, F. Sandoval, Identification of noisy dynamical systems with parameter estimation based on Hopfield neural networks. Neurocomputing **121**, 14–24 (2013)
110. X. Chen, L. Huang, Z. Guo, Finite time stability of periodic solution for Hopfield neural networks with discontinuous activations. Neurocomputing **103**, 43–49 (2013)
111. M.A. Cohen, S. Grossberg, Absolute stability of global pattern formation and parallel memory storage by competitive neural networks. IEEE Trans. Syst. Man Cybern. SMC **13**, 815–826 (1983)
112. T. Liang, Y. Yang, Y. Liu, L. Li, Existence and global exponential stability of almost periodic solutions to Cohen-Grossberg neural networks with distributed delays on time scales. Neurocomputing **123**, 207–215 (2014)
113. J.-L. Wang, H.-N. Wun, L. Guo, Stability analysis of reaction-diffusion Cohen-Grossberg neural networks under impulsive control. Neurocomputing **106**, 21–30 (2013)
114. C. Zhou, H. Zhang, H. Zhang, C. Dang, Global exponential stability of impulsive fuzzy Cohen-Grossberg neural networks with mixed delays and reaction-diffusion terms. Neurocomputing **91**, 67–76 (2012)
115. S. Arik, An analysis of exponential stability of delayed neural networks with time varying delays. Neural Netw. **17**, 1027–1031 (2004)
116. Z. Wang, H. Zhang, B. Jiang, LMI-based approach for global asymptotic stability analysis of recurrent neural networks with various delays and structures. IEEE Trans. Neural Netw. **22**, 1032–1045 (2011)
117. Z. Wang, H. Zhang, Synchronization stability in complex interconnected neural networks with nonsymmetric coupling. Neurocomputing **108**, 84–92 (2013)
118. E. Yucel, S. Arik, New exponential stability results for delayed neural networks with time varying delays. Phys. D **191**, 314–322 (2004)
119. J. Hale, H. Koçak, *Dynamics and Bifurcations* (Springer, New York, 1991)
120. L.O. Chua, Cellular neural networks: theory. IEEE Trans. Circuits Syst. **35**, 1257–1272 (1988)
121. Y. Li, J. Shu, Anti-periodic solutions to impulsive shunting inhibitory cellular neural networks with distributed delays on time scales. Commun. Nonlinear Sci. Numer. Simul. **16**, 3326–3336 (2011)
122. S. Ahmad, I.M. Stamova, Global exponential stability for impulsive cellular neural networks with time-varying delays. Nonlinear Anal. Theory, Methods Appl **69**, 786–795 (2008)
123. K. Li, X. Zhang, Z. Li, Global exponential stability of impulsive cellular neural networks with time-varying and distributed delay. Chaos Solitons Fractals **41**, 1427–1434 (2009)
124. L. Pan, J. Cao, Anti-periodic solution for delayed cellular neural networks with impulsive effects. Nonlinear Anal. Real World Appl. **12**, 3014–3027 (2011)

125. X. Song, X. Xin, W. Huang, Exponential stability of delayed and impulsive cellular neural networks with partially Lipschitz continuous activation functions. Neural Netw. **29–30**, 80–90 (2012)
126. I.M. Stamova, R. Ilarionov, On global exponential stability for impulsive cellular neural networks with time-varying delays. Comput. Math. Appl. **59**, 3508–3515 (2010)
127. Q. Wang, X. Liu, Exponential stability of impulsive cellular neural networks with time delay via Lyapunov functionals. Appl. Math. Comput. **194**, 186–198 (2007)
128. B. Wu, Y. Liu, J. Lu, New results on global exponential stability for impulsive cellular neural networks with any bounded time-varying delays. Math. Comput. Model. **55**, 837–843 (2012)
129. Y. Yang, J. Cao, Stability and periodicity in delayed cellular neural networks with impulsive effects. Nonlinear Anal. Real World Appl. **8**, 362–374 (2007)
130. M. Akhmet, *Nonlinear Hybrid Continuous/Discrete-Time Models* (Atlantis Press, Paris, 2011)
131. M. Akhmet, *Principles of Discontinuous Dynamical Systems* (Springer, New York, 2010)
132. M. Akhmet, M.O. Fen, Chaotification of impulsive systems by perturbations. Int. J. Bifurc. Chaos **24**, 1450078 (2014)
133. Y. Yao, W.J. Freeman, Model of biological pattern recognition with spatially chaotic dynamics. Neural Netw. **3**(2), 153–170 (1990)
134. M.U. Akhmet, M.O. Fen, Entrainment by chaos. J. Nonlinear Sci. **24**, 411–439 (2014)
135. E.M. Izhikevich, Weakly connected quasi-periodic oscillators, FM interactions, and multiplexing in the brain. SIAM J. Appl. Math. **59**, 2193–2223 (1999)
136. B.L. McNaughton, C.A. Barnes, P. Andersen, Synaptic efficacy and EPSP summation in granule cells of rat fascia dentata studied in vitro. J. Neurophysiol. **46**, 952–966 (1981)
137. M. Abeles, Neural codes for higher brain functions, in *Information Processing by the Brain*, ed. by H.J. Markowitsch (Hans Huber Publishers, Toronto, 1988)
138. A.M. Samoilenko, N.A. Perestyuk, *Impulsive Differential Equations* (World Scientific, Singapore, 1995)
139. Y. Huang, X.S. Yang, Hyperchaos and bifurcation in a new class of four-dimensional Hopfield neural networks. Neurocomputing **69**, 1787–1795 (2006)
140. W.-Z. Huang, Y. Huang, Chaos, bifurcation and robustness of a class of Hopfield neural networks. Int. J. Bifurc. Chaos **21**, 885–895 (2011)
141. Q. Li, X.-S. Yang, F. Yang, Hyperchaos in Hopfield-type neural networks. Neurocomputing **67**, 275–280 (2005)
142. P.C. Rech, Chaos and hyperchaos in a Hopfield neural network. Neurocomputing **74**, 3361–3364 (2011)
143. X.-S. Yang, Q. Yuan, Chaos and transient chaos in simple Hopfield neural networks. Neurocomputing **69**, 232–241 (2005)
144. Q. Yuan, Q. Li, X.-S. Yang, Horseshoe chaos in a class of simple Hopfield neural networks. Chaos Solitons Fractals **39**, 1522–1529 (2009)
145. M. Cartwright, J. Littlewood, On nonlinear differential equations of the second order I: the equation $\ddot{y} - k(1 - y^2)'y + y = bkcos(\lambda t + a)$, k large. J. Lond. Math. Soc. **20**, 180–189 (1945)
146. M. Levi, *Qualitative Analysis of the Periodically Forced Relaxation Oscillations* (Memoirs of the American Mathematical Society, United States of America, 1981)
147. N. Levinson, A second order differential equation with singular solutions. Ann. Math. **50**, 127–153 (1949)
148. Y. Ueda, Randomly transitional phenomena in the system governed by Duffing's equation. J. Stat. Phys. **20**, 181–196 (1979)
149. Y. Ueda, Steady motions exhibited by Duffing's equation: a picture book of regular and chaotic motions, in *New Approaches to Nonlinear Problems in Dynamics*, ed. by P.J. Holmes (SIAM, Philadelphia, 1980)
150. V.V. Nemytskii, V.V. Stepanov, *Qualitative Theory of Differential Equations* (Princeton University Press, Princeton, 1960)
151. L. Shilnikov, Bifurcations and strange attractors, in *Proceedings of the International Congress of Mathematicians*, vol. III (Higher Education Press, Beijing, 2002), pp. 349–372

152. S. Smale, Differentiable dynamical systems. Bull. Am. Math. Soc. **73**, 747–817 (1967)
153. K. Pyragas, Continuous control of chaos by self-controlling feedback. Phys. Rev. A **170**, 421–428 (1992)
154. H. Alonso, T. Mendonça, P. Rocha, Hopfield neural networks for on-line parameter estimation. Neural Netw. **22**, 450–462 (2009)
155. A.C. Mathias, P.C. Rech, Hopfield neural network: the hyperbolic tangent and the piecewise-linear activation functions. Neural Netw. **34**, 42–45 (2012)
156. J. Peng, Z.-B. Xu, H. Qiao, B. Zhang, A critical analysis on global convergence of Hopfield-type neural networks. IEEE Trans. Circuits Syst.-I: Regul. Pap. **52**, 804–814 (2005)
157. X. Huang, Z. Zhao, Z. Wang, Y. Li, Chaos and hyperchaos in fractional-order cellular neural networks. Neurocomputing **94**, 13–21 (2012)
158. B. Kosko, Bidirectional associative memories. IEEE Trans. Syst. Man Cybern. **18**, 49–60 (1988)
159. J. Xiao, Z. Zeng, A. Wu, New criteria for exponential stability of delayed recurrent neural networks. Neurocomputing **134**, 182–188 (2014)
160. Z. Zhang, K. Liu, Y. Yang, New LMI-based condition on global asymptotic stability concerning BAM neural networks of neutral type. Neurocomputing **81**, 24–32 (2012)
161. S. Zhu, Y. Shen, Robustness analysis for connection weight matrix of global exponential stability recurrent neural networks. Neurocomputing **101**, 370–374 (2013)
162. F.C. Hoppensteadt, E.M. Izhikevich, *Weakly Connected Neural Networks* (Springer, New York, 1997)
163. F. Pasemann, M. Hild, K. Zahedi, SO(2)-networks as neural oscillators. Comput. Methods Neural Model. Lect. Notes Comput. Sci. **2686**, 144–151 (2003)
164. C.K. Ahn, M.K. Song, New sets of criteria for exponential $L_2 - L_\infty$ stability of Takagi-Sugeno fuzzy systems combined with Hopfield neural networks. Int. J. Innov. Comput. Inf. Control **9**, 2979–2986 (2013)
165. S. Arik, A new condition for robust stability of uncertain neural networks with time delays. Neurocomputing **128**, 476–482 (2014)
166. X. Li, J. Jia, Global robust stability analysis for BAM neural networks with time-varying delays. Neurocomputing **120**, 499–503 (2013)
167. F. Li, Global stability at a limit cycle of switched Boolean networks under arbitrary switching signals. Neurocomputing **133**, 63–66 (2014)
168. M. Pulido, O. Castillo, P. Melin, Genetic optimization of ensemble neural networks for complex time series prediction of the Mexican exchange. Int. J. Innov. Comput. Inf. Control **9**, 4151–4166 (2013)
169. X. Su, Z. Li, Y. Feng, L. Wu, New global exponential stability criteria for interval-delayed neural networks. Proc. Inst. Mech. Eng. Part I: J. Syst. Control Eng. **225**, 125–136 (2011)
170. X. Xu, J. Zhang, J. Shi, Exponential stability of complex-valued neural networks with mixed delays. Neurocomputing **128**, 483–490 (2014)
171. R. Yang, Z. Zhang, P. Shi, Exponential stability on stochastic neural networks with discrete interval and distributed delays. IEEE Trans. Neural Netw. **21**, 169–175 (2010)
172. E. Sander, J.A. Yorke, Connecting period-doubling cascades to chaos. Int. J. Bifurc. Chaos **22**, 1–16 (2012)
173. M. Farkas, *Periodic Motions* (Springer, New York, 2010)
174. E. Ott, C. Grebogi, J.A. Yorke, Controlling chaos. Phys. Rev. Lett. **64**, 1196–1199 (1990)
175. H.G. Schuster, *Handbook of Chaos Control* (Wiley-VCH, Weinheim, 1999)
176. H. Haken, *Advanced Synergetics: Instability Hierarchies of Self-Organizing Systems and Devices* (Springer, Berlin, 1983)
177. G. Nicolis, I. Prigogine, *Exploring Complexity: An Introduction* (W.H. Freeman, New York, 1989)
178. A. Damasio, Synchronous activation in multiple cortical areas: a mechanism for recall. Sem. Neurosci. **2**, 287–296 (1990)
179. M.M. Mesulam, Large-scale neurocognitive networks and distributed processing for attention, language, and memory. Ann. Neurol. **28**, 597–613 (1990)

180. F. Varela, J.-P. Lachaux, E. Rodriguez, J. Martinerie, The brainweb: phase synchronization and large-scale integration. Nat. Rev. Neurosci. **2**, 229–239 (2001)
181. M.U. Akhmet, M.O. Fen, Attraction of Li-Yorke chaos by retarded SICNNs. Neurocomputing **147**, 330–342 (2015)
182. P. Balasubramaniam, P. Muthukumar, Synchronization of chaotic systems using feedback controller: an application to Diffie-Hellman key exchange protocol and ElGamal public key cryptosystem. J. Egypt. Math. Soc. **22**, 365–372 (2014)
183. A. Khadra, X. Liu, X. Shen, Application of impulsive synchronization to communication security. IEEE Trans. Circuits Syst.-I, Fundam. Theory Appl. **50**, 341–351 (2003)
184. P. Muthukumar, P. Balasubramaniam, Feedback synchronization of the fractional order reverse butterfly-shaped chaotic system and its application to digital cryptography. Nonlinear Dyn. **74**, 1169–1181 (2013)
185. P. Muthukumar, P. Balasubramaniam, K. Ratnavelu, Synchronization of a novel fractional order stretch-twist-fold (STF) flow chaotic system and its application to a new authenticated encryption scheme (AES). Nonlinear Dyn. **77**, 1547–1559 (2014)
186. T. Yang, L.O. Chua, Impulsive control and synchronization of nonlinear dynamical systems and application to secure communication. Int. J. Bifurc. Chaos **7**, 645–664 (1997)
187. K. M. Akhmet, M.O. Fen, Impulsive SICNNs with chaotic postsynaptic currents (submitted)
188. M. Akhmet, M.O. Fen, Generation of cyclic/toroidal chaos by Hopfield neural networks. Neurocomputing **145**, 230–239 (2014)
189. K. Kaneko, I. Tsuda, *Complex Systems: Chaos and Beyond, A Constructive Approach with Applications in Life Sciences* (Springer, Berlin, 2000)

Chapter 9
The Prevalence of Weather Unpredictability

It is found that Lorenz systems can be unidirectionally coupled such that the chaos expands from the drive system. This is true if the response system is not chaotic, but admits a global attractor—an equilibrium or a cycle. Lorenz in his genius study explains the unpredictability for a regional weather and considered the phenomenon also globally. We make an effort to provide a mechanism which may help understand the lack of weather forecasting better on the basis of Lorenz systems. Our suggestions concern an application of meteorological regional models.

9.1 Introduction

Significant investigations for weather forecasting in the history started with the numerical studies of Richardson [1]. The first successful numerical weather forecast was announced by Charney et al. [2]. They proposed a method for the numerical solution of barotropic vorticity equation over a limited area of the Earth's surface. Although this oversimplified model was unable to describe the baroclinic instability process, one can say that their study gave a light for the future applied computer modelings for weather forecasting. A fairly realistic atmospheric circulation was deduced in the study [3], which deals with the essential aspect of nonlinear interactions through the analysis of a simple set of equations, that is, low-order models.

In the zero-dimensional modelings of the atmosphere, one essentially attempts to follow the evolution of global surface-air temperature as a result of changes in global radiative balance [4–6], where dimension refers to the number of independent space variables that are used to describe the model domain, that is, to physical-space dimensions [7]. On the other hand, there are two kinds of one-dimensional atmospheric models, for which the single spatial variable is latitude or height. The former are called energy-balance models [8, 9], while the models in which the details of radiative equilibrium are investigated with respect to a height coordinate are named as radiative-convective models [10–12], since convection plays a key role in vertical heat

M. Akhmet and M.O. Fen, *Replication of Chaos in Neural Networks, Economics and Physics*, Nonlinear Physical Science,
DOI 10.1007/978-3-662-47500-3_9

transfer. Considering the third space coordinate that is not explicitly included, two-dimensional atmospheric models can also be characterized mainly in two classes—models that resolve explicitly two horizontal coordinates [13–16] or explicitly a meridional coordinate and height [17–19]. Another class of two-dimensional models can be considered as the extension of energy-balance models to resolve zonal and meridional surface features [20–22]. Nowadays, the modeling of the atmosphere continues to be a remarkable research area for scientists [23–25].

In his famous study, to investigate the dynamics of the atmosphere, Lorenz [26] built a mathematical model consisting of a system of three differential equations in the following form

$$\begin{aligned}\frac{dx_1}{dt} &= -\sigma x_1 + \sigma x_2\\ \frac{dx_2}{dt} &= -x_1x_3 + rx_1 - x_2\\ \frac{dx_3}{dt} &= x_1x_2 - bx_3,\end{aligned} \tag{9.1.1}$$

where σ, r, and b are constants.

System (9.1.1) is a simplification of a model, derived by Saltzman [27], to study finite amplitude convection. The studies of Saltzman originate from the Rayleigh–Bénard convection which describes heat flow through a fluid, like air or water. In this modeling, one considers a fluid between two horizontal plates where the gravity is assumed to be in the downward direction and the temperature of the lower plate is maintained at a higher value than the temperature of the upper one. Rayleigh [28] found that if the temperature difference is kept at a constant value, then the system possesses a steady-state solution in which there is no motion and convection should take place if this solution becomes unstable. In other words, depending on the temperature difference between the plates, heat can be transferred by conduction or by convection. Assuming variations in only $x_1 - x_3$ plane, Saltzman [27] considered the equations

$$\begin{aligned}&\frac{\partial}{\partial t}\nabla^2\psi + \frac{\partial\left(\psi, \nabla^2\psi\right)}{\partial(x_1, x_3)} - g\varepsilon\frac{\partial\theta}{\partial x_1} - \nu\nabla^4\psi = 0\\ &\frac{\partial\theta}{\partial t} + \frac{\partial\left(\psi, \theta\right)}{\partial(x_1, x_3)} - \frac{\Delta T_0}{H}\frac{\partial\psi}{\partial x_1} - \kappa\nabla^2\theta = 0,\end{aligned} \tag{9.1.2}$$

where ψ is a stream function for the two-dimensional motion, θ is the departure of temperature from that occurring in the state of no convection and the constants ΔT_0, H, g, ε, ν, and κ denote, respectively, the temperature contrast between the lower and upper boundaries of the fluid, the height of the fluid under consideration, the acceleration of gravity, the coefficient of thermal expansion, the kinematic viscosity, and the thermal conductivity [26, 27]. In his study, Saltzman [27] achieved an infinite system by means of applying Fourier series methods to system (9.1.2), and then used the simplification procedure proposed by Lorenz [29] to obtain a system with finite

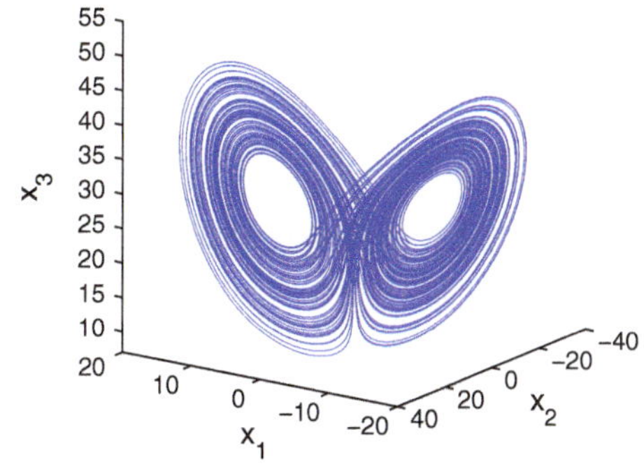

Fig. 9.1 The chaotic trajectory of system (9.1.1) with $\sigma = 10$, $r = 28$ and $b = 8/3$

number of terms. Lorenz [26] sets all but three Fourier coefficients equal to zero and as a consequence attained system (9.1.1), which describes an idealized model of a fluid.

In system (9.1.1), the variable x_1 is proportional to the circulatory fluid flow velocity, while the variable x_2 is proportional to the temperature difference between the ascending and descending currents. Positive x_1 values indicate clockwise rotations of the fluid and negative x_1 values mean counterclockwise motions. The variable x_3, on the other hand, is proportional to the distortion of the vertical temperature profile from linearity, a positive value indicating that the strongest gradients occur near the boundaries. The parameters σ and r are called the Prandtl and Rayleigh numbers, respectively [26, 30, 31].

The dynamics of the Lorenz system (9.1.1) is very rich. For instance, with different values of the parameters σ, r, and b, the system can exhibit stable periodic orbits, homoclinic explosions, period-doubling bifurcations, and chaotic attractors [31]. Figure 9.1 depicts the chaotic trajectory of system (9.1.1) with $\sigma = 10$, $r = 28$, and $b = 8/3$ corresponding to the initial data $x_1(0) = -12.89$, $x_2(0) = -8.91$, $x_3(0) = 36.59$.

According to Lorenz [32], the butterfly effect is possible and he understands this phenomenon as a positive answer to the following question: "Does the flap of a butterfly's wings in Brazil set off a tornado in Texas?" *From this question one can immediately decide that the butterfly effect is a global phenomenon, and consequently, the underlying mathematics of the globalization has to be investigated. This is what we do in the present chapter.* Lorenz was the first who discovered sensitivity with the aid of system (9.1.1) and then made the conclusion on the effect. Nowadays, there is an agreement that the butterfly effect exists, if we mean *sensitivity=unpredictability*, but the tornado caused by a butterfly's flap is questionable. It seems that Lorenz himself believed that sensitivity discovered in his equation is a strong indicator of the butterfly effect in its original meteorological sense. Possibly his intuition is based on the idea that the system of ordinary differential equations is derived from a system of partial differential equations. There should be a deeper interpretation for the effect of chaotic dynamics in the three-dimensional system on the infinite-dimensional one. We also believe that the opinion of Lorenz, who considered his results as an evidence of the meteorological butterfly effect (tornado), is very reasonable. Moreover, his claim has to be considered as a challenging problem for mathematicians. Apparently, we are only in the beginning of the answer of the Lorenz's question, and if one thinks positively on the subject, then by our opinion several next questions emerge.

The first one is whether sensitivity in meteorological models is a reflection of the butterfly effect. Definitely, this question needs a thorough investigation. Possibly it requests a deep analysis on the basis of ordinary and partial differential equations. The problem is not solved in this chapter at all. We axiomatize somehow the state assuming that the butterfly effect is sensitivity in mathematical sense or, more generally, chaos. We can also reduce the question by considering the problem of unpredictability through sensitivity. Consequently, the following questions are reasonable: Can one explain the *global* unpredictability of weather by applying models similar to the Lorenz system? How Lorenz systems can be utilized for a *global* description of the weather? These and adjoint questions are on the agenda.

The physical properties of the atmosphere are not the same throughout the Earth. The tropical atmosphere possesses considerably a different behavior from those in the temperate and polar latitudes, as if it was a different fluid [32]. Taking inspiration from its multifaceted structure, we propose to consider that the atmosphere divided into subregions such that the dynamic properties of each region differ considerably from the others. In this case, we suppose that the dynamics of each subregion of the atmosphere subjects to its own Lorenz system. That is, for different subregions, the coefficients of the corresponding Lorenz system are different. Since for some parameter values, chaos can take place in the Lorenz system and for some not; such chaotic or non-chaotic motions should have prolonged forever, conflicting the realistic dynamics of the atmosphere, where global unpredictability is present. To extend our attitude for the butterfly effect, we propose that instability, which may occur in a subregion, can be imported to neighbor subregions of the atmosphere, such that chaos occurs not only endogenously, but also exogenously. In other words, exterior perturbations influencing a part of the atmosphere may cause a chaotic behavior to occur in that region. In addition to this, we suppose that these perturbations most probably originate through the neighboring regions within the atmosphere, and the dynamics of the connected Lorenz systems help analyze this. We understand that our results give a light only on one among many questions how the weather processes have to be described through mathematical models. For example, in our discussion, we consider unidirectional connection of subsystems which is not true in reality. The question whether the overlapping of two chaotic dynamics may produce regularity can also be considered in future investigations. We guess that it is not possible, but an analysis has to be made. One can also investigate synchronization of chaos in neighbor regions, etc. Nevertheless, we issue from the point of view that what we have done is a one more small step in the mathematical approach to the complexity of the weather. This is not a modeling of the atmosphere, but rather an effort to explain how the weather unpredictability can be arranged over the Earth on the basis of the Lorenz's meteorological model. In fact, this is also true for other meteorological models, since mathematical properties of stability, attraction, and chaos attractors are common for all models.

Although we are in the beginning of the theory, we suppose that our rigorously approved idea for the global unpredictability and for the extension of chaotic behavior from one Lorenz system to another will give a light for the justification of the erratic behavior observed in dynamical systems of meteorology. Generally, analysis of chaotic dynamics in atmospheric models is rather numerical [33–37] or depends on the observation of time-series [38, 39].

There are many published papers which have results about chaos considering first of all its mathematical meaning. This is true either for differential equations [40, 41] or data analysis [42]. Apparently, there are still few articles with meteorological interpretation of chaos ingredients. In this chapter, the main attention is given to global chaos over the Earth in its mathematical sense. Moreover, what is not less important is the discussion of the chaos presence as unpredictability of the weather, which is the reflection of sensitivity in the chaos definition. Thus, not only the readers who are interested in the strong mathematical basics of global chaos, but also the ones who are looking for an evidence of unpredictability of weather as a global phenomenon may find their interests in the present chapter.

The appearance of chaos in differential/discrete equations may be either endogenous or exogenous. As the first type of chaos birth, one can take into account the irregular motions that occur in Lorenz, Rössler, Chua systems, the logistic map, Duffing and Van der Pol equations [43–48]. To indicate the endogenous irregularity, we use (i) ingredients of Devaney and Li-Yorke chaos, (ii) period-doubling route to chaos, (iii) intermittency, and (iv) positive Lyapunov exponents. Symbolic dynamics and Smale horseshoes were widely used for that purpose [41, 44, 46, 49–53]. While the endogenous chaos production is widespread and historically unique, the exogenous chaos as generated by irregular perturbations has not been intensively investigated yet. In the present chapter, we will appeal to endogenous chaos, but mostly to exogenous chaos.

The assumptions used in this chapter are as follows:

(i) The whole atmosphere of the Earth is partitioned in a finite number of subregions.
(ii) In each of the subregions, the dynamics of the weather is governed by the Lorenz system with certain coefficients.
(iii) There are subregions for which the corresponding Lorenz systems admit a chaos with main ingredient as sensitivity, which means unpredictability of weather in the meteorological sense and there are subregions, where Lorenz systems are non-chaotic and with equilibriums or cycles as global attractors.
(iv) The Lorenz systems are connected unidirectionally.

The main goal of this chapter is to show that under conditions (i)–(iv) not only regions mentioned in (iii) are subdued to unpredictability, but the Earth's global weather also is unpredictable, that is in each of its subregions.

The principal novelty of our investigation is that we create exogenous chaotic perturbations by means of the solutions of a chaotic Lorenz system, plug it into a regular Lorenz system, and find that chaos is inherited by the solutions of the latter.

Such an approach has been widely used for differential equations before, but for regular disturbance functions. That is, it has been shown that an (almost) periodic perturbation function implies the existence of an (almost) periodic solution of the system. While the literature on chaos synchronization [54–61] has also produced methods of generating chaos in a system by plugging in terms that are chaotic, it relies on the asymptotic convergence between the chaotic exogenous terms and the solution of the response system for the proof of chaos creation. Instead, we provide a direct verification of the ingredients of chaos for the perturbed system [62–70]. Moreover, in Sect. 9.5 we represent the appearance of cyclic chaos, which cannot be reduced to generalized synchronization. Very interesting examples of applications of discrete dynamics to continuous chaos analysis were provided in the papers [71–74]. In these studies, the general technique of dynamical synthesis [71] was developed, and this technique was used in the paper [65].

9.2 Coupling Mechanism for Unpredictability

In order to describe the chaos expansion, let us localize the global process by taking into account only two adjacent subregions of the atmosphere, labeled A and B. In the beginning, the subregion A is assumed to be chaotic, while the subregion B is non-chaotic. By the phrase "chaotic subregion," we mean that the coefficients of the corresponding Lorenz system are such that the system possesses a chaotic attractor. In a similar way, one should understand from the phrase "non-chaotic subregion" that the corresponding Lorenz system does not exhibit chaotic motions such that it admits a global asymptotically stable equilibrium or a globally attracting limit cycle.

In our discussion, we couple the Lorenz systems corresponding to subregions A and B unidirectionally such that the existing chaos propagates from one to another. We suppose that the dynamics of the subregion A is described by the Lorenz system (9.1.1) in which the coefficients σ, r, and b are chosen in such a way that the system is chaotic. In addition to this, we consider the Lorenz system

$$\begin{aligned} \frac{du_1}{dt} &= -\overline{\sigma}u_1 + \overline{\sigma}u_2 \\ \frac{du_2}{dt} &= -u_1u_3 + \overline{r}u_1 - u_2 \\ \frac{du_3}{dt} &= u_1u_2 - \overline{b}u_3, \end{aligned} \tag{9.2.3}$$

where the parameters $\overline{\sigma}$, $\overline{r}$, and $\overline{b}$ are such that the system is non-chaotic. To actualize the chaos transmission process, we perturb system (9.2.3) with the solutions of (9.1.1) to set up the system

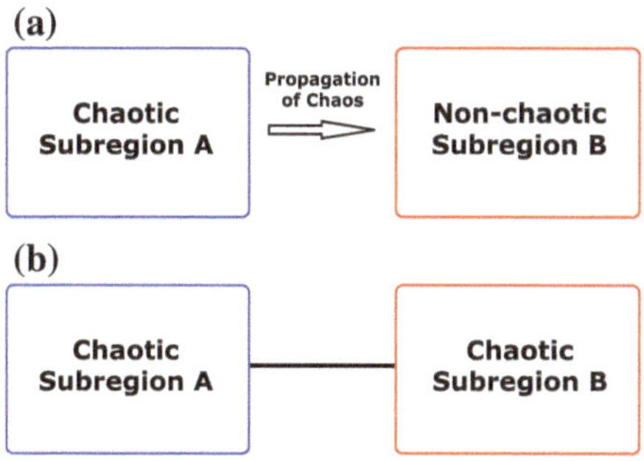

Fig. 9.2 Schematic representation of the chaos extension mechanism. **a** The dynamics during the transmission of chaos, **b** The state of the weather after the transmission of chaos. The unidirectional coupling of Lorenz systems gives rise to the chaotification of the initially non-chaotic system such that as a result the unpredictability has been propagated from subregion A to subregion B

$$\begin{aligned}
\frac{dy_1}{dt} &= -\overline{\sigma} y_1 + \overline{\sigma} y_2 + g_1(x(t)) \\
\frac{dy_2}{dt} &= -y_1 y_3 + \overline{r} y_1 - y_2 + g_2(x(t)) \\
\frac{dy_3}{dt} &= y_1 y_2 - \overline{b} y_3 + g_3(x(t)),
\end{aligned} \tag{9.2.4}$$

where $x(t) = (x_1(t), x_2(t), x_3(t))$. The conditions on the continuous function $g(x) = (g_1(x), g_2(x), g_3(x))$ will be mentioned in Sect. 9.8. System (9.2.4) represents the dynamics of the subregion B after the transmission of chaos, whereas system (9.2.3) represents the dynamics before the process is carried out. The propagation mechanism is represented schematically in Fig. 9.2. By chaos propagation, we mean the process of unidirectional coupling of Lorenz systems. Figure 9.2a illustrates the dynamics during the transmission of chaos. After the transmission of unpredictability is achieved, the dynamics of both subregions, A and B, exhibit chaotic behavior as shown in Fig. 9.2b.

It is worth noting that the mentioned local process can be maintained by considering more subregions, whose dynamics are also described by Lorenz systems.

9.3 Extension of Lorenz Unpredictability

To demonstrate the extension of sensitivity, we will show numerically on an example that the divergence of two nearby solutions in the driving chaotic Lorenz system (9.1.1) leads to the presence of the same feature in system (9.2.4). Additionally, we will apply our method on a third Lorenz system, in order to show the maintainability of the process. The mathematical description of sensitivity and a theoretical proof for its extension are presented in Sect. 9.8.

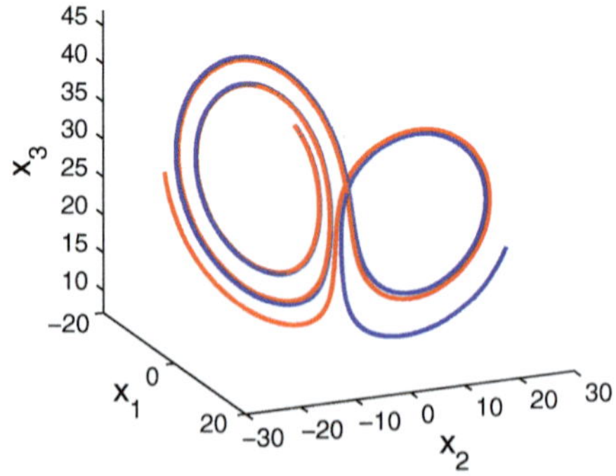

Fig. 9.3 Sensitivity in the Lorenz system (9.3.5). The figure represents the divergence of two initially nearby trajectories of system (9.3.5), which are shown in *blue* and *red* colors

Let us begin by considering the chaotic system [26]

$$\begin{aligned}\frac{dx_1}{dt} &= -10x_1 + 10x_2\\ \frac{dx_2}{dt} &= -x_1x_3 + 28x_1 - x_2\\ \frac{dx_3}{dt} &= x_1x_2 - (8/3)x_3,\end{aligned} \tag{9.3.5}$$

which is in the form of (9.1.1) with the coefficients $\sigma = 10$, $r = 28$ and $b = 8/3$. To observe sensitivity of system (9.3.5), in Fig. 9.3, we represent two initially nearby trajectories for $t \in [0, 3]$, corresponding to the initial data $x_1(0) = -8.57$, $x_2(0) = -2.39$, $x_3(0) = 33.08$, and $x_1(0) = -8.53$, $x_2(0) = -2.47$, $x_3(0) = 33.05$ which are shown in blue and red colors, respectively.

Next, we take into account the Lorenz system

$$\begin{aligned}\frac{du_1}{dt} &= -10u_1 + 10u_2\\ \frac{du_2}{dt} &= -u_1u_3 + 0.35u_1 - u_2\\ \frac{du_3}{dt} &= u_1u_2 - (8/3)u_3,\end{aligned} \tag{9.3.6}$$

which possesses a stable equilibrium point [31]. We perturb (9.3.6) with the solutions of (9.3.5) to set up the following system:

$$\begin{aligned}\frac{dy_1}{dt} &= -10y_1 + 10y_2 + 0.3x_1(t) - 0.15\sin(x_1(t))\\ \frac{dy_2}{dt} &= -y_1y_3 + 0.35y_1 - y_2 + 1.6x_2(t)\\ \frac{dy_3}{dt} &= y_1y_2 - (8/3)y_3 + 0.1\tan(x_3(t)/65).\end{aligned} \tag{9.3.7}$$

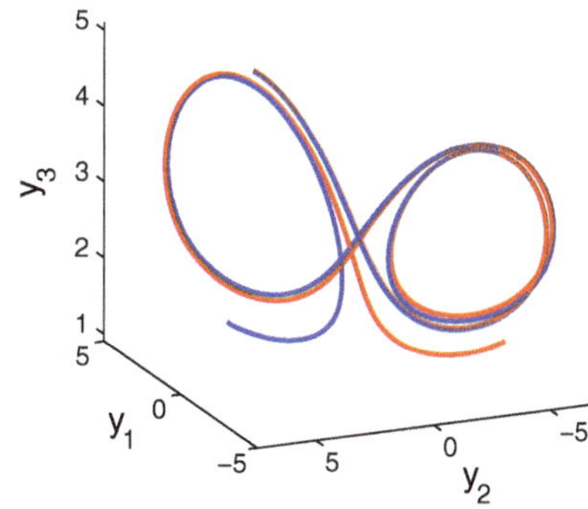

Fig. 9.4 Extension of sensitivity in system (9.3.7). The divergence of the initially nearby solutions of system (9.3.7) is observable in the figure

System (9.3.7) is in the form of (9.2.4) with $\overline{\sigma} = 10$, $\overline{r} = 0.35$, $\overline{b} = 8/3$, $g_1(x(t)) = 0.3x_1(t) - 0.15\sin(x_1(t))$, $g_2(x(t)) = 1.6x_2(t)$, and $g_3(x(t)) = 0.1\tan(x_3(t)/65)$.

The applied perturbations may not be in accordance with realistic air flows in the atmosphere. However, the exemplification reveals the propagation of unpredictability and indicate the possibility for the usage of different types of perturbations in the systems. We make use of "toy" perturbations because of the lack of preexisting ones, which should be found through experimental investigations.

To reveal numerically the extension of sensitivity in system (9.3.7), we represent in Fig. 9.4 the projections of two initially nearby trajectories of the unidirectionally coupled system (9.3.5) + (9.3.7) on the $y_1 - y_2 - y_3$ space for $t \in [0, 3]$. The trajectory with blue color corresponds to the initial data $x_1(0) = -8.57$, $x_2(0) = -2.39$, $x_3(0) = 33.08$, $y_1(0) = 3.91$, $y_2(0) = 1.86$, $y_3(0) = 4.39$, and the one with red color corresponds to the initial data $x_1(0) = -8.53$, $x_2(0) = -2.47$, $x_3(0) = 33.05$, $y_1(0) = 3.91$, $y_2(0) = 1.87$, $y_3(0) = 4.40$. The divergence of the initially nearby trajectories seen in Fig. 9.4 manifests the sensitivity feature in system (9.3.7).

Now, we consider the system

$$\begin{aligned}\frac{dz_1}{dt} &= -10z_1 + 10z_2 + 12y_1(t)\\ \frac{dz_2}{dt} &= -z_1z_3 + 0.1z_1 - z_2 + 20[y_2(t) + 2\arctan(y_2(t)/5)]\\ \frac{dz_3}{dt} &= z_1z_2 - (8/3)z_3 - 8y_3(t).\end{aligned} \tag{9.3.8}$$

System (9.3.8) is also in the form of (9.2.4), but this time the perturbations $h_1(y(t)) = 12y_1(t)$, $h_2(y(t)) = 20[y_2(t) + 2\arctan(y_2(t)/5)]$ and $h_3(y(t)) = -8y_3(t)$ are provided by the solutions of system (9.3.7).

Figure 9.5 shows the projections of two trajectories, which are initially nearby, of the 9-dimensional system (9.3.5) + (9.3.7) + (9.3.8) on the $z_1 - z_2 - z_3$ space. The trajectory with blue color has the initial data $x_1(0) = -8.57$, $x_2(0) = -2.39$, $x_3(0) = 33.08$, $y_1(0) = 3.91$, $y_2(0) = 1.86$, $y_3(0) = 4.39$, $z_1(0) = 6.92$, $z_2(0) = -6.18$, $z_3(0) = 10.48$, whereas the one with red color has the initial data

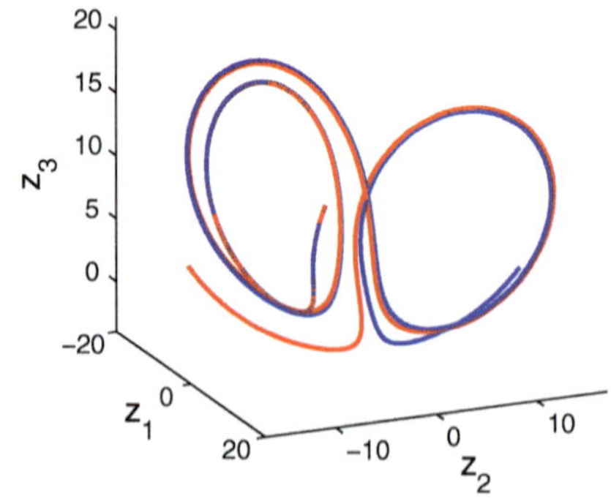

Fig. 9.5 Extension of sensitivity in system (9.3.8)

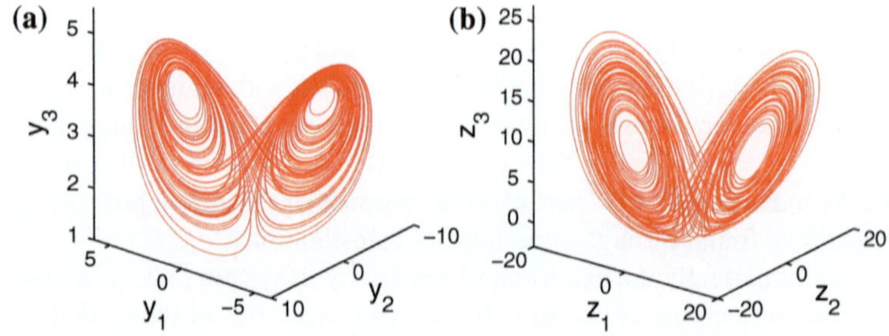

Fig. 9.6 3-dimensional projections of the chaotic trajectory of system (9.3.5)+(9.3.7)+(9.3.8). **a** The projection on the $y_1 - y_2 - y_3$ space. **b** The projection on the $z_1 - z_2 - z_3$ space. The picture represented in (**a**) illustrates the chaotic trajectory of the perturbed Lorenz system (9.3.7), while the picture in (**b**) corresponds to the perturbed Lorenz system (9.3.8). The projection on the $x_1 - x_2 - x_3$ space is the classical Lorenz attractor corresponding to system (9.3.5). The pictures represented in (**a**) and (**b**) confirm both the extension of chaos and the existence of a chaotic attractor in the 9-dimensional phase space

$x_1(0) = -8.53$, $x_2(0) = -2.47$, $x_3(0) = 33.05$, $y_1(0) = 3.91$, $y_2(0) = 1.87$, $y_3(0) = 4.40$, $z_1(0) = 6.89$, $z_2(0) = -6.18$, $z_3(0) = 10.47$. The utilized time interval is the same with Figs. 9.3 and 9.4. It is seen in Fig. 9.5 that although the depicted trajectories are initially nearby, later they diverge from each other. In other words, it is demonstrated that the sensitivity of system (9.3.7) is extended to (9.3.8). Moreover, one can conclude from the simulations that the system (9.3.5)+(9.3.7)+(9.3.8) is also sensitive.

In the next simulation, the trajectory of system (9.3.5)+(9.3.7)+(9.3.8) with $x_1(0) = -12.89$, $x_2(0) = -8.91$, $x_3(0) = 36.59$, $y_1(0) = -4.21$, $y_2(0) = -4.96$, $y_3(0) = 3.07$, $z_1(0) = -14.06$, $z_2(0) = -8.38$, $z_3(0) = 16.93$ is considered. The 3-dimensional projections of the trajectory on the $y_1 - y_2 - y_3$ and $z_1 - z_2 - z_3$ spaces are depicted in Fig. 9.6. Both of the pictures represented in Fig. 9.6a and b manifest not only the chaos extension but also the existence of a chaotic attractor in the 9-

dimensional phase space. It is worth noting that the projection on the $x_1 - x_2 - x_3$ space is the classical Lorenz attractor, which is shown in Fig. 9.1.

To illustrate the extension of chaos in large collections of interconnected Lorenz systems, let us introduce the following 27-dimensional system consisting of the subsystems $S_1, S_2, \ldots, S_9$:

$$\left.\begin{aligned}\frac{dx_1}{dt} &= -10x_1 + 10x_2\\ \frac{dx_2}{dt} &= -x_1x_3 + 28x_1 - x_2\\ \frac{dx_3}{dt} &= x_1x_2 - (8/3)x_3\end{aligned}\right\} S_1$$

$$\left.\begin{aligned}\frac{dy_1}{dt} &= -10y_1 + 10y_2 + 8x_1(t)\\ \frac{dy_2}{dt} &= -y_1y_3 + 0.21y_1 - y_2 + x_2(t) + 0.001x_2^3(t)\\ \frac{dy_3}{dt} &= y_1y_2 - (8/3)y_3 + 2x_3(t)\end{aligned}\right\} S_2$$

$$\left.\begin{aligned}\frac{dz_1}{dt} &= -10z_1 + 10z_2 + 4x_2(t)\\ \frac{dz_2}{dt} &= -z_1z_3 + 0.02z_1 - z_2 + 3x_3(t)\\ \frac{dz_3}{dt} &= z_1z_2 - (8/3)z_3 + x_1(t) + 0.1\cos(x_1(t))\end{aligned}\right\} S_3$$

$$\left.\begin{aligned}\frac{dw_1}{dt} &= -10w_1 + 10w_2 - x_1(t)\\ \frac{dw_2}{dt} &= -w_1w_3 + 0.34w_1 - w_2 + 4\tanh(x_2(t))\\ \frac{dw_3}{dt} &= w_1w_2 - (8/3)w_3 - 5x_3(t)\end{aligned}\right\} S_4$$

$$\left.\begin{aligned}\frac{d\zeta_1}{dt} &= -10\zeta_1 + 10\zeta_2 + \tan(y_1(t)/20)\\ \frac{d\zeta_2}{dt} &= -\zeta_1\zeta_3 + 0.12\zeta_1 - \zeta_2 + 2.5y_2(t)\\ \frac{d\zeta_3}{dt} &= \zeta_1\zeta_2 - (8/3)\zeta_3 - 10y_3(t)\end{aligned}\right\} S_5$$

$$\left.\begin{aligned}\frac{d\eta_1}{dt} &= -10\eta_1 + 10\eta_2 + 8y_1(t)\\ \frac{d\eta_2}{dt} &= -\eta_1\eta_3 + 0.29\eta_1 - \eta_2 + 4.5y_3(t)\\ \frac{d\eta_3}{dt} &= \eta_1\eta_2 - (8/3)\eta_3 - e^{y_2(t)/30}\end{aligned}\right\} S_6$$

$$\left.\begin{aligned}\frac{d\kappa_1}{dt} &= -10\kappa_1 + 10\kappa_2 + 4z_1(t)\\ \frac{d\kappa_2}{dt} &= -\kappa_1\kappa_3 + 0.19\kappa_1 - \kappa_2 + 9z_2(t)\\ \frac{d\kappa_3}{dt} &= \kappa_1\kappa_2 - (8/3)\kappa_3 + 6z_3(t)\end{aligned}\right\} S_7$$

$$\left.\begin{aligned}\frac{d\rho_1}{dt} &= -10\rho_1 + 10\rho_2 + 4w_1(t)\\ \frac{d\rho_2}{dt} &= -\rho_1\rho_3 + 0.17\rho_1 - \rho_2 + 7w_2(t)\\ \frac{d\rho_3}{dt} &= \rho_1\rho_2 - (8/3)\rho_3 - 3\tanh(w_3(t))\end{aligned}\right\} S_8$$

$$\left.\begin{aligned}\frac{d\tau_1}{dt} &= -10\tau_1 + 10\tau_2 + \arctan(w_1(t))\\ \frac{d\tau_2}{dt} &= -\tau_1\tau_3 + 0.32\tau_1 - \tau_2 + 9w_2(t)\\ \frac{d\tau_3}{dt} &= \tau_1\tau_2 - (8/3)\tau_3 + w_3(t).\end{aligned}\right\} S_9$$

The coefficients of S_1 are chosen in such a way that the system is chaotic [26]. The systems $S_2, S_3, \ldots, S_9$ are designed such that if the corresponding perturbations $x(t)$, $y(t)$, $z(t)$, $w(t)$ are chaotic, then the systems possess chaos. However, in the absence of the perturbations, $S_2, S_3, \ldots, S_9$ admit stable equilibria and they are all non-chaotic. The connection topology of the systems $S_1, S_2, \ldots, S_9$ is represented in Fig. 9.7. On the other hand, Fig. 9.8 depicts the chaotic attractors corresponding the each S_i, $i = 1, 2, \ldots, 9$, such that collectively the picture can be considered as the chaotic attractor of the whole 27-dimensional system. One can confirm that Fig. 9.8 supports our ideas such that the chaos of S_1 generates chaos in the remaining subsystems even if they are non-chaotic in the absence of the perturbations.

We shall continue in the next section by the extension of chaos obtained through period-doubling cascade.

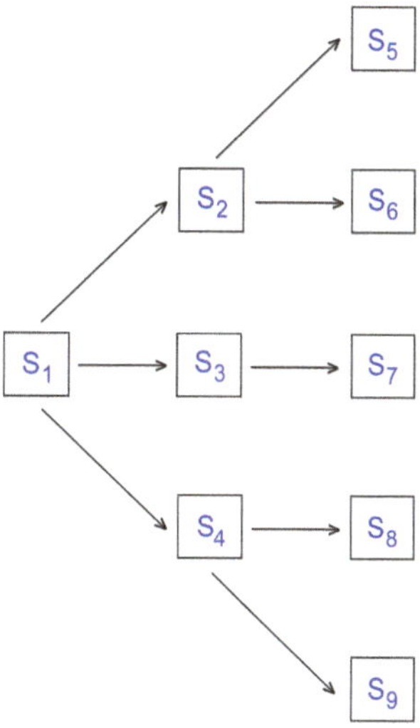

Fig. 9.7 The topology of the unpredictability extension through Lorenz systems $S_1, S_2, \ldots, S_9$

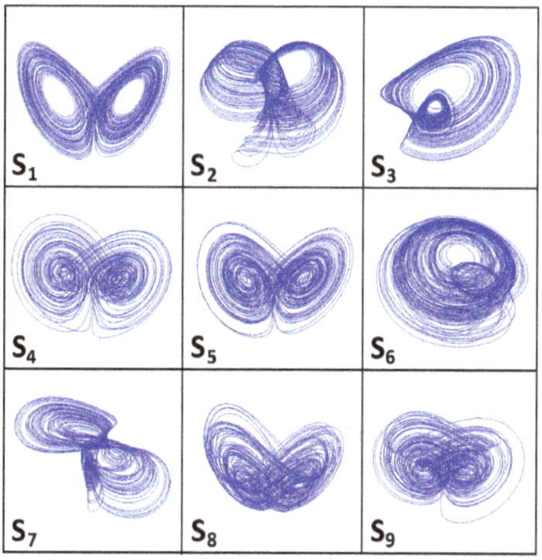

Fig. 9.8 The chaotic attractors of the Lorenz systems $S_1, S_2, \ldots, S_9$, an illustration of the butterfly effect

9.4 Period-Doubling Cascade

Consider the Lorenz system [31, 75]

$$\begin{aligned}\frac{dx_1}{dt} &= -10x_1 + 10x_2\\ \frac{dx_2}{dt} &= -x_1x_3 + rx_1 - x_2\\ \frac{dx_3}{dt} &= x_1x_2 - (8/3)x_3,\end{aligned} \tag{9.4.9}$$

where r is a parameter.

For the values of r between 99.98 and 100.795, system (9.4.9) possesses two symmetric stable periodic orbits such that one of them spirals round twice in $x_1 > 0$ and once in $x_1 < 0$, whereas another spirals round twice in $x_1 < 0$ and once in $x_1 > 0$. The book [31] calls such periodic orbits as x^2y and y^2x, respectively. That is, "x" is written every time when the orbit spirals round in $x_1 > 0$, while "y" is written every time when it spirals round in $x_1 < 0$. As r decreases towards 99.98, a period-doubling bifurcation occurs in the system such that two new symmetric stable periodic orbits (x^2yx^2y and y^2xy^2x) appear and the previous periodic orbits lose their stability [31, 75]. According to Franceschini [75], system (9.4.9) undergoes infinitely many period-doubling bifurcations at the parameter values 99.547, 99.529, 99.5255, and so on. The sequence of bifurcation parameter values accumulates at $r_\infty = 99.524$. For values of r smaller than r_∞ infinitely many unstable periodic orbits take place in the dynamics of system (9.4.9) [31, 75].

To extend the period-doubling cascade of (9.4.9), we perturb the system

$$\begin{aligned} \frac{du_1}{dt} &= -10u_1 + 10u_2 \\ \frac{du_2}{dt} &= -u_1u_3 + 0.27u_1 - u_2 \\ \frac{du_3}{dt} &= u_1u_2 - (8/3)u_3 \end{aligned} \tag{9.4.10}$$

with the solutions of (9.4.9), and set up the system

$$\begin{aligned} \frac{dy_1}{dt} &= -10y_1 + 10y_2 + 1.8x_1(t) \\ \frac{dy_2}{dt} &= -y_1y_3 + 0.27y_1 - y_2 + x_2(t) \\ \frac{dy_3}{dt} &= y_1y_2 - (8/3)y_3 + 0.3x_3(t). \end{aligned} \tag{9.4.11}$$

Using Theorem 15.8 [76], one can verify that for each periodic $x(t)$, system (9.4.11) admits a periodic solution with the same period.

In Fig. 9.9, we represent the stable periodic orbits of systems (9.4.9) and (9.4.11). Figure 9.9a shows the y^2x periodic orbit of system (9.4.9) for $r = 100.36$, while Fig. 9.9b depicts the corresponding periodic orbit of system (9.4.11). Similarly, Fig. 9.9c and d represent the y^2xy^2x periodic orbit of system (9.4.9) with $r = 99.74$ and the corresponding periodic orbit of system (9.4.11), respectively.

Now, we demonstrate the extension of period-doubling cascade in Fig. 9.10. The projection of the trajectory of system (9.4.9) with $r = 99.51$ corresponding to the initial data $x_1(0) = 10.58$, $x_2(0) = 28.19$, $x_3(0) = 53.32$ on the $x_1 - x_3$ plane is shown in Fig. 9.10a. Making use of the initial data $y_1(0) = 2.23$, $y_2(0) = 1.26$, $y_3(0) = 9.64$, the projection of the corresponding trajectory of system (9.4.11) on the $y_1 - y_3$ plane is depicted in Fig. 9.10b. Moreover, the irregular behavior of the y_3 coordinate over time is illustrated in Fig. 9.11. The simulation results reveal that the period-doubling cascade of system (9.4.9) is extended to system (9.4.11).

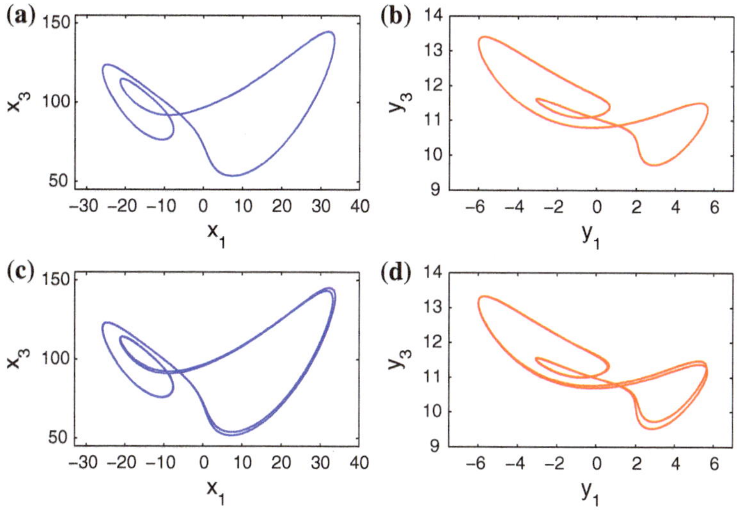

Fig. 9.9 The stable periodic orbits of systems (9.4.9) and (9.4.11)

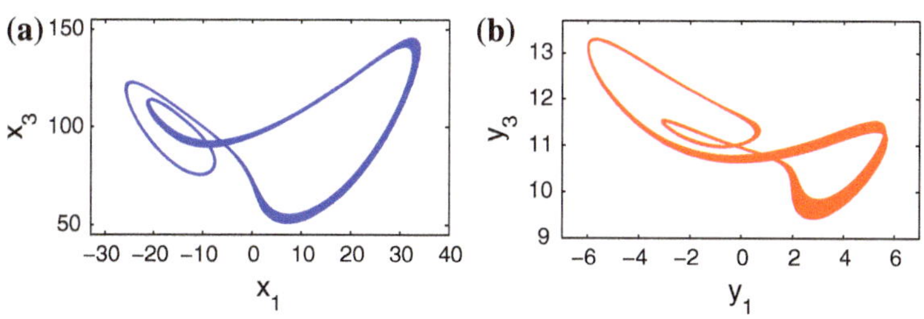

Fig. 9.10 Extension of period-doubling cascade in the unidirectionally coupled Lorenz systems (9.4.9)+(9.4.11)

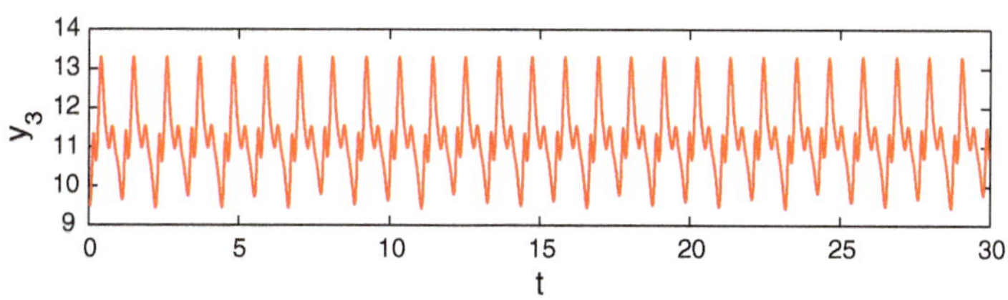

Fig. 9.11 The irregular behavior of the y_3 coordinate of system (9.4.11) with $r = 99.51$

A theoretical investigation of the extension of period-doubling cascade is provided in Sect. 9.8.

For some parameter values, the Lorenz system can exhibit limit cycles [31]. In the next section, we will consider the Lorenz system (9.2.3) with a globally attracting limit cycle and verify numerically how to achieve a motion that behaves chaotically and cyclically in the same time.

9.5 Cyclic Chaos in Lorenz Systems

In our previous illustrations, we considered system (9.2.3) with a stable equilibrium point. Now, we shall consider the model with a limit cycle. The numerical simulations represented in this part are theoretically based on the paper [70] in which the main result is about the existence of infinitely many unstable periodic solutions and extension of sensitivity, which is understood as unpredictability for weather investigation.

According to Sparrow [31], the Lorenz system

$$\begin{aligned}\frac{du_1}{dt} &= -10u_1 + 10u_2\\ \frac{du_2}{dt} &= -u_1u_3 + 350u_1 - u_2\\ \frac{du_3}{dt} &= u_1u_2 - (8/3)u_3\end{aligned} \tag{9.5.12}$$

possesses a globally attracting limit cycle. We perturb system (9.5.12) with the solutions of (9.3.5), and set up the following system:

$$\begin{aligned}\frac{dy_1}{dt} &= -10y_1 + 10y_2 + 2.3x_1(t)\\ \frac{dy_2}{dt} &= -y_1y_3 + 350y_1 - y_2 + 2x_2(t)\\ \frac{dy_3}{dt} &= y_1y_2 - (8/3)y_3 + 1.5x_3(t).\end{aligned} \tag{9.5.13}$$

Making use of the solution of system (9.3.5) with $x_1(0) = 5.71$, $x_2(0) = 9.01$, $x_3(0) = 17.06$, we depict the trajectory of (9.5.13) corresponding to the initial data $y_1(0) = -21.67$, $y_2(0) = 34.33$, $y_3(0) = 346.38$ in Fig. 9.12a. The projection of the same trajectory on the $y_1 - y_2$ plane is shown in Fig. 9.12b. Both figures reveal that the trajectory behaves chaotically around the limit cycle of (9.5.12).

To confirm one more time that the considered trajectory is essentially chaotic, the graph of the y_2 coordinate of system (9.5.13) is illustrated in Fig. 9.13.

Although the system (9.5.12) possesses a globally attracting limit cycle, the simulations seen in Figs. 9.12 and 9.13 indicate that the applied perturbation makes the system behave chaotically. In other words, chaotic behavior is seized by the limit cycle of system (9.5.12) and as a result a motion which behaves both chaotically and cyclically appears.

To make comparison of our approach with that of generalized synchronization [54–56, 58, 61], let us apply the auxiliary system approach [54, 55] to the couple (9.3.5)+(9.5.13).

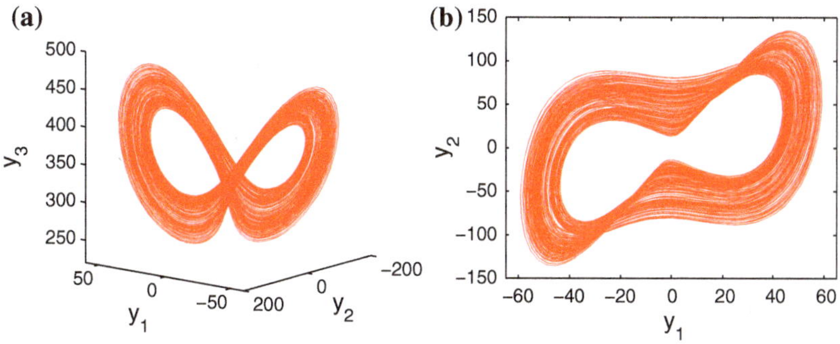

Fig. 9.12 The projections of the chaotic trajectory produced by the coupled system (9.3.5) + (9.5.13). **a** The 3-dimensional projection on the $y_1 - y_2 - y_3$ space; **b** The 2-dimensional projection on the $y_1 - y_2$ plane. The pictures in (**a**) and (**b**) represent a motion which behaves both chaotically and cyclically around the stable limit cycle of system (9.5.12)

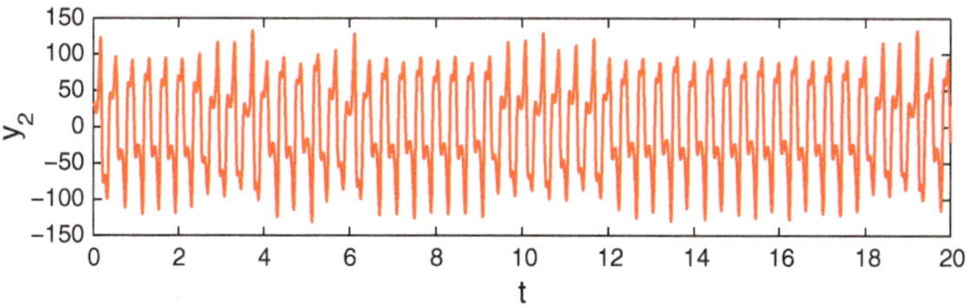

Fig. 9.13 The time-series for the y_2-coordinate of system (9.5.13). The picture confirms the chaotic behavior of the motion. The remaining coordinates of system (9.5.13), which are not just pictured here, behave also chaotically

The corresponding auxiliary system is

$$\begin{aligned}\frac{dz_1}{dt} &= -10z_1 + 10z_2 + 2.3x_1(t)\\ \frac{dz_2}{dt} &= -z_1z_3 + 350z_1 - y_2 + 2x_2(t)\\ \frac{dz_3}{dt} &= z_1z_2 - (8/3)z_3 + 1.5x_3(t).\end{aligned} \tag{9.5.14}$$

The projection of the stroboscopic plot of the 9-dimensional system (9.3.5) + (9.5.13) + (9.5.14) on the $y_2 - z_2$ plane is depicted in Fig. 9.14. The figure is obtained by marking the trajectory with the initial data $x_1(0) = 5.71$, $x_2(0) = 9.01$, $x_3(0) = 17.06$, $y_1(0) = -21.67$, $y_2(0) = 34.33$, $y_3(0) = 346.38$, $z_1(0) = -46.26$, $z_2(0) = -49.73$, $z_3(0) = 415.87$, and by omitting the first 200 iterations. It is observable in Fig. 9.14 that the stroboscopic plot is not on the line $z_2 = y_2$. Therefore, we conclude that generalized synchronization does not take place in the dynamics of the couple (9.3.5) + (9.5.13).

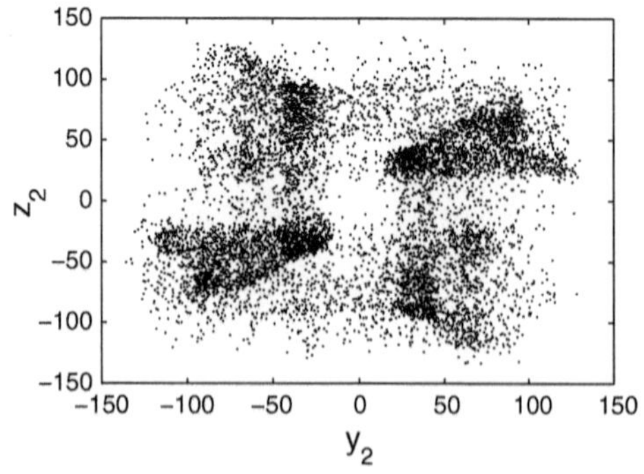

Fig. 9.14 Application of the auxiliary system approach to system (9.3.5)+(9.5.13) indicates that generalized synchronization does not exist for the couple

Another approach to investigate the presence or absence of generalized synchronization is the evaluation of conditional Lyapunov exponents [55, 58, 60].

To determine the conditional Lyapunov exponents, we take into account the following variational equations for system (9.5.13):

$$\begin{aligned}\frac{d\xi_1}{dt} &= -10\xi_1 + 10\xi_2\\ \frac{d\xi_2}{dt} &= (-y_3(t) + 350)\xi_1 - \xi_2 - y_1(t)\xi_3 \\ \frac{d\xi_3}{dt} &= y_2(t)\xi_1 + y_1(t)\xi_2 - (8/3)\xi_3.\end{aligned} \tag{9.5.15}$$

Using the solution $y(t)$ of system (9.5.13) corresponding to the initial data $x_1(0) = 5.71$, $x_2(0) = 9.01$, $x_3(0) = 17.06$, $y_1(0) = -21.67$, $y_2(0) = 34.33$, $y_3(0) = 346.38$, we evaluated the largest Lyapunov exponent of system (9.5.15) as 0.0226. That is, system (9.5.13) possesses a positive conditional Lyapunov exponent, and this result reveals one more time the absence of generalized synchronization.

In the next part, we will continue with the extension of intermittency.

9.6 Intermittency in the Weather Dynamics

Pomeau and Manneville [52] observed intermittency in the Lorenz system

$$\begin{aligned}\frac{dx_1}{dt} &= -10x_1 + 10x_2\\ \frac{dx_2}{dt} &= -x_1x_3 + rx_1 - x_2 \\ \frac{dx_3}{dt} &= x_1x_2 - (8/3)x_3,\end{aligned} \tag{9.6.16}$$

where r is slightly larger than the critical value $r_c \approx 166.06$.

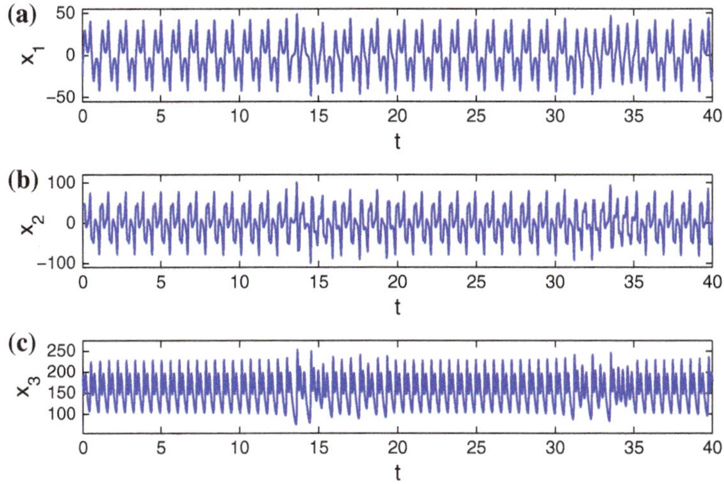

Fig. 9.15 Intermittency in the Lorenz system (9.6.16) with $r = 166.3$. **a** The graph of the x_1 coordinate. **b** The graph of the x_2 coordinate. **c** The graph of the x_3 coordinate

To illustrate the intermittent behavior of system (9.6.16) with $r = 166.3$, let us consider the solution corresponding to the initial data $x_1(0) = -23.3$, $x_2(0) = 38.3$, $x_3(0) = 193.4$. The graphs of the x_1, x_2, and x_3 coordinates of the solution are shown in Fig. 9.15, where one can see that regular oscillations are interrupted by irregular ones.

Now, we use solutions of (9.6.16) with $r = 166.3$ to perturb system (9.2.3), where $\overline{\sigma} = 10$, $\overline{r} = 0.35$, $\overline{b} = 8/3$, and constitute the following system:

$$\begin{aligned} \frac{dy_1}{dt} &= -10y_1 + 10y_2 + 0.7x_1(t) \\ \frac{dy_2}{dt} &= -y_1y_3 + 0.35y_1 - y_2 - x_2(t) \\ \frac{dy_3}{dt} &= y_1y_2 - (8/3)y_3 + 0.2x_3(t). \end{aligned} \tag{9.6.17}$$

The graphs of the y_1, y_2, and y_3 coordinates of system (9.6.17) are represented in Fig. 9.16. The initial data $x_1(0) = -23.3$, $x_2(0) = 38.3$, $x_3(0) = 193.4$, $y_1(0) = 1.3$, $y_2(0) = 5.5$, $y_3(0) = 12.1$ are used in the simulation. It is revealed in Fig. 9.16 that the intermittent behavior of the prior Lorenz system is extended.

9.7 Self-Organization and Synergetics

We have to say that the idea of the transition of chaos from one system to another as well as the arrangement of chaos in an ordered way can be considered as another level of self-organization [77, 78]. Durrenmatt [79] described that "... a system is

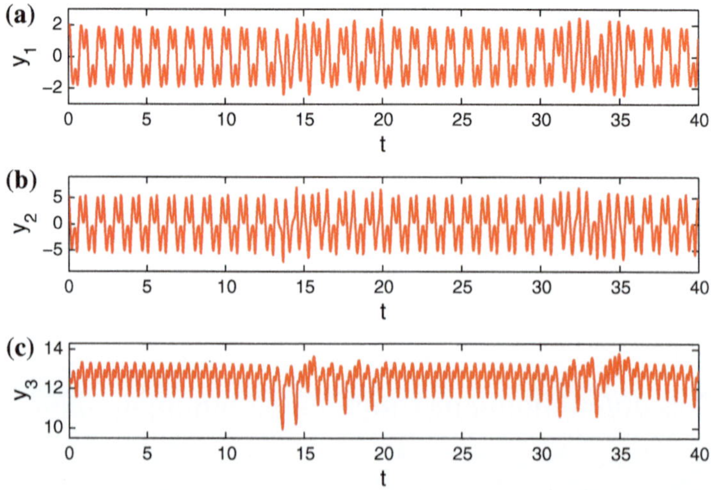

Fig. 9.16 Extension of intermittency in system (9.6.17). The behaviors of the y_1, y_2, and y_3 coordinates are shown in pictures (**a**), (**b**) and (**c**), respectively. The extension of the intermittent behavior is observable such that regular motions are interrupted by irregular ones

self-organizing if it acquires a spatial, temporal or functional structure without specific interference from the outside. By "specific" we mean that the structure of functioning is not impressed on the system, but the system is acted upon from the outside in a nonspecific fashion." There are three approaches to self-organization, namely thermodynamic (dissipative structures), synergetic, and the autowave. For the theory of dynamical systems (e.g., differential equations), the phenomenon means that an autonomous system of equations admits a regular and stable motion (periodic, quasi-periodic, almost periodic). This is what in the literature is called autowaves processes [80] or self-excited oscillations [81]. We are inclined to add to the list another phenomenon, which is a consequence of the butterfly effect. Consider the collection of subsystems, $S_1, S_2, \ldots, S_9$, (the number 9 can be replaced by any natural number n_0, in general) which is introduced in Sect. 9.3 once again. Because of the connections and the conditions discovered in our analysis, likewise S_1 all the other subsystems, S_i, $i = 2, 3, \ldots, 9$, are also chaotic. We suppose that this is a self-organization. This phenomenon can be restricted only for autonomous systems or it can be even interpreted for non-autonomous systems, too. So, we can say that the butterfly effect is an example of self-organization, that is a coherent behavior of a large number of systems [77].

In his fascinating paper, the German theoretical physicist Haken [77] introduced a new interdisciplinary field of science, synergetics, which deals with the origins and the evolution of spatio-temporal structures. The profound part of synergetics is based on the dynamical systems theory. Depending on the discussion of the present chapter, it is natural that we concentrate on the differential equations, and everything that will be said below about synergetics concerns first of all dynamical systems with mathematical approach. One of the main features of systems in synergetics is self-organization, which has been discussed above, and we approved that the phenomenon is present in the butterfly effect. According to Haken [77], the central question in synergetics is whether there are general principles which govern the self-organized formation of structures and/or functions. The main principles by the founder of the theory are instability, order parameters, and slaving [77]. Instability is understood as the formation or collapse of structures (patterns). This is very common in fluid dynamics, lasers, chemistry, and biology [77, 82–85]. A number of examples of instability can be found in the literature about morphogenesis [86] and the pattern formation examples can be found in fluid dynamics. The phenomenon is called as instability because the former state of fluid transforms to a new one, loses its ability to persist, and becomes unstable. We see instability in the butterfly effect, as consecutive chaotification of systems $S_2, S_3, \ldots$ joined to the source S_1 of chaos. The concepts of the order parameter and slaving are strongly connected in synergetics. For differential equations theory, order parameters mean those phase variables, whose behavior formate the main properties of a macroscopic structure, which dominate over all other variables in the formation such that they can even depend on the order parameters functionally. The dependence that is proved (discovered) mathematically is what we call as slaving. It is not difficult to see in the butterfly effect that the variables of the system S_1 are order parameters, and they determine the chaotic behavior of the joined systems' variables. That is, the slaving principle is present here.

9.8 The Mathematical Background

In our theoretical discussions, we consider more general coupled systems, which are not necessarily Lorenz systems. We will denote by $\mathbb{R}$ and $\mathbb{N}$ the sets of real numbers and natural numbers, respectively, and we will make use of the usual Euclidean norm for vectors.

Let us consider the autonomous systems

$$\frac{dx}{dt} = F(x), \tag{9.8.18}$$

and

$$\frac{du}{dt} = f(u), \tag{9.8.19}$$

where $t \geq 0$ and the functions $F : \mathbb{R}^m \to \mathbb{R}^m$ and $f : \mathbb{R}^n \to \mathbb{R}^n$ are continuous in their arguments.

We perturb system (9.8.19) with the solutions of (9.8.18) and obtain the system in the form,

$$\frac{dy}{dt} = f(y) + \mu g(x(t)), \tag{9.8.20}$$

where the number μ is nonzero and the function $g : \mathbb{R}^m \to \mathbb{R}^n$ is continuous. It is worth noting that the systems (9.1.1), (9.2.3), and (9.2.4) are in the form of (9.8.18), (9.8.19), and (9.8.20), respectively.

We mainly assume that system (9.8.18) possesses a chaotic attractor, let us say a set in $\mathbb{R}^m$. Fix x_0 from the attractor and take a solution $x(t)$ of (9.8.18) with $x(0) = x_0$. Since we use the solution $x(t)$ as a perturbation in system (9.8.20), we call it as *chaotic function*. Chaotic functions may be irregular as well as regular (periodic and unstable) [26, 41, 49, 53, 87, 88].

Our purpose is to prove rigorously the extension of chaos from system (9.8.18) to system (9.8.20). In our theoretical discussions, we request the existence of a bounded positively invariant region for system (9.8.20). Such an invariant region can be achieved by different methods and one of them is mentioned in the next part. We will show the extension of sensitivity and the existence infinitely many unstable periodic solutions in Sects. 9.8.2 and 9.8.3, respectively.

In the following parts, for a given solution $x(t)$ of system (9.8.18), we will denote by $\phi_{x(t)}(t, t_0, y_0)$ the unique solution of system (9.8.20) satisfying the initial condition $\phi_{x(t)}(t_0, t_0, y_0) = y_0$.

9.8.1 Bounded Positively Invariant Region

Making benefit of Lyapunov functions and uniform ultimate boundedness [76, 89], we present a method in Theorem 9.1 for the existence of a bounded positively invariant set of system (9.8.20). Then, we will apply this technique to the Lorenz system.

Solutions of system (9.8.20) are uniformly ultimately bounded if there exists a number $B_0 > 0$ and corresponding to any number $\alpha > 0$ there exists a number $T(\alpha) > 0$ such that $\|y_0\| \leq \alpha$ implies that for each solution $x(t)$ of system (9.8.18) and $t_0 \geq 0$ we have $\left\|\phi_{x(t)}(t, t_0, y_0)\right\| < B_0$ for all $t \geq t_0 + T(\alpha)$.

The following condition is required:

(A1) There exists a positive number M_g such that $\sup_{x \in \mathbb{R}^m} \|g(x)\| \leq M_g$.

Theorem 9.1 *Suppose that condition* $(A1)$ *is fulfilled and there exists a Lyapunov function* $V(x)$ *defined on* $\mathbb{R}^n$ *such that* $V(x)$ *has continuous first-order partial derivatives. Additionally, assume that there exists a number* $B \geq 0$ *such that the following conditions are satisfied on the region* $\|x\| \geq B$:

(i) $V(x) \geq a(\|x\|)$, *where* $a(r)$ *is a continuous, increasing function defined for* $r \geq B$ *which satisfies* $a(B) > 0$ *and* $a(r) \to \infty$ *as* $r \to \infty$;
(ii) $V'_{(9.8.19)}(x) \leq -b(\|x\|)$, *where* $b(r)$ *is an increasing function defined for* $r \geq B$ *which satisfies* $b(B) > 0$;

(iii) $\left\| \frac{\partial V}{\partial x}(x) \right\| \leq c\left(\|x\|\right)$, *where $c(r)$ is a function defined for $r \geq B$ and there exists a positive number M_0 such that $0 < c(r) \leq M_0 b(r)$ for all $r \geq B$.*

Then, for sufficiently small $|\mu|$, the solutions of system (9.8.20) are uniformly ultimately bounded.

Proof Fix arbitrary numbers $t_0 \geq 0$, $\alpha > 0$ and a solution $x(t)$ of system (9.8.18). Take a number β satisfying $0 < \beta < b(B)$. We consider system (9.8.20) with a nonzero number μ which satisfies the inequality

$$|\mu| \leq \frac{1}{M_0 M_g}\left(1 - \frac{\beta}{b(B)}\right).$$

Our aim is to show the existence of numbers $B_0 > B$ and $T(\alpha) \geq 0$, independent of t_0, such that if $\|y_0\| \leq \alpha$, then $\left\|\phi_{x(t)}(t, t_0, y_0)\right\| < B_0$ for all $t \geq t_0 + T(\alpha)$.

Consider an arbitrary $y_0 \in \mathbb{R}^n$ such that $\|y_0\| \leq \alpha$. For the sake of brevity, let us denote $y(t) = \phi_{x(t)}(t, t_0, y_0)$. In the proof, both of the possibilities $\|y_0\| < B$ and $\|y_0\| \geq B$ will be considered. We start with the former.

Let $M_V = \max\limits_{\|x\|=B} V(x)$. Since $a(r) \to \infty$ as $r \to \infty$, there exists a number $B_0 > B$ such that $a(B_0) \geq M_V$.

Now, suppose that there exists a moment $s_1 > t_0$ such that $\|y(s_1)\| \geq B_0$. It is possible to find a moment s_2 satisfying $t_0 < s_2 < s_1$ such that $\|y(s_2)\| = B$ and $\|y(t)\| \geq B$ for all $t \in [s_2, s_1]$.

Assumptions (ii) and (iii) imply for $s_2 \leq t \leq s_1$ that

$$\begin{aligned}
\frac{dV(y(t))}{dt} &= \frac{\partial V}{\partial x}(y(t)) \cdot (f(y(t)) + \mu g(x(t))) \\
&\leq -b\left(\|y(t)\|\right) + |\mu| M_g c\left(\|y(t)\|\right) \\
&\leq (|\mu| M_0 M_g - 1) b(B) \\
&\leq -\beta,
\end{aligned}$$

where "$\cdot$" denotes the scalar product.

The last inequality implies that $V(y(s_1)) < V(y(s_2))$. On the other hand, by the help of assumption (i), we have $V(y(s_2)) \leq M_V \leq a(B_0) \leq V(y(s_1))$. This is a contradiction. Therefore, for all $t \geq t_0$, the inequality $\|y(t)\| < B_0$ is valid.

Next, we consider the possibility $\|y_0\| \geq B$. Since the function $V(x)$ is continuous and $\|y_0\| \leq \alpha$, one can find a number $K(\alpha) > 0$ such that $V(y_0) \leq K(\alpha)$. By means of condition (i) used together with the inequality $\|y_0\| \geq B$, we have that $K(\alpha) \geq a(B)$.

Assume that there exists a moment $\bar{t} > t_0 + \dfrac{K(\alpha) - a(B)}{\beta}$ such that $\left\|y(\bar{t})\right\| \geq B$.

If there exists $t_1 \in [t_0, \bar{t}]$ such that $\|y(t_1)\| < B$, then by means of uniqueness of solutions, using a similar discussion to the case $\|y_0\| < B$ considered above, one can show that for all $t \geq t_1$ the inequality $\|y(t)\| < B_0$ holds. On the other hand, if for all $t \in [t_0, \overline{t}]$ the inequality $\|y(t)\| \geq B$ is valid, then one can verify that the inequality

$$V(y(\overline{t})) \leq V(y_0) - \beta(\overline{t} - t_0)$$

holds. Under the circumstances we attain that

$$a(B) \leq V(y_0) - \beta(\overline{t} - t_0) \leq K(\alpha) - \beta(\overline{t} - t_0) < a(B).$$

This is a contradiction. Hence, for all $t > t_0 + T(\alpha)$, where $T(\alpha) = \dfrac{K(\alpha) - a(B)}{\beta}$, we have $\|y(t)\| < B_0$. Consequently, the solutions of system (9.8.20) are uniformly ultimately bounded.

Next, we shall verify the conditions of Theorem 9.1 for the Lorenz model. Let us consider the system (9.2.3) with the parameters $\overline{\sigma} > 0$, $0 < \overline{r} < \sqrt{2} - 1$, $\overline{b} > 0$, and take into account the Lyapunov function

$$V(u) = \frac{1}{\overline{\sigma}} u_1^2 + u_2^2 + u_3^2,$$

where $u = (u_1, u_2, u_3) \in \mathbb{R}^3$.

Set $\gamma_1 = \min\left\{1, \dfrac{1}{\overline{\sigma}}\right\}$ and define the function $a(r)$ through the formula $a(r) = \gamma_1 r^2$. In that case, the relation $V(u) \geq \gamma_1(u_1^2 + u_2^2 + u_3^2) = a\left(\|u\|\right)$ holds. On the other hand, one can verify that

$$\begin{aligned} &V'_{(9.2.3)}(u) = \frac{2}{\overline{\sigma}} u_1 u_1' + 2u_2 u_2' + 2u_3 u_3' \\ &= \frac{2}{\overline{\sigma}} u_1[\overline{\sigma}(-u_1 + u_2)] + 2u_2(-u_1 u_3 + \overline{r} u_1 - u_2) + 2u_3(u_1 u_2 - \overline{b} u_3) \\ &= 2(\overline{r} + 1) u_1 u_2 - 2u_1^2 - 2u_2^2 - 2\overline{b} u_3^2. \end{aligned}$$

Now, let $\gamma_2 = \min\left\{1, 2 - (\overline{r} + 1)^2, 2\overline{b}\right\}$. Making use of the identity

$$2(\overline{r} + 1) u_1 u_2 = u_1^2 + (\overline{r} + 1)^2 u_2^2 - [u_1 - (\overline{r} + 1) u_2]^2$$

we attain the inequality

$$\begin{aligned} &V'_{(9.2.3)}(u) = -[u_1 - (\overline{r} + 1) u_2]^2 - u_1^2 - \left[2 - (\overline{r} + 1)^2\right] u_2^2 - 2\overline{b} u_3^2 \\ &\leq -u_1^2 - \left[2 - (\overline{r} + 1)^2\right] u_2^2 - 2\overline{b} u_3^2 \\ &\leq -b\left(\|u\|\right), \end{aligned}$$

where the function $b(r)$ is defined through the formula $b(r) = \gamma_2 r^2$. The last inequality validates the condition (ii) of Theorem 9.1.

Furthermore, one can obtain that

$$\left\| \frac{\partial V}{\partial u}(u) \right\| = 2\sqrt{\frac{1}{\overline{\sigma}^2}u_1^2 + u_2^2 + u_3^2} \leq c\,(\|u\|)\,,$$

where $c(r) = 2\gamma_3 r$ and $\gamma_3 = \max\left\{1, \frac{1}{\overline{\sigma}}\right\}$. If we take $M_0 = \frac{2\gamma_3}{\gamma_2}$, then the inequality $c(r) \leq M_0 b(r)$ holds for all $r \geq 1$. Consequently, for $B = 1$, the conditions of Theorem 9.1 are satisfied for system (9.2.3) with the coefficients $\overline{\sigma} > 0$, $0 < \overline{r} < \sqrt{2} - 1$ and $\overline{b} > 0$.

In the next section, we will continue with the extension of sensitivity, which can be considered as the unique ingredient of chaos for a set of bounded solutions [26, 41, 90].

9.8.2 *Unpredictability Analysis*

Extension of the sensitivity feature through system (9.8.20) will be handled in the present part. We shall begin with the meaning of the aforementioned property for systems (9.8.18) and (9.8.20). The main result will be stated in Theorem 9.2.

System (9.8.18) is called sensitive if there exist positive numbers ε_0 and Δ such that for an arbitrary positive number δ_0 and for each chaotic solution $x(t)$ of system (9.8.18), there exist a chaotic solution $\overline{x}(t)$ of the same system and an interval $J \subset [0, \infty)$, with a length no less than Δ, such that $\|x(0) - \overline{x}(0)\| < \delta_0$ and $\|x(t) - \overline{x}(t)\| > \varepsilon_0$ for all $t \in J$.

Our main assumption is the existence of a bounded positively invariant set $\mathscr{K}$ for system (9.8.20). The existence of such an invariant set can be shown, for example, using Theorem 9.1.

We say that system (9.8.20) is sensitive if there exist positive numbers ε_1 and $\overline{\Delta}$ such that for an arbitrary positive number δ_1, each $y_0 \in \mathscr{K}$ and a chaotic solution $x(t)$ of (9.8.18), there exist $y_1 \in \mathscr{K}$, a chaotic solution $\overline{x}(t)$ of (9.8.18) and an interval $J^1 \subset [0, \infty)$, with a length no less than $\overline{\Delta}$, such that $\|y_0 - y_1\| < \delta_1$ and $\left\|\phi_{x(t)}(t, 0, y_0) - \phi_{\overline{x}(t)}(t, 0, y_1)\right\| > \varepsilon_1$ for all $t \in J^1$.

The following assumptions are needed:

(A2) There exists a positive number M_F such that $\sup\limits_{x \in \mathbb{R}^m} \|F(x)\| \leq M_F$;

(A3) There exists a positive number L_f such that $\|f(y_1) - f(y_2)\| \leq L_f \|y_1 - y_2\|$ for all $y_1, y_2 \in \mathbb{R}^n$;

(A4) There exists a positive number L_g such that $\|g(x_1) - g(x_2)\| \geq L_g \|x_1 - x_2\|$ for all $x_1, x_2 \in \mathbb{R}^m$.

In the next theorem, the extension of sensitivity from system (9.8.18) to system (9.8.20) is considered.

Theorem 9.2 *Suppose that conditions* $(A1) - (A4)$ *hold. If system (9.8.18) is sensitive, then the same is true for system (9.8.20).*

Proof Fix arbitrary $\delta_1 > 0$, $y_0 \in \mathscr{K}$ and a chaotic solution $x(t)$ of (9.8.18). Since system (9.8.18) is sensitive, one can find $\varepsilon_0 > 0$ and $\Delta > 0$ such that for arbitrary $\delta_0 > 0$ both of the inequalities $\|x(0) - \overline{x}(0)\| < \delta_0$ and $\|x(t) - \overline{x}(t)\| > \varepsilon_0$, $t \in J$, hold for some chaotic solution $\overline{x}(t)$ of (9.8.18) and for some interval $J \subset [0, \infty)$, whose length is not less than Δ.

Take an arbitrary $y_1 \in \mathscr{K}$ such that $\|y_0 - y_1\| < \delta_1$. For the sake of brevity, let us denote $y(t) = \phi_{x(t)}(t, 0, y_0)$ and $\overline{y}(t) = \phi_{\overline{x}(t)}(t, 0, y_1)$.

It is worth noting that there exist positive numbers K_0 and H_F such that $\|y(t)\|$, $\|\overline{y}(t)\| \leq K_0$ for all $t \geq 0$ and $\sup\limits_{t \in \mathbb{R}_+} \|x(t)\| \leq H_F$ for each chaotic solution $x(t)$ of system (9.8.18).

Our aim is to determine positive numbers ε_1, $\overline{\Delta}$ and an interval $J^1 \subset [0, \infty)$ with length $\overline{\Delta}$ such that the inequality $\|y(t) - \overline{y}(t)\| > \varepsilon_1$ holds for all $t \in J^1$.

Since the derivative of each chaotic solution $x(t)$ of (9.8.18) lies inside the tube with radius M_F, the collection of chaotic solutions of system (9.8.18) is an equicontinuous family on $[0, \infty)$. Suppose that $g(x) = (g_1(x), g_2(x), \ldots, g_n(x))$, where each g_j, $1 \leq j \leq n$, is a real valued function. Making use of the uniform continuity of the function $\overline{g} : \mathbb{R}^m \times \mathbb{R}^m \to \mathbb{R}^n$, defined as $\overline{g}(v_1, v_2) = g(v_1) - g(v_2)$, on the compact region $\mathscr{R} = \{(v_1, v_2) \in \mathbb{R}^m \times \mathbb{R}^m : \|v_1\| \leq H_F, \|v_2\| \leq H_F\}$ together with the equicontinuity of the collection of chaotic solutions of (9.8.18), one can verify that the collection $\mathscr{F}$ consisting of the functions of the form $g_j(x_1(t)) - g_j(x_2(t))$, $1 \leq j \leq n$, where $x_1(t)$ and $x_2(t)$ are chaotic solutions of system (9.8.18), is an equicontinuous family on $[0, \infty)$.

According to the equicontinuity of the family $\mathscr{F}$, one can find a positive number $\tau < \Delta$, which is independent of $x(t)$ and $\overline{x}(t)$, such that for any $t_1, t_2 \in [0, \infty)$ with $|t_1 - t_2| < \tau$, the inequality

$$\left|\left(g_j\left(x(t_1)\right) - g_j\left(\overline{x}(t_1)\right)\right) - \left(g_j\left(x(t_2)\right) - g_j\left(\overline{x}(t_2)\right)\right)\right| < \frac{L_g \varepsilon_0}{2n} \quad (9.8.21)$$

holds for all $1 \leq j \leq n$.

Condition $(A4)$ implies that $\|g(x(t)) - g(\overline{x}(t))\| \geq L_g \|x(t) - \overline{x}(t)\|$, $t \in J$. Therefore, for each $t \in J$, there exists an integer j_0, $1 \leq j_0 \leq n$, which possibly depends on t, such that

$$\left|g_{j_0}(x(t)) - g_{j_0}(\overline{x}(t))\right| \geq \frac{L_g}{n} \|x(t) - \overline{x}(t)\|.$$

Otherwise, if there exists $s \in J$ such that for all $1 \leq j \leq n$ the inequality

$$\left|g_j\left(x\left(s\right)\right) - g_j(\overline{x}(s))\right| < \frac{L_g}{n} \left\|x(s) - \overline{x}(s)\right\|$$

holds, then one encounters with a contradiction since

$$\left\|g(x(s)) - g(\overline{x}(s))\right\| \leq \sum_{j=1}^{n} \left|g_j(x(s)) - g_j(\overline{x}(s))\right| < L_g \left\|x(s) - \overline{x}(s)\right\|.$$

Denote by s_0 the midpoint of the interval J, and let $\theta = s_0 - \tau/2$. There exists an integer j_0, $1 \leq j_0 \leq n$, such that

$$\left|g_{j_0}(x(s_0)) - g_{j_0}(\overline{x}(s_0))\right| \geq \frac{L_g}{n} \left\|x(s_0) - \overline{x}(s_0)\right\| > \frac{L_g \varepsilon_0}{n}. \tag{9.8.22}$$

On the other hand, making use of the inequality (9.8.21) we acquire for all $t \in [\theta, \theta + \tau]$ that

$$\begin{aligned}
&\left|g_{j_0}\left(x(s_0)\right) - g_{j_0}\left(\overline{x}(s_0)\right)\right| - \left|g_{j_0}\left(x(t)\right) - g_{j_0}\left(\overline{x}(t)\right)\right| \\
&\leq \left|\left(g_{j_0}\left(x(t)\right) - g_{j_0}\left(\overline{x}(t)\right)\right) - \left(g_{j_0}\left(x(s_0)\right) - g_{j_0}\left(\overline{x}(s_0)\right)\right)\right| \\
&< \frac{L_g \varepsilon_0}{2n}.
\end{aligned}$$

Therefore, by means of (9.8.22), we have that the inequality

$$\left|g_{j_0}\left(x(t)\right) - g_{j_0}\left(\overline{x}(t)\right)\right| > \left|g_{j_0}\left(x(s_0)\right) - g_{j_0}\left(\overline{x}(s_0)\right)\right| - \frac{L_g \varepsilon_0}{2n} > \frac{L_g \varepsilon_0}{2n} \tag{9.8.23}$$

is valid for $t \in [\theta, \theta + \tau]$.

One can find numbers $s_1, s_2, \ldots, s_n \in [\theta, \theta + \tau]$ such that

$$\begin{aligned}
\int_{\theta}^{\theta+\tau} \left[g(x(s)) - g(\overline{x}(s))\right] ds = \Big(&\tau \left[g_1(x(s_1)) - g_1(\overline{x}(s_1))\right], \\
&\tau \left[g_2(x(s_2)) - g_2(\overline{x}(s_2))\right], \ldots, \tau \left[g_n(x(s_n)) - g_n(\overline{x}(s_n))\right]\Big).
\end{aligned}$$

By using the inequality (9.8.23), we attain that

$$\left\|\int_{\theta}^{\theta+\tau} \left[g(x(s)) - g(\overline{x}(s))\right] ds\right\| \geq \tau \left|g_{j_0}(x(s_{j_0})) - g_{j_0}(\overline{x}(s_{j_0}))\right| > \frac{\tau L_g \varepsilon_0}{2n}.$$

The relation

$$y(t) - \overline{y}(t) = (y(\theta) - \overline{y}(\theta)) + \int_{\theta}^{t} [f(y(s)) - f(\overline{y}(s))]\, ds$$
$$+ \int_{\theta}^{t} \mu[g(x(s)) - g(\overline{x}(s))]ds,$$

where $t \in [\theta, \theta + \tau]$, yields

$$\|y(\theta + \tau) - \overline{y}(\theta + \tau)\| \geq |\mu| \left\| \int_{\theta}^{\theta+\tau} [g(x(s)) - g(\overline{x}(s))]ds \right\|$$
$$- \|y(\theta) - \overline{y}(\theta)\| - \int_{\theta}^{\theta+\tau} L_f \|y(s) - \overline{y}(s)\|\, ds$$
$$> \frac{|\mu|\, \tau L_g \varepsilon_0}{2n} - \|y(\theta) - \overline{y}(\theta)\| - \int_{\theta}^{\theta+\tau} L_f \|y(s) - \overline{y}(s)\|\, ds.$$

The last inequality implies that

$$\max_{t\in[\theta,\theta+\tau]} \|y(t) - \overline{y}(t)\| \geq \|y(\theta + \tau) - \overline{y}(\theta + \tau)\|$$
$$> \frac{|\mu|\, \tau L_g \varepsilon_0}{2n} - (1 + \tau L_f) \max_{t\in[\theta,\theta+\tau]} \|y(t) - \overline{y}(t)\|\, .$$

Therefore, $\max\limits_{t\in[\theta,\theta+\tau]} \|y(t) - \overline{y}(t)\| > \dfrac{|\mu|\, \tau L_g \varepsilon_0}{2n(2 + \tau L_f)}$.

Suppose that $\max\limits_{t\in[\theta,\theta+\tau]} \|y(t) - \overline{y}(t)\| = \|y(\xi) - \overline{y}(\xi)\|$ for some $\xi \in [\theta, \theta + \tau]$. Define

$$\overline{\Delta} = \min \left\{ \frac{\tau}{2}, \frac{|\mu|\, \tau L_g \varepsilon_0}{8n(K_0 L_f + M_g |\mu|)(2 + \tau L_f)} \right\}$$

and let

$$\theta^1 = \begin{cases} \xi, & \text{if } \xi \leq \theta + \tau/2 \\ \xi - \overline{\Delta}, & \text{if } \xi > \theta + \tau/2 \end{cases}.$$

For $t \in [\theta^1, \theta^1 + \overline{\Delta}]$, by favor of the equation

$$y(t) - \overline{y}(t) = (y(\xi) - \overline{y}(\xi)) + \int_{\xi}^{t} [f(y(s)) - f(\overline{y}(s))]\, ds$$
$$+ \int_{\xi}^{t} \mu[g(x(s)) - g(\overline{x}(s))]ds,$$

one can obtain that

$$\begin{aligned}\|y(t)-\overline{y}(t)\| &\geq \|y(\xi)-\overline{y}(\xi)\| - \left|\int_\xi^t L_f \|y(s)-\overline{y}(s)\|\, ds\right| \\ &- |\mu| \left|\int_\xi^t \|g(x(s))-g(\overline{x}(s))\|\, ds\right| \\ &> \frac{|\mu|\,\tau L_g \varepsilon_0}{2n(2+\tau L_f)} - 2\overline{\Delta}\left(K_0 L_f + M_g |\mu|\right) \\ &\geq \frac{|\mu|\,\tau L_g \varepsilon_0}{4n(2+\tau L_f)}.\end{aligned}$$

The length of the interval $J^1 = [\theta^1, \theta^1 + \overline{\Delta}]$ does not depend on $x(t)$, $\overline{x}(t)$, and for $t \in J^1$ the inequality $\|y(t) - \overline{y}(t)\| > \varepsilon_1$ holds, where $\varepsilon_1 = \dfrac{|\mu|\,\tau L_g \varepsilon_0}{4n(2+\tau L_f)}$. Consequently, system (9.8.20) is sensitive.

9.8.3 Unstable Cycles and Unpredictability

Assume that system (9.8.18) admits a period-doubling cascade. That is, there exists an equation

$$x' = G(x, \lambda), \tag{9.8.24}$$

where λ is a parameter and the function $G : \mathbb{R}^m \times \mathbb{R} \to \mathbb{R}^m$ is such that for some finite number λ_∞, $G(x, \lambda_\infty)$ is equal to the function $F(x)$ in the right-hand side of system (9.8.18).

System (9.8.18) is said to admit a period-doubling cascade [49, 87, 88, 91] if there exists a sequence of period-doubling bifurcation values $\{\lambda_j\}_{j\in\mathbb{N}}$ satisfying $\lambda_j \to \lambda_\infty$ as $j \to \infty$ such that as the parameter λ increases or decreases through λ_j system (9.8.24) undergoes a period-doubling bifurcation for each $j \in \mathbb{N}$. As a consequence, at the parameter value $\lambda = \lambda_\infty$, there exist infinitely many unstable periodic solutions of system (9.8.24), and hence of system (9.8.18), all lying in a bounded region.

Now, let us introduce the following definition [76]. We say that the solutions of the non-autonomous system (9.8.20), with a fixed $x(t)$, are ultimately bounded if there exists a number $B > 0$ such that for every solution $y(t)$, $y(t_0) = y_0$, of system (9.8.20), there exists a positive number R such that the inequality $\|y(t)\| < B$ holds for all $t \geq t_0 + R$.

We say that system (9.8.20) replicates the period-doubling cascade of system (9.8.18) if for each periodic solution $x(t)$ of (9.8.18), system (9.8.20) admits a periodic solution with the same period.

The following condition is required in the next theorem, which can be verified using Theorem 15.8 [76].

(A5) Solutions of system (9.8.20) are ultimately bounded by a bound common for all $x(t)$.

Theorem 9.3 *If conditions* $(A1) - (A5)$ *hold, then system (9.8.20) replicates the period-doubling cascade of system (9.8.18).*

It is worth noting that the instability of the infinite number of periodic solutions of system (9.8.20) is ensured by Theorem 9.2.

9.9 Notes

The question *"Does the flap of a butterfly's wings in Brazil set off a tornado in Texas?"* is very impressive and it has done a lot to popularize chaos for both mathematicians and non-mathematicians. Some of the authors say that the question relates sensitive dependence on initial conditions in dynamical systems considered as unpredictability for meteorological observations. Lorenz himself, in successive his talks and the book [32], was obsessed by the question and sincerely believed its possibility. He also supposed that his system can give a key for the positive answer of the question. We rely on the intuition of the great mathematician and meteorologist.

What we have done is a small step in the mathematical approach to the complexity of the weather. Our suggestions are not about a modeling, but rather an effort to answer the question why the weather is unpredictable at each point of the Earth, on the basis of the Lorenz's meteorological model and other models.

We should recognize that all our discussions can be considered as a "toy object" in the theory and according to the complexity phenomenon in meteorological investigations, one can say that the investigation of chaos in meteorology still remains as a "toy object" [26, 92]. We do not take into account changes which may happen because of the day light evolution, variety of land forms, seasonal differences in the region, etc., but what we propose is to connect regional mathematical models into a global net so that understanding the unpredictability becomes possible. The results of the present chapter were published in the paper [93].

Since the chaotification principles proposed in this chapter are not specific for the Lorenz system, they can be applied to other meteorological models as well, without any restrictions on the dimension and the number of the coupled systems. For instance, one can consider the Lorenz model of general circulation of the atmosphere [94]

$$\begin{aligned}\frac{dX_1}{dt} &= -X_2^2 - X_3^2 - \tilde{a}X_1 + \tilde{a}F\\ \frac{dX_2}{dt} &= X_1X_2 - \tilde{b}X_1X_3 - X_2 + G\\ \frac{dX_3}{dt} &= \tilde{b}X_1X_2 + X_1X_3 - X_3,\end{aligned} \tag{9.9.25}$$

where X_1 represents the strength of a large-scale westerly wind current, X_2 and X_3 represent the cosine and sine phases of a chain of superposed large-scale eddies, the parameter F represents the external-heating contrast, and G represents the heating contrast between oceans and continents. The coefficient $\tilde{b}$, if greater than unity, allows the displacement to occur more rapidly than the amplification, and the coefficient $\tilde{a}$, if less than unity, allows the westerly current to damp less rapidly than the eddies [94–97].

For $\tilde{a} > 0$ and $\tilde{b} > -1$, let us take into account the Lyapunov function $V(X) = X_1^2 + X_2^2 + X_3^2$, and set $\overline{a} = \min\{\tilde{a}, 1\}$, $\overline{b} = \sqrt{\tilde{a}^2 F^2 + G^2}$. One can verify that

$$V'_{(9.9.25)}(X) \leq -2\overline{a}\left(X_1^2 + X_2^2 + X_3^2\right) + 2\overline{b}\sqrt{X_1^2 + X_2^2 + X_3^2}$$

and

$$\left\| \frac{\partial V}{\partial X} \right\| = 2\sqrt{X_1^2 + X_2^2 + X_3^2}.$$

Therefore, the conditions of Theorem 9.1 are satisfied with $a(r) = r^2$, $b(r) = 2\overline{a}r^2 - 2\overline{b}r$, $c(r) = 2r$, $M_0 = 1$ and $B = \dfrac{1+\overline{b}}{\overline{a}}$. Consequently, our theory is also applicable to the Lorenz model of general circulation of the atmosphere.

References

1. L.F. Richardson, *Weather Prediction By Numerical Process* (Cambridge University Press, Cambridge, 1922)
2. J.G. Charney, R. Fjörtoft, J. Von Neumann, Numerical integration of the barotropic vorticity equations. Tellus **2**, 237–254 (1950)
3. J.G. Charney, On the theory of the general circulation of the atmosphere, in *The Atmosphere and Sea in Motion*, ed. by B. Bolin (Rockefeller Institute Press and Oxford University Press, New York, 1959), pp. 178–193
4. C. Crafoord, E. Källen, A note on the condition for existence of more than one steady-state solution in Budyko-Sellers type models. J. Atmos. Sci. **35**, 1123–1125 (1978)
5. M. Ghil, S. Childress, *Topics in Geophysical Fluid Dynamics: Atmospheric Dynamics, Dynamo Theory and Climate Dynamics* (Springer, New York, 1987)
6. M. Ghil, Atmospheric modeling, in *Natural Climate Variability on Decade-to-Century Time Scales*, ed. by D.G. Martinson, K. Bryan, M. Ghil, M.D. Hall, T.R. Karl, E.S. Sarachik, S. Sorooshian, L.D. Talley (National Academy Press, Washington, 1995), pp. 164–168
7. M. Ghil, A.W. Robert, Solving problems with GCMs: general circulation models and their role in the climate modeling hierarchy, in *General Circulation Model Development: Past, Present and Future*, ed. by D.A. Randall (Academic Press, New York, 2000), pp. 285–325
8. M.L. Budyko, The effect of solar radiation variations on the climate of the earth. Tellus **21**, 611–619 (1969)
9. W.D. Sellers, A climate model based on the energy balance of the earth-atmosphere system. J. Appl. Meteorol **8**, 392–400 (1969)
10. T.P. Charlock, W.D. Sellers, Aerosol effects on climate: calculations with time-dependent and steady-state radiative-convective model. J. Atmos. Sci. **38**, 1327–1341 (1980)
11. S. Manabe, R.F. Strickler, Thermal equilibrium of the atmosphere with a convective adjustment. J. Atmos. Sci. **21**, 361–385 (1964)

12. V. Ramanathan, J.A. Coakley, Climate modeling through radioactive convective models. Rev. Geophys. Space Phys. **16**, 465–489 (1978)
13. J.G. Charney, J.G. DeVore, Multiple flow equilibria in the atmosphere and blocking. J. Atmos. Sci. **36**, 1205–1216 (1979)
14. B. Legras, M. Ghil, Persistent anomalies, blocking and variations in atmospheric predictability. J. Atmos. Sci. **42**, 433–471 (1985)
15. E.N. Lorenz, The mechanics of vacillation. J. Atmos. Sci. **20**, 448–464 (1963b)
16. B.B. Reinhold, R.T. Pierrehumbert, Dynamics of weather regimes: quasi-stationary waves and blocking. Mon. Weather Rev. **110**, 1105–1145 (1982)
17. H. Gallée, J.P. van Ypersele, T. Fichefet, C. Tricot, A. Berger, Simulation of the last glacial cycle by a coupled, sectorially averaged climate-ice-sheet model, I. The climate model. J. Geophys. Res. **96**, 139–161 (1991)
18. M.C. MacCracken, S.J. Ghan, Design and use of zonally-averaged models, in *Physically-Based Modelling and Simulation of Climate and Climatic Change, Part 2*, ed. by M.E. Schlesinger (Springer, Netherlands, 2000), pp. 755–809
19. B. Saltzman, A.D. Vernekar, Global equilibrium solutions for the zonally averaged macroclimate. J. Geophys. Res. **77**, 3936–3945 (1972)
20. J. Adem, Incorporation of advection of heat by mean winds and by ocean currents in a thermodynamic model for long-range weather prediction. Mon. Weather Rev. **98**, 776–786 (1970)
21. F. Chen, M. Ghil, Interdecadal variability in a hybrid coupled ocean-atmosphere model. J. Phys. Oceanogr. **26**, 1561–1578 (1996)
22. G.R. North, J.G. Mengel, D.A. Short, Simple energy balance model resolving the seasons and the continents: application to the astronomical theory of the ice ages. J. Geophys. Res. **88**, 6576–6586 (1983)
23. H. Gao, J. Duan, Dynamics of a coupled atmosphere-ocean model. Nonlinear Anal. Real World Appl. **5**, 667–693 (2004)
24. J.L.D. Rodriguez, M. Thompson, A balanced atmospheric model of Lorenz. Nonlinear Anal. Real World Appl. **11**, 3251–3271 (2010)
25. O.S. Rozanova, J.-L. Yu, C.-K. Hu, On the position of a vortex in a two-dimensional model of atmosphere. Nonlinear Anal. Real World Appl. **13**, 1941–1954 (2012)
26. E.N. Lorenz, Deterministic nonperiodic flow. J. Atmos. Sci. **20**, 130–141 (1963)
27. B. Saltzman, Finite amplitude free convection as an initial value problem. J. Atmos. Sci. **19**, 329–341 (1962)
28. L. Rayleigh, On convective currents in a horizontal layer of fluid when the higher temperature is on the under side. Philos. Mag. **32**, 529–546 (1916)
29. E.N. Lorenz, Maximum simplification of the dynamic equations. Tellus **12**, 243–254 (1960)
30. K.T. Alligood, T.D. Sauer, J.A. Yorke, *Chaos: An Introduction to Dynamical Systems* (Springer, New York, 1996)
31. C. Sparrow, *The Lorenz Equations: Bifurcations, Chaos and Strange Attractors* (Springer, New York, 1982)
32. E.N. Lorenz, *The Essence of Chaos* (UCL Press, London, 1993)
33. K. Fraedrich, Estimating the dimensions of weather and climate attractors. J. Atmos. Sci. **43**, 419–432 (1993)
34. H. Itoh, M. Kimoto, Multiple attractors and chaotic itinerancy in a quasigeostrophic model with realistic topography: implications for weather regimes and low-frequency variability. J. Atmos. Sci. **53**, 2217–2231 (1996)
35. V. Krishnamurthy, A predictability study of Lorenz's 28-variable model as a dynamical system. J. Atmos. Sci. **50**, 2215–2229 (1993)
36. E.N. Lorenz, Designing chaotic models. J. Atmos. Sci. **62**, 1574–1587 (2005)
37. H. Mukougawa, M. Kimoto, S. Yoden, A relationship between local error growth and quasi-stationary states: case study in the Lorenz system. J. Atmos. Sci. **48**, 1231–1237 (1991)
38. P. Grassberger, I. Procaccia, Characterization of strange attractors. Phys. Rev. Lett. **50**, 346–349 (1983a)

39. P. Grassberger, I. Procaccia, Measuring the strangeness of strange attractors. Phys. D: Nonlinear Phenom. **9**, 189–208 (1983b)
40. H. Li, X. Liao, S. Ullah, L. Xiao, Analytical proof on the existence of chaos in a generalized duffing-type oscillator with fractional-order deflection. Nonlinear Anal. Real World Appl. **13**, 2724–2733 (2012)
41. S. Wiggins, *Global Bifurcations and Chaos: Analytical Methods* (Springer, New York, 1988)
42. Y. Feliks, Nonlinear dynamics and chaos in the sea and land breeze. J. Atmos. Sci. **61**, 2169–2187 (2004)
43. L.O. Chua, M. Komuro, T. Matsumoto, The double scroll family, parts I and II. IEEE Trans. Circuit Syst. **CAS-33**, 1072–1118 (1986)
44. R.L. Devaney, *An Introduction to Chaotic Dynamical Systems* (Addison-Wesley, Menlo Park, 1989)
45. J. Guckenheimer, P.J. Holmes, *Nonlinear Oscillations, Dynamical Systems and Bifurcations of Vector Fields* (Springer, New York, 1997)
46. T.Y. Li, J.A. Yorke, Period three implies chaos. Am. Math. Mon. **82**, 985–992 (1975)
47. O.E. Rössler, An equation for continuous chaos. Phys. Lett. **57A**, 397–398 (1976)
48. J.M.T. Thompson, H.B. Stewart, *Nonlinear Dynamics and Chaos* (Wiley, Chichester, 2002)
49. M.J. Feigenbaum, Universal behavior in nonlinear systems. Los Alamos Sci./Summer 4-27 (1980)
50. J. Kennedy, J.A. Yorke, Topological horseshoes. Trans. Am. Math. Soc. **353**, 2513–2530 (2001)
51. J. Kennedy, S. Koçak, J.A. Yorke, A chaos lemma. Am. Math. Mon. **108**, 411–423 (2001)
52. Y. Pomeau, P. Manneville, Intermittent transition to turbulence in dissipative dynamical systems. Commun. Math. Phys. **74**, 189–197 (1980)
53. H.G. Schuster, W. Just, *Deterministic Chaos, An Introduction* (Wiley-VCH, Federal Republic of Germany, 2005)
54. H.D.I. Abarbanel, N.F. Rulkov, M.M. Sushchik, Generalized synchronization of chaos: the auxiliary system approach. Phys. Rev. E **53**, 4528–4535 (1996)
55. J.M. Gonzáles-Miranda, *Synchronization and Control of Chaos* (Imperial College Press, London, 2004)
56. B.R. Hunt, E. Ott, J.A. Yorke, Differentiable generalized synchronization of chaos. Phys. Rev. E **55**(4), 4029–4034 (1997)
57. T. Kapitaniak, Synchronization of chaos using continuous control. Phys. Rev. E **50**, 1642–1644 (1994)
58. L. Kocarev, U. Parlitz, Generalized synchronization, predictability, and equivalence of unidirectionally coupled dynamical systems. Phys. Rev. Lett. **76**(11), 1816–1819 (1996)
59. E.E.N. Macau, C. Grebogi, Y.-C. Lai, Active synchronization in nonhyperbolic hyperchaotic systems. Phys. Rev. E **65**, 027202 (2002)
60. L.M. Pecora, T.L. Carroll, Synchronization in chaotic systems. Phys. Rev. Lett. **64**, 821–825 (1990)
61. N.F. Rulkov, M.M. Sushchik, L.S. Tsimring, H.D.I. Abarbanel, Generalized synchronization of chaos in directionally coupled chaotic systems. Phys. Rev. E **51**(2), 980–994 (1995)
62. M. Akhmet, *Principles of Discontinuous Dynamical Systems* (Springer, New York, 2010)
63. M. Akhmet, *Nonlinear Hybrid Continuous/Discrete-Time Models* (Atlantis Press, Paris, 2011)
64. M.U. Akhmet, Devaney's chaos of a relay system. Commun. Nonlinear Sci. Numer. Simul. **14**, 1486–1493 (2009)
65. M.U. Akhmet, Dynamical synthesis of quasi-minimal sets. Int. J. Bifurc. Chaos **19**, 2423–2427 (2009)
66. M.U. Akhmet, Li-Yorke chaos in the system with impacts. J. Math. Anal. Appl. **351**, 804–810 (2009)
67. M.U. Akhmet, M.O. Fen, Chaos generation in hyperbolic systems. Interdiscip. J. Discont. Nonlinear. Complex. **1**, 367–386 (2012)
68. M.U. Akhmet, M.O. Fen, Chaotic period-doubling and OGY control for the forced duffing equation. Commun. Nonlinear. Sci. Numer. Simul. **17**, 1929–1946 (2012)

69. M.U. Akhmet, M.O. Fen, Replication of chaos. Commun. Nonlinear. Sci. Numer. Simul. **18**, 2626–2666 (2013)
70. M.U. Akhmet, M.O. Fen, Entrainment by chaos. J. Nonlinear Sci. **24**, 411–439 (2014)
71. R. Brown, L. Chua, Dynamical synthesis of poincaré maps. Int. J. Bifurc. Chaos **3**, 1235–1267 (1993)
72. R. Brown, L. Chua, From almost periodic to chaotic: the fundamental map. Int. J. Bifurc. Chaos **6**, 1111–1125 (1996)
73. R. Brown, L. Chua, Chaos: generating complexity from simplicity. Int. J. Bifurc. Chaos **7**, 2427–2436 (1997)
74. R. Brown, R. Berezdivin, L. Chua, Chaos and complexity. Int. J. Bifurc. Chaos **11**, 19–26 (2001)
75. V. Franceschini, A Feigenbaum sequence of bifurcations in the Lorenz model. J. Stat. Phys. **22**, 397–406 (1980)
76. T. Yoshizawa, *Stability Theory and the Existence of Periodic Solutions and Almost Periodic Solutions* (Springer, Berlin, 1975)
77. H. Haken, *Advanced Synergetics: Instability Hierarchies of Self-Organizing Systems and Devices* (Springer, Berlin, 1983)
78. G. Nicolis, I. Prigogine, *Exploring Complexity: An Introduction* (W.H. Freeman, New York, 1989)
79. F. Durrenmatt, *The Physicists* (Grove, New York, 1964)
80. A.A. Andronov, A.A. Vitt, C.E. Khaikin, *Theory of Oscillations* (Pergamon Press, Oxford, 1966)
81. F.C. Moon, *Chaotic Vibrations: An Introduction for Applied Scientists and Engineers* (Wiley, Hoboken, 2004)
82. H. Haken, *Information and Self-Organization: A Macroscopic Approach to Complex Systems* (Springer, Berlin, 1988)
83. H. Haken, *Brain Dynamics, Synchronization and Activity Patterns in Pulse-Coupled Neural Nets with Delays and Noise* (Springer, Berlin, 2002)
84. J.D. Murray, *Mathematical Biology II: Spatial Models and Biomedical Applications* (Springer, New York, 2003)
85. M.A. Vorontsov, W.B. Miller, *Self-organization in Optical Systems and Applications in Information Technology* (Springer, Berlin, 1998)
86. A.M. Turing, The chemical basis of morphogenesis. Philos. Trans. R. Soc. Lond. Ser. B Biol. Sci. **237**, 37–72 (1952)
87. E. Sander, J.A. Yorke, Period-doubling cascades galore. Ergod. Theory Dyn. Syst. **31**, 1249–1267 (2011)
88. E. Sander, J.A. Yorke, Connecting period-doubling cascades to chaos. Int. J. Bifurc. Chaos **22**, 1–16 (2012)
89. N. Rouche, P. Habets, M. Laloy, *Stability Theory by Liapunov's Direct Method* (Springer, Berlin, 1977)
90. C. Robinson, *Dynamical Systems: Stability, Symbolic Dynamics, and Chaos* (CRC Press, Boca Raton, 1995)
91. I. Zelinka, S. Celikovsky, H. Richter, G. Chen (eds.), *Evolutionary Algorithms and Chaotic Systems* (Springer, Berlin, 2010)
92. C. Grebogi, J.A. Yorke, *The Impact of Chaos on Science and Society* (United Nations University Press, Tokyo, 1997)
93. M. Akhmet, M.O. Fen, Extension of Lorenz unpredictability. Int. J. Bifurc. Chaos (in press)
94. E.N. Lorenz, Irregularity: a fundamental property of the atmosphere. Tellus **36A**, 98–110 (1984)
95. H. Broer, C. Simó, R. Vitolo, Bifurcations and strange attractors in the Lorenz-84 climate model with seasonal forcing. Nonlinearity **15**, 1205–1267 (2002)
96. C. Masoller, A.C. Sicardi, Schifino, L. Romanelli, Characterization of strange attractors of Lorenz model of general circulation of the atmosphere. Chaos Solitons Fractals **6**, 357–366 (1995)
97. P.J. Roebber, Climate variability in a low-order coupled atmosphere-ocean model. Tellus **47A**, 473–494 (1995)

Chapter 10
Spatiotemporal Chaos in Glow Discharge-Semiconductor Systems

Spatiotemporal chaos is one of the complicated structures observed in spatially extended dynamical systems and it is characterized by chaotic properties both in time and space coordinates. The existence of a positive Lyapunov exponent can be used to detect spatiotemporal complexity, which can be observed, for example, in liquid crystal light valves, electroconvection, cardiac fibrillation, chemical reaction-diffusion systems, and fluidized granular matter. Spatially extended dynamical systems often serve as standard models for the investigation of complex phenomena in electronics. A special interest is directed toward pattern formation phenomena in electronic media, mainly the nonlinear gas discharge systems. It is clear that chaos can appear as an intrinsic property of systems as well as through couplings. The interaction of spatially extended systems is important for neural networks, reentry initiation in coupled parallel fibers, thermal convection in multilayered media, and for systems consisting of several weakly coupled spatially extended systems such as the electrohydrodynamical convection in liquid crystals. In the present chapter, we numerically verify the appearance of cyclic chaotic behavior in unidirectionally coupled glow discharge-semiconductor systems. The chaos in the response system is obtained through period-doubling cascade of the drive system such that it admits infinitely many unstable periodic solutions and sensitivity is present. Previously, the extension of chaos through couplings has been considered by synchronization [1–7]. The task is difficult for partial differential equations because of the choice of connecting parameters [8–10]. Kocarev et al. [8] suggested a useful time-discontinuous monitoring for synchronization, but our choice is based on a finite-dimensional connection. It is demonstrated that the present results cannot be reduced to anyone in the theory of synchronization of chaos. The technique of chaos extension suggested in the present chapter can be related to technical problems [11, 12], where collectives of microdischarge systems are considered and in models which appear in neural networks, hydrodynamics, optics, chemical reactions, and electrical oscillators. Stabilization of multidimensional periodical regimes can be useful in the applications of the glow discharge systems in conventional and energy saving lamps, beamers, flat TV screens, etc.

M. Akhmet and M.O. Fen, *Replication of Chaos in Neural Networks, Economics and Physics*, Nonlinear Physical Science,
DOI 10.1007/978-3-662-47500-3_10

10.1 Introduction

The investigations of chaos theory for continuous-time dynamics started due to the needs of real-world applications, especially with the studies of Poincaré [13], Cartwright and Littlewood [14], Levinson [15], Lorenz [16], and Ueda [17]. Chaotic dynamics has high effectiveness in the analysis of electrical processes of neural networks [18, 19] and can be used for optimization and self-organization problems in robotics [20]. The reason for that is the opportunities provided by the dynamical structure of chaos.

Starting from the primary investigations [14–17], chaos has been found as an internal property of systems, and studies in this sense have prolonged until today, for example, by the construction of discrete maps [21–24]. At the very beginning of the chaos analysis, one has to mention the Smale Horseshoes technique [25] and symbolic dynamics [26]. Another opportunity to reveal chaotic dynamics is the usage of bifurcation diagrams [27, 28].

If one considers a mechanical or electrical system and perturb it by an external force which is bounded, periodic or almost periodic, then the forced system can produce a behavior with a similar property, boundedness/periodicity/almost periodicity [29–33]. A reasonable question appears whether it is possible to use a chaotic force to obtain the same type of complexity in physical systems.

To meet the challenge, we introduced rigorous description of chaotic force as a function or *a set of functions* and described *the input–output mechanism* for ordinary differential equations in the studies [34–46]. It was rigorously proved that an irregular behavior can follow the chaotic force very likely as regular motions do. We have applied the machinery to mechanical and electrical systems with a finite number of freedom [34–42, 44, 46] as well as to neural networks [43, 45]. In the present chapter, we apply the theory to unidirectionally coupled glow discharge-semiconductor (GDS) systems.

10.2 Preliminaries

Chaotic dynamics can appear in systems as an intrinsic property and it can be extended through interactions. In the literature, an effective and unique way of the chaos extension from one system to another has been suggested within the scope of generalized synchronization [1–5, 7], which characterizes the dynamics of a response system that is driven by the output of a chaotic driving system. Suppose that the dynamics of the drive and response systems are governed by the following systems with a skew product structure

$$x' = F(x) \tag{10.2.1}$$

and

$$y' = G(y, H(x)), \tag{10.2.2}$$

respectively, where $x \in \mathbb{R}^m$, $y \in \mathbb{R}^n$. Generalized synchronization is said to occur if there exist sets I_x, I_y of initial conditions and a transformation ϕ, defined on the chaotic attractor of (10.2.1), such that for all $x(0) \in I_x$ and $y(0) \in I_y$ the relation $\lim_{t\to\infty} \|y(t) - \phi(x(t))\| = 0$ holds. In that case, a motion which starts on $I_x \times I_y$ collapses onto a manifold $M \subset I_x \times I_y$ of synchronized motions. The transformation ϕ is not required to exist for the transient trajectories. When ϕ is the identity, the identical synchronization takes place [3, 6].

The synchronization of a large class of unidirectionally coupled chaotic partial differential equations was deeply investigated in [8, 9], where the synchronization was achieved by applying the driving signals only at a finite number of space points. The synchronization of spatiotemporal chaos in a pair of complex Ginzburg–Landau equations was performed in [10] for the case when all space points are continuously driven. In the present chapter, we use perturbations to a single coordinate of an infinite dimensional response system, which is nonchaotic in the absence of driving, to obtain chaotic motions in the system.

It has not been investigated whether the response system admits the same type of chaos with the drive system in the theory of chaos synchronization yet. The replication of chaos with specific types such as Devaney [47], Li-Yorke [23], and period-doubling cascade [48–50] was investigated for drive-response couples for the first time in the papers [34–46].

In the study [44], we considered a system of the form

$$u' = K(u), \tag{10.2.3}$$

where $K : \mathbb{R}^n \to \mathbb{R}^n$ is a continuously differentiable function. We supposed that system (10.2.3) possesses an orbitally stable limit cycle and perturbed it with solutions of a chaos generating system, in the form of (10.2.1), and set up the system

$$y' = K(y) + \mu M(x), \tag{10.2.4}$$

where μ is a nonzero number and $M : \mathbb{R}^m \to \mathbb{R}^n$ is a continuous function. The extension of sensitivity and chaos through period-doubling cascade for the coupled system (10.2.1)–(10.2.4) were rigorously proved in the paper [44]. As a result, we achieved *chaotic cycles*, that is, motions which behave cyclically and chaotically simultaneously.

The rich experience of chaos expansion in finite-dimensional spaces provides a confidence that our approach mentioned in [44] has to work also in infinite-dimensional spaces. In this chapter, we numerically observe the presence of orbitally stable limit cycles in the 2-dimensional projections of the infinite-dimensional space as well as their deformation to chaotic cycles under chaotic perturbations. By using

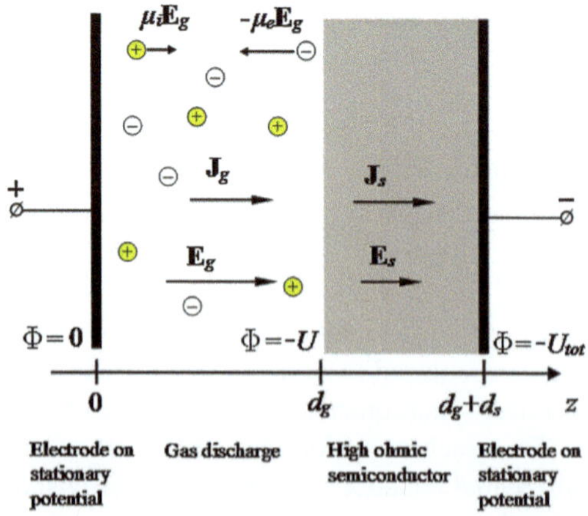

Fig. 10.1 A cross section of a planar discharge cell: it consists of a metal anode, a gas layer, a high-ohmic cathode, and another metal contact. The subscripts g and s refer to the gas and semiconductor regions

the technique presented in [44], one can elaborate the results of the present chapter from the theoretical point of view. Although couplings of GDS systems have not been performed in the literature yet, our results reveal the opportunity of chaos extension in such systems.

Summarizing, electronic systems are important tools for synchronization and chaos extension. In the present chapter, we make use of our previous approach [44] to extend chaos in unidirectionally coupled GDS systems.

10.2.1 Description of the GDS Model

Our GDS was previously studied both theoretically and experimentally in [51–66]. It represents a planar plasma layer coupled to a planar semiconductor layer, which are sandwiched between two planar electrodes to which a DC voltage is applied (see Fig. 10.1). We used a one-dimensional fluid model for this system, where any pattern formation in the transversal direction is excluded and only the single dimension normal to the layers is resolved. For the gas discharge, the model takes into account electron and ion drift in the electric field, bulk impact ionization, and secondary emission from the cathode as well as space charge effects. The semiconductor is approximated with a constant conductivity.

The gas discharge part of the model consists of continuity equations for two charged species, namely, electrons and positive ions with particle densities n_e and n_i:

$$\partial_t n_\mathrm{e} + \nabla \cdot \Gamma_\mathrm{e} = S_e, \qquad (10.2.5)$$

$$\partial_t n_i + \nabla \cdot \Gamma_i = S_i, \qquad (10.2.6)$$

which are coupled to Poisson's equation for the electric field in electrostatic approximation:

$$\nabla \cdot \mathbf{E} = \frac{e}{\varepsilon_0}(n_i - n_e), \quad \mathbf{E} = -\nabla \Phi. \tag{10.2.7}$$

Here, Φ is the electric potential, $\mathbf{E}$ is the electric field in the gas discharge, e is the elementary charge, and ε_0 is the dielectric constant. The vector fields Γ_e and Γ_i are the particle flux densities, that in simplest approximation are described by drift only. (In general, particle diffusion $D_{e,i}\nabla n_{e,i}$ could be included.) The drift velocities are assumed to depend linearly on the local electric field with mobilities $\mu_e \gg \mu_i$:

$$\Gamma_e = -\mu_e n_e \mathbf{E}, \quad \Gamma_i = \mu_i n_i \mathbf{E}, \tag{10.2.8}$$

hence the total electric current in the discharge is

$$\mathbf{J} = \varepsilon_0 \partial_t \mathbf{E} + e\left(\Gamma_i - \Gamma_e\right) = \varepsilon_0 \partial_t \mathbf{E} + e\left(\mu_i n_i + \mu_e n_e\right)\mathbf{E}. \tag{10.2.9}$$

Two types of ionization processes are taken into account: the α process of electron impact ionization in the bulk of the gas, and the γ process of electron emission by ion impact onto the cathode. In a local field approximation, the α process determines the source terms in the continuity equations (10.2.5) and (10.2.6):

$$S_e = S_i = |\Gamma_e|\, \alpha_0\, \alpha\left(|\mathbf{E}|/E_0\right), \tag{10.2.10}$$

where we use the classical Townsend approximation

$$\alpha\left(|\mathbf{E}|/E_0\right) = \exp\left(-E_0/|\mathbf{E}|\right). \tag{10.2.11}$$

The effect of the semiconductor layer with thickness d_s, conductivity σ_s, dielectric constant ε_s is described by the external circuit equation

$$\partial_t U = \frac{U_{tot} - U - R_s J}{T_s}, \tag{10.2.12}$$

where U_{tot} is the applied voltage, $U = \int_0^{d_g} E\, dZ$ is the voltage over the gas discharge which is the electric field E integrated over the height d_g of the discharge, $R_s = d_s/\sigma_s$ is the resistance of the semiconductor layer, where σ_s is its conductivity, and $T_s = \varepsilon_s \varepsilon_0/\sigma_s$ is the Maxwell relaxation time of the semiconductor with dielectric constant ε_s.

Following the traditions of the synchronization of chaotic systems, we will call the coupled GDS systems as the *drive* and *response* systems.

The goal of our investigation is to extend the spatiotemporal chaos of a drive GDS system to a response GDS system by means of a special connection mechanism between the systems. In order to make the present chapter self-sufficient, we complete the chaos analysis of the GDS system, which was initiated in the papers [63, 64]. The method of the analysis, as well as the connection mechanism are our theoretical suggestions [34–46].

The chaos obtained through period-doubling cascade [48–50] is under investigation in this chapter. In other words, the existence of infinitely many unstable periodic solutions and the presence of sensitivity [47] are considered. One of the advantages of our approach is the controllability of the extended chaos [3, 41, 42, 44, 67]. It is possible to stabilize an unstable periodic solution of the response GDS system by controlling the chaos of the drive system. The presented technique is applicable to large number of interconnected GDS systems and the control of the global chaos can also be achieved. This approach can be useful for applications of the gas discharge systems in conventional and energy saving lamps, beamers, flat TV screens, etc. [11, 12].

10.2.2 The Model in Dimensionless Form

The dimensional analysis is performed essentially as in [63, 64]. In dimensional units, Z parametrizes the direction normal to the layers. The anode of the gas discharge is at $Z = 0$, the cathode end of the discharge is at $Z = d_g$, and the semiconductor extends up to $Z = d_g + d_s$.

When diffusion is neglected, the ion current and the ion density at the anode vanish. This is described by the boundary condition on the anode $Z = 0$:

$$\Gamma_i(0, t) = 0 \quad \Rightarrow \quad n_i(0, t) = 0. \tag{10.2.13}$$

The boundary condition at the cathode, $Z = d_g$, describes the γ-process of secondary electron emission:

$$\left|\Gamma_e\left(d_g, t\right)\right| = \gamma \left|\Gamma_i\left(d_g, t\right)\right| \quad \Rightarrow \quad \mu_e n_e\left(d_g, t\right) = \gamma \mu_i n_i\left(d_g, t\right). \tag{10.2.14}$$

Finally, a DC voltage U_{tot} is applied to the system determining the electric potential on the boundaries

$$\Phi(0, t) = 0, \quad \Phi\left(d_g + d_s, t\right) = -U_{tot}. \tag{10.2.15}$$

Here, the first potential vanishes due to gauge freedom. We denote the potential at the interface between the semiconductor and the gas discharge by $-U$ so that $\Phi\left(d_g, t\right) = -U$.

Let us introduce the intrinsic parameters of the system as $t_0 = \dfrac{1}{\alpha_0 \mu_e E_0}$, $Z_0 = \dfrac{1}{\alpha_0}$, $n_0 = \dfrac{\varepsilon_0 \alpha_0 E_0}{e}$. In the studies [63, 64], the problem was reduced to one spatial dimension z such that the GDS system takes the following dimensionless form:

$$\begin{aligned}
&\partial_\tau \sigma - \partial_z (\mathscr{E}\sigma) = \sigma \mathscr{E} \alpha (\mathscr{E}), \\
&\partial_\tau \rho + \mu \partial_z (\mathscr{E}\rho) = \sigma \mathscr{E} \alpha (\mathscr{E}), \\
&\partial_z \mathscr{E} = \rho - \sigma, \quad \mathscr{E} = -\partial_z \phi,
\end{aligned} \tag{10.2.16}$$

where the dimensionless time, coordinates, and fields are $z = \frac{Z}{Z_0}$, $\tau = \frac{t}{t_0}$, $\sigma(z,\tau) = \frac{n_e(Z,t)}{n_0}$, $\rho(z,\tau) = \frac{n_i(Z,t)}{n_0}$, $\mathscr{E}(z,\tau) = \frac{E(Z,t)}{E_0}$, $\phi(z,\tau) = \frac{\Phi(Z,t)}{E_0 Z_0}$ and $\alpha(\mathscr{E}) = e^{-1/|\mathscr{E}|}$.

The intrinsic dimensionless parameters of the gas discharge are the mobility ratio μ of electrons and ions and the length ratio L of discharge gap width and impact ionization length. That is, $\mu = \frac{\mu_i}{\mu_e}$ and $L = \frac{d_g}{Z_0}$. The boundary conditions become

$$\begin{aligned}
&\rho(0,\tau) = 0, \\
&\sigma(L,\tau) = \gamma \mu \rho(L,\tau), \\
&\phi(0,\tau) = 0, \quad \phi(L,\tau) = -\mathscr{U},
\end{aligned} \tag{10.2.17}$$

and the external circuit is described by

$$\partial_\tau \mathscr{U} = \frac{\mathscr{U}_{tot} - \mathscr{U} - \mathscr{R}_s j}{\tau_s}, \tag{10.2.18}$$

where the total applied voltage is rescaled as $\mathscr{U}_{tot} = U_{tot}/(E_0 Z_0)$, dimensionless voltage $\mathscr{U}(\tau) = \int_0^L \mathscr{E}\, dz$, time scale $\tau_s = T_s/t_0$, resistance $\mathscr{R}_s = R_s\, e\mu_e n_0/Z_0$, and spatially conserved total current $j(\tau) = \partial_\tau \mathscr{E} + \mu\rho\mathscr{E} + \sigma\mathscr{E}$.

We consider a regime corresponding to a transition between Townsend and glow discharge. The parameters are taken as in the experiments [66] and in our previous work [63]. The discharge is in nitrogen at 40 mbar, in a gap of 1.4 mm. We used the ion mobility $\mu_i = 23.33\ \mathrm{cm}^2/(\mathrm{V\,s})$ and electron mobility $\mu_e = 6666.6\ \mathrm{cm}^2/(\mathrm{V\,s})$, therefore the mobility ratio is $\mu = \mu_i/\mu_e = 0.0035$. The secondary emission coefficient was taken as $\gamma = 0.08$. The applied voltages U_{tot} are in the range of 513–570 V. For $\alpha_0 = Ap = [27.8\ \mu\mathrm{m}]^{-1}$ and for $E_0 = Bp = 10.3\ \mathrm{kV/cm}$, we used values from [58]. The semiconductor layer consists of 1.5 mm of GaAs with dielectric constant $\varepsilon_s = 13.1$ and conductivity $\sigma_s = (2.6 \times 10^5\ \Omega\,\mathrm{cm})^{-1}$. Corresponding dimensionless parameters are $L = 50$, $\mathscr{R}_s = 30597$, $\tau_s = 7435$, and a total voltage range $\mathscr{U}_{tot}$ between 17.67 and 20.03.

10.3 Chaotically Coupled GDS Systems

In the present section, we will extend the spatiotemporal chaos of a drive GDS system through utilizing its voltage over the gas discharge as a chaotic control applied to the electric circuit of a response GDS system. In the coupling, the voltage over the discharge of the drive system is applied as a perturbation to the circuit equation of the response system. The presence of chaos in the response system will be shown numerically. Moreover, we will compare our results with generalized synchronization.

The full analysis of the spatiotemporal chaos in the GDS system (10.2.16)–(10.2.18) is provided in Sect. 10.4, where the bifurcation diagram as well as the chaotic behaviors in the voltage, electric field, electron density, and ion density of the system are represented. According to these results, the GDS system

$$\begin{aligned} &\partial_\tau \sigma - \partial_z (\mathscr{E}\sigma) = \sigma \mathscr{E} \alpha(\mathscr{E}), \\ &\partial_\tau \rho + \mu \partial_z (\mathscr{E}\rho) = \sigma \mathscr{E} \alpha(\mathscr{E}), \\ &\partial_z \mathscr{E} = \rho - \sigma, \quad \mathscr{E} = -\partial_z \phi, \\ &\partial_\tau \mathscr{U} = \frac{20 - \mathscr{U} - \mathscr{R}_s j}{\tau_s}, \end{aligned} \tag{10.3.19}$$

is chaotic, and it will be accompanied by the boundary conditions

$$\begin{aligned} &\rho(0,\tau) = 0, \\ &\sigma(L,\tau) = \gamma \mu \rho(L,\tau), \\ &\phi(0,\tau) = 0, \quad \phi(L,\tau) = -\mathscr{U}. \end{aligned}$$

We will take into account (10.3.19) as the drive system.

The solutions of (10.3.19) will be used as a perturbation for the response GDS system in the form,

$$\begin{aligned} &\partial_\tau \widetilde{\sigma} - \partial_z (\widetilde{\mathscr{E}}\widetilde{\sigma}) = \widetilde{\sigma} \widetilde{\mathscr{E}} \alpha(\widetilde{\mathscr{E}}), \\ &\partial_\tau \widetilde{\rho} + \mu \partial_z (\widetilde{\mathscr{E}}\widetilde{\rho}) = \widetilde{\sigma} \widetilde{\mathscr{E}} \alpha(\widetilde{\mathscr{E}}), \\ &\partial_z \widetilde{\mathscr{E}} = \widetilde{\rho} - \widetilde{\sigma}, \quad \widetilde{\mathscr{E}} = -\partial_z \widetilde{\phi}, \\ &\partial_\tau \mathscr{V} = \frac{\mathscr{V}_{tot} - \mathscr{V} - \mathscr{R}_s \widetilde{j} + \delta \mathscr{U}(\tau)}{\tau_s}, \end{aligned} \tag{10.3.20}$$

with the boundary conditions

$$\begin{aligned} &\widetilde{\rho}(0,\tau) = 0, \\ &\widetilde{\sigma}(L,\tau) = \gamma \mu \widetilde{\rho}(L,\tau), \\ &\widetilde{\phi}(0,\tau) = 0, \quad \widetilde{\phi}(L,\tau) = -\mathscr{V}. \end{aligned}$$

In system (10.3.20), δ is a nonzero number and the term $\delta \mathscr{U}(\tau)/\tau_s$ is the perturbation from the drive system (10.3.19).

It is shown in Sect. 10.4 for the parameter value $\mathscr{U}_{tot} = 17.7$ that the projection of the attractor of system (10.2.16)–(10.2.18) on the domain of Eq. (10.2.18) is a stable limit cycle (see Fig. 10.7). That is, in the absence of driving, the response system (10.3.20) with $\mathscr{V}_{tot} = 17.7$ does not possess chaos. We will numerically show that the response GDS system possesses chaotic motions near the limit cycle, provided

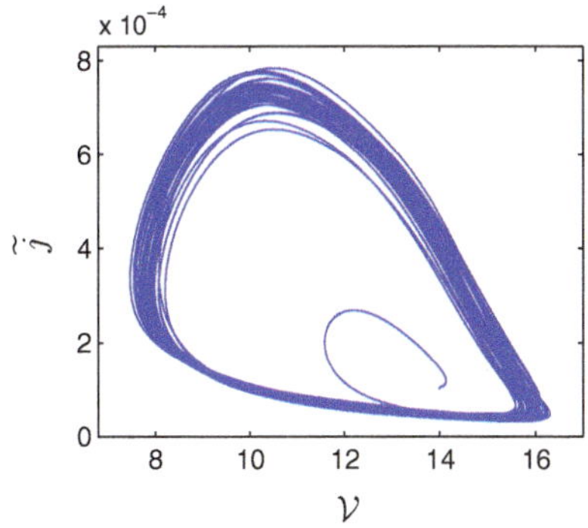

Fig. 10.2 The trajectory of the response system (10.3.20) in the $\mathcal{V} - \tilde{j}$ plane manifests the chaotic cycle

that the driving effect is included. Our results are theoretically based on the study [44], where we have proved that if the drive system admits infinitely many unstable periodic solutions and sensitivity, then the response system does the same. Since the attractor exists in system (10.2.16)–(10.2.18) with $\mathcal{U}_{tot} = 17.7$, one can conclude by the extension of our results presented in [44] that if the number $|\delta|$ in Eq. (10.3.20) is sufficiently small, then system (10.3.20) possesses cyclic chaos on the $\mathcal{V} - \tilde{j}$ plane.

Let us take $\mathcal{V}_{tot} = 17.7$ and $\delta = 0.047$ in the response GDS system (10.3.20). Using the solution of the drive system shown in Figs. 10.8, 10.9 and 10.10, we depict in Fig. 10.2 the projection of a chaotic solution of (10.3.20) on the $\mathcal{V} - \tilde{j}$ plane. The figure reveals that the response GDS system possesses motions that behave chaotically around the limit cycle of system (10.2.16)–(10.2.18) with $\mathcal{U}_{tot} = 17.7$. Moreover, to support the presence of chaos in the response system, we depict in Fig. 10.3 the time series of the $\mathcal{V}$ coordinate. The amplitude ranges 15–16.6 and 7.4–8.6 are used in Fig. 10.3b, c, respectively, to increase the visibility of chaotic behavior.

Figure 10.4a–c, depict, respectively, the chaotic behaviors in the electric field, electron density, and ion density of system (10.3.20). The figure supports the presence of chaos in the response GDS system such that it is the expansion of the one which takes place on the $\mathcal{V} - \tilde{j}$ plane.

Now, let us compare our results with generalized synchronization (GS) [1–5, 7]. According to Kocarev and Parlitz [5], GS occurs for the coupled systems (10.2.1) and (10.2.2) if and only if for all $x_0 \in I_x$, y_{10}, $y_{20} \in I_y$, the asymptotic stability criterion $\lim_{t\to\infty} \|y(t, x_0, y_{10}) - y(t, x_0, y_{20})\| = 0$ holds, where $y(t, x_0, y_{10})$ and $y(t, x_0, y_{20})$ denote the solutions of (10.2.2) with the initial data $y(0, x_0, y_{10}) = y_{10}$, $y(0, x_0, y_{20}) = y_{20}$ and the same $x(t)$, $x(0) = x_0$. This criterion is a mathematical formulation of the auxiliary system approach [1, 3]. We shall make use of the auxiliary system approach to demonstrate the absence of generalized synchronization in the coupled system (10.3.19)–(10.3.20).

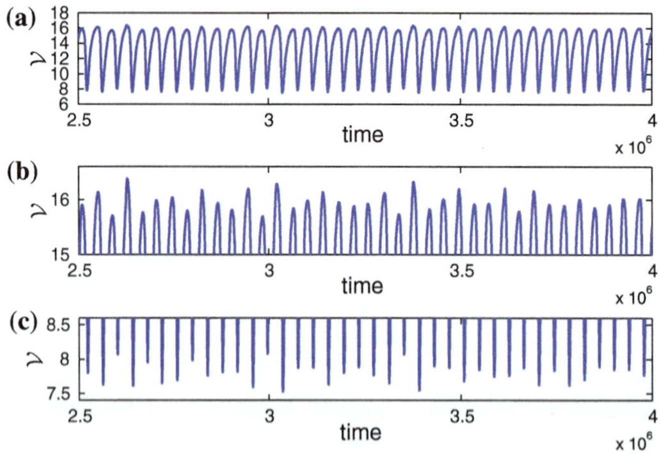

Fig. 10.3 The behavior of the $\mathcal{V}$ coordinate of system (10.3.20) is shown in (**a**). In **b**, **c**, where the chaotic behavior is observable, the amplitudes are restricted to the ranges 15–16.6 and 7.4–8.6, respectively

We introduce the auxiliary system

$$\begin{aligned}
&\partial_\tau \overline{\sigma} - \partial_z \left(\overline{\mathscr{E}\sigma}\right) = \overline{\sigma}\overline{\mathscr{E}}\alpha\left(\overline{\mathscr{E}}\right),\\
&\partial_\tau \overline{\rho} + \mu\partial_z \left(\overline{\mathscr{E}\rho}\right) = \overline{\sigma}\overline{\mathscr{E}}\alpha\left(\overline{\mathscr{E}}\right),\\
&\partial_z \overline{\mathscr{E}} = \widetilde{\rho} - \overline{\sigma}, \quad \overline{\mathscr{E}} = -\partial_z \overline{\phi},\\
&\partial_\tau \mathscr{W} = \frac{17.7 - \mathscr{W} - \mathscr{R}_s \overline{j} + 0.047\mathscr{U}(\tau)}{\tau_s}
\end{aligned} \tag{10.3.21}$$

with the boundary conditions

$$\begin{aligned}
&\overline{\rho}(0,\tau) = 0,\\
&\overline{\sigma}(L,\tau) = \gamma\mu\overline{\rho}(L,\tau),\\
&\overline{\phi}(0,\tau) = 0, \quad \overline{\phi}(L,\tau) = -\mathscr{W}.
\end{aligned}$$

Making use of the solution $\mathscr{U}(\tau)$ whose graph is represented in Fig. 10.9 in both of the systems (10.3.20) and (10.3.21), we depict in Fig. 10.5 the projection of the stroboscopic plot of system (10.3.20)–(10.3.21) on the $\mathcal{V} - \mathscr{W}$ plane. The first 500 iterations are omitted in the simulation. The time interval $[0, 80 \times 10^6]$ is used and the time step is taken as $\Delta\tau = 5000$. Since the plot does not take place on the line $\mathscr{W} = \mathcal{V}$, we conclude that *generalized synchronization is not achieved* in the dynamics of the coupled system (10.3.19)–(10.3.20).

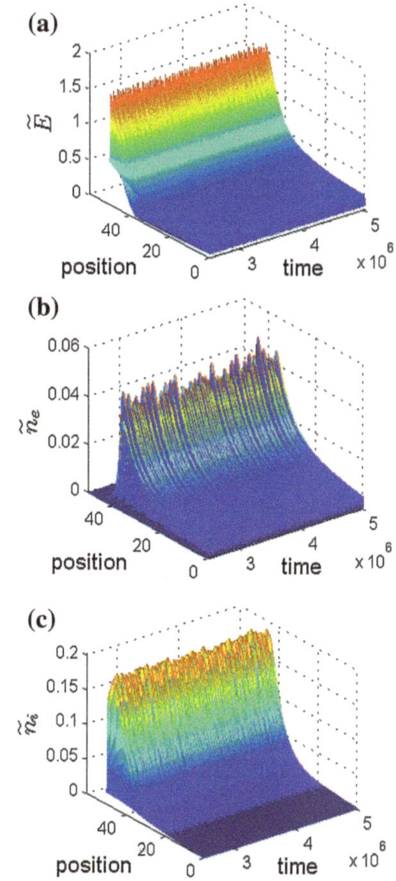

Fig. 10.4 Time evolution of profiles of the **a** electric field $\widetilde{E}$, **b** electron density $\widetilde{n}_e$, and **c** ion density $\widetilde{n}_i$ support the existence of chaotic motions around the periodic solution

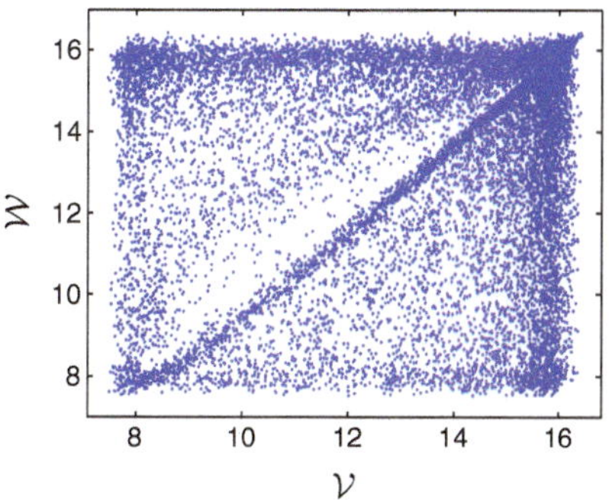

Fig. 10.5 Application of the auxiliary system approach reveals that the coupled systems (10.3.19) and (10.3.20) are not synchronized

10.4 The Chaos in the Drive GDS System

In this part, we will extend the results of [63] about the presence of chaos in GDS systems. In the paper [63], only a finite number of period-doubling bifurcations were indicated. However, in the present section, we represent the occurrence of infinitely many period-doubling bifurcations by means of a bifurcation diagram and we definitely reveal the regions of regularity and chaoticity.

The bifurcation diagram corresponding to the $\mathscr{U}$ coordinate of system (10.2.16)–(10.2.18) with the boundary conditions (10.2.17) is pictured in Fig. 10.6. Here, $\mathscr{U}_{tot}$ is the bifurcation parameter. Supporting the results of [63], it is observable in the figure that the system displays period-doubling bifurcations and leads to chaos. The period-doubling bifurcations occur approximately at the $\mathscr{U}_{tot}$ values 18.315, 18.782, 18.902, 18.939, etc., and a period-six window appears near $\mathscr{U}_{tot} = 19.073$ in the bifurcation diagram.

One can conclude from Fig. 10.6 that the system (10.2.16)–(10.2.18) possesses a stable periodic solution for $\mathscr{U}_{tot} = 17.7$. The projection of a solution that approaches to the stable limit cycle, which is the projection of the attractor of the global system (10.2.16)–(10.2.18) on the domain of (10.2.18) with $\mathscr{U}_{tot} = 17.7$, is depicted in Fig. 10.7. This result confirms the existence of an attractor as a periodic solution in the spatiotemporal equation.

The bifurcation diagram shown in Fig. 10.6 confirms that the drive GDS system (10.3.19) is chaotic. The projection of a chaotic solution of (10.3.19) on the $\mathscr{U} - j$ plane is represented in Fig. 10.8. Moreover, the time series of the $\mathscr{U}$ coordinate of the same solution is shown Fig. 10.9, where one can see the chaotic behavior.

The profiles of the electric field E, electron density n_e, and ion density n_i of (10.3.19) are pictured in Fig. 10.10a–c, respectively. Figure 10.10 also confirms the presence of chaos in the drive system.

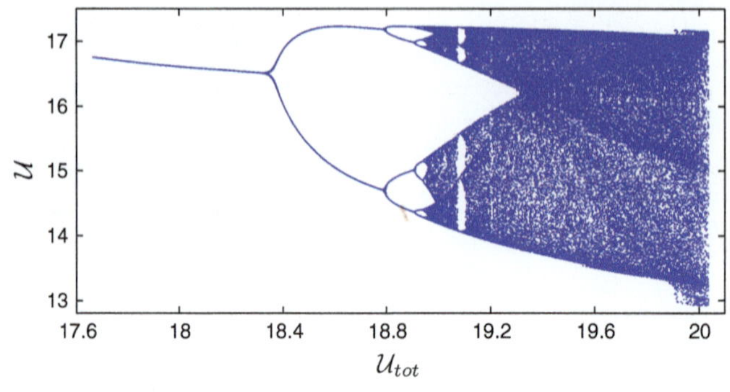

Fig. 10.6 The bifurcation diagram of system (10.2.16)–(10.2.18) for the values of the parameter $\mathscr{U}_{tot}$ between 17.67 and 20.03

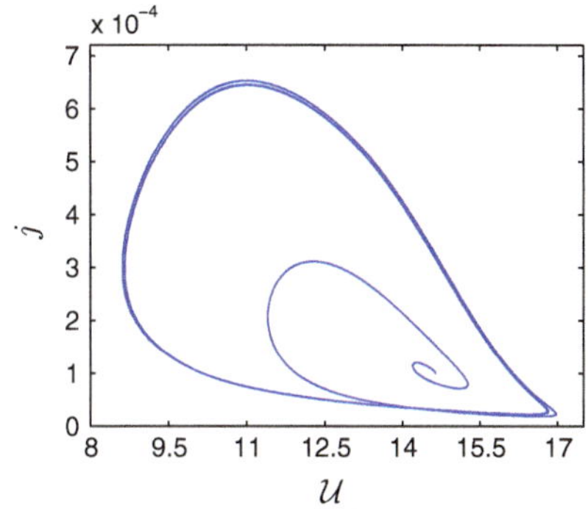

Fig. 10.7 The figure reveals a limit cycle, the projection of the attractor of the global system on the domain of Eq. (10.2.18) with $\mathscr{U}_{tot} = 17.7$

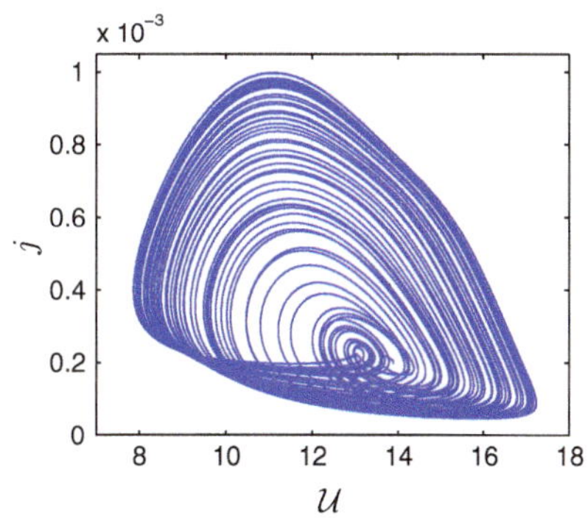

Fig. 10.8 The projection of the chaotic solution of the drive GDS system (10.3.19) on the $\mathscr{U} - j$ plane

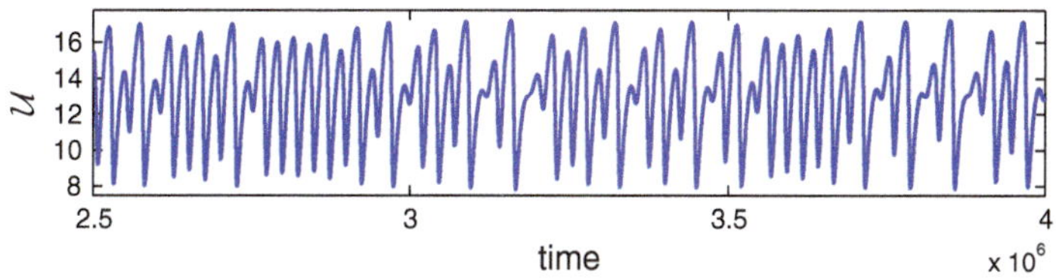

Fig. 10.9 The chaotic behavior of the $\mathscr{U}$ coordinate of system (10.3.19)

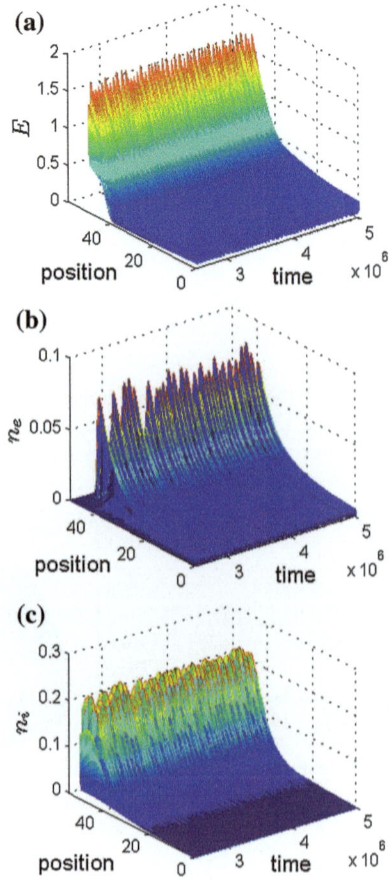

Fig. 10.10 Profiles of the **a** electric field E, **b** electron density n_e, and **c** ion density n_i as functions of time support the existence of spatiotemporal chaos

10.5 Notes

In the studies [34–46], we applied the input–output mechanism to systems that admit stable equilibrium points as well as limit cycles. It is theoretically proved in [44] that weak forcing of systems with stable limit cycles leads to the deformation of limit cycles to chaotic cycles, that is, motions that behave chaotically around the limit cycle. This phenomenon cannot be explained by the theory of generalized synchronization [1–5, 7], and it is also used in the present chapter. In the electrical sense, the chaotification of limit cycles is much more preferable than that procedure for asymptotic equilibria, because of the role of oscillations for electronics.

In this chapter, we utilize GDS systems as drive and response electrical models. GDS systems were analyzed for a chaos presence in [63]. We complete the analysis by constructing the full period-doubling bifurcation diagram to demonstrate that the drive system admits infinitely many unstable periodic solutions as well as sensitivity. However, this is only an auxiliary result. The main novelty of the present chapter with respect to the previous studies [57, 63, 64] is that we consider these systems which are coupled in a unidirectional way and prove that the chaos can be extended through couplings of GDS systems as well as in their arbitrary large collectives. This type of chaos extension may give benefits in further applications, for example, in economic lamps and flat TV screens [11, 12]. We suggest that our way of numerical analysis and special design of complexity can be further verified experimentally. It is worth noting that our approach is not generalized synchronization of chaos at all. This is demonstrated through the special method of auxiliary system approach [1, 3]. The results of Chap. 10 were published in the paper [68].

References

1. H.D.I. Abarbanel, N.F. Rulkov, M.M. Sushchik, Generalized synchronization of chaos: the auxiliary system approach. Phys. Rev. E **53**, 4528–4535 (1996)
2. U.S. Freitas, E.E.N. Macau, C. Grebogi, Using geometric control and chaotic synchronization to estimate an unknown model parameter. Phys. Rev. E **71**, 047203 (2005)
3. J.M. Gonzáles-Miranda, *Synchronization and Control of Chaos* (Imperial College Press, London, 2004)
4. B.R. Hunt, E. Ott, J.A. Yorke, Differentiable generalized synchronization of chaos. Phys. Rev. E **55**(4), 4029–4034 (1997)
5. L. Kocarev, U. Parlitz, Generalized synchronization, predictability, and equivalence of unidirectionally coupled dynamical systems. Phys. Rev. Lett. **76**(11), 1816–1819 (1996)
6. L.M. Pecora, T.L. Carroll, Synchronization in chaotic systems. Phys. Rev. Lett. **64**, 821–825 (1990)
7. N.F. Rulkov, M.M. Sushchik, L.S. Tsimring, H.D.I. Abarbanel, Generalized synchronization of chaos in directionally coupled chaotic systems. Phys. Rev. E **51**(2), 980–994 (1995)
8. L. Kocarev, Z. Tasev, U. Parlitz, Synchronizing spatiotemporal chaos of partial differential equations. Phys. Rev. Lett. **79**, 51–54 (1997)
9. L. Kocarev, Z. Tasev, T. Stojanovski, U. Parlitz, Synchronizing spatiotemporal chaos. Chaos **7**, 635–643 (1997)
10. M.M. Sushchik, Ph.D. dissertation, University of California, San Diego, (1996)
11. U. Kogelschatz, Filamentary, patterned, and diffuse barrier discharges. IEEE Trans. Plasma Sci. **30**, 1400–1408 (2002)
12. U. Kogelschatz, Dielectric-barrier discharges: their history, discharge physics, and industrial applications. Plasma Chem. Plasma Process. **23**, 1–46 (2003)
13. K.G. Andersson, Poincaré's discovery of homoclinic points. Arch. Hist. Exact Sci. **48**, 133–147 (1994)
14. M. Cartwright, J. Littlewood, On nonlinear differential equations of the second order I: the equation $\ddot{y} - k(1 - y^2)'y + y = bkcos(\lambda t + a)$, k large. J. Lond. Math. Soc. **20**, 180–189 (1945)
15. N. Levinson, A second order differential equation with singular solutions. Ann. Math. **50**, 127–153 (1949)
16. E.N. Lorenz, Deterministic nonperiodic flow. J. Atmos. Sci. **20**, 130–141 (1963)
17. Y. Ueda, Random phenomena resulting from non-linearity in the system described by Duffing's equation. Trans. Inst. Electr. Eng. Jpn. **98A**, 167–173 (1978)
18. C.A. Skarda, W.J. Freeman, How brains make chaos in order to make sense of the world. Behav. Brain Sci. **10**(2), 161–173 (1987)
19. M. Watanabe, K. Aihara, S. Kondo, Self-organization dynamics in chaotic neural networks. Control Chaos Math. Model. **8**, 320–333 (1997)

20. S. Steingrube, M. Timme, F. Wörgötter, P. Manoonpong, Self-organized adaptation of a simple neural circuit enables complex robot behaviour. Nat. Phy. **6**, 224–230 (2010)
21. E. Akin, S. Kolyada, Li-Yorke sensitivity. Nonlinearity **16**, 1421–1433 (2003)
22. P. Li, Z. Li, W.A. Halang, G. Chen, Li-Yorke chaos in a spatiotemporal chaotic system. Chaos Solitons Fractals **33**(2), 335–341 (2007)
23. T.Y. Li, J.A. Yorke, Period three implies chaos. Am. Math. Mon. **82**, 985–992 (1975)
24. F.R. Marotto, Snap-back repellers imply chaos in $\mathbb{R}^n$,. J. Math. Anal. Appl. **63**, 199–223 (1978)
25. S. Smale, Differentiable dynamical systems. Bull. Am. Math. Soc. **73**, 747–817 (1967)
26. S. Wiggins, *Global Bifurcations and Chaos: Analytical Methods* (Springer, New York, 1988)
27. H.L.D. de S. Cavalcante and J.R.R. Leite, Experimental bifurcations and homoclinic chaos in a laser with a saturable absorber, Chaos **18**, 023–107 (2008)
28. A. Coillet, Y.K. Chembo, Routes to spatiotemporal chaos in Kerr optical frequency combs. Chaos **24**, 013113 (2014)
29. P. Hagedorn, A. DasGupta, *Vibrations and Waves in Continuous Mechanical Systems* (Wiley, Hoboken, 2007)
30. J.F. Rodrigues, G. Seregin, J.M. Urbano, *Trends in Partial Differential Equations of Mathematical Physics* (Birkhäuser, Germany, 2005)
31. M. Roseau, *Vibrations in Mechanical Systems* (Springer, Berlin, 1987)
32. D.S. Steinberg, *Vibration Analysis for Electronic Equipment* (Wiley, Hoboken, 2000)
33. N.E. Tovmasyan, *Boundary Value Problems for Partial Differential Equations and Applications in Electrodynamics* (World Scientific, Singapore, 1994)
34. M.U. Akhmet, Creating a chaos in a system with relay. Int. J. Qual. Theory Differ. Equ. Appl. **3**, 3–7 (2009)
35. M.U. Akhmet, Devaney's chaos of a relay system. Commun. Nonlinear Sci. Numer. Simul. **14**, 1486–1493 (2009)
36. M.U. Akhmet, Dynamical synthesis of quasi-minimal sets. Int. J. Bifur. Chaos **19**, 2423–2427 (2009)
37. M.U. Akhmet, Li-Yorke chaos in the system with impacts. J. Math. Anal. Appl. **351**, 804–810 (2009)
38. M.U. Akhmet, Shadowing and dynamical synthesis. Int. J. Bifur. Chaos **19**, 3339–3346 (2009)
39. M.U. Akhmet, Homoclinical structure of the chaotic attractor. Commun. Nonlinear Sci. Numer. Simul. **15**, 819–822 (2010)
40. M.U. Akhmet, M.O. Fen, Chaos generation in hyperbolic systems. Interdiscip. J. Discontin. Nonlinearity Complex. **1**, 367–386 (2012)
41. M.U. Akhmet, M.O. Fen, Chaotic period-doubling and OGY control for the forced Duffing equation. Commun. Nonlinear Sci. Numer. Simul. **17**, 1929–1946 (2012)
42. M.U. Akhmet, M.O. Fen, Replication of chaos. Commun. Nonlinear Sci. Numer. Simul. **18**, 2626–2666 (2013)
43. M.U. Akhmet, M.O. Fen, Shunting inhibitory cellular neural networks with chaotic external inputs. Chaos **23**, 023112 (2013)
44. M.U. Akhmet, M.O. Fen, Entrainment by chaos. J. Nonlinear Sci. **24**, 411–439 (2014)
45. M. Akhmet, M.O. Fen, Generation of cyclic/toroidal chaos by Hopfield neural networks. Neurocomputing **145**, 230–239 (2014)
46. M.U. Akhmet, M.O. Fen, A new method of chaos generation. Nonlinear Stud. **21**, 195–203 (2014)
47. R.L. Devaney, *An Introduction to Chaotic Dynamical Systems* (Addison-Wesley, Boston, 1989)
48. M.J. Feigenbaum, Universal behavior in nonlinear systems. Los Alamos Sci. Summer **1**, 4–27 (1980)
49. E. Sander, J.A. Yorke, Period-doubling cascades galore. Ergod. Theory Dyn. Syst. **31**, 1249–1267 (2011)
50. E. Sander, J.A. Yorke, A period-doubling cascade precedes chaos for planar maps. Chaos **23**, 033113 (2013)
51. E. Ammelt, Yu. A. Astrov, H.G. Purwins, Stripe Turing structures in a two-dimensional gas discharge system. Phys. Rev. E **55**, 6731–6740 (1997)

52. Yu.A. Astrov, E. Ammelt, H.G. Purwins, Experimental evidence for zigzag instability of solitary stripes in a gas discharge system. Phys. Rev. Lett. **78**, 3129–3132 (1997)
53. Yu.A. Astrov, Yu.A. Logvin, Formation of clusters of localized states in a gas discharge system via a self-completion scenario. Phys. Rev. Lett. **79**, 2983–2986 (1997)
54. A. von Engel, M. Steenbeck, *Elektrische Gasentladungen* (Springer, Berlin, 1934)
55. E.L. Gurevich, A.S. Moskalenko, A.L. Zanin, Yu.A. Astrov, H.G. Purwins, Rotating waves in a planar DC-driven gas-discharge system with semi-insulating GaAs cathode. Phys. Lett. A **307**, 299–303 (2003)
56. M.S. Mokrov, Yu.P. Raizer, Simulation of current filamentation in a DC-driven planar gas discharge-semiconductor system. J. Phy. D-Appl. Phys. **44**(42), 425202 (2011)
57. I. Rafatov, D.D. Šijačić, U. Ebert, Spatiotemporal patterns in a dc semiconductor-gas-discharge system: stability analysis and full numerical solutions. Phys. Rev. E **76**, 036206 (2007)
58. Y.P. Raizer, *Gas Discharge Physics* (Springer, Berlin, 1997). 2nd corrected printing
59. YuP Raizer, U. Ebert, D.D. Šijačić, Dependence of the transition from Townsend to glow discharge on secondary emission. Phys. Rev. E **70**, 017401 (2004)
60. Yu.P. Raizer, E.L. Gurevich, M.S. Mokrov, Self-sustained oscillations in a low-current discharge with a semiconductor serving as a cathode and ballast resistor: II. Theory Tech. Phys. **51**, 185–197 (2006)
61. Yu.P. Raizer, M.S. Mokrov, A simple physical model of hexagonal patterns in a Townsend discharge with a semiconductor cathode. J. Phys. D-Appl. Phys. **43**, 25520 (2010)
62. Yu.P. Raizer, M.S. Mokrov, Physical mechanisms of self-organization and formation of current patterns in gas discharges of the Townsend and glow types. Phys. Plasmas **20**, 101604 (2013)
63. D.D. Šijačić, U. Ebert, I. Rafatov, Period doubling cascade in glow discharges: local versus global differential conductivity. Phys. Rev. E **70**, 056220 (2004)
64. D.D. Šijačić, U. Ebert, I. Rafatov, Oscillations in DC driven barrier discharges: numerical solutions, stability analysis, and phase diagram. Phys. Rev. E **71**, 066402 (2005)
65. D.D. Šijačić, U. Ebert, Transition from Townsend to glow discharge: subcritical, mixed, or supercritical characteristics. Phys. Rev. E **66**, 066410 (2002)
66. C. Strümpel, Y.A. Astrov, H.-G. Purwins, Nonlinear interaction of homogeneously oscillating domains in a planar gas discharge system. Phys. Rev. E **62**, 4889–4897 (2000)
67. K. Pyragas, Continuous control of chaos by self-controlling feedback. Phys. Rev. A **170**, 421–428 (1992)
68. M. Akhmet, I. Rafatov, M.O. Fen, Extension of spatiotemporal chaos in glow discharge-semiconductor systems. Chaos **24**, 043127 (2014)